Joseph John Bevelacqua
Health Physics in the 21st Century

Related Titles

Turner, J. E.

Atoms, Radiation, and Radiation Protection

2007
ISBN: 978-3-527-40606-7

Martin, J. E.

Physics for Radiation Protection

A Handbook

2006
ISBN: 978-3-527-40611-1

Lieser, K. H.

Nuclear and Radiochemistry

Fundamentals and Applications

2001
ISBN: 978-3-527-30317-5

Attix, F. H.

Introduction to Radiological Physics and Radiation Dosimetry

1986
ISBN: 978-0-471-01146-0

Bevelacqua, J. J.

Basic Health Physics

Problems and Solutions

approx. 568 pages
1999
Hardcover
ISBN: 978-0-471-29711-6

Bevelacqua, J. J.

Contemporary Health Physics

Problems and Solutions

approx. 449 pages
1995
Hardcover
ISBN: 978-0-471-01801-8

Joseph John Bevelacqua

Health Physics in the 21st Century

WILEY-VCH Verlag GmbH & Co. KGaA

The Author

Dr. Joseph John Bevelacqua
Bevelacqua Resources
Richland, WA USA
bevelresou@aol.com

Cover Picture:
Beam Eye View
A Schematic by Frank Tecker, CERN
Summer School 2007

All books published by Wiley-VCH are carefully produced. Nevertheless, authors, editors, and publisher do not warrant the information contained in these books, including this book, to be free of errors. Readers are advised to keep in mind that statements, data, illustrations, procedural details or other items may inadvertently be inaccurate.

Library of Congress Card No.:
applied for

British Library Cataloguing-in-Publication Data
A catalogue record for this book is available from the British Library.

Bibliographic information published by the Deutsche Nationalbibliothek
Die Deutsche Nationalbibliothek lists this publication in the Deutsche Nationalbibliografie; detailed bibliographic data are available on the Internet at <http://dnb.d-nb.de>.

© 2008 WILEY-VCH Verlag GmbH & Co. KGaA, Weinheim

All rights reserved (including those of translation into other languages). No part of this book may be reproduced in any form – by photoprinting, microfilm, or any other means – nor transmitted or translated into a machine language without written permission from the publishers. Registered names, trademarks, etc. used in this book, even when not specifically marked as such, are not to be considered unprotected by law.

Composition   Thomson Digital, Noida, India
Printing   Betz-Druck GmbH, Darmstadt
Bookbinding   Litges & Dopf GmbH, Heppenheim

Printed in the Federal Republic of Germany
Printed on acid-free paper

ISBN: 978-3-527-40822-1

This book is dedicated to my wife Terry
and
Sammy and Chelsea
and
Lucy, Anna, Samuel, and Matthew
and
David and Hanna
and
Isaiah

Contents

Preface *XIX*
Acknowledgments *XXI*
A Note on Units *XXIII*

I	**Overview of Volume I** *1*	
1	**Introduction** *3*	
	References *5*	
II	**Fission and Fusion Energy** *7*	
2	**Fission Power Production** *9*	
2.1	Overview *9*	
2.2	Basic Health Physics Considerations *9*	
2.3	Fission Reactor History *13*	
2.4	Generation II Power Reactors *13*	
2.4.1	Pressurized Water Reactors *14*	
2.4.1.1	Core *15*	
2.4.1.2	Reactor Vessel *15*	
2.4.1.3	Primary Coolant System *15*	
2.4.1.4	Steam System *16*	
2.4.1.5	Reactor Control and Protection Systems *16*	
2.4.1.6	Engineered Safety Features *17*	
2.4.2	Boiling Water Reactors *17*	
2.4.2.1	BWR Reactor Assembly *18*	
2.4.2.2	BWR Reactor Core *18*	
2.4.3	CANDU Reactors *18*	
2.4.3.1	General Description *18*	
2.4.3.2	Control Systems *19*	
2.4.3.3	Steam System *19*	

Health Physics in the 21st Century. Joseph John Bevelacqua
Copyright © 2008 WILEY-VCH Verlag GmbH & Co. KGaA, Weinheim
ISBN: 978-3-527-40822-1

2.4.3.4	Safety Systems 19	
2.4.4	High-Temperature Gas-Cooled Reactors 19	
2.4.5	Liquid Metal Fast Breeder Reactors 20	
2.4.6	Generation II Summary 20	
2.5	Generation III and IV Radiological Design Characteristics 21	
2.6	Generation III 22	
2.6.1	Safety Objectives and Standards 25	
2.6.2	PWRs 26	
2.6.3	BWRs 27	
2.6.4	Advanced CANDU 27	
2.6.5	HTGRs 28	
2.6.6	Generation III Safety System Examples 28	
2.6.6.1	Emergency Condenser System 29	
2.6.6.2	Containment Cooling Condensers 29	
2.6.6.3	Core Flooding System 29	
2.6.6.4	Passive Pressure Pulse Transmitters 29	
2.7	Generation IV 30	
2.7.1	Gas-Cooled Fast Reactors 31	
2.7.2	Lead–Bismuth-Cooled Fast Reactors 32	
2.7.3	Molten Salt Epithermal Reactors 36	
2.7.4	Sodium-Cooled Fast Reactors 37	
2.7.5	Supercritical Water-Cooled Reactors 37	
2.7.6	Very High Temperature Reactors (VHTR) 38	
2.7.7	Radionuclide Impacts 38	
2.7.8	Hydrogen Production 39	
2.7.9	Deployment of Generation IV Reactors 40	
2.8	Generic Health Physics Hazards 40	
2.9	Specific Health Physics Hazards 41	
2.9.1	Buildup of Activity in Filters, Demineralizers, and Waste Gas Tanks 41	
2.9.2	Activation of Reactor Components 44	
2.9.3	Fuel Damage 45	
2.9.4	Reactor Coolant System Leakage 45	
2.9.5	Hot Particle Dose 46	
2.9.6	Effluent Releases 47	
2.9.6.1	Light Water and Heavy Water Reactor Effluents 47	
2.9.6.2	Gas-Cooled Reactor Effluents 47	
2.9.6.3	CO_2-Cooled Reactor Effluents 48	
2.9.6.4	Helium-Cooled Reactor Effluents 48	
2.10	Advanced Reactor ALARA Measures 49	
2.11	Radiological Considerations During Reactor Accidents 49	
2.12	Beyond Design Basis Events 53	
2.13	Other Events 56	
2.14	Probabilistic Risk Assessment 56	
2.15	Semi-Infinite Cloud Model 57	

2.16	Normal Operations *58*	
2.16.1	Health Physics *59*	
2.16.2	Maintenance *59*	
2.16.3	Operators *60*	
2.17	Outage Operations *60*	
2.18	Abnormal Operations *61*	
2.19	Emergency Operations *61*	
	Problems *62*	
	References *67*	
3	**Fusion Power Production** *71*	
3.1	Overview *71*	
3.2	Fusion Process Candidates *72*	
3.3	Physics of Plasmas *73*	
3.4	Plasma Properties and Characteristics *75*	
3.5	Plasma Confinement *79*	
3.6	Overview of an Initial Fusion Power Facility *81*	
3.7	ITER *83*	
3.8	ITER Safety Characteristics *84*	
3.9	General Radiological Characteristics *85*	
3.10	Accident Scenarios/Design Basis Events *87*	
3.10.1	Loss-of-Coolant Accidents *87*	
3.10.2	Loss-of-Flow Accidents *87*	
3.10.3	Loss-of-Vacuum Accidents *88*	
3.10.4	Plasma Transients *88*	
3.10.5	Magnet Fault Transients *89*	
3.10.6	Loss of Cryogen *89*	
3.10.7	Tritium Plant Events *89*	
3.10.8	Auxiliary System Accidents *90*	
3.11	Radioactive Source Term *90*	
3.12	Beyond Design Basis Events *90*	
3.13	Assumptions for Evaluating the Consequences of Postulated ITER Events *90*	
3.14	Caveats Regarding the ITER Technical Basis *92*	
3.15	Overview of Fusion Energy Radiation Protection *94*	
3.16	D-T Systematics *95*	
3.17	Ionizing Radiation Sources *97*	
3.18	Nuclear Materials *100*	
3.19	External Ionizing Radiation Hazards *100*	
3.19.1	Alpha Particles *100*	
3.19.2	Beta Particles *101*	
3.19.3	Photons *101*	
3.19.4	Neutrons *102*	
3.19.4.1	Vanadium Activation – Vacuum Vessel Liner *103*	
3.19.4.2	Activation of Stainless Steel – Vacuum Vessel Structural Material *104*	

3.19.5	Heavy Ions *106*	
3.20	Uncertainties in Health Physics Assessments Associated with External Ionizing Radiation *106*	
3.21	Internal Ionizing Radiation Hazards *108*	
3.21.1	Tritium *108*	
3.21.2	Particulates *109*	
3.22	Measurement of Ionizing Radiation *109*	
3.22.1	Measurement of External Radiation *110*	
3.22.2	Tritium Measurement *112*	
3.22.2.1	Ion Chamber Tritium-in-Air Monitors *112*	
3.22.2.2	Tritium Bubbler *112*	
3.22.2.3	Composition Measurements *113*	
3.22.2.4	Thermal Methods *113*	
3.23	Maintenance *113*	
3.23.1	Vacuum Vessel Maintenance *114*	
3.23.2	Vacuum Vessel-Cooling Water System Maintenance *114*	
3.23.3	Routine Maintenance *114*	
3.24	Accident Scenarios *117*	
3.25	Regulatory Requirements *117*	
3.25.1	ALARA-Confinement Methods and Fusion Process Types *118*	
3.25.2	ALARA – Design Features *120*	
3.26	Other Radiological Considerations *120*	
3.27	Other Hazards *121*	
3.28	Other Applications *121*	
3.28.1	Cold Fusion *122*	
3.28.2	Sonoluminescence *122*	
3.29	Conclusions *123*	
	Problems *124*	
	References *128*	
III	**Accelerators** *131*	
4	**Colliders and Charged Particle Accelerators** *133*	
4.1	Introduction *133*	
4.2	Candidate Twenty-First Century Accelerator Facilities *133*	
4.2.1	Radiation Characteristics of Low-Energy Accelerators *134*	
4.3	Types of Twenty-First Century Accelerators *137*	
4.3.1	Spallation Neutron Source *138*	
4.3.1.1	Machine Overview *138*	
4.3.1.2	Ion Source *138*	
4.3.1.3	LINAC *138*	
4.3.1.4	Accumulator Ring *139*	
4.3.1.5	Hg Target *139*	
4.3.1.6	Applications *139*	

4.3.1.7	SNS Design Decisions	*139*
4.3.1.8	Radiation Protection Regulations	*139*
4.3.1.9	Health Physics Considerations	*140*
4.3.2	Electron–Positron Colliders – Existing Machines	*140*
4.3.2.1	Overview	*140*
4.3.2.2	Electromagnetic Cascade Showers	*143*
4.3.2.3	External Bremsstrahlung	*145*
4.3.2.4	Photoneutron Production	*146*
4.3.2.5	Muons	*146*
4.3.2.6	Synchrotron Radiation	*147*
4.3.2.7	Radiation Levels at the Large Electron–Positron Collider	*149*
4.3.2.8	LEP Radiation Levels Outside the Shielding	*149*
4.3.2.9	Radiation Levels Inside the LEP Machine Tunnel	*149*
4.3.3	Hadron Colliders	*150*
4.3.3.1	Large Hadron Collider	*150*
4.3.3.1.1	CMS	*151*
4.3.3.1.2	ATLAS	*151*
4.3.3.1.3	LHCb	*151*
4.3.3.1.4	TOTEM	*152*
4.3.3.1.5	ALICE	*152*
4.3.3.1.6	Antiprotons	*152*
4.3.3.1.7	Proton Reactions	*154*
4.3.3.1.8	Neutrons	*154*
4.3.3.1.9	Muons	*154*
4.3.3.1.10	Hadronic (Nuclear) Cascade	*154*
4.3.3.1.11	Heavy Ions	*156*
4.3.3.1.12	Synchrotron Radiation	*156*
4.3.3.1.13	High-Power Beam Loss Events	*157*
4.3.4	Heavy-Ion Colliders	*157*
4.3.4.1	Examples of RHIC Radiological Hazards	*159*
4.3.4.2	Radiation Protection Philosophy	*159*
4.3.4.3	Personnel Safety Envelope	*159*
4.3.4.4	Collider Safety Envelope Parameters	*159*
4.3.4.5	Beam Loss Control	*160*
4.3.4.6	Particle Accelerator Safety System	*160*
4.4	Planned Accelerator Facilities	*160*
4.4.1	International Linear Collider	*161*
4.4.1.1	Electron Source/LINAC	*161*
4.4.1.2	Positron Source/LINAC	*161*
4.4.1.3	Electron-Damping Ring	*162*
4.4.1.4	Positron-Damping Ring	*162*
4.4.1.5	Main LINACs	*162*
4.4.1.6	Interaction Area	*162*
4.4.1.7	Evolving ILC Design	*162*
4.4.1.8	ILC Health Physics	*163*

4.4.2	Muon Colliders *163*	
4.4.2.1	Neutrino Characteristics *164*	
4.4.2.2	Neutrino Beam Characteristics at a Muon Collider *165*	
4.4.2.3	Neutrino Interaction Model *167*	
4.4.2.4	Neutrino Effective Dose *167*	
4.4.2.5	Bounding Neutrino Effective Dose – Linear Muon Collider *168*	
4.4.2.6	Bounding Neutrino Effective Dose – Circular Muon Collider *171*	
4.4.2.7	ALARA Impacts of Muon Colliders *177*	
4.4.2.8	Other Radiation Protection Issues *178*	
4.4.3	Very Large Hadron Collider *181*	
4.5	Common Health Physics Issues in Twenty-First Century Accelerators *181*	
4.5.1	Sources of Radiation *182*	
4.5.2	Activation *184*	
4.5.3	Radiation Shielding *184*	
4.5.4	Radiation Measurements *184*	
4.5.5	Environment *187*	
4.5.6	Operational Radiation Safety *189*	
4.5.7	Safety Systems *189*	
4.6	Other Applications *190*	
	Problems *190*	
	References *195*	
5	**Light Sources** *199*	
5.1	Overview *199*	
5.2	Physical Basis *200*	
5.2.1	Bremsstrahlung *200*	
5.2.2	Synchrotron Radiation *201*	
5.3	Overview of Photon Light Sources – Insertion Devices *201*	
5.4	X-Ray Tubes *202*	
5.5	Overview of Synchrotron Radiation Sources and Their Evolution *203*	
5.6	X-Ray Radiation from Storage Rings *204*	
5.6.1	Bending Magnets *204*	
5.6.2	Insertion Devices *205*	
5.6.3	Wigglers *205*	
5.6.4	Undulators *205*	
5.7	Brightness Trends *206*	
5.8	Physics of Photon Light Sources *206*	
5.8.1	Brightness of a Synchrotron Radiation Source *206*	
5.9	Motion of Accelerated Electrons *209*	
5.10	Insertion Device Radiation Properties *211*	
5.10.1	Power and Power Density *213*	
5.11	FEL Overview *215*	
5.12	Physical Model of a FEL *216*	
5.12.1	FEL Physics *218*	

5.13	FEL Characteristics	*220*
5.14	Optical Gain	*220*
5.14.1	Cavity Design	*221*
5.14.2	Optical Klystron	*223*
5.15	Accessible FEL Output	*224*
5.16	X-Ray Free-Electron Lasers	*224*
5.17	Threshold X-Ray Free Electron	*225*
5.18	Near-Term X-Ray FELs	*226*
5.19	Gamma-Ray Free-Electron Lasers (GRFEL)	*226*
5.20	Other Photon-Generating Approaches	*227*
5.20.1	Compton Backscattering	*228*
5.20.2	Laser Accelerators	*229*
5.20.2.1	Basic Theory	*229*
5.20.3	Laser Wake-Field Acceleration (LWFA)	*229*
5.20.4	Laser Ion Acceleration (LIA)	*230*
5.20.5	Future Possibilities	*230*
5.21	X-Ray Induced Isomeric Transitions	*231*
5.22	Gamma-Ray Laser/Fission-Based Photon Sources	*232*
5.23	Photon Source Health Physics and Other Hazards	*234*
5.23.1	Ionizing Radiation	*234*
5.23.2	NonIonizing Radiation	*235*
5.23.3	Activation of Accelerator Components	*236*
5.23.4	Shielding Design and Safety Analysis	*236*
5.24	Evaluation of Radiation Dose	*237*
5.25	General Safety Requirements	*238*
5.26	Radioactive and Toxic Gases	*238*
5.27	Laser Safety Calculations	*239*
5.27.1	Limiting Aperture	*239*
5.27.2	Exposure Time/Maximum Permissible Exposure	*239*
	Problems	*240*
	References	*245*
IV	**Space**	*249*
6	**Manned Planetary Missions**	*251*
6.1	Overview	*251*
6.2	Introduction	*251*
6.3	Terminology	*252*
6.4	Basic Physics Overview	*253*
6.5	Radiation Protection Limitations	*255*
6.6	Overview of the Space Radiation Environment	*255*
6.6.1	General Characterization	*257*
6.6.2	Trapped or van Allen Belt Radiation	*258*
6.6.3	Galactic Cosmic Ray Radiation	*259*

6.6.4	Solar Flare Radiation or Solar Particle Events 259
6.7	Calculation of Absorbed and Effective Doses 260
6.8	Historical Space Missions 260
6.8.1	Low-Earth Orbit Radiation Environment 260
6.8.2	The Space Radiation Environment Outside Earth's Magnetic Field 261
6.8.3	Radiation Data from Historical Missions 263
6.8.4	Gemini 263
6.8.5	Skylab 265
6.8.6	Space Transport Shuttle 265
6.8.7	Mir Space Station 266
6.8.8	International Space Station 266
6.8.9	Apollo Lunar Missions 266
6.8.10	Validation of LEO and Lunar Mission Absorbed Dose Rates 267
6.9	LEO and Lunar Colonization 268
6.10	GCR and SPE Contributions to Manned Planetary Missions 269
6.10.1	GCR Doses 269
6.10.2	SPE Doses 270
6.10.3	Planetary Mission to Mars 275
6.10.4	Mars Orbital Dynamics 275
6.10.5	Overview of Mars Mission Doses 278
6.10.6	Oak Ridge National Laboratory (ORNL) Mars Mission 278
6.10.7	Trapped Radiation Contribution 278
6.10.8	GCR Contribution 278
6.10.9	SPE Contribution 279
6.10.10	Mars Mission Doses 279
6.11	Other Planetary Missions 280
6.11.1	Planetary Atmospheric Attenuation 285
6.12	Mars and Outer Planet Mission Shielding 286
6.13	Electromagnetic Deflection 288
6.13.1	EM Field Deflector Physics 289
6.13.2	Case I – Deflection Using a Static Magnetic Field 291
6.13.3	Case II – Deflection Using a Static Electric Field 291
6.13.4	Engineering Considerations for EM Field Generation 295
6.14	Space Radiation Biology 295
6.15	Final Thoughts 296
	Problems 296
	References 300
7	**Deep Space Missions** 303
7.1	Introduction 303
7.2	Stellar Radiation 303
7.2.1	Origin of Stars 304
7.2.2	Low Mass Stars 304
7.2.3	High Mass Stars 305

7.2.4	Star Types	*307*
7.2.5	MS Star Health Physics Considerations	*309*
7.2.6	Supernovas	*309*
7.2.7	White Dwarfs, Pulsars, and Black Holes	*311*
7.2.8	Dark Matter/Dark Energy	*311*
7.2.9	Gamma-Ray Bursts	*312*
7.3	Galaxies	*314*
7.3.1	Distance Scales	*314*
7.3.2	Characteristics of Galaxies	*315*
7.4	Deep Space Radiation Characteristics	*317*
7.5	Overview of Deep Space Missions	*319*
7.6	Trajectories	*319*
7.6.1	Spacetime and Geodesics	*320*
7.7	Candidate Missions	*321*
7.8	Propulsion Requirements for Deep Space Missions	*322*
7.9	Candidate Propulsion Systems Based on Existing Science and Technology	*323*
7.9.1	Antimatter Propulsion	*323*
7.9.2	Fission Driven Electric Propulsion	*323*
7.9.3	Fusion Propulsion	*324*
7.9.4	Interstellar Ramjet	*324*
7.9.5	Unique Nuclear Reactions	*325*
7.10	Technology Growth Potential	*325*
7.10.1	Dyson Spheres	*326*
7.11	Sources of Radiation in Deep Space	*327*
7.12	Mission Doses	*327*
7.12.1	Trapped Radiation	*328*
7.12.2	Galactic Cosmic Radiation	*329*
7.12.3	SPE Radiation	*330*
7.12.4	Radiation from a Fusion Reactor Propulsion System	*330*
7.12.4.1	Distance	*331*
7.12.4.2	Shielding	*331*
7.13	Time to Reach Alpha Centauri	*333*
7.14	Countermeasures for Mitigating Radiation and Other Concerns During Deep Space Missions	*334*
7.15	Theoretical Propulsion Options	*335*
7.15.1	Modifying (Warping) Spacetime	*336*
7.15.2	Wormholes	*338*
7.15.3	Folding Spacetime	*338*
7.15.4	Mapping Spacetime	*338*
7.16	Spatial Anomalies	*339*
7.17	Special Considerations	*339*
7.18	Point Source Relationship	*340*
	Problems	*344*
	References	*348*

V	**Answers and Solutions** *351*	
	Solutions *353*	
	Solutions for Chapter 2	*353*
	Solutions for Chapter 3	*368*
	Solutions for Chapter 4	*384*
	Solutions for Chapter 5	*403*
	Solutions for Chapter 6	*421*
	Solutions for Chapter 7	*435*
VI	**Appendixes** *453*	
A	**Significant Events and Important Dates in Physics and Health Physics** *455*	
	References *463*	
B	**Production Equations in Health Physics** *465*	
B.1	Introduction *465*	
B.2	Theory *465*	
B.3	Examples *468*	
B.3.1	Activation *468*	
B.3.2	Demineralizer Activity *469*	
B.3.3	Surface Deposition *469*	
B.3.4	Release of Radioactive Material into a Room *470*	
B.4	Conclusions *471*	
	References *471*	
C	**Key Health Physics Relationships** *473*	
	References *482*	
D	**Internal Dosimetry** *483*	
D.1	Introduction *483*	
D.2	Overview of Internal Dosimetry Models *483*	
D.3	MIRD Methodology *485*	
D.4	ICRP Methodology *487*	
D.5	Biological Effects *487*	
D.6	ICRP 26/30 and ICRP 60/66 Terminology *490*	
D.7	ICRP 26 and ICRP 60 Recommendations *490*	
D.8	Calculation of Internal Dose Equivalents Using ICRP 26/30 *491*	
D.9	Calculation of Equivalent and Effective Doses Using ICRP 60/66 *493*	
D.10	Model Dependence *495*	
D.11	Conclusions *495*	
	References *495*	

E	**The Standard Model of Particle Physics** *497*
E.1	Overview *497*
E.2	Particle Properties and Supporting Terminology *497*
E.2.1	Terminology *497*
E.3	Basic Physics *498*
E.3.1	Basic Particle Properties *498*
E.3.2	Fundamental Interactions *501*
E.4	Fundamental Interactions and Their Health Physics Impacts *503*
E.4.1	Conservation Laws *504*
E.4.2	Consequences of the Conservation Laws and the Standard Model *506*
E.5	Cross-Section Relationships for Specific Processes *508*
	References *508*

F	**Special Theory of Relativity** *509*
F.1	Length, Mass, and Time *509*
F.1.1	Cosmic Ray Muons and Pions *510*
F.2	Energy and Momentum *512*
	References *513*

G	**Muon Characteristics** *515*
G.1	Overview *515*
G.2	Stopping Power and Range *515*
	References *518*

H	**Luminosity** *519*
H.1	Overview *519*
H.2	Accelerator Physics *519*
	References *521*

I	**Dose Factors for Typical Radiation Types** *523*
I.1	Overview *523*
I.2	Dose Factors *523*
I.3	Dose Terminology *524*
	References *524*

J	**Health Physics Related Computer Codes** *525*
J.1	Code Overview *525*
J.1.1	EGS Code System *525*
J.1.2	ENDF *525*
J.1.3	FLUKA *526*
J.1.4	JENDL *526*
J.1.5	MARS *526*
J.1.6	MCNP *526*
J.1.7	MCNPX *526*

J.1.8	MicroShield® 527
J.1.9	MicroSkyshine® 527
J.1.10	SCALE 5 527
J.1.11	SKYSHINE-KSU 528
J.1.12	SPAR 528
J.2	Code Utilization 528
	References 529

K	**Systematics of Heavy Ion Interactions with Matter** 531
K.1	Introduction 531
K.2	Overview of External Radiation Sources 531
K.3	Physical Basis for Heavy Ion Interactions with Matter 532
K.4	Range Calculations 535
K.5	Tissue Absorbed Dose from a Heavy Ion Beam 536
K.6	Determination of Total Reaction Cross Section 537
	References 537

L	**Curvature Systematics in General Relativity** 539
L.1	Introduction 539
L.2	Basic Curvature Quantities 539
L.3	Tensors and Connection Coefficients 541
L.3.1	Flat Spacetime Geometry 541
L.3.2	Schwarzschild Geometry 543
L.3.3	MT Wormhole Geometry 545
L.3.4	Generalized Schwarzchild Geometry 547
L.3.5	Friedman–Robertson–Walker (FRW) Geometry 549
L.4	Conclusions 552
	References 552

Index 553

Preface

Health Physics in the 21st Century is intended to bridge the gap between existing health physics textbooks and reference materials needed by a practicing health physicist as the twenty-first century progresses. This material necessarily encompasses emerging radiation-generating technologies, advances in existing technology, and applications of existing technology to new areas. As the twenty-first century unfolds, this gap will rapidly broaden. It is unlikely that the present text will be a definitive health physics work, but it will hopefully encourage other authors to present material that will advance the field and prepare health physicists for upcoming challenges.

The topics selected for inclusion in this text are based on the author's judgment of areas that merit presentation and development. Some areas involve incremental steps in existing health physics knowledge including aspects of Generation III and IV fission reactors. Other topics, such as deep space missions and muon colliders, require the development of concepts that will be relatively new to some health physicists. Additional knowledge regarding the nature and application of fundamental interactions is also required. In addition, paradigm shifts in thinking are necessary. For example, health physicists are currently trained to disregard neutrino effective dose values owing to their trivial magnitude. Although this is certainly true for fission reactors and twentieth-century and early twenty-first century accelerators, it is not a valid assumption as muon energies reach the PeV energy range.

Additional skill development is needed in areas such as the planetary and deep space radiation environment with the need to apply special and general relativity to health physics problems. The possibility also exists for the alteration of fundamental radiological properties such as the gamma-ray energy emitted during a nuclear transition. The modification of established physical phenomena is primarily of interest to theoretical physicists, but it could become an important consideration in health physics evaluations of the deep space radiation environment as well. Spatial abnormalities, such as the *Pioneer Anomaly*, may be an early indication of the complexities of deep space travel that could have a profound impact on health physics evaluations.

As a means to facilitate the transition to new concepts, over 200 problems with solutions are provided. These problems are an integral part of the text, and they serve

Health Physics in the 21st Century. Joseph John Bevelacqua
Copyright © 2008 WILEY-VCH Verlag GmbH & Co. KGaA, Weinheim
ISBN: 978-3-527-40822-1

to integrate and amplify the chapter and appendix information. Students are encouraged to work carefully on each problem to maximize the benefit of this text.

If in the first few years of the twentieth century a health physicist were to have written the textbook *Health Physics in the 20th Century*, the individual would have invariably missed numerous developments and would have failed to predict a wide range of phenomena that emerged to dominate the twentieth-century field. However, it is the author's view that such an effort would have been worthwhile because it would serve to stimulate the field and prompt additional publications to correct the incorrect perceptions advanced by *Health Physics in the 20th Century*. This book is written with a similar desire.

This text is primarily intended for upper level undergraduate and graduate health physics courses. *Health Physics in the 21st Century* is also written for advanced undergraduate and graduate science and engineering courses. It will also be a useful reference for scientists and engineers participating in the evolving technologies, including Generation IV fission reactors, fusion technology, advanced accelerators, light sources, free-electron lasers, and space technology and exploration. The author also hopes that this text will be used by the various health physics certification boards (e.g., the American Board of Health Physics) in developing examination questions.

The success of *Health Physics in the 21st Century* will be judged by history. It is hoped that this text will be worthy of at least a footnote by a future author when she writes *Health Physics in the 22nd Century*.

I wish the health physicists of the twenty-first century well in upholding the highest traditions of the field that have been advanced by the founders of the various national and international Health Physics and Radiation Protection Societies. *Bonne chance*.

Joseph John Bevelacqua, PhD, CHP
Bevelacqua Resources
Richland, WA USA

Acknowledgments

Numerous individuals and organizations assisted the author in the development of this text by providing information, computer codes, or technical assistance. The author apologizes in advance to any individual or organization that was inadvertently omitted.

The author is pleased to acknowledge the support of

- Atomic Energy of Canada Limited for ACR-1000 information (Ron Oberth).
- Brookhaven National Laboratory, National Nuclear Data Center for cross-section and other nuclear data.
- Dartmouth College for fusion publications applicable to the International Thermonuclear Experimental Reactor (Dr David C. Montgomery).
- European Organization for Nuclear Research (CERN) for reports and information.
- European Space Agency for reports and information.
- Federal State Unitary Enterprise, OKB Mechanical Engineering for information about Russian gas cooled reactor (Dr Nikolay Kodochigov).
- Fermi National Accelerator Laboratory for providing access to the MARS Code (Dr Nikolai Mokhov) and for reports and information.
- International Atomic Energy Agency for information related to Generation III and IV fission reactors (Drs John Cleveland, M. Mabrouk, A. Stanculescu, M. Zafirakis-Gomez, and J. Segota) and for nuclear and dosimetry data.
- Japan Atomic Energy Research Institute for nuclear and dosimetry data.
- Los Alamos National Laboratory for providing access to MCNPX (Dr Greg McKinney and Diane Pelowitz) and for reports and information.
- National Aeronautics and Space Administration for reports and information.
- National Research Council of Canada for providing access to the EGS4 transport code.
- Oak Ridge National Laboratory's Radiation Safety Information Computational Information Center (Barbara Snow and Alice Rice) for providing access to numerous computer codes.
- Research Center for Charged Particle Data, National Institute of Radiological Sciences, Chiba, Japan (Dr N. Mafsufugi) for heavy ion reaction model information.

- Research Reactor Safety Analysis Services (Dr Robert C. Nelson) for space launch safety analysis information.
- Stanford Linear Accelerator Center for reports and information.

As one of the purposes of this text is to maintain the technical basis for evolving American Board of Health Physics Certification Examinations, some of the problems were derived from questions that appeared on previous examinations. As a prior panel member, vice chair, and chair of the Part II Examination Panel, I would like to thank my panel and all others whose exam questions have been consulted in formulating questions for this textbook.

The author is also fortunate to have worked with colleagues, students, mentors, and teachers who have shared their wisdom and knowledge, provided encouragement or otherwise influenced the content of this text. The following individuals are acknowledged for their assistance during the author's career: Dick Amato, John Auxier, Lee Booth, Ed Carr, Paul Dirac, Bill Halliday, Tom Hess, Gordon Lodde, Bob Nelson, John Philpott, Lew Pitchford, John Poston, John Rawlings, Don Robson, Bob Rogan, Mike Slobodien, Jim Tarpinian, Jim Turner, and George Vargo.

The continuing encouragement of my wife Terry is gratefully acknowledged.

I would also like to thank the staff of Wiley-VCH with whom I have enjoyed working, particularly Anja Tschörtner, Ulrike Werner, Hans-Jochen Schmitt, and Dr Alexander Grossmann. The advice and encouragement of George Telecki of John Wiley & Sons, Inc. is also acknowledged.

A Note on Units

The author breaks from his previous tendency to utilize the conventional units encountered in the United States. Although these traditional units are a source of comfort to the author and many applied health physicists of United States, their use is not appropriate for this text. As the US regulations become harmonized with international recommendations and regulations, there is an emerging preference for the SI System of Units. This trend will likely to continue into the twenty-first century, and it is reasonable to expect that SI units will eventually become the standard in the United States as it is throughout the world.

For those readers who feel more comfortable with conventional units, the following conversion factors are provided:

SI unit	Traditional US unit
Bq	2.7×10^{-11} Ci
Gy	100 rad
C/kg air	3881 R
Sv	100 rem

As the reader can note the choice of units is more a matter of familiarity and comfort. However, uniformity and clear communication between various scientific and engineering fields and nations suggest the need to adopt the SI System. Accordingly, the SI System is adopted in this text.

1
Overview of Volume I

The first volume of *Health Physics in the 21st* provides a tie between twentieth century and twenty-first century health physics. Chapters dealing with fission reactors, high-energy accelerators, light sources, and low-Earth orbit missions bridge these two centuries and provide logical ties between these time periods. Other topics involving muon colliders, X-ray and γ-ray lasers, planetary space travel, interstellar travel, and fusion power reactors push the time line deeper into or beyond the twenty-first century.

Volume I introduces new topics and basic knowledge required to understand the anticipated evolution of the health physics field. Background information is provided in 12 appendixes to smooth the transition of information needed to comprehend the emerging radiation generating technologies. The reader should consult these appendixes as they are referenced in the main text.

1
Introduction

As the end of nineteenth century approached, science was in a contented state. Man's understanding of space and time, matter and energy, and the basic physics principles appeared to be fundamentally correct. The basic physical laws were established. Newton's laws described the motion of objects, their interaction, and fundamental characteristics such as momentum and energy. Maxwell equations explained known electric and magnetic phenomena. Dalton's ideas revealed the atomic nature of matter, and Mendeleev devised a periodic system for chemical elements.

It was generally believed that the basic laws of the universe were known and that the essential interactions involving these laws had been determined. The remaining challenges involved filling in the details regarding the known interactions and fundamental laws. Little did mankind know that radical revisions to the nineteenth-century view would soon be required with the emergence of new physics, which coincided with the birth of health physics profession.

The discovery of X-rays by William's Röntgen in 1895 was an unexpected event. Röntgen's discovery initiated the birth of new science including the field of health physics, and signaled the first of many events that shook the foundation of nineteenth century physics. These events led to an improved view of matter and its interaction properties. For example, atomic energy levels were described by a series of improving models and theories; the nucleus, its energy levels, and interactions were revealed; an increasing, large set of fundamental particles was discovered and their number continued to grow; additional radiation types were found and their interactions characterized and quantified.

Although Maxwell equations survived, the new physics replaced most of the nineteenth century physics. The new physics was satisfactorily described through quantum mechanics, quantum electrodynamics, special and general relativity, the nuclear shell model, nuclear optical models, and the Standard Model of Particle Physics.

Although Röntgen's discovery of X-rays was the genesis for the health physics field, the Manhattan Project cemented health physics as a profession. Other events such as the Hiroshima and Nagasaki atomic bomb attacks, weapons production activities, development of nuclear fission technology, medical and industrial applications of radioisotopes, manned space missions, nuclear fusion studies, high-energy accelerator operation, and studies of the biological effects of ionizing radiation significantly

Health Physics in the 21st Century. Joseph John Bevelacqua
Copyright © 2008 WILEY-VCH Verlag GmbH & Co. KGaA, Weinheim
ISBN: 978-3-527-40822-1

influenced the health physics field. From these beginnings, the health physics field continued to grow.

The twentieth century saw a maturation of the health physics profession and its scientific basis. A standard set of units evolved and the various radiation types were characterized. National and international organizations were formed to foster sustained development and standardization. Instrumentation advances permitted the detection of a variety of ionizing radiation types over a wide range of energies.

A summary of the key events and dates associated with the physics and health physics professions is provided in Appendix A. Although the events selected for inclusion in Appendix A are somewhat subjective, they are representative of significant events influencing the development of the health physics profession.

The events of Appendix A will also influence the development of twenty-first century radiation generating technologies, and the associated development of health physics practices to protect workers from the radiation hazards of these technologies. Although the evolution of these technologies is uncertain, they will certainly involve energies and radiation types that will require careful management.

Health Physics in the 21st Century reviews emerging and maturing radiation generating technologies that will affect the health physics profession. It is hoped that this review will foster additional research into these areas and into areas not yet imagined.

Health physics is a dynamic and vital field and has an exciting future. However, significant challenges will likely arise as new physics emerges, new particles and radiation types are discovered, and old paradigms fall. For example, the Standard Model will be superseded, as was Classical Mechanics, and be replaced by an improved theory. It is unclear what theories will emerge, but the diversity of current approaches (e.g., supersymmetry, quantum gravity, twistor theory, string theory, grand unification theories, and higher dimensional theories) offer insight into an exciting future.

There is an intimate linkage between the health physics profession and emerging physics. New physics produces new energy regimes that lead to the production of a diversity of radiation types. In some cases, the magnitude of known hazards will increase, but new hazards may also emerge. For example, neutrinos do not present a significant hazard at light water reactors or twentieth-century accelerators. However, when accelerator energies reach the PeV energy range or when the particle fluences significantly increase, neutrino effective doses can no longer be ignored.

Although a large number of emerging radiation generating technologies exist, it is practical to only consider a representative subset. The following are judged by the author to be representative of future health physics challenges and these are further explored in this book:

- Generation III and IV fission reactors
- Fusion power facilities
- High-energy accelerators including muon colliders
- Photon light sources including free electron lasers

- Manned planetary missions
- Deep space missions
- Advanced nuclear fuel cycles including laser isotope separation and actinide transmutation
- Radiation therapy using heavy ions, exotic particles, and antimatter
- Radioactive dispersal devices and improvised nuclear devices
- Evolving regulatory considerations

The first six of these listed topics are covered in Volume I of *Health Physics in the 21st Century*. The remaining topics and additional areas are the subject of Volume II.

The twenty-first century should be an exciting time for the health physics profession. It is the author's desire that this book contributes in some small measure to the education of twenty-first century health physicists and their understanding of emerging radiation generating technologies. The author also hopes that this book will foster additional effort to improve upon and further develop the topics covered in this book and additional emerging areas.

Good luck and best wishes in advancing our proud profession.

References

Bevelacqua, J.J. (1995) *Contemporary Health Physics: Problems and Solutions*, John Wiley & Sons, Inc., New York.

Bevelacqua, J.J. (1999) *Basic Health Physics: Problems and Solutions*, John Wiley & Sons, Inc., New York.

Cember, H. (1996) *Introduction to Health Physics*, 3rd edn, McGraw-Hill, New York.

Glasstone, S. (1982) *Energy Deskbook*, United States Department of Energy Technical Information Center, Oak Ridge, TN.

Griffiths, D. (1987) *Introduction to Elementary Particles*, John Wiley & Sons, Inc., New York.

Jackson, J.D. (1999) *Classical Electrodynamics*, 3rd edn, John Wiley & Sons, Inc., New York.

Penrose, R. (2005) *The Road to Reality: A Complete Guide to the Laws of the Universe*, Alfred A Knopf, New York.

Turner, J.E. (1995) *Atoms, Radiation, and Radiation Protection*, 2nd edn, John Wiley and Sons, Inc., New York.

II
Fission and Fusion Energy

Part Two examines the health physics impacts of power generating technologies. Chapter 2 reviews fission power reactors with an emphasis on Generation III and IV reactors, which will provide electricity generating capacity for at least the first half of the twenty-first century. The fusion power capacity will be phased into the electrical power mix during the second half of the twenty-first century.

2
Fission Power Production

2.1
Overview

The fission power reactor health physics in the twenty-first century will be a combination of familiar as well as new challenges. Pressurized water reactors (PWRs) and boiling water reactors (BWRs) will still be operating and dominating the reactor fleet at least for the first quarter of the twenty-first century. High-temperature gas-cooled reactors (HTGRs) may become an important vehicle for hydrogen production and liquid metal fast breeder reactors (LMFBRs) may also have renewed life in the twenty-first century as a means to eliminate actinides in the nuclear fuel cycle. In addition, new reactor types will likely emerge.

The diversity of fission reactor types and concepts is governed by national and international policies and the manner in which the nuclear fuel cycle – reviewed in Volume II – is operated. This chapter examines the health physics aspects of twenty-first century fission reactors.

Chapter 2 presents basic health physics considerations for fission reactor operations. The operational characteristics of Generation II, III, and IV reactors are outlined to facilitate an understanding of specific health physics concerns in both normal and off-normal circumstances.

2.2
Basic Health Physics Considerations

Power reactor health physicists continue to show responsibility toward protecting the health and safety of the public, including the nuclear facilities' workers, from a variety of radiation hazards associated with the fission process used to produce electricity. These include external radiation sources (e.g., gamma, beta, and neutron radiation types) as well as internal sources of beta and alpha radiation. The situation often gets complicated when mixed radiation fields resulting from a combination of these radiation types occur.

Power reactors consist of a fuel core in which fissions occur and the fission energy is utilized to heat the reactor coolant. The nuclear fuel core contains fissile material (^{233}U, ^{235}U, or ^{239}Pu) and fertile material (^{238}U or ^{232}Th). At present, most commercial reactors use uranium fuel enriched to about 3–5 wt% in ^{235}U. The fuel consists of uranium dioxide pellets contained within zirconium alloy rods that are arranged in a fuel bundle. The core is composed of hundreds of fuel bundles or assemblies.

A fuel core may be cooled by a variety of materials including light water, heavy water, helium gas, and liquid metal that may also serve to modify the fission neutron energy spectrum to enhance reactor performance. The primary coolant that directly cools the core contains dissolved and suspended radionuclides that present an external as well as internal radiation hazard. The coolant system consists of a reactor vessel (containing the reactor core), cooling fluid (water or gas) piping, heat exchanger equipment (including steam generators), and a multitude of support pumps, valves, instrumentation and associated lines, and system control components.

The radiation hazards at a fission reactor consist of the direct gamma and neutron radiation from the fission process, fission products, and activation products resulting from the neutron irradiation of reactor components and structures. A summary of typical fission and activation products is provided in Table 2.1.

Radionuclides are products of the fission process or are activation products generated by the fission neutron spectrum such as a thermal neutron spectrum used in light water reactors (LWRs) and a fast neutron spectrum used in liquid metal reactors. Common fission products are radioactive krypton and xenon, radioiodine, radiocesium, and radiostrontium; activation products in light water and heavy water reactor coolants include ^{3}H and ^{16}N; structural activation products include isotopes of iron, cobalt, and nickel. Activation products are discussed in detail in subsequent sections.

In water-cooled reactors, effective doses inside the containment building during power operations are dominated by ^{16}N and neutron radiation. Effective dose values for maintenance and repair activities are dominated by activation products (usually ^{60}Co or ^{58}Co). Events involving mechanical fuel damage or failure result in the release of isotopes of krypton, xenon, and iodine. Specific accident release characteristics are discussed further in the text.

Typical production modes for selected activation products are provided in Table 2.2. Such tables are important because advanced reactors vary in terms of their construction and component configurations, but the basic driving force remains the fission of uranium and plutonium. For this reason, Tables 2.1 and 2.2 may serve as guidelines for the health physics considerations at a fission reactor regardless of its specific operational characteristics.

Before reviewing specific health physics aspects of fission reactors, a brief review of the various reactor types is provided. Details regarding Generation I, II, III, and IV reactors are also presented.

Table 2.1 Typical radionuclides at a fission power reactor.

Nuclide	Half-life	Decay mode and significant radiation types	Health physics consideration
^{3}H	12.3 yr	$\beta^- E_{max} = 18.6$ keV@100%	Internal hazard particularly during refueling and primary component maintenance operations
^{16}N	7.14 s	$\beta^- E_{max} = 4.29$ MeV@68% $= 10.4$ MeV@26% $\gamma = 2.74$ MeV@0.76% $= 6.13$ MeV@69% $= 7.12$ MeV@5%	Water and air activation product Significantly contributes to containment radiation doses and impacts the design of primary shielding
^{60}Co	5.27 yr	$\beta^- E_{max} = 0.318$ MeV@100% $\gamma = 1.17$ MeV@100% $= 1.33$ MeV@100%	Activation product that generally dominates personnel dose considerations at an operating reactor
^{85}Kr	10.7 yr	$\beta^- E_{max} = 0.687$ MeV@100% $\gamma = 0.514$ MeV@0.43%	Long-lived fission gas that dominates release considerations for spent fuel after about 1 year following discharge from the reactor Fission gases also present an off-site release concern (submersion hazard)
^{88}Kr	2.84 h	$\beta^- E_{max} = 0.52$ MeV@67% $= 2.91$ MeV@14% $\gamma = 0.20$ MeV@26% $= 0.83$ MeV@13% $= 1.53$ MeV@11% $= 2.20$ MeV@13% $= 2.39$ MeV@35%	Example of short-lived fission gas Fission gases also present an off-site release concern (submersion hazard)
^{88}Rb	17.8 min	$\beta^- E_{max} = 2.58$ MeV@13% $= 5.32$ MeV@78% $\gamma = 0.90$ MeV@14% $= 1.84$ MeV@21% $= 3.01$ MeV@2.4%	Daughter of ^{88}Kr Detection of ^{88}Rb is often an indication of fuel cladding degradation ^{88}Rb is a particulate that is detected as low-level skin or clothing contamination
^{90}Sr	28.6 yr	$\beta^- E_{max} = 0.546$ MeV@100%	Long-lived fission product that presents an external as well as internal concern ^{90}Sr is not normally an operational concern unless fuel damage occurs

(continued)

Table 2.1 (Continued)

Nuclide	Half-life	Decay mode and significant radiation types	Health physics consideration
^{90}Y	64.1 h	$\beta^- E_{max} = 2.28$ MeV @ 100%	Daughter of ^{90}Sr. Fission product that presents an external as well as internal concern
^{131}I	8.04 d	$\beta^- E_{max} = 0.606$ MeV @ 89% $\gamma = 0.28$ MeV @ 6.1% $= 0.36$ MeV @ 81% $= 0.64$ MeV @ 7.3%	Fission product that presents an internal hazard to the thyroid and an external radiation concern ^{131}I also is an off-site release concern
^{133}Xe	5.245 d	$\beta^- E_{max} = 0.35$ MeV @ 99% $\gamma = 0.031$ MeV @ 39% $= 0.035$ MeV @ 9.1% $= 0.081$ MeV @ 36%	Example of short-lived fission gas Fission gases also present an off-site release concern (submersion hazard)
^{135}Xe	9.11 h	$\beta^- E_{max} = 0.91$ MeV @ 96% $\gamma = 0.25$ MeV @ 90% $= 0.61$ MeV @ 2.9%	Example of short-lived fission gas. As a fission gas, it is a submersion concern Fission gases also present an off-site release concern
^{137}Cs	30.2 yr	$\beta^- E_{max} = 0.51$ MeV @ 95% $= 1.17$ MeV @ 5.4% $\gamma = 0.66$ MeV @ 90%	Long-lived fission product that presents an external as well as internal concern ^{137}Cs is not normally an operational concern unless fuel damage occurs
^{147}Pm	2.62 yr	$\beta^- E_{max} = 0.22$ MeV @ 100% $\gamma = 0.12$ MeV @ 0.0029%	Example of intermediate half-life fission product. These fission products are not normally an operational concern unless fuel damage occurs
^{238}U	4.47×10^9 yr	$\alpha = 4.15$ MeV @ 23% $= 4.20$ MeV @ 77% X-rays	Dominant uranium isotope in light water fuel. Neutron capture leads to the production of ^{239}Pu
^{239}Pu	2.41×10^4 yr	$\alpha = 5.10$ MeV @ 12% $= 5.14$ MeV @ 15% $= 5.16$ MeV @ 73% X-rays	Internal radiation hazard if fuel damage occurs ^{239}Pu is not normally an operational concern unless fuel damage occurs

Table 2.2 Typical fission power reactor activation products and their production modes.

Nuclide	Neutron energy region	Component activated	Production reaction
^{3}H	Thermal	Lithium hydroxide[a]	^{6}Li(n, α)^{3}H
	Thermal	Boric acid[b]	^{10}B(n, 2α)^{3}H
	Thermal	Primary coolant	^{2}H(n, γ)^{3}H
^{54}Mn	Fast	Stainless steel[c]	^{54}Fe(n, p)^{54}Mn
^{59}Fe	Thermal	Stainless steel[c]	^{58}Fe(n, γ)^{59}Fe
^{58}Co	Thermal	Stainless steel[c] and stellite[d]	^{57}Co(n, γ)^{58}Co
	Fast	Stainless steel[c] and stellite[d]	^{59}Co(n, 2n)^{58}Co
	Fast	Stainless steel[c] and stellite[d]	^{58}Ni(n, p)^{58}Co
^{60}Co	Thermal	Stainless steel[c] and stellite[d]	^{59}Co(n, γ)^{60}Co
^{95}Zr	Thermal	Zirconium fuel cladding	^{94}Zr(n, γ)^{95}Zr

[a] Primary coolant corrosion control (PWR).
[b] Primary coolant reactivity control (PWR).
[c] Reactor coolant system structures and corrosion and wear products.
[d] Valve seats and hard-facing components.

2.3
Fission Reactor History

The progression of fission reactor development is characterized in terms of four generations. The Generation I reactors were developed in 1950–1960s and none are operational today. Generation II reactors are the facilities from 1970s to mid-1990s and include all US commercial reactors and most reactors currently operating in the world. Generation III are advanced reactors with the first becoming operational in 1996 in Japan and others are under construction or ready to be commissioned. Generation IV designs are still in development stage and are expected to arrive in another decade or two. The focus of this book is on Generation III and IV reactors. Before we review Generation III and IV reactors, let us discuss the essential operating characteristics of Generation II systems.

2.4
Generation II Power Reactors

The most common Generation II power reactor designs are PWRs and BWRs. A Canadian-Deuterium-Uranium (CANDU) reactor is a variant of the PWR. HTGRs and FBRs are also part of the Generation II reactor fleet. A summary of the overall characteristics of Generation II reactors, including the reactor type, coolant type, coolant temperature, and coolant pressure, is outlined in Table 2.3. Comments that affect health physics considerations are also provided.

Table 2.3 Characteristics of various Generation II power reactors.

Reactor type	Coolant			Comments
	Type	Pressure (MPa)	Temperature (°C)	
PWR	Light water	13.8	260–316	Steam generators transfer primary heat to the secondary coolant
CANDU	Heavy water	10.0	288	Steam generators transfer primary heat to the secondary coolant Selective isolation of individual pressure tubes permits online refueling
BWR	Light water	6.9	260	The primary coolant is directly converted to steam that drives the turbine
HTGR	Helium	5.2	704	Helium coolant is pumped to a steam generator, and heat is transferred to the secondary coolant
LMFBR	Sodium	a	a	Fast neutrons are not moderated and primary sodium heats secondary sodium, which then heats water to produce steam

^{a}Varies but subject to the constraint that the temperature remains above the melting point of sodium (98 °C).

2.4.1
Pressurized Water Reactors

A pressurized water reactor is a thermal power system in which ordinary or light water (H_2O) functions as the moderator, coolant, and neutron reflection material. PWRs operate at higher pressures than BWRs to inhibit boiling within the reactor core. At the turn of this century, about two-thirds of operating reactors were PWRs.

The health physics concerns and operating characteristics of a PWR are most easily understood by discussing its major components. These components include the fuel core, reactor vessel, primary coolant system, steam system, control systems, and engineered safety features.

2.4.1.1 Core
In a power reactor, the fission process occurs within the core. The PWR reactor core is assembled from rectilinear fuel assemblies and is approximately cylindrical in shape with a height of about 4 m and a diameter of about 4 m. The core consists of 20 000 or more fuel rods combined into fuel assemblies each containing 200 or more rods. The fuel rods are loaded with cylindrical uranium dioxide fuel pellets contained within a zirconium alloy fuel rod.

One-fourth to one-third of the fuel assemblies are replaced during refueling outages. The time between refueling outages varies between 12 and 24 months depending on the initial enrichment of the fuel and the operating philosophy of the owner. With sound preventive maintenance, longer cycles are viable and lead to improved economics and higher capacity factors.

In a PWR, control rods are inserted through the top of the reactor vessel into the core. However, not all fuel assemblies contain control rods, and typically 30–70 control rods are utilized in the core design. A number of stainless steel or zirconium alloy tubes compose a control rod assembly. Each of these tubes contains neutron poisons, normally a combination of silver, indium, and cadmium.

2.4.1.2 Reactor Vessel
The reactor core is enclosed by a steel core barrel and is supported in a large cylindrical reactor vessel. Reactor vessels are on the order of 10.7–13.7 m high and have an internal diameter of about 4.6 m. Carbon steel forms the reactor vessel, 20–23 cm thick, and the interior surface is stainless steel clad.

2.4.1.3 Primary Coolant System
Water flows upward through the reactor core where it is heated to about 329 °C before flowing to the primary side of the steam generator. After transferring a portion of the reactor coolant's energy to water on the secondary side of the steam generator, the high-pressure primary water is pumped back to the reactor vessel. This water enters a nozzle just above the top of the core, and then flows down through the annular region between the vessel and the core barrel. At the bottom of the core, the water reverses direction and flows upward into the bottom of the core. As it flows through the core, the core heat is removed and the coolant is heated.

The letdown system removes a portion of the primary coolant, typically, 115–380 Lpm and monitors it for radioactivity. Increasing letdown radiation levels may be an indication of fuel defects or failures that permit fission products to escape from the fuel matrix and cladding and into the primary coolant; these may also be attributed to releases of trapped corrosion deposits often referred to as "CRUD bursts" caused by mechanical or thermal cycles. CRUD is a historical term for *corrosion undetermined products*. The letdown system also connects with filters; radioactive material clean-up systems such as demineralizers; systems to add chemicals for pH control, lithium for corrosion control, and boric acid for reactivity control; gas clean-up systems; and waste gas holdup tanks.

2.4.1.4 Steam System

PWRs typically have two, three, or four independent steam generators. Most steam generators consist of thousands of inverted U-shaped tubes enclosed in a shell casing. Other designs have straight tubes and are referred to as "once-through steam generators." The high-pressure, high-temperature water from the primary coolant system flows inside the steam generator tubes. Heat is transferred to the lower pressure secondary water outside the tubes and that secondary coolant boils to produce steam. The resultant steam flows to the PWR steam turbine system that drives the electric generator. The exhaust steam from the low-pressure turbines drains to condensers. The low-pressure turbine condensate is heated and than returned to the secondary side of the steam generator.

Steam line radiation detectors are designed to indicate the presence of ^{16}N in the secondary system. Detection of ^{16}N activity is an indication that a breach of a steam generator tube has occurred, and that primary coolant is entering the secondary system.

Air ejectors (eductors) take their suction on the condenser vapor space to maintain a vacuum that is used to facilitate the efficient transport of steam to the turbine. An air ejector removes noncondensable gases (e.g., air and fission noble gases) and transports them past a radiation monitor where they are characterized prior to release to the environment.

A portion of the secondary water in the PWR steam generator is removed to reduce chemical contaminants that promote corrosion of the steam generator tubes. Corrosion is a significant factor in the degradation of steam generator tubes requiring tube repair, tube plugging, or replacement of the steam generator. Steam generator tube damage presents a pathway for radioactive material in the primary system to enter the secondary system.

This liquid pathway intentionally removed from the steam generator is commonly known as steam generator blowdown. The blowdown liquid is discharged to the environment after being monitored for radiation. Recall that the condenser's air ejector radiation monitor measures the secondary system's noble gas activity. A positive indication of air ejector or blowdown activity is an indication of primary to secondary leakage via leaking or cracked steam generator tubes. Normally, concurrent, elevated reading on steam line monitors, the air ejector monitor, and the blowdown monitor indicate steam generator tube damage.

2.4.1.5 Reactor Control and Protection Systems

A dissolved neutron absorber or poison usually in the form of boric acid is used for reactivity control in a PWR. During power operation, the boron concentration in the primary coolant is periodically (e.g., at least once a shift) reduced to compensate for the loss of reactivity in the fuel as fissions occur and fission product poisons build up in the core. Control rods are used for fine reactivity adjustments during reactor operation, but their use is infrequent. Reactor start-up occurs either through rod withdrawal or through dilution of the boric acid concentration in the primary coolant.

2.4.1.6 Engineered Safety Features

Engineered safety systems provide ultimate protection to the core and safe shutdown capability. The emergency core cooling system consists of independent safety systems that are initiated as a function of predetermined reactor coolant system parameters including pressure. Generation III and IV reactors rely on passive safety systems, not the active systems used in Generation II reactors. For example, the Generation III reactors use passive safety systems to mitigate most design basis events (DBEs). Rather than employing active components, Generation III systems are passively actuated from stored energy derived from batteries or compresses gases, or use natural phenomena such as gravity, evaporation, and condensation. Passive safety systems perform a number of functions in Generation III systems including reactor core decay heat removal, ensuring reactor cooling water supply, maintaining containment cooling and control room habitability.

Another safety feature of a PWR is its containment building. The PWR containment structure encloses the reactor vessel and its concrete radiation shielding, the pressurizer, the reactor coolant piping connecting the reactor vessel and steam generators, steam generators, reactor coolant pumps, and safety injection accumulators. PWR containments are cylindrical in shape and constructed from reinforced concrete with a metal liner. The concrete thickness is on the order of 1.1 m with a 3.8-cm steel liner plate. Typical containment volumes are between 2.8×10^4 and $5.7 \times 10^4 \, \text{m}^3$.

An additional safety system, known as reactor building spray, is housed within the containment structure. The spray system functions to reduce containment pressures by condensing the steam accompanying a loss-of-coolant accident (LOCAs) and also serves to reduce the associated iodine source term. The reduction in iodine concentration is accomplished by spraying a mixture containing water and sodium hydroxide into the containment atmosphere that enhances the retention of radioiodine in the condensed steam. This same principle forms the basis for administering radioiodine in a basic pH solution in therapeutic nuclear medicine applications to minimize the amount of radioactive material that volatilizes.

2.4.2 Boiling Water Reactors

The boiling water reactor is a direct-cycle steam generating system. In a BWR, water is boiled using core fission heat energy within the reactor vessel. The resultant steam exits the reactor vessel and is transported to the turbine generator. After flowing through the turbine generator, the steam is condensed and collected in the condenser. From the condenser, the condensate is pumped back to the reactor vessel. As there is no secondary cooling system, the steam contains radioactive material including activation products such as ^{16}N. The presence of ^{16}N in the steam system leads to secondary system dose rates that are larger than the corresponding PWR doses rates.

2.4.2.1 BWR Reactor Assembly

The BWR reactor assembly consists of the reactor vessel, internals (such as the steam separator, steam dryer, and core barrel), fuel elements, and control rods system. The BWRs control rods enter through the bottom of the core to avoid the engineering challenges associated with their transit through the steam separation and drying components.

2.4.2.2 BWR Reactor Core

The BWR core is composed of a group of fuel assemblies arranged to approximate a right circular cylinder. An individual fuel assembly has fuel rods arranged in a square array that is similar to the construction of, but smaller than, a PWR fuel assembly.

In a PWR, a control rod is inserted into a single fuel assembly. BWR control rods service four assemblies and have a cruciform cross section. A fuel assemble resides in each of the four quadrants of the BWR cruciform control rod.

The cruciform BWR control blade is comprised of stainless steel tubes, containing compacted boron carbide power (neutron poison), arranged in a stainless steel sheath. BWR control rod blades are control devices that are used to control the reactor core's neutron flux profile and provide shutdown capability.

2.4.3
CANDU Reactors

The CANDU reactor uses heavy water (99.8% deuterium dioxide) as the moderator, primary coolant, and reflector material. Although most CANDU reactors are operating in Canada, these systems are also deployed in Europe and Asia. A typical Generation-II CANDU system is on the order of 600 MW (electric), but advanced CANDU systems come with larger power output.

CANDU reactors carry all the associated external radiation hazards associated with PWRs and BWRs. However, the heavy water coolant and CANDU neutron spectrum are very effective in producing tritium via the $^2H(n, \gamma)^3H$ reaction.

Tritium exposures represent a significant fraction of the total worker effective dose in a CANDU facility. Historically, 30–40% of the effective dose received by CANDU workers is because of the exposure of tritium. The dose control and the expense of the deuterium oxide coolant require primary system leakage be kept to near zero levels.

2.4.3.1 General Description

The Generation-II CANDU reactor vessel is a cylindrical steel vessel oriented horizontally. The vessel is 6.1 m in length, has a diameter of 7.9 m, and is penetrated by about 400 horizontal channels (pressure tubes) designed to withstand primary system pressure. These channels contain the fuel elements, and the pressurized primary coolant flows through the channels and around the fuel elements to remove the fission heat.

The CANDU reactor fuel is composed of natural uranium dioxide pellets packed into a corrosion-resistant zirconium alloy tube about 0.5-m long and 1.3-cm diameter to form a cylindrical fuel rod. Each fuel assembly contains 37 rods and 12 assemblies are packed end-to-end in each pressure tube.

CANDU reactors are unique as refueling or the removal of spent fuel and its replacement by new fuel is performed while the reactor is operating. The refueling machine inserts a fresh fuel bundle into one end of a horizontal pressure tube that is temporarily isolated from the main coolant channel. A spent fuel bundle is then displaced at the other end and is removed.

2.4.3.2 Control Systems

The CANDU reactor has several types of vertical control elements that are strong neutron absorbing materials. These control systems include cadmium rods used during reactor start-up and shutdown, absorbing rods to control power and ensure a uniform power distribution within the core, and emergency shutdown injection of gadolinium nitrate into the core.

2.4.3.3 Steam System

The ends of the pressure tubes are connected to manifolds that direct the heated heavy water coolant to conventional U-tube steam generators. After exiting the steam generator, the primary coolant is pumped back to the inlet plenum and then into the pressure tubes.

2.4.3.4 Safety Systems

A break in a single pressure tube results in the loss of primary coolant. Once the break is detected, the damaged tube is disconnected and reactor operation proceeds with the other tubes. A more serious loss-of-coolant accident can result in fuel damage with a subsequent release of radioactivity. It would develop from a break in one of the coolant headers or in the piping connecting the reactor and steam generators. For these larger LOCA events, an emergency core cooling system supplies additional cooling capability. The separate moderator system provides a significant heat sink.

A concrete containment structure encloses the primary coolant system including the reactor vessel, steam generators, and interconnecting piping. A water spray system within the CANDU containment condenses steam and reduces containment pressures resulting from a large break LOCA.

2.4.4 High-Temperature Gas-Cooled Reactors

The high-temperature gas-cooled reactor is a thermal nuclear power system that utilizes graphite as the moderator material and helium gas as the coolant medium. Graphite becomes physically stronger with increasing temperature up to about 2480 °C. The chemically inert property of helium permits reactor operation at high temperatures. Superheated steam, derived from the energy transfer from the helium gas to water in a heat exchanger, drives the turbine generator to produce electricity. Although other fuel cycles are possible, the basic HTGR design utilizes the effective conversion of fertile ^{232}Th into fissile ^{233}U. Fuel reprocessing is required to recover the ^{233}U produced from ^{232}Th for use in subsequent cycles.

Other reactor designs use pressurized CO_2 gas as the coolant. The Magnox and advanced gas-cooled reactors (AGRs) are graphite-moderated reactors that use uranium fuel. AGRs are a second-generation of gas-cooled reactors that have higher fuel and gas temperatures than Magnox reactors. These AGR system characteristics yield a higher steam cycle efficiency than the Magnox reactor.

2.4.5
Liquid Metal Fast Breeder Reactors

A liquid metal fast breeder reactor is designed to produce more fuel than it consumes, and to produce electrical energy. As liquid sodium is utilized as the coolant, the LMFBR has no moderator. Consequently, fast neutrons produce most of the fission events. Usually, ^{238}U is used to produce ^{239}Pu via neutron capture. In addition, some fissionable ^{241}Pu is produced. The LMFBR is advantageous in extending uranium reserves to enhance the long-term use of nuclear power generation.

A primary activation product of the sodium coolant is ^{24}Na produced through the ^{23}Na(n, γ) reaction. The resultant ^{24}Na has a half-life of about 15 h and decays with the emission of two gamma rays, 1.369 and 2.754 MeV, each with a yield of 100%.

The heat from the primary sodium coolant is transferred to an intermediate heat exchanger with sodium on its secondary side. After transferring its heat, the primary sodium coolant is returned to the reactor. The heated secondary sodium is pumped to a steam generator where it heats water in the tertiary system and produces steam. As in other reactor designs, the steam is directed to a turbine for the production of electricity.

Periodically, the LMFBR fuel rods are removed from the reactor vessel for recovery of plutonium and uranium. In principle, the plutonium produced by an LMFBR would be sufficient to refuel that reactor and provide part of the fuel for another reactor. Reprocessing and applications of fast reactors in future fuel cycles are addressed in Volume II.

2.4.6
Generation II Summary

Generation II reactors provide reasonably reliable power and safety performance. However, their safety performance has been criticized, with the March 28, 1979 Three Mile Island Accident serving as the focal point of that criticism. Generation III reactors are envisioned to improve the safety performance and economic viability of fission reactors. Their design and operational characteristics are summarized in the following section of this chapter. Prior to reviewing these characteristics, an overview of the key radiological design considerations of Generation III and IV reactors is presented.

2.5
Generation III and IV Radiological Design Characteristics

From a radiological perspective, the Generation III and IV facility design should ensure that effective doses to plant workers and members of the public are as low as reasonably achievable (ALARA). This is achieved through the design of systems, structures, and components that

- attain optimal reliability and maintainability to reduce maintenance frequency and duration requirements for radioactive components;
- reduce radiation fields to allow operations, maintenance, and inspection activities to be performed in a manner that leads to optimum effective doses. The design should accommodate the use of robotic technology to perform maintenance and surveillance in high radiation areas;
- reduce access, repair, and equipment removal times to limit the time spent in radiation fields. Adequate equipment spacing and job preparation areas facilitate access for maintenance, repair, and inspection;
- utilize modularized components to facilitate their replacement or removal to a lower radiation area for repair;
- accommodate remote and semiremote operation, maintenance, and inspection to reduce the time spent in radiation fields;
- reduce concentrations of cobalt and nickel for materials in contact with the primary coolant to minimize the production of the ^{58}Co and ^{60}Co activation products. These isotopes are the major sources of radiation exposure during shutdown, maintenance, and inspection activities at light water reactors. Exceptions to this design specification may be necessary to enhance component or system reliability. However, the decision to utilize materials that produce ^{58}Co and ^{60}Co activation products must be made in a deliberate manner.
- separate radioactive and nonradioactive systems. High radiation sources should be located in separate shielded cubicles. In addition, equipment requiring periodic servicing or maintenance (e.g., pumps, valves, and control panels) should be separated from sources with higher radioactivity such as tanks and demineralizers;
- incorporate reach rods or motor operators into valves located in high radiation areas;
- reduce the accumulation of radioactive materials in equipment and piping. Flushing connections facilitate the removal of radioactive materials from system components. Locating drains at low points enhances the achievement of this design aspect. Piping should be seamless, and the number of fittings minimized to reduce the accumulation of radioactive materials at seams and welds;
- locate systems that generate radioactive waste close to waste processing systems to minimize the length of piping carrying these materials;

- minimize the potential for pipe plugging by routing lines that carry resin slurries vertically. Large radius bends should be used instead of elbows to minimize the potential for pipe plugging.

These radiological design characteristics are common to a variety of organizations involved in advanced reactor regulation and standards. This includes the US Nuclear Regulatory Commission (NRC) and the International Atomic Energy Agency (IAEA). With these general design considerations established, the specific characteristics of Generation III and IV fission reactors are considered.

2.6
Generation III

The Generation II reactors are characterized by a wide variety of design concepts within the various reactor types (e.g., PWRs). These varied design concepts complicated the licensing process and the lack of standardization hindered the effective communication of operating experience. As illustrated by the Three Mile Island Unit 2 accident, Generation II reactors could be vulnerable to off-normal operating conditions and utilized active safety systems. Active systems require either electrical or mechanical operation to occur during an off-normal event, and these systems are potentially vulnerable to mechanical or electrical failures. A more reliable safety system operates passively and uses inherent physical properties as the basis for their design. These inherent properties include physical phenomena such as gravity and convection as the basis for their functionality.

Incorporating the lessons learned from operating Generation II reactors, Generation III reactors utilize standardized designs to facilitate the licensing process, reduce capital cost, and reduce construction time. Generation III designs are also less complex, have larger safety margins, and should be more reliable than Generation II reactors. As a result, Generation III reactors minimize their maintenance requirements, simplify operations, and are less vulnerable to operational upsets. These characteristics should lead to higher capacity factors and longer operating lifetimes. In addition, the Generation III designs focus on passive safety systems. Examples of Generation III reactors are summarized in Table 2.4 which include advanced boiling water reactors (ABWR), advanced pressurized water reactors (APR or APWR), advanced passive (AP) PWRs, European pressurized water reactor (EPWR), simplified boiling water reactors (SBWR), economic and simplified boiling water reactor (ESBWR), CANDU reactors, advanced Canadian deuterium reactors (ACR), pebble bed modular reactor (PBMR), and gas turbine–modular helium reactors (GT-MHR).

The reactor types noted in Table 2.4 include light water reactors, heavy water reactors, and high-temperature reactors (HTRs). Light water and heavy water reactors operate predominantly with uranium fuel. Operation with uranium and plutonium fuels is also possible. High-temperature reactors operate with a variety of fuel types including highly enriched uranium and thorium, ^{233}U and thorium, and plutonium

Table 2.4 Generation III reactors.

Country (vendor)	Reactor	Power rating (MWe)	Design progress	Main features
US–Japan (General Electric–Hitachi–Toshiba)	ABWR	1300	Commercial operation in Japan since 1996	Light water reactor with a 4-year construction period, improved efficiency, simplified operation, and evolutionary design
South Korea (derived from Westinghouse designs)	APR-1400	1400	Developed for new South Korean Shin Kori Units 3 and 4, expected to be operating about 2010	Light water reactor with simplified operation, increased reliability, projected 4-year construction period, and evolutionary design
US (Westinghouse)	AP-600 (PWR)	600	United States Nuclear Regulatory Commission design certified in 1999	Light water reactor with passive safety features, projected 3-year construction period, and 60-year plant life
US (Westinghouse)	AP-1000 (PWR)	1100	NRC design approval 2004	Light water reactor with passive safety features, projected 3-year construction period, 60-year plant life, and capable of operating with a mixed oxide core
Japan (Westinghouse and Mitsubishi)	APWR	1500	Basic design in progress with an APWR planned at Tsuruga, Japan	Light water reactor with hybrid active and passive safety systems, and simplified design, construction, and operation
France–Germany (Framatome ANP)	EPWR	1600	Future French standard with French design approval	Light water reactor with improved safety features, high fuel efficiency, and low projected costs
Germany (Framatome ANP)	SBWR-1000	1200	Under development with pre-certification in the United States	Light water reactor with high fuel efficiency and passive safety features

(continued)

Table 2.4 (*Continued*)

Country (vendor)	Reactor	Power rating (MWe)	Design progress	Main features
USA (General Electric)	ESBWR	1390	Developed from ABWR with precertification in the United States	Light water reactor with short construction time and enhanced safety features
Canada (Atomic Energy of Canada, Limited)	CANDU-9	925–1300	Canadian licensing approval 1997	Heavy water reactor with flexible fuel requirements and passive safety features
Canada (Atomic Energy of Canada Limited)	ACR	700 1000	ACR-700: precertification in the United States, ACR-1000 proposed for United Kingdom	Light water reactor with low-enriched fuel and passive safety features
South Africa (Eskom, BNFL)	PBMR	165 (module)	Prototype construction planned for circa 2006, precertification in the United States	High-temperature gas-cooled reactor designed to be a low-cost, modular plant. It features a direct-cycle gas turbine, high fuel efficiency, and passive safety features
USA–Russia–multinational (General Atomics–Minatom)	GT-MHR	285 (module)	Under development in Russia by a multinational joint venture	High-temperature gas-cooled reactor designed to be a low-cost, modular plant. It features a direct-cycle gas turbine, high fuel efficiency, and passive safety features

Derived from Uranium Information Centre Nuclear Issues Briefing Paper 16 (2005)

and thorium. Thorium fuels have been used most prominently in HTRs. Other reactor development is possible using fast neutron systems.

The Generation III reactors represent incremental enhancements of the basic Generation II design. Therefore, their overall operating characteristics and accident consequences are similar to the Generation II designs. The major improvements come in enhanced safety performance and gains in operability and maintainability.

2.6.1
Safety Objectives and Standards

Although there are a large number of national and international organizations that provide safety objectives and standards for advanced reactors, IAEA guidelines are representative of these standards and sufficient for the purpose of this book. Additional safety objectives and design criteria are provided in subsequent sections.

IAEA safety standards address five areas: (1) safety of nuclear facilities, (2) radiation protection and safety of radiation sources, (3) safe management of radioactive waste, (4) safe transport of nuclear materials, and (5) general safety. The IAEAs International Nuclear Safety Advisory Group (INSAG) also provides advice and guidance.

In the radiation protection area, INSAG recommendations take into account recommendations made by a number of international organizations including the United Nations Scientific Committee on the Effects of Atomic Radiation (UNSCEAR), the International Commission on Radiological Protection (ICRP), and the International Commission on Radiation Units and Measurements (ICRU). The IAEA proposes an overall radiation protection objective "to ensure that in all operational states radiation exposure within the installation or due to any planned release of radioactive material from the installation is kept below prescribed limits and as low as reasonably achievable, and to ensure mitigation of the radiological consequences of any accident."

To further foster public safety, the IAEA considers improved accident mitigation to be an essential complementary means to ensure public safety. Associated with this goal is the need to demonstrate that for accidents without core melt, no protective actions (evacuation or sheltering) are required for population within the vicinity of the plant. For severe design basis accidents, only protective actions that are limited in area and time are needed. This includes restrictions on food consumption after the release of radioactive material from the facility.

The INSAG and IAEA Safety Standards Series documents offer a number of safety goals for future nuclear plants. These include reducing the core damage frequency (CDF) relative to current plants, considering severe accidents in the design of the plants, ensuring that severe accident releases to the environment are maintained as low as practicable with a goal of providing simplification of emergency planning, providing human factors to minimize operator activity during an emergency event, adopting digital instrumentation and control systems, and utilizing passive systems and components.

To reduce the probability of an accident and to mitigate their radiological consequences, system design improvements will be incorporated into future reactors. Examples of improvements in the Generation II design concept include the following:

- Maintaining larger safety design margins. Examples include larger water inventories in the pressurizer to provide primary system surge and transient control and in the steam generators to maintain a heat sink during transients. Additional safety margins include a lower core power density and negative reactivity coefficients to reduce safety system challenges.

- Providing reliable, redundant, and diverse safety systems. These systems should also incorporate greater physical separation to ensure that events or failures do not impact multiple trains of a safety system.
- Incorporating passive cooling and condensing systems.
- Providing more robust containment buildings that are sufficiently large to withstand the temperatures and pressures from a design basis accident. These containments would not require fast acting pressure reduction systems, but would utilize passive systems (e.g., hydrogen control systems) to ensure their integrity in the event of a severe accident. The incorporation of an outer, second containment should also be considered to provide additional protection against external events and allow for detection and filtration of radioactive material that leaked from the inner containment.

To present the health physics aspects of Generation III reactors, a discussion of the operating characteristics of these reactors is presented. These operating characteristics provide a basis for developing the health physics characteristics of the Generation III systems. The health physics characteristics are not specific to any of the reactor types of Table 2.4, and represent generic descriptions for a particular reactor type. For simplicity, the Generation III reactors are broadly classified as PWRs, BWRs, CANDUs, and HTGRs. As the Generation III designs are incremental improvements over their Generation II counterparts, only differences between these two designs are addressed.

2.6.2
PWRs

The Generation III pressurized water reactor systems utilize the basic Generation II structures including steam generators, but have enhanced safety systems and operating characteristics. In advanced PWRs, the operational safety of the plant is improved by increasing the ability of the design to accommodate abnormal plant conditions without activating safety systems and to provide plant operators sufficient time to address these conditions before the activation of automated safety systems. In addition, new safety systems are added and the reliability of existing systems is increased. For example, the control room and information processing systems have been revamped to facilitate operator understanding and response to plant conditions.

The reactor is housed in an enhanced containment designed to withstand credible events. These changes lead to increased safety, improved operability, improved reliability and availability, and reduced cost. Another feature of advanced PWRs is that they can be configured to operate with plutonium fuel in a mixed oxide fuel mode.

The Generation III PWRs have reduced occupational doses as a result of improvements in reliability and maintainability. Additional dose savings are achieved by improving steam generator performance that minimizes dose intensive inspection and surveillance activities. In addition, eliminating steam generator replacement outages not only leads to significant dose savings but also contributes to the improved economic viability of advanced PWRs. Steam generator reliability improvements will be the key indicator of the success of Generation III PWRs.

2.6.3
BWRs

The design for the ABWR differs from today's BWR in a number of ways. Advanced boiling water reactors offer a variety of innovations including natural circulation in the core region, no recirculation pumps, and, for most designs, passive decay heat removal systems. The safety improvements incorporated in the ABWR result in a more compact design than the BWR. For example, the ABWR's facility volume is only about 70% of the volume of contemporary BWRs. This size reduction decreases the construction time and cost, and makes the design more robust and less susceptible to Earthquakes. In Generation II BWRs, the control rods are hydraulic. In the ABWR, they are electrohydraulic. Having an additional drive mechanism reduces the probability of failure and enhances the safety performance. All major equipment and components are engineered for reliability and ease of maintenance. These design considerations minimize downtime and reduce worker radiation doses. Most of the ABWR improvements are based on proven BWR technology, and do not need extensive research and development (R&D) for their deployment. ABWR R&D should focus on demonstrating the economic viability of the design enhancements.

Given the improvements in reliability and maintenance, Generation III BWRs should have reduced occupational doses compared to their Generation II counterparts. Minimizing high dose activities involving contaminated systems is the key element of dose reduction at a Generation III BWR.

Both Generation III PWR and BWR units also need to demonstrate source term control including minimizing primary system corrosion, primary coolant system activity, and source term buildup in primary system components. Source term control depends on maintaining rigorous chemistry specifications, ensuring replacement components minimize the generation of activation products, and maintaining good fuel performance with minimal fuel damage.

2.6.4
Advanced CANDU

The advanced CANDU reactor incorporates a compact core design, improved thermal efficiency through higher pressure steam turbines, a light water primary cooling system, a heavy water moderator completely contained within the calandria, online refueling, extended fuel life of three to six times over natural uranium by using slightly enriched uranium (SEU) 2% oxide fuel, and prefabricated structures and

Table 2.5 Comparison of Generation II and III CANDU core systems.

Core parameter	Generation II (CANDU-6)	Generation III (ACR-700)
Power (MWe)	728	731
Fuel channels	380	284
Diameter (m)	7.59	5.21

systems. A comparison of the Generation II and III CANDU Core Systems is summarized in Table 2.5.

The compact core size in the Generation III CANDU has a number of health physics implications. Maintenance and surveillance requirements are reduced because there are fewer fuel channels and the core is smaller. These changes result in reduced work around the calandria that translates into less occupational dose. The tritium dose is also reduced because the coolant has been changed from heavy to light water. This is primarily the result of the reaction,

$$^{2}H + n \rightarrow {}^{3}H + \gamma, \tag{2.1}$$

that occurs to a lesser degree because of the switch from heavy to light water coolant. As noted earlier, 30–40% of the anticipated occupational dose in a Generation II CANDU is attributed to tritium intakes. Eliminating the heavy water coolant reduces the tritium dose. Other Generation III CANDU options using thorium fuel are also under consideration.

2.6.5
HTGRs

Proponents of HTGRs suggest their proliferation resistance, inherent safety, fuel construction that retains fission products, and ability to produce high temperatures for hydrogen production are positive considerations for their deployment. Improved safety and reliability depends on high-quality fuel manufacturing and development of an industrial infrastructure.

From a health physics perspective, the HTGR design offers a number of advantages. The coated fuel concept provides an enhanced fission product barrier that minimizes releases of fission products into the gas coolant and subsequently to the environment. Compared to PWRs and BWRs, the HTGR gas coolant produces minimal activation products with tritium being the major product. In addition, the elimination of ^{16}N produced in light water coolant has definite benefits from both radiation protection and design simplification perspectives.

2.6.6
Generation III Safety System Examples

To illustrate the passive safety system philosophy in a Generation III facility, aspects of Framatome's SBWR-1000 reactor are presented. The SBWR-1000 partially replaces active safety systems with passive systems. Examples include (1) the emergency condenser system, (2) containment cooling condenser, (3) core flooding system, and (4) pressure pulse transmitters. Each of these systems is briefly addressed because they mitigate the radiological source term during an abnormal or emergency event by protecting one or more fission product barriers (i.e., the fuel/clad, the primary coolant system piping, and the containment structure).

2.6.6.1 Emergency Condenser System

One of the key innovations of an advanced BWR is the passive design of the emergency condensers that remove heat from the reactor core upon a drop in reactor pressure vessel water level. The tubes of the emergency condensers are submerged in the core flooding pool and filled with water when the reactor vessel's water level is in the normal operating range. As reactor vessel water level decreases, water drains from the condenser tubes. Steam from the reactor then enters the drained condenser tubes and condenses. The condensed steam inside the tubes flows by gravity into the reactor vessel and maintains core cooling. The action of the emergency condensers, which maintains the integrity of the fuel and minimizes the radiological source term available for release to the plant/environment, occurs automatically and requires no electrical power, control logic, or switching operations.

2.6.6.2 Containment Cooling Condensers

During a LOCA, steam is released into the BWR drywell (containment) causing drywell temperatures to increase, and action is required to mitigate this increase. The cooling condensers remove heat from the drywell and transfer it to the water in the storage pool located above the reactor. In the SBWR-1000, this is a passive heat transfer that requires no electric power or switching operations, and relies on no active systems. Removing containment heat minimizes the potential for a loss of fission product barriers and limits the release of radioactive material to the environment.

2.6.6.3 Core Flooding System

When reactor coolant system pressure drops below a preset value, self-actuating check valves open and permit the gravity flow of water from the core flooding system to the reactor vessel. Check valves open on differential pressure between the core flooding tank and the primary coolant system. Both pressure- and gravity-induced flow are passive features requiring no electric power or active switching operations. As a result of these passive features, water covers the core, provides core cooling, and minimizes the breach of the fuel and cladding. Preserving the fuel/clad fission product barrier minimizes the release of radioactive material from the primary system. Passive core flooding systems are also incorporated into a number of Generation II designs.

2.6.6.4 Passive Pressure Pulse Transmitters

Passive pressure pulse transmitters are small heat exchangers. When reactor water level decreases, pressure increases on the secondary side of the heat exchanger. This pressure increase changes the position of a pilot valve connected to the secondary side of the heat exchanger. The change in valve position initiates action to trip (shutdown) the reactor and initiate containment isolation without the need for electrical power or logic signals. The pilot value function is similar to the Generation II Turbine/Generator Trip design involving pilot valves that change position in the electrohydraulic system. This passive reactor trip protects the fuel/clad and primary coolant piping fission product barriers. Containment isolation ensures the integrity of the third fission product barrier.

2.7 Generation IV

The Generation IV International Forum (GIF), initiated in the year 2000, represents countries having a vested interest in nuclear energy. The forum, including the United States, Argentina, Brazil, Canada, France, Japan, South Korea, South Africa, Switzerland, and the United Kingdom, are committed to the joint development of the next generation of nuclear technology. These 10 nations agreed on six Generation IV nuclear reactor technologies for deployment between 2015 and 2025, and the characteristics of these reactors are summarized in Table 2.6. Some of these technologies operate at higher temperatures than the Generation II and III reactors, and four are designated for hydrogen production.

The six design concepts represent the potential for improved economics, safety, reliability, and proliferation resistance. The Generation IV technologies maximize the utilization of fissile resources and minimize high-level waste. The specific Generation IV reactors addressed by the GIF are gas-cooled fast reactors (GFRs), lead–bismuth-cooled fast reactors (LFRs), molten salt epithermal reactors (MSRs), sodium-cooled fast reactors (SFRs), supercritical water-cooled reactors (SWCR), and very high-temperature, helium-cooled, graphite-moderated thermal reactors (VHTRs). An advantage of the Generation IV design is the capability for full actinide recycling using a closed fuel cycle concept.

Table 2.6 summarizes the Generation IV systems that include a variety of reactor types and design philosophies. The basic characteristics of these systems are summarized in subsequent sections of the text. Before reviewing the Generation IV reactor types, an examination of expected activation products is provided. These activation products are summarized in Table 2.7 and focus on the unique Generation IV specific isotopes. The production mode for the activation product is also provided. Subsequent discussion outlines the unique Generation IV materials that lead to radionuclides not normally associated with Generation II and III reactors. The Generation IV reactors also produce the fission related activation products provided in Table 2.1.

Generation IV graphite-moderated reactors may also produce other isotopes owing to the impurities in this moderator. These isotopes include ^{38}Cl [$^{37}Cl(n, \gamma)$], ^{46}Sc [$^{45}Sc(n, \gamma)$], ^{82}Br [$^{81}Br(n, \gamma)$], and ^{152}Eu [$^{151}Eu(n, \gamma)$].

Table 2.6 Generation IV reactor concept characteristics.

Reactor technology	Power rating (MWe)	Operating temperature (°C)	Fuel	Economic justification
Gas-cooled fast reactors	288	850	^{238}U, other fertile materials, and fissile materials	Electricity and hydrogen are produced
Lead–bismuth-cooled fast reactors	50–1200	550–800	Fuel is ^{238}U metal or nitride	Electricity and hydrogen are produced
Molten salt epithermal reactors	1000	700–800	The uranium fuel is dissolved in a salt coolant	Electricity and hydrogen are produced
Sodium-cooled fast reactors	150–1500	550	Metal or mixed oxide (MOX)	Electricity is produced
Supercritical water-cooled reactors (thermal and fast versions)	1500	510–550	UO_2	Electricity is produced
Very high-temperature, helium-cooled, graphite-moderated thermal reactors	250	1000	UO_2 Can also utilize PBMR or GT-MHR fuel	Electricity and hydrogen are produced

Derived from Uranium Information Centre, Briefing Paper # 77 (2005).

2.7.1
Gas-Cooled Fast Reactors

The reference Generation IV GFR utilizes a fast neutron spectrum, and has a helium-cooled reactor core in conjunction with a direct Brayton cycle. The Brayton or Joule cycle is an ideal, closed cycle process that incorporates a direct-cycle helium turbine for electricity production and uses process heat for the thermochemical production of hydrogen. With a fast neutron spectrum and full actinide recycle, the GFR minimizes the production of long-lived radionuclides.

The GFR design produces a high thermal efficiency of about 48%. This is considerably higher than the efficiency of about 33% encountered in Generation II light water systems. Operating in conjunction with a closed fuel cycle, GFRs enhance the utilization of uranium and minimize waste generation. Actinide waste is limited by incorporating full actinide recycle as part of the closed fuel cycle. To improve efficiency, the GFR facility is colocated with other fuel cycle facilities for on-site spent fuel processing and fuel refabrication.

As part of the GFR development process, different fuels will be tested for their compatibility to operate with a fast neutron spectrum and at high temperatures. GFR fuel incorporates a number of enhancements including advanced coatings and ceramic fuel composites to facilitate fission product retention.

In view of its operating characteristics, GFRs require unique materials for their successful implementation. These include ceramic materials used for incore structures. A variety of materials are under consideration including SiC, ZrC, TiC, NbC, ZrN, TiN, MgO, and $Zr(Y)O_2$. Intermetallic compounds such as Zr_3Si_2 are promising candidates for the neutron reflector material in a GFR. A representative set of these materials utilized in the Generation IV Industrial Forum reactors are reflected in Table 2.7.

The GFR concept has a number of characteristics that support health physics design objectives. These include the fuel composition and actinide recycle that permit operation of a closed fuel cycle. The fuel composition fosters fission product retention and limits the release of radioactive materials from the fuel fission product barrier.

Full actinide recycle eliminates long-term waste disposal and associated radiation dose concerns. In addition, the closed nature of the fuel cycle limits the occupational doses associated with waste disposal and storage. The extent of these health physics advantages depends on GFR fuel performance and the development of actinide recycling technology.

2.7.2
Lead–Bismuth-Cooled Fast Reactors

Liquid metal reactors, using both lead and bismuth coolants, are inherently safe systems. These reactors have the potential for significant waste volume reduction compared to advanced light water reactors. The key advantage of the liquid metal reactors is their potential to recycle essentially all of the actinides. Challenges include proliferation resistance and economic viability. While their economics can be improved through design simplification including modularization of the design, proliferation issues can be minimized if the design is successful in efficiently recycling actinides, particularly ^{239}Pu.

The LFR system utilizes a fast neutron spectrum, and its core is cooled with a lead or lead–bismuth eutectic coolant. The system operating characteristics offer considerable flexibility. Core lifetimes can be as long as 15–20 years. Planned power ratings include 50–150 MWe (fabricated reactor module), 300–400 MWe (modular design), and 1200 MWe (base load facility).

The LFR concept utilizes a closed fuel cycle with the supporting fuel cycle facilities residing in a central or regional location. Within the closed fuel cycle, LFR facilities provide efficient utilization of uranium resources and management of actinides.

LFR reactors are cooled passively through natural convection with an outlet temperature of 550 °C. Outlet temperatures could reach 800 °C depending on the

Table 2.7 Activation products in Generation IV fission power reactors.

Nuclide	Half-life	Decay mode	Production mode
^{3}H	12.3 yr	β^-	GFRs and VHTRs (^{4}He gas coolant and fuel): tertiary fission, ^{4}He$(\gamma, p)^3$H, and ^{4}He$(n, d)^3$H MSRs (fluoride salt coolant): ^{6}Li$(n, \alpha)^3$H
^{10}Be	1.6×10^6 yr	β^-	MSRs (beryllium fluoride salt coolant): ^{9}Be$(n, \gamma)^{10}$Be
^{14}C	5730 yr	β^-	GFRs and VHTRs (graphite): ^{14}N$(n, p)^{14}$C and ^{13}C$(n, \gamma)^{14}$C GFRs (gas coolant): ^{17}O$(n, \alpha)^{14}$C
^{15}O	122 s	β^+ γ	GFRs (gas coolant): ^{16}O$(n, 2n)^{15}$O
^{16}N	7.14 s	β^- γ	GFRs (gas coolant): ^{16}O$(n, p)^{16}$N MSRs (fluoride salt coolant): ^{19}F$(n, \alpha)^{16}$N
^{17}N	4.2 s	β^- γ n	GFRs (gas coolant): ^{17}O$(n, p)^{17}$N
^{18}F	110 m	β^+ γ	MSRs (fluoride salt coolant): ^{19}F$(n, 2n)^{18}$F
^{19}O	26.9 s	β^- γ	GFRs (gas coolant): ^{18}O$(n, \gamma)^{19}$O MSRs (fluoride salt coolant): ^{19}F$(n, p)^{19}$O
^{20}F	11.0 s	β^- γ	MSRs (fluoride salt coolant): ^{19}F$(n, \gamma)^{20}$F SFRs (liquid sodium coolant): ^{23}Na$(n, \alpha)^{20}$F
^{22}Na	2.60 yr	β^+ γ	MSRs (sodium salt coolant) and SFRs (liquid sodium coolant): ^{23}Na$(n, 2n)^{22}$Na and ^{23}Na$(\gamma, n)^{22}$Na
^{23}Ne	37.2 s	β^- γ	MSRs (sodium salt coolant) and SFRs (liquid sodium coolant): ^{23}Na$(n, p)^{23}$Ne
^{24}Na	15.0 d	β^- γ	MSRs (sodium salt coolant) and SFRs (liquid sodium coolant): ^{23}Na$(n, \gamma)^{24}$Na GFRs (in core materials): ^{24}Mg$(n, p)^{24}$Na
^{25}Na	59.3 s	β^- γ	GFRs (in core materials): ^{25}Mg$(n, p)^{25}$Na
^{27}Mg	9.46 m	β^- γ	GFRs (in core materials): ^{26}Mg$(n, \gamma)^{27}$Mg GFRs (in core materials): ^{30}Si$(n, \alpha)^{27}$Mg

(continued)

Table 2.7 (Continued)

Nuclide	Half-life	Decay mode	Production mode
^{28}Al	2.25 m	β^-, γ	GFRs (in core materials) and VHTRs (fuel coating): ^{28}Si(n, p)^{28}Al
^{29}Al	6.5 m	β^-, γ	GFRs (in core materials) and VHTRs (fuel coating): ^{29}Si(n, p)^{29}Al
^{31}Si	2.62 h	β^-, γ	GFRs (in core materials) and VHTRs (fuel coating): ^{30}Si(n, γ)^{31}Si
^{45}Ca	165 d	β^-, γ	GFRs (in core materials): ^{48}Ti(n, α)^{45}Ca
^{45}Ti	3.08 h	β^+, γ	GFRs (in core materials): ^{46}Ti(n, 2n)^{45}Ti
^{46}Sc	83.8 d	β^-, γ	GFRs (in core materials): ^{46}Ti(n, p)^{46}Sc
^{47}Ca	4.54 d	β^-, γ	GFRs (in core materials): ^{50}Ti(n, α)^{47}Ca
^{47}Sc	3.42 d	β^-, γ	GFRs (in core materials): ^{47}Ti(n, p)^{47}Sc
^{48}Sc	43.7 h	β^-, γ	GFRs (in core materials): ^{48}Ti(n, p)^{48}Sc
^{51}Ti	4.80 m	β^-, γ	GFRs (in core materials): ^{50}Ti(n, γ)^{51}Ti
^{88}Y	107 d	β^+, γ	GFRs (in core materials): ^{89}Y(n, 2n)^{88}Y
^{89}Sr	50.5 d	β^-, γ	GFRs (in core materials): ^{89}Y(n, p)^{89}Sr GFRs (in core materials), MSRs (coolant component) and VHTRs (fuel coating): ^{92}Zr(n, α)^{89}Sr
^{89m}Y	15.7 s	γ	GFRs (in core materials): ^{89}Y(n, n')^{89m}Y
^{89}Zr	78.4 h	β^+, γ	GFRs (in core materials), MSRs (coolant component) and VHTRs (fuel coating): ^{90}Zr(n, 2n)^{89}Zr
^{90}Y	64.1 h	β^-	GFRs (in core materials): ^{89}Y(n, γ)^{90}Y and ^{93}Nb(n, α)^{90}Y GFRs (in core materials), MSRs (coolant component), and VHTRs (fuel coating): ^{90}Zr(n, p)^{90}Y

Table 2.7 (Continued)

Nuclide	Half-life	Decay mode	Production mode
^{90m}Y	3.19 h	β^-, γ	GFRs (in core materials): ^{89}Y(n, γ)^{90m}Y
^{92}Nb	3.7×10^7 yr	γ	GFRs (in core materials): ^{93}Nb(n, 2n)^{92}Nb
^{92m}Nb	10.1 d	γ	GFRs (in core materials): ^{93}Nb(n, 2n)^{92m}Nb
^{93}Zr	1.5×10^6 yr	β^-, γ	GFRs (in core materials), MSRs (coolant component), and VHTRs (fuel coating): ^{92}Zr(n, γ)^{93}Zr
^{93m}Nb	14.6 yr	γ	GFRs (in core materials): ^{93}Nb(n, n')^{93m}Nb
^{94}Nb	2.0×10^4 yr	β^-, γ	GFRs (in core materials): ^{93}Nb(n, γ)^{94}Nb
^{94m}Nb	6.2 m	β^-, γ	GFRs (in core materials): ^{93}Nb(n, γ)^{94m}Nb
^{95}Zr	64.0 d	β^-, γ	MSRs (coolant component) and VHTRs (fuel coating): ^{94}Zr(n, γ)^{95}Zr
^{97}Zr	16.8 h	β^-, γ	MSRs (coolant component) and VHTRs (fuel coating): ^{96}Zr(n, γ)^{97}Zr
^{203}Pb	51.9 h	γ	LFRs (lead coolant): ^{204}Pb(n, 2n)^{203}Pb
^{204m}Pb	66.9 m	γ	LFRs (lead coolant): ^{204}Pb(n, n')^{204m}Pb
^{205}Pb	1.4×10^7 yr	EC, X-rays	LFRs (lead coolant): ^{204}Pb(n, γ)^{205}Pb
^{209}Pb	3.3 h	β^-	LFRs (lead coolant): ^{208}Pb(n, γ)^{209}Pb
^{210}Pb	22.3 yr	β^-, γ, α	LFRs (bismuth coolant): ^{209}Bi(n, p)^{209}Pb + n $\rightarrow$ ^{210}Pb
^{210}Bi	5.01 d	β^-, γ, α	LFRs (bismuth coolant): ^{209}Bi(n, γ)^{210}Bi
^{210m}Bi	3.5×10^6 yr	α, γ	LFRs (bismuth coolant): ^{209}Bi(n, γ)^{210m}Bi
^{210}Po	138 d	α, γ	LFRs (bismuth coolant): ^{209}Bi(n, γ)^{210}Bi$\xrightarrow{\beta^-}$^{210}Po

success of fuel and materials research and development. LFR applications include the generation of electricity, hydrogen production, and desalination of seawater.

The LFR system incorporates a number of heat transport innovations. Natural circulation, lift pumps, and in-vessel steam generators enhance the heat transfer characteristics of the design. However, chemistry controls must be developed to facilitate the control of oxygen and ^{210}Pb. If issues concerning the development of fuel and reactor materials and achieving acceptable corrosion control properties for these materials can be resolved, then the professed advantages of the LFR can be realized.

2.7.3
Molten Salt Epithermal Reactors

The molten salt reactor utilizes an epithermal neutron spectrum to fission uranium. MSRs have potential advantages in terms of conversion efficiency, proliferation resistance attributable to the lower fuel inventory and plutonium buildup, and a reduced source term with online separation and removal of fission products. The circulating molten salt fuel is a mixture of zirconium, sodium, and uranium fluorides. Other options include lithium and beryllium fluoride with dissolved thorium and ^{233}U. The molten salt/fuel flows in channels through the core's graphite moderator. The MSR reference power level is 1000 MWe, the primary system operates at low pressures (<0.5 MPa), and the coolant outlet temperature is $\geq$700 °C. The fission heat is transferred through an intermediate heat exchanger, to a secondary coolant loop, and then to a final heat exchanger to the power conversion system.

As the fuel is in a liquid state, fuel processing is performed while the reactor is operating. The actinides and fission products form fluorides in the liquid coolant. This chemistry permits the fuel cycle to be tailored for the destruction (burnup) of minor actinides and plutonium and the removal of fission products. As the MSR fuel cycle allows full actinide recycle, high-level waste consists of fission products only. However, the entire MSR concept requires further refinement and development particularly of high-temperature structural materials, development of appropriate fuel characteristics, development of a molten salt-to-water heat exchanger, and resolution of nuclear and hydrogen safety issues.

As the fuel is dissolved in the coolant, the MSR design only has two fission product barriers. This is a significant reduction in the number of currently accepted fission product barriers. Any primary coolant leakage leads to the release of fission products into the plant and the containment building is the only remaining barrier. Auxiliary building (AB) leakage merits special attention because no containment exists in Generation II and III AB structures.

Given the level of development required for the MSR design to become fully mature, additional health physics issues may emerge. Potential areas of concern include the capability of the liquid fuel/coolant to retain fission and activation products following primary coolant leakage, and the capability of facility structures, systems, and components to contain fission and activation products following primary coolant leakage.

2.7.4
Sodium-Cooled Fast Reactors

SFRs operate with a fast neutron spectrum and utilize a liquid sodium coolant. Core heat is transferred through an intermediate sodium-to-sodium heat exchanger to a steam generating system for electricity generation. A variety of fuel options and power levels are envisioned for the SFR.

Plant options include a 150–1500-MWe system with metal alloy fuel and on-site processing. Mixed oxide fuel is utilized in a higher power version (500–1500 MWe), which requires reprocessing in off-site facilities. Both versions facilitate a closed fuel cycle with full actinide recycle.

Given the existing experience with Generation II sodium-cooled reactors, the health physics concerns are better defined than in other Generation IV systems. These concerns are similar to those of Generation II reactors, but are complicated by the potential for the sodium–water chemical reaction to mobilize fission and activation products. Health physics issues could also arise from the implementation of the closed fuel cycle with full actinide recycle. Experience with twentieth-century reprocessing suggests waste storage, environmental concerns, maintenance of heavily contaminated equipment, and decommissioning issues merit thorough evaluation.

2.7.5
Supercritical Water-Cooled Reactors

The supercritical water-cooled reactor operates above the critical point of water (22.1 MPa, 374 °C). The baseline SWCR is a 1500-MWe facility that operates at 25 MPa with a core outlet temperature of 510 °C. This baseline has a thermal efficiency of about 44%. The SWCR has a number of possible operating options. Higher power and temperature options are also feasible.

The two broad operating possibilities include a system based on either a thermal spectrum or a fast spectrum. The thermal neutron version uses once-through uranium dioxide fuel. The use of a single-pass fuel cycle negates the positive benefits of actinide recycle. From a health physics perspective, a single pass fuel cycle without actinide recycle is not a desirable Generation IV alternative.

A fast spectrum version permits actinide recycle using conventional reprocessing technology. However, the fast reactor version must overcome materials development issues. Both options utilize a direct-cycle system without a phase change, passive safety features, and operational characteristics similar to those of the simplified BWR. The secondary system is based on supercritical turbine technology utilized in advanced fossil power facilities.

The SWCR facility should have health physics issues that are similar to Generation II and III BWRs. Additional health physics issues may arise if fast reactor materials are not developed with sufficient lifetime and desired operational characteristics.

2.7.6
Very High Temperature Reactors (VHTR)

The VHTR is a graphite-moderated, helium-cooled reactor that operates with a thermal neutron spectrum. It operates with an outlet temperature of 1000 °C that permits high-temperature applications (e.g., hydrogen production and process heat for the petrochemical industry) to be accomplished. VHTR fuel consists of coated particles using materials such as SiC and ZrC that are formed into pebble elements or prismatic blocks. The plant uses once-through uranium fuel or U/Pu fuel to produce electricity or process heat through an intermediate heat exchanger. The VHTRs open fuel cycle does not address the waste disposal issues associated with long-term spent fuel storage. As noted earlier, an open fuel cycle is not desirable from a health physics perspective.

The VHTR requires technology advancements in fuel performance and high-temperature materials development. Shortcomings in either of these areas would potentially weaken the fuel and the primary coolant system fission product barriers. If these issues are resolved, the health physics issues will resemble those at a Generation II and III HTGR facility.

2.7.7
Radionuclide Impacts

The extent to which the radionuclides of Table 2.7 dominate effective doses at a Generation IV facility depends on operational characteristics that are not yet fully defined. On the basis of the Generation II and III experience, the following conclusions appear to apply to Generation IV systems:

(1) ^{3}H in the HTO form and ^{14}C as CO_2 present an internal radiation hazard particularly during refueling operations and primary system maintenance. The extent of the hazard depends on allowable leakage and primary system performance characteristics. A portion of the activation products of Table 2.1 including ^{131}I also present an internal radiation hazard.

(2) Submersion hazards exist for radioactive gases (e.g., ^{15}O, ^{16}N, ^{17}N, and ^{19}O). The noble gases produced in the fission process also present a submersion hazard, and these are primarily comprised of isotopes of Kr and Xe.

(3) External hazards exist for a variety of nuclides including the coolant activation products ^{16}N and ^{24}Na. The extent of the external radiation hazard depends on the magnitude of the production of fission and activation products (see Tables 2.1 and 2.7) that decay via beta and gamma emission.

(4) Off-site releases from a Generation IV reactor are expected to be similar to those from Generation II and III facilities. The off-site release source term is dominated by radioiodine and noble gas activity. MSRs present potential health physics issues because there are only two fission product barriers after the fuel is dissolved in the coolant.

(5) Open vice closed fuel cycles present additional health physics concerns. The open fuel cycle associated with SWCRs (thermal option) and VHTRs have negative waste storage and associated effective dose impacts. These concerns include the long-term storage of actinides with the potential for their release into the environment. Closed fuel cycle options have positive nuclear proliferation and waste disposal aspects as all actinides are destroyed during the recycling process. However, the effective dose profile merits careful evaluation.

2.7.8
Hydrogen Production

Interest in hydrogen production is growing in the United States and other countries, and four of the six Generation IV design concepts have hydrogen production as a design goal. The production of hydrogen using a nuclear reactor is at an early stage of development. Hydrogen production technology is at a level that is roughly equivalent to the nuclear power industry in the early 1960s with respect to the production of electricity.

Three basic approaches have been advanced for the nuclear energy production of hydrogen. The first (nuclear-assisted steam reforming of natural gas) uses nuclear heat to reduce the amount of natural gas needed to produce hydrogen. Hot electrolysis is the second approach and it produces oxygen and hydrogen from water using heat, not electricity. Finally, thermochemical cycles use a series of chemical reactions and high temperatures to convert water into hydrogen and oxygen. All three of these processes use heat as the basis for hydrogen production. Of these three, thermochemical hydrogen production is viewed as the most cost effective approach.

The leading thermochemical sequence is the sulfur–iodine process that consists of three chemical reactions:

$$\text{Heat input at } 800°C: \quad 2H_2SO_4 \rightarrow 2SO_2 + 2H_2O + O_2, \tag{2.2}$$

$$\text{Heat input at } 450°C: \quad 2HI \rightarrow I_2 + H_2, \tag{2.3}$$

$$\text{Heat rejection at } 120°C: \quad I_2 + SO_2 + 2H_2O \rightarrow 2HI + H_2SO_4. \tag{2.4}$$

The net result of these three reactions is that heat and water yield hydrogen and oxygen. The other chemical compounds (H_2SO_4 and HI) are recycled.

The health physics aspects of hydrogen production depend on the reactor design that drives the hydrogen production requirements. Only high-temperature reactor designs are candidates for hydrogen production. The optimum design matches the electrochemical generation and hydrogen generation requirements. In addition, the nuclear heat production reactor and chemical hydrogen production facility must be separated. Preliminary design studies suggest that a separation distance of at least a kilometer may be necessary to ensure that potential accidents in one facility do not affect the other.

Table 2.8 Generation IV deployment summary.

Generation IV concept	Base case deployment date	Type of fuel cycle
GFR	2025	Closed with actinide recycling
LFR	2025	Closed with actinide recycling
MSR	2025	Closed with actinide recycling
SFR	2015	Closed with actinide recycling
SWCR	2025	Closed with actinide recycling or open with once-through fuel
VHTR	2020	Open with once-through fuel

Derived from GIF-002-00 (2002).

2.7.9
Deployment of Generation IV Reactors

Table 2.8 provides a summary of the current estimate of deployment dates for various Generation IV reactor types. Generation IV reactors are projected to be deployed in the 2015–2025 time frame. The sodium-cooled fast reactor has the most optimistic deployment date of 2015. This is somewhat expected as there is scalable operating experience from Generation II SFR designs. These deployment dates are contingent on the development of the Generation IV reactors and resolution of the open issues previously identified.

The Generation IV concepts vary in their ability to close the fuel cycle and address the disposition of high-level waste. The VHTR and thermal SWCR designs do not advance the long-term high-level waste disposal issue because they utilize an open fuel cycle with no reprocessing or actinide recycle. The long-term management of high-level waste is a significant health physics issue that impacts worker doses, off-site doses, and environmental protection.

2.8
Generic Health Physics Hazards

The presentation to this point provides an overall characterization of existing and planned fission reactors. This information provides the basis for a discussion of specific health physics hazards associated with these fission reactors.

The power reactor health physicist must deal with a wide variety of issues. Issues, such as internal and external dose control, are not unique to the power reactor field, but their application is unique to the reactor environment. To illustrate the power reactor health physics issues, examples of dose concerns associated with commercial reactors are presented. The focus is on expected hazards encountered in Generation III and IV reactors.

Table 2.9 summarizes health physics hazards associated with generic power reactor activities. Examples of these work activities include primary component maintenance during outages and power operations, steam generator surveillance and repair, spent fuel pool work activities, refueling operations, containment at power inspections, waste-processing operations, component decontamination, and spill clean-up. Although these activities are typical of the work that is part of a power reactor environment, they are not a complete listing of the challenging tasks faced by health physics personnel. The activities of Table 2.9 involve both internal and external exposure pathways.

2.9
Specific Health Physics Hazards

The generic descriptions of Table 2.9 provide an overview of the radiation hazards that affect task performance at a Generation III and IV fission reactor. Knowledge of these generic hazards facilitates the introduction of specific hazards. For specificity, selected tasks and facility conditions are chosen to illustrate the health physics hazards. These tasks and conditions are the buildup of radioactive material in components such as filters, demineralizers, and waste gas decay tanks; activation of reactor components; fuel damage; reactor coolant system leakage; hot particle doses; and effluent releases.

2.9.1
Buildup of Activity in Filters, Demineralizers, and Waste Gas Tanks

Air filters trap airborne radioactive material, liquid filters remove suspended particulates, demineralizers retain radioactive material from liquid streams, and waste gas decay tanks collect fission gases and iodine removed from the primary coolant. All power reactor types benefit from minimizing their radioactive source terms. The activity that accumulates in filters and demineralizers are primarily activation products. Fission products accumulate if fuel damage occurred during previous operations. MSR designs must contend with fission products and actinides as the fuel is dissolved in the salt coolant.

Filters are commonly used to reduce effluent concentrations. A variety of air filter types (e.g., high-efficiency particulate air and charcoal) remove airborne activation products, fission products, and iodine. Liquid filters vary in construction and composition, but all types mechanically remove radioactive material suspended in liquid streams.

Demineralizers remove radioactivity from fluid systems using an ion-exchange process. The radiation levels inside demineralizer cubicles associated with spent fuel clean-up systems can exceed the criteria for very high radiation areas. After fuel damage, demineralizer radiation levels increase dramatically because of the increase in the influent source term.

MSR demineralizer systems will be unique because the fuel is dissolved in the salt coolant. Activation and fission products are removed from the coolant as part of

Table 2.9 Generation III and IV power reactor generic work activities and associated health physics hazards.

Work activity	Reactor types	Hazards[a]
Primary component maintenance during an outage	All	Activation products Fission products (depending on the fuel integrity)[b] Hot particles
Primary component maintenance during power operations	All	Activation products Fission products (depending on the fuel integrity)[b] Hot particles Fission neutrons Fission gamma rays Fission betas ^{16}N photons
Steam generator Eddy current surveillance and tube repair during an outage	All reactors with steam generators (independent of type)	Activation products Fission products (depending on the fuel integrity)[b] Hot particles
Spent fuel pool activities including fuel rearrangement, control rod replacement, fuel assembly reconstitution, and clean-up activities	All except MSR	Activation products Fission products (depending on the fuel integrity)[b] Hot particles Criticality
Refueling operations	All	Activation products Fission products (depending on the fuel integrity)[b] Hot particles Tritium Criticality
Containment at power inspections[c]	All	Noble gases Tritium Iodine Neutrons ^{16}N photons

Table 2.9 (Continued)

Work activity	Reactor types	Hazards[a]
Online radioactive waste processing and actinide recycle	MSR	Activation products Criticality Fission products Hot particles Tritium

[a]Tritium is a hazard for all activities at a CANDU reactor.
[b]MSRs have no fuel fission product barrier as the fuel is dissolved in the coolant. Refueling occurs while the reactor is operating.
[c]This is a Generation II activity that improved maintenance and outage planning. Operating experience and operating policy determine if it will be utilized at Generation III and IV facilities.

the facility's design. Demineralizer loading and change out are unique aspects of the MSR, and the selection of ion-exchange media requires careful selection to avoid radiation degradation of the media.

Waste gas decay tanks accumulate fission gases and iodine. The radioactive material accumulates and is retained until it meets the criteria for release into the environment.

The accumulation of activity in a system is described in terms of production equations. Production equations are important in a number of health physics applications and are described in Appendix B.

The buildup of activity of isotope i (A_i) on a filter, in a demineralizer bed, or in a waste gas decay tank, is determined by the system properties and isotopes present in the fluid entering these components. Important parameters impacting the buildup of the activity in filters and demineralizers include the concentration of the isotope in the fluid entering the device (C_i), the system flow rate (F), the time the filter or demineralizer is operating or processing influent (t_{op}), and the time the system is isolated (t_{decay}) from the influent stream,

$$A_i = \frac{C_i e_i F}{\lambda}(1-\exp(-\lambda t_{op}))\exp(-\lambda t_{decay}), \quad (2.5)$$

where e_i is the efficiency of the filter or demineralizer for removal of isotope i and λ_i is the radioactive decay constant of isotope i. Fluids containing multiple isotopes require the application of Equation 2.5 for each nuclide present in the influent stream.

The types of radioactive material deposited in filters and demineralizers vary with the specific design. Activation products are design specific as noted in Table 2.7. They vary considerably and depend on the coolant type, materials used in the construction of the primary system, fuel type, and the reactor's neutron spectrum (i.e., thermal or fast). Generic activation products are provided in Table 2.1.

Fission product generation depends on the specific fuel composition and neutron spectrum incorporated into the design. For example, fission products are derived from a variety of fissile nuclides including ^{233}U, ^{235}U, ^{239}Pu, and ^{241}Pu for thermal fission and ^{232}Th and ^{238}U for fast fission.

2.9.2
Activation of Reactor Components

Another source of activity is the direct irradiation of reactor components and the activation of corrosion products. Corrosion or wear products dissolved or suspended in the primary coolant pass through the core region and are subjected to the core's neutron fluence. Activation occurs by a variety of neutron-induced reactions, and the nuclides produced depend on the neutron spectrum and fluence impinging upon the material in the core region. Specific activation mechanisms are illustrated in Tables 2.2 and 2.7.

The activity derived from the irradiation is given by the production equations of Appendix B and has the specific form,

$$A_i = N_i \sigma \phi (1 - \exp(-\lambda t_{\text{irr}})) \exp(-\lambda t_{\text{decay}}), \tag{2.6}$$

where N_i is the number of target atoms that are activated, σ is the cross section for the activation reaction, ϕ is the fluence rate or flux inducing the activation reaction, t_{irr} is the time the target is irradiated or exposed to the core flux, and t_{decay} is the decay time or time the target was removed from the reactor's core region or activating flux.

Once the activity of a source is known, its radiological impact is determined from the knowledge of its basic geometry. Common geometries include the point, line, disk, and slab sources. Relationships for calculating dose information from these geometries are summarized in Appendix C, which also provides a summary of other important health physics relationships that are useful in external dosimetry applications. Internal dosimetry relationships are described in Appendix D.

For example, the dose rate at a distance r from a small source is obtained from a point source approximation. The point source approximation is accurate to about 1% whenever the distance from the source is at least three times the largest source dimension.

A second useful relationship encountered in a power reactor environment is the line source approximation. The line source equation is useful when assessing the dose from sample lines or piping carrying primary coolant or other radioactive fluids. Fuel rods, resin columns, and irradiated rods are also credibly approximated using line sources.

The third useful relationship for estimating the dose rate from typical power reactor components is the thin disk source approximation. A disk source provides a reasonable approximation to the dose rate from a radioactive spill or contaminated surfaces.

Slab sources are useful in approximating the dose rates from contaminated soil, contaminated pools, or demineralzer beds. Dose rates from a spent fuel pool, from contaminated concrete floors or walls, and from the activity deposited in demineralizer beds is addressed with reasonable accuracy with a slab source approximation.

2.9.3
Fuel Damage

A nuclear reactor contains a number of barriers designed to prevent fission products from escaping from the reactor core to the environment. These barriers include the fuel matrix and fuel element cladding or coating, the reactor coolant system and included piping, and the containment building. A breach of any of these barriers warrants serious attention to prevent the release of radioactive material into the environment.

The nature of the fuel fission product barrier depends on the fuel construction. For example, in PWRs, BWRs, and CANDUs, the fuel fission product barriers consist of UO_2 pellets enclosed within a stainless steel or zirconium alloy tube. In VHTRs, the fuel is coated in a ceramic, and the fuel fission product barrier consists of the SiC or ZrC fuel coating and the fuel material.

Fuel barrier damage facilitates the release of fission products contained between the fuel pellet and cladding (gap activity) or between the ceramic coating and fuel and increases the primary coolant activity. Noble gas activity entering the primary coolant is either released into the containment atmosphere via leakage paths or to off-gas systems. These gaseous fission products are an early indication that a fuel cladding/coating failure or mechanical damage to the cladding/coating has occurred. BWRs normally detect fuel failure by detection of fission gases in the off-gas system. However, PWRs normally monitor the primary coolant line or letdown filter lines for these fission products or monitor the containment atmosphere for released noble gases (e.g., xenon and krypton) and their daughter products. The analysis of primary coolant samples by gamma spectroscopy is a routine confirmatory action in either type of reactor.

The MSR concept does not have a defined fuel fission product barrier because the fuel is dissolved in the molten salt coolant. If the salt effectively retains the fission products, then the salt solution could be considered a type of barrier. However, the retention of fission gases in the salt coolant appears to be an open issue for MSRs.

2.9.4
Reactor Coolant System Leakage

Leakage from the primary coolant system is an undesirable but inevitable problem at a power reactor. Value seams, pump seals, value packing, and instrument line connections provide pathways for small leaks that contaminate local areas. This contamination must be controlled to limit station external and internal doses. In addition to primary system leaks, health physicists must address leakage from the primary to secondary systems for reactors using steam generators.

Leakage from steam generator tubes to the secondary system presents a health physics concern, because additional plant areas become contaminated. As the secondary components are considered clean systems, the presence of contamination has a negative impact on facility operations and expands areas requiring stringent radiological controls.

Secondary coolant contamination has a number of negative health physics aspects. The secondary activity tends to concentrate in components such as the main steam isolation valves and high-pressure turbine piping resulting in surface contamination areas and local hot spots. Secondary ion-exchange resins and filters become contaminated, which adds to the facility's contamination problems and increases the volume of radioactive waste generated. Steam generators clean-up systems also become contaminated. Contaminated secondary system areas increase health physics survey requirements and decontamination activities.

Primary to secondary leakage also increases the likelihood of a release of noble gases and iodine to the environment. The most likely release pathways are through a secondary system relief valve or through the condenser air ejector.

2.9.5
Hot Particle Dose

Particulate matter is produced within the primary coolant system through a variety of processes and power reactor activities. The maintenance of pumps, valves, and piping create small particles during the process of cutting, grinding, and welding. Operation of valves and pumps leads to wearing of active surfaces, which produces small particulate materials. Cladding erosion and failures or erosion of control rod surfaces contribute additional matter to the reactor coolant system. This material is often too small to be removed by the reactor coolant system's filters. Therefore, it passes through the core and is activated by the neutron fluence, which leads to the creation of highly activated, microscopic material called a "hot particle."

Hot particles are composed of activation products and possibly fission fragments depending upon the integrity of the fuel fission product barrier. Particles may contain either single isotopes or a large number of radioisotopes. Hot particles present a skin dose hazard. Beta radiation is the dominant contributor to the skin dose, but gamma contributions can approach about 30% of the beta dose contribution.

The dose from a hot particle residing on the surface of the skin is given by the relationship,

$$D = \frac{t}{S}\sum_i A_i F_i, \tag{2.7}$$

where D is the absorbed dose to the skin from the hot particle, A_i is the particle activity for radionuclide i, F_i is the dose factor for radionuclide i ($Gy\,m^2/MBq\,h$), t is the residence time on the skin, S is the area over which the dose is averaged, and i is the number of radionuclides in the hot particle.

Following NCRP 130, the skin dose is averaged over $10\,cm^2$ and evaluated at a depth of $7\,mg/cm^2$ at the basal cell layer depth. As the dose from a point source falls off rapidly as 1 over r-squared, the dose from a hot particle is highly localized. Hot particles also attach to the eye, enter the lungs through inhalation, and irradiate the gastrointestinal tract after ingestion. NCRP 130 provides specific guidelines for addressing these specific conditions.

2.9.6 Effluent Releases

The effluents that characterize a facility depend on the core materials, reactor materials, and specific design aspects of the Generation III or IV system. Examples of the nuclides that may appear in a facility are summarized in Tables 2.1 and 2.7.

2.9.6.1 Light Water and Heavy Water Reactor Effluents

Although off-gas systems are designed to trap most gaseous effluents, quantities of noble gases, ^{3}H, ^{14}C, and iodine isotopes are available for release. These isotopes are generated through the activation and fission processes. Their release is facilitated by defects in the fuel clad/coating.

Production mechanisms are design specific. For example, tritium arises from the neutron activation of the primary coolant [^{2}H(n, γ)^{3}H] and from tertiary fission. In a PWR, tritium is also produced from neutron capture in ^{10}B used for reactivity control [^{10}B(n, 2α)^{3}H] and from neutron capture in ^{6}Li used for chemistry control [^{6}Li(n, α)^{3}H]. ^{14}C is usually produced from the ^{14}N(n, p)^{14}C or ^{17}O(n, α)^{14}C reactions.

Liquid effluents include fission and activation products as well as tritium. Tritium is the dominant liquid effluent in PWRs.

The quantity of fission products in liquid waste depends on the integrity of the fuel fission product barrier. Liquid waste clean-up systems, including filtration and demineralization, remove most of these radionuclides from the effluent stream.

Fission product radionuclides generated from binary fission include ^{85}Kr, ^{87}Kr, ^{88}Kr, ^{133}Xe, ^{135}Xe, ^{137}Xe, ^{131}I, ^{137}Cs, ^{137}Ba, ^{141}Ce, ^{144}Ce, ^{103}Ru, ^{106}Ru, ^{103}Rh, ^{106}Rh, ^{90}Sr, and ^{90}Y. Activation products are produced by neutron capture by materials in the vicinity of the nuclear core including chemical control agents dissolved in the primary coolant, stainless steel or stellite corrosion, and wear products resulting from system maintenance or operation, primary coolant system piping, the reactor vessel, and core structural material.

Activation products are produced from a variety of reactions including ^{54}Fe(n, p)^{54}Mn, ^{58}Fe(n, γ)^{59}Fe, ^{57}Co(n, γ)^{58}Co, ^{58}Ni(n, p)^{58}Co, ^{59}Co(n, γ)^{60}Co, and ^{94}Zr(n, γ)^{95}Zr. The aforementioned (n, γ) reactions are normally induced by thermal neutrons, and the (n, p) reactions are initiated by fast neutrons.

2.9.6.2 Gas-Cooled Reactor Effluents

The gas-cooled reactors have different materials of construction than water-cooled reactors. Accordingly, different radionuclides inventories and effluents are expected. The following discussion describes isotopes unique to specific gas-cooled reactors. To make the discussion complete, both CO_2 and ^{4}He coolant versions are also presented. Expected radionuclides such as fission gases are not listed as they have already been discussed.

2.9.6.3 CO_2-Cooled Reactor Effluents

Advanced CO_2 gas-cooled reactors (AGCRs) developed in the United Kingdom are graphite-moderated facilities. In the CO_2 AGCRs, isotopes of significance include ^{3}H, ^{14}C, ^{16}N, ^{35}S, and ^{41}Ar. Activation of the CO_2 primary coolant produces ^{14}C, ^{16}N, and ^{41}Ar, and activation of the graphite moderator yields ^{3}H, ^{14}C, and ^{35}S. Fission products similar to those noted for PWRs, BWRs, and CANDUs are also produced. Their possibility of release depends on the status of the fuel fission product barrier.

The CO_2 coolant activation products ^{14}C and ^{16}N arise from the ^{13}C(n, γ)^{14}C and ^{16}O(n, p)^{16}N reactions, respectively. ^{40}Ar is an impurity in the CO_2 coolant and ^{41}Ar is derived from the ^{40}Ar(n, γ)^{41}Ar reaction. ^{14}C is also produced from nitrogen impurities in the coolant through the ^{14}N(n, p)^{14}C reaction.

The graphite moderator may contain trace sulfur and chlorine impurities that lead to ^{35}S via the ^{34}S(n, γ)^{35}S and ^{35}Cl(n, p)^{35}S reactions. Graphite may also contain lithium impurities that upon the capture of thermal neutrons produce tritium through the ^{6}Li(n, α)^{3}H reaction.

In addition to ^{3}H, ^{14}C, ^{16}N, ^{35}S, and ^{41}Ar, the irradiated AGCR fuel is a source of activation products. Following disassembly, irradiated AGCR fuel is stored in pools until its activity is reduced to a level that permits its off-site transport. During this cooling period, gradual dissolution of radionuclides in the graphite sleeves and stainless steel cladding are released into the pool water. These radionuclides involve both fission and activation products including ^{46}Sc, ^{54}Mn, ^{55}Fe, ^{58}Co, ^{60}Co, ^{124}Sb, ^{134}Cs, ^{137}Cs, and ^{182}Ta.

2.9.6.4 Helium-Cooled Reactor Effluents

One of the key features affecting the effluent release in helium-cooled reactors is the concentration of impurities in the graphite moderator. The impurities vary considerably with the type of graphite used in the design. It is expected that a variety of elements will be found in the graphite moderator including boron, cesium, calcium, carbon, chlorine, cobalt, helium, iron, lithium, nickel, nitrogen, niobium, and uranium. The concentrations of these elements directly affect the effluent concentrations of their activation products such as ^{3}H and ^{14}C. For example, for French nuclear grade graphite following three years of irradiation, the maximal activation activity is a few MBq/g of graphite with about 40% of the activity attributed to ^{3}H, and 15–20% each to ^{55}Fe and ^{60}Co. Other isotopes produced in graphite include ^{14}C, ^{36}Cl, ^{41}Ca, and ^{59}Ni.

Helium-cooled reactor metallic materials are dominated by chromium, iron, and nickel with smaller quantities of cobalt and molybdenum. These elements lead to activation products including ^{55}Fe, ^{59}Ni, ^{60}Co, and ^{63}Ni. The concentrations of these radionuclides depend on the concentrations of their parent isotopes in the original metallic material.

The previous discussion illustrates the complexity involved in gas-cooled reactor effluents. The specific design requirements including material specifications govern the radionuclides produced and their abundance. For example, the graphite specification controls the impurities and their concentrations. The allowance for impurities in the graphite has a significant impact on the resulting activation products. Therefore, identical Generation IV helium-cooled reactors have different effluent radionuclide characteristics if their graphite specifications are not the same.

2.10
Advanced Reactor ALARA Measures

One of the radiological considerations in advanced reactor design is minimizing worker radiation doses. Components are designed to be nearly maintenance free or requiring infrequent attention, but when required, repairs should occur quickly to reduce radiation doses. Thermal insulation is designed to be reusable, to be quickly removed, and reinstalled.

Component materials are selected to minimize the production of activation products. In particular, cobalt alloys are restricted. This minimizes a major source term (^{58}Co and ^{60}Co) of activation products that occurs in Generation II facilities.

The component arrangement and accessibility are optimized in advanced Generation III and IV reactors. These features enhance task completion, minimize radiation doses, and facilitate operability testing of reactor components.

Maintenance and in-service inspection are also optimized. Heat exchangers, tanks, and vessels are designed to minimize the collection of radioactive material and to facilitate the removal of any radioactive material collecting within their boundaries. Components are arranged to allow adequate room for maintenance and inspection activities.

2.11
Radiological Considerations During Reactor Accidents

Reactor accidents are broadly classified as design basis events and beyond design basis events (BDBEs). Design basis events are caused by a component failure such a break in primary system piping or steam generator tubes. Beyond design basis events include multiple failures such as a loss of all power (off-site and on-site emergency power) or ruptures of tubes in multiple steam generators. These events are discussed below in greater detail. Plant procedures exist to address both design and beyond design basis events.

There are four generic types of design basis accidents that occur at a fission reactor. These are loss-of-coolant accidents that involve a loss of core coolant, steam generator tube ruptures (SGTRs) resulting from breaches in the tubes forming a boundary between the primary and secondary coolants, fuel handling accidents (FHAs) that result in damage to the fuel cladding, and waste gas decay tank ruptures (WGDTRs) involving a loss of integrity in structures containing fission gases. These events are significant because they permit radioactive material to escape from engineered systems and enter plant areas or the environment in an uncontrolled manner.

Reactor accidents vary in severity, but the most significant radiological event involves core damage that leads to a significant radioactive release. Other events, including failure of waste gas decay tanks or spent fuel element breaches, are less severe, but more likely scenarios. Before discussing the basic reactor accident types, a brief review of fission product barriers is warranted.

The following discussion applies to all Generation IV reactor types except MSRs. As the fuel is dissolved in the coolant, molten salt reactors only have two fission product barriers.

Most reactor types define three fission product barriers: the fuel, the reactor coolant system, and the containment structure. Each of these forms a barrier to the movement of fission products contained within the fuel matrix into plant areas or the environment. As such, preserving the integrity of the fission product barriers is crucial to both maintaining the control of radioactive material and maintaining an effective health physics program.

The first fission product barrier includes the fuel pellet or fuel material. The fuel material and its associated coating or cladding retains solid and gaseous fission products. For pellet/clad configuration fuel, fission product activity is often classified as either gap activity or total fuel pin activity. Gap activity is fission product activity residing in the gaps between the fuel pellets and the gap between the fuel pellet and the cladding. The total fuel pin activity is the total activity that includes the gap activity plus the activity contained within the fuel material. The fuel fission product barrier is absent in MSRs.

The second fission product barrier is the primary coolant system boundary including its piping and components. Any break in primary piping permits radioactive material to be released into the facility.

The third fission product barrier is the containment structure that encloses the primary coolant system. Any breach of the containment structure creates a pathway for radioactive material to reach the environment. Penetration of any of the three fission product barriers facilitates the release of radioactive material in an uncontrolled manner. If the event is severe and involves a breach of multiple barriers, a major reactor accident occurs.

The four reactor accident categories are further defined as

(1) *Loss-of-coolant accidents* – In a containment building LOCA, the reactor's primary piping is breached and cooling flow is reduced or lost. As a result, the temperature of the nuclear fuel increases. As the fuel temperature increases, the fuel fission product barrier degrades and fission products are released into the primary coolant. The loss of the fuel barrier is significant because it is the second failed fission product barrier, and the loss facilitates the release of radioactive material to uncontrolled areas including the environment. If the LOCA is severe, the fuel will eventually melt with the subsequent release of additional radioactive material to the primary coolant.

Fuel cladding degradation can occur even without fuel melting. Breaches in the clad caused by impacts of foreign material or localized heating release fission

2.11 Radiological Considerations During Reactor Accidents

radionuclides into the primary coolant. Subsequent breaches in the primary coolant system or containment building offer a release path to the environment. With fuel damage and a breach in primary piping, only the containment barrier prevents a release of radioactive material into the environment.

If the primary piping breach occurs in the auxiliary building (e.g., in the letdown line), a release pathway to the environment exists. The auxiliary building is not a containment barrier.

Containment building failures facilitate a release into the environment. Examples of containment failures include malfunctions of purge valves, air supply valves, containment hatch valves, penetrations, and containment isolation valves.

(2) *Steam generator tube ruptures* – Steam generator tubes form a barrier between the primary and secondary coolants. If a tube rupture or leak occurs, a pathway is created that mixes the primary (radioactive) and secondary (nonradioactive) fluids. The secondary (clean) part of the plant becomes contaminated and its radiation levels exhibit a significant increase. Failures of atmospheric or steam generator safety values or other secondary system piping, valves, or components provide a direct release pathway for the primary coolant's radioactive material to reach the environment.

An SGTR is a special class of LOCA with the primary system leak occurring through the steam generator tubes. In addition to the secondary impacts, the primary system experiences the radiological consequences of a LOCA with the severity depending on the magnitude of the primary to secondary leakage.

(3) *Fuel handling accidents* – The nuclear fuel residing in the fuel storage pool or the reactor core is periodically moved during refueling operations or operations involving fuel inspection or control rod maintenance. Accidents during these evolutions damage the fuel fission product barrier and lead to a release of radionuclides into the radiologically controlled plant areas or the environment. Fission gases dominate the radiological release. Iodine may also be released depending on the age and extent of the damage to the fuel assembly. If spent fuel is involved in the FHA and it has been out of the reactor for about 1 year, the short-lived noble gas activity will have decayed and the dominant isotope in the release is ^{85}Kr that has a half-life of 10.7 years.

(4) *Waste gas decay tank ruptures* – Waste gas decay tanks store fission gases and possibly radioiodine to permit their decay before the release of these radioactive materials into the environment. Failures of the tank structure, valves, or associated components release fission gases into the plant. As these tanks reside in the auxiliary building, a release into the environment is likely to take place.

The extent to which these design basis accidents lead to radiological consequences depends largely on the integrity of the reactor fuel. If the fuel fission product barrier remains intact, the releases will be characterized by the steady state activity of the primary coolant. The radiological hazards increase proportionally with the degree to which the fuel fission product barrier degrades and releases fission products into the primary coolant. Tables 2.10 summarizes the various power reactor accident types, the types of radiological releases that could occur, plant systems that mitigate the release, and methods that could be utilized to mitigate the release. Table 2.10 information is derived from operating Generation II and III facilities and their operating characteristics.

Table 2.10 is generic in that it does not focus on a specific design type. As an example of a specific Generation III reactor design, the design basis events and their radiological consequences for the AP-1000 are presented.

The NRC Certification Review for the AP-1000 provides an assessment of the design basis events for this reactor type. On the basis of the currently available information, the NRCs estimate of the radiological consequences of AP-1000 design basis events are summarized in Table 2.11. The radiological consequences are provided for the AP-1000 control room, the exclusion area boundary (EAB), and low-population zone (LPZ).

The EAB is the area surrounding the reactor, in which the reactor licensee has the authority to determine all activities including exclusion or removal of personnel and property from the area. The US Nuclear Regulatory Commission (USNRC)

Table 2.10 Design basis event accident mitigation.

Accident type	Release type	Mitigation	Termination
LOCA	Iodine[a] Noble gas Particulate	NaOH spray (PWRs) Suppression pool (BWRs) Ice condensers Filtration ECCS[b]	In-plant repairs Reestablish core cooling Isolate leak (e.g., letdown line)
SGTR	Iodine[a] Noble gas	Filtration Release via the condenser ECCS[b] Protect intact steam generators	Cool and depressurize the primary coolant system In-plant repairs
FHA (<1 yr old fuel)	Iodine[a] Noble gas	Filtration	Fuel assembly depressurizes
FHA (>1 yr old fuel)	^{85}Kr	Filtration	Fuel assembly depressurizes
WGDTR	Iodine[a] Noble gas	Filtration	In-plant repairs (e.g., tank isolation) Tank depressurizes

[a]Depends on the extent of fuel barrier defects.
[b]Emergency core cooling system.

Table 2.11 AP-1000 radiological consequences of design basis accidents (total effective dose equivalent (TEDE)).[a]

Postulated accident	EAB (mSv)	LPZ (mSv)	Control room (mSv)
Loss-of-coolant accident	190	150	34
Main steamline break outside containment with an accident-initiated iodine spike	2	8	13
Reactor coolant pump shaft seizure without feedwater available	<1	<1	12
Rod ejection accident	15	24	11
Fuel-handling accident	24	10	29
Small line break accident	10	4	14
Steam generator tube rupture with accident-initiated iodine spike	5	7	26
Spent fuel pool boiling	n/a[b]	<0.1	<0.1

[a] Derived from NUREG-1793 (2006).
[b] n/a: not applicable.

quantifies the EAB as the perimeter of a 2760-ft (841-m) radius circle from the circumference of a 630-ft (192-m) circle encompassing the proposed power block housing the reactor containment structure. LPZ is similarly defined. The LPZ is the area immediately surrounding the EAB. The USNRC quantifies the LPZ as a 2-mile (3.2-km) radius circle from the circumference of a 630-ft (192-m) circle encompassing the proposed power block housing the reactor containment structure.

The radiological impact of the events described in Table 2.11 is less severe than the corresponding Generation II accidents. This result is expected on the basis of the Generation III design criteria with its enhanced safety performance. Similar improvements are expected for Generation IV facilities.

2.12
Beyond Design Basis Events

Beyond design basis events are situations that involve failure of multiple safety barriers, and are potentially significant from a radiological perspective. As noted in Table 2.12, these events involve combinations of design basis events.

The BDBEs include loss of power events in which all on-site and off-site power is lost, tube ruptures in more than one of the facility's steam generators, faults in multiple steam generators, combinations of steam generator faults and ruptures, and LOCAs coinciding with the loss of power. A steam generator fault is a break in the secondary system piping or secondary system component failure such as a relief valve

Table 2.12 Beyond design basis event accident mitigation.

Accident type	Release type	Mitigation	Termination
Loss of all on-site and off-site power	Iodine[a] Noble gas Particulate	Station batteries Reflux cooling Steam-driven auxiliary feed pumps Establish on-site emergency power Protect primary piping integrity Filtration ECCS[b]	In-plant repairs to equipment damaged by the loss of power Reestablish core cooling with electric-driven pumps Restore off-site electric power Stabilize the primary coolant system
LOCA with loss of power	Iodine[a] Noble gas Particulate	Station batteries Reflux cooling Steam-driven auxiliary feed pumps Establish on-site emergency power Protect primary piping integrity NaOH spray (PWRs) Suppression pool (BWRs) Ice condensers Filtration ECCS[b]	In-plant repairs to equipment damaged by the loss of power Reestablish core cooling with electric-driven pumps Restore off-site electric power Isolate source of primary leakage Stabilize the primary coolant system
Ruptures in multiple SGs	Iodine[a] Noble gas	Protect intact steam generators Filtration Release via the condenser ECCS[b]	Cool and depressurize the primary coolant system In-plant repairs
Faults in multiple SGs	Iodine[a] Noble gas	Protect intact steam generators Isolate fault locations Protect primary piping by avoiding overcooling resulting from the fault ECCS[b]	Primary system pressure and temperature trending to acceptable values In-plant repairs
Combination of faulted and ruptured SGs	Iodine[a] Noble gas	Protect intact steam generators Isolate fault locations Protect primary piping by avoiding overcooling resulting from the fault Filtration Release via the condenser ECCS[b]	Primary system pressure and temperature trending to acceptable values In-plant repairs

[a]Depends on the extent of fuel barrier defects.
[b]Emergency core coolant system.

that opens and does not reseat (close). The fault provides a pathway for a release of secondary coolant into the environment.

The loss of off-site and on-site electrical power disables a number of safety systems that jeopardizes the ability to cool the core and mitigate the release of radioactive material. These systems provide core cooling or supply feedwater to the secondary side of the steam generators. Some feedwater is provided by steam-driven auxiliary pumps, but their flow rate decreases as the core heat decreases.

Upon losing power, primary flow to the core ceases, but a process known as reflux cooling provides some core cooling. Reflux cooling is primary coolant converted to steam by core decay heat. The steam condenses inside primary piping and then flows back into the core. Maintaining the integrity of primary and secondary piping systems minimizes fuel damage and the release of radioactive material from the reactor coolant system. However, a prolonged loss of power will increase core temperatures and eventually damage or even melt the fuel.

Station batteries provide a source of direct current (DC) that can be converted into alternating current to power safety system pumps and valves. The lifetime and capability of the DC system is limited (on the order of a few hours) so expeditious recovery is essential.

LOCAs coinciding with the loss of electric power have the characteristics of the DBE loss of coolant events. With the loss of power, the LOCA severity is increased because core cooling is limited. Without electric-driven pumps to provide cooling water the likelihood of fuel damage and melting is increased. Any degradation of the secondary system restricts the core's heat sink that facilitates an increase in core damage and possibly melting of the core or selected fuel assembles. Fuel damage, including fuel melting, releases fission products into the reactor coolant system and possibly into the environment.

Multiple ruptured steam generators are a more severe version of an SGTR. As such, it has radiological consequences that are similar to, but more severe than, an SGTR.

With a single ruptured steam generator, the intact steam generators can be used to provide long-term core cooling. With multiple ruptures emphasis will be placed on using any intact steam generators and to preserve their integrity.

Steam generator faults are breaks in steam generator secondary piping. Faults lead to a rapid loss of secondary coolant that result in overcooling the primary system. Overcooling is significant because primary system pressure and temperature limits could be exceeded resulting in stressing the primary piping and components. This stress increases the potential for primary system damage including component rupture that would lead to a LOCA.

The overcooling condition exists as long as the faulted steam generator receives feedwater (secondary coolant). Recovery from a fault condition includes feedwater isolation and subsequent restoration of the primary system to acceptable pressure and temperature conditions.

Rupture/fault combinations have characteristics of both types of events. If both events occur in the same steam generator, the combination of the loss of heat sink with the loss of primary coolant to the secondary system presents an energetic pathway for the release of primary coolant. Fuel damage exacerbates the consequence of the rupture/fault event.

The off-site consequences of a BDBE result in the release of iodine and noble gases into the environment. The consequences of an environmental release depend on the release rate of radioactive material, the particular radioactive material released, the meteorological conditions during the event, and the release duration. The variation in release consequences as a function of these quantities is complex and scenario specific.

2.13
Other Events

The September 11, 2001 attacks in the United States caused regulators to review the design basis for nuclear power plants. Proposals to establish no-fly zones near reactors or install lattice-like barriers to protect nuclear reactors from an aircraft attack have been forwarded.

The US Nuclear Regulatory Commission recently determined that making nuclear power plants crash proof to an airliner attack by terrorists is impractical. Protection against an air attack is the responsibility of the military and the Federal Aviation Agency. The NRC has directed that the operators of nuclear plants focus on preventing radioactive material from escaping in the event of an air attack and to improve evacuation plans to protect the health and safety of the public.

Additional discussion regarding power reactor radiological events caused by intentional human intervention is addressed in Volume II. These reactor events are caused by a number of initiators and are not limited to aircraft events.

2.14
Probabilistic Risk Assessment

The US regulatory environment utilizes probabilistic risk assessments (PRAs) as a tool to evaluate severe accidents. As such, the NRC utilizes a goal of less than 1×10^{-4}/year for core damage frequency and less than 1×10^{-6}/year for a large release frequency (LRF). PRAs are also used to uncover design and operational vulnerabilities; strengthen programs and activities in areas such as training, emergency operations, reliability assurance, and safety evaluations; and evaluate maintenance and surveillance frequencies.

Although the previous discussion on severe accidents was generic, the certification of the AP-1000 design by the USNRC provides specific severe accident examples for a Generation III facility. Results of the AP-1000 analysis, summarized in Table 2.13, support the previous discussion on the safety performance of Generation III reactors. The AP-1000 total CDF values are a factor of 17–1250 lower than the total CDF range for Generation II reactors. This CDF reduction represents a significant improvement in safety performance.

Table 2.13 Comparison of AP-1000 core damage frequency contributions by initiating event (internal events and power operation).[a]

Initiating event	CDF (/yr)	Operating PWR[b] results (CDF range/yr)
LOCAs (total)	2.1×10^{-7}	1×10^{-6} to 8×10^{-5}
–Large	4.5×10^{-8}	—
–Spurious automatic depressurization system actuation	3.0×10^{-8}	—
–Safety injection line break	9.5×10^{-8}	—
–Medium	1.6×10^{-8}	—
–Small	1.8×10^{-8}	—
–Core makeup tank line break	4.0×10^{-9}	—
–RCS leak	3.0×10^{-9}	—
Steam generator tube rupture	7.0×10^{-9}	9.0×10^{-9} to 3.0×10^{-5}
Transients	8.0×10^{-9}	5.0×10^{-7} to 3.0×10^{-4}
Loss of off-site power/station blackout	1.0×10^{-9}	1.0×10^{-8} to 7.0×10^{-5}
Anticipated transient without scram	5.0×10^{-9}	1.0×10^{-8} to 4.0×10^{-5}
Interfacing system LOCA	5.0×10^{-11}	1.0×10^{-9} to 8.0×10^{-6}
Vessel rupture	1.0×10^{-8}	1.0×10^{-7}
Total	2.4×10^{-7}	4.0×10^{-6} to 3.0×10^{-4}

[a] Derived from NUREG-1793 (2006).
[b] Generation II Reactors, NUREG-1560 (1996).

2.15
Semi-Infinite Cloud Model

In each of the design basis events, radioactive gas can be released into plant areas and then into the environment. The gamma absorbed dose rate ($\dot{D}_K$) from nuclide K in the radioactive gas cloud is often assessed using a semi-infinite cloud model,

$$\dot{D}_K = k \sum_K E_K \chi_K(r, t), \tag{2.8}$$

where k is a conversion factor for the semi-infinite cloud, E_K is the average gamma energy per disintegration for nuclide K, and $\chi_K(r, t)$ is the air concentration of nuclide K at distance r and time t from the release point. In Equation 2.8, k is defined for a semi-infinite gas cloud as

$$k = \frac{1 \text{ Gy}}{\text{J/kg}} \frac{1.6 \times 10^{-13} \text{ J}}{\text{MeV}} \frac{10^6 \text{ dis/s}}{\text{MBq}} \frac{1 \text{ m}^3}{1.293 \text{ kg}} (0.5) = 6.19 \times 10^{-8} \frac{\text{Gy dis m}^3}{\text{s MeV MBq}}. \tag{2.9}$$

It is often more convenient to measure the release source term at the nuclear facility rather than at the receptor location. This is particularly true in the early stages of an accident when field measurements are not practical. The air concentration is related to the release rate (Q) from the facility

$$x_K(r,t) = \frac{x_K(r,t)}{Q_K} Q_K, \quad (2.10)$$

where χ/Q is the atmospheric dispersion parameter. With this relationship, the absorbed dose rate equation becomes

$$\dot{D} = k \sum_K E_K \frac{x_K(r,t)}{Q_K} Q_K = \sum_K Q_K (\text{DRCF}_K) \frac{x_K(r,t)}{Q_K}, \quad (2.11)$$

where the dose rate correction factor DRCF_K is given by

$$\text{DRCF}_K = k E_K. \quad (2.12)$$

Equation 2.11 provides the total absorbed dose rate ($\dot{D}$), which is the sum of the dose rates from each released radionuclide. In Equation 2.11, the release time is assumed to be short relative to the released radionuclide's half-life. Accordingly, Equation 2.11 contains no radioactive decay or other removal terms.

The semi-infinite cloud model assumes that the release rate is constant and that the atmospheric conditions, as described by the dispersion parameter, are also constant. Accident events are not likely to meet either of these conditions for extended periods of time. Another assumption is that the plume dimensions are large compared to the distance the gamma rays travel in air. This assumption is not valid close to the source, but it is more easily achieved further from the source. In addition, the model does not account for radiation buildup factors caused by the Compton scattering of the gamma ray photons, and excludes the air attenuation of photons.

This semi-infinite cloud model applies to whole body, thyroid, bone, and other organ doses. The dose conversion factors provide organ or whole body doses and reflect the type of radiological releases that occur.

Radioactive releases have an impact on plant accessibility and facility workgroup activities. The impact of the radiological environment on various workgroups is summarized in subsequent discussion.

2.16
Normal Operations

Normal operational activities typically lead to low levels of radiation that are carefully monitored. A good operational philosophy is "Every µSv counts." To illustrate normal operational activities, typical activities performed by health physics, maintenance, and operations staff are reviewed. The activities performed by individual work groups vary from plant to plant.

The doses received by the various groups depend on the planned work activity with the largest station doses received in years in which the facility is off-line for refueling or maintenance outages. In the twentieth century, operating cycles between outages varied between 12 and 24 months. It is possible that operating cycle durations will increase in the twenty-first century.

2.16.1
Health Physics

During normal operations, most of the radiation dose received by the health physics group is attributed to health physics and decontamination technician activities. These activities include routine radiation surveys; sample counting; repair, maintenance, and calibration of health physics survey instrumentation; calibration of station radiation monitors; collecting environmental samples; collecting radioactive trash; radioactive waste processing; performing ALARA reviews; writing radiation work permits; and job coverage activities. The highest effective doses are received in the coverage of high dose activities on primary system components. In general, health physics personnel receive doses less than 10 mSv during outage years and a few mSv during nonoutage years.

Health physics doses include both beta–gamma and neutron radiation components. The beta–gamma dose is derived from coverage of auxiliary building tasks and the neutron doses arise from job coverage of containment repair tasks and coverage of containment at power inspection activities. Neutron doses arise from the fission process and the beta–gamma doses result from the decay of fission and activation products.

2.16.2
Maintenance

Maintenance personnel are responsible for preventive and corrective activities including valve and pump repair, refurbishing electrical and mechanical components, performing surveillances, equipment repair, and instrument calibration. During normal operating conditions, work is normally performed in low-dose areas of the plant. Although some repair activity occurs in the containment building, most work is in lower dose areas that are shielded from primary system piping. If maintenance is to be performed in a high dose area, shielding and other ALARA measures are warranted. If the component resides in a containment area receiving neutron and ^{16}N gamma radiation, the reactor power level may require reduction to permit the repair to occur while the reactor remains online. Most of the dose received by maintenance personnel is because of the beta–gamma activity. This reflects the fact that most nonoutage maintenance is not performed in the vicinity of the reactor's neutron radiation. Maintenance doses are similar to those received by health physics personnel. ALARA measures are warranted to ensure maintenance activities are performed in an efficient radiological manner.

2.16.3
Operators

Equipment operators receive most of the operations group's dose, but reactor operators also receive dose in the performance of their duties. Operational activities vary with the philosophy of the utility responsible for the facility license.

Equipment operators or auxiliary operators (AOs) perform a variety of activities including valve lineups, equipment tag-outs, surveillance testing, in-plant equipment operation, radioactive waste processing, chemical additions, spent fuel activities, logging plant conditions, and containment at power inspections. AOs also support control room activities.

Licensed reactor operators (ROs) are qualified to perform all auxiliary operator activities and also manipulate controls related to core reactivity and reactor safety systems. For ROs, equipment operation normally occurs from control room panels. ROs also perform surveillance testing of reactor components with the support of AOs. Normally, RO radiation doses are less than AO doses.

Licensed senior reactor operators (SROs) direct the activities of AOs and ROs. They may perform any activity performed by AOs and ROs. However, their role is supervisory and their doses are normally lower than RO doses. SROs must be present whenever fuel is manipulated in either containment or at the spent fuel pool.

Operator doses vary but are similar in magnitude to health physics and maintenance doses. As operator activities occur in the containment as well as auxiliary building, their doses usually involve more neutron radiation than received by health physics and maintenance personnel.

2.17
Outage Operations

Outage activities occur in a frequency of about 24 months and these activities dominate the station collective dose. Most outage work occurs on systems that are impractical to schedule during power operations. For specificity, PWR outage activities are defined. This is reasonable because two thirds of currently operating reactors are PWRs. BWR activities are similar, but do not include steam generator work.

Outage activities include refueling, work on primary system components, primary coolant system surveillance and testing, operational testing, electrical tasks, safety system maintenance, and secondary system activities. Outage activities are dominated by beta–gamma-emitting radionuclides from activated components and radioactive material deposited on the interior surfaces of primary system components.

Refueling activities occurring during an outage include reactor head removal, fuel removal, fuel rearrangement, new fuel addition, control rod latching and unlatching, and head reinstallation. Refueling activities expose operations personnel to hot particles and present the possibility of tritium intakes. Criticality is also an issue whenever fuel is moved.

Primary component work involves a variety of activities including reactor coolant pump seal replacement and motor repair, valve and pump electrical and mechanical maintenance, instrument calibration, control rod drive repair, and component replacement. ALARA planning and dose tracking are important considerations in ensuring that high dose jobs do not exceed the planned dose budgets.

Steam generator surveillance and testing include eddy current testing of steam generator tubes to ensure their integrity. Tubes not meeting the integrity requirements are either repaired or plugged. Requiring that primary to secondary leakage is at a minimum ensures the secondary system contains minimal quantities of radioactive material. Steam generator surveillance and testing are a dose intensive activity.

Outage testing is performed on both primary and secondary components, but primary testing leads to the larger doses. Safety systems and reactor protection systems are tested during outages because their testing during power operations is not practical.

2.18
Abnormal Operations

Abnormal operations involve off-normal events that offer the potential for effective doses that exceed those occurring during normal operations. These events include natural phenomena (e.g., high winds, abnormal precipitation, high or low temperatures, and seismic events) as well as plant-specific events (loss of power, reactor coolant leakage, secondary system leakage, radioactive releases, elevated radiation levels, elevated airborne activity, and fuel handling events). Abnormal events are important because they may be a precursor to an escalating condition that leads to an emergency event.

An increased level of health physics response occurs for abnormal events that are associated with radioactive material. Elevated radiation levels and airborne radioactive material concentrations should be anticipated. The health physics response to an abnormal event must proceed in an ALARA manner.

During an abnormal event, maintenance and operations effective doses depend on the specific event sequence. For events leading to a reactor trip, the doses are dominated by beta–gamma activity. If the reactor remains online during the abnormal event, personnel may receive neutron as well as beta–gamma effective doses.

2.19
Emergency Operations

Emergency operations indicate a degraded plant condition that has the potential for releasing radioactive material into the environment or plant areas. Emergency events proceed from relatively minor events (i.e., unusual events) to more significant situations. In increasing order of severity these are the alert, site area emergency, and general emergency. This is the emergency classification scheme adopted in the United States.

The health physics response involves both in-plant and environmental monitoring. ALARA planning is essential because the radiation levels exceed those encountered during abnormal events. Good communication between operations, maintenance, and health physics personnel must be maintained to ensure task completion occurs in an ALARA manner.

Emergency events include the design basis events of Table 2.10 and the beyond design basis events of Table 2.12. The severity of these events suggests that safety systems have actuated, the reactor has tripped, and releases of radioactive material or elevated plant radiation levels exist. Both noble gas and iodine source terms should be considered in any repair activity. Repair activities must consider ALARA principles in both task planning and implementation.

Problems

02-01 The Green Bay Sodium-Cooled Fast Reactor Power Station operated without incident during its first 5 years of service. During year 6, an instrument line fails and primary sodium coolant leaks onto the containment floor. The leak assumes a 10-m diameter shape, but has not reacted with the air or the surface supporting the spill. The spill assumes a thin disk geometry, has a total activity 4×10^7 MBq, and the dose factor (gamma constant) for the activated coolant is 5.7×10^{-7} Sv m^2/MBq h.

(a) What is the effective dose rate at 10 m above the spill on the axis of the disk?
(b) If the metal coolant is released into the environment as a respirable aerosol at a rate of 10^5 MBq/s, what is the effective dose rate received by an off-site individual submerged in a semi-infinite cloud of the aerosol? Assume the dose rate conversion factor for the metal aerosol is 3×10^{-8} Sv m^3/MBq s, the applicable wind speed is 2.0 m/s, and the dispersion coefficient is 5×10^{-4} m^{-2}.

02-02 You are responsible for managing the health physics aspects for work on irradiated fuel assemblies in a spent fuel pool at an advanced pressurized water reactor. A fuel assembly recently removed from the core has a slowly leaking fuel pin (rod) that must be replaced prior to the assembly's return to the reactor vessel. The assembly has been out of the core for 42 h. You must determine the off-site dose consequences should the cladding on the leaking fuel rupture while reconstituting the assembly (e.g., replacing the damaged fuel pin).

For the APWR facility, the iodine decontamination factor (DF) through water is 100. At the time of the postulated clad rupture, the mean wind speed is 2.5 m/s. Data from station meteorological towers indicate that the air temperature at 60 m is 21 °C and the air temperature at 10 m is 21.5 °C. The following table provides additional meteorological data:

Meteorological parameters	
Stability class	Temperature gradient [ΔT (°C)/50 m]
A	<-0.95
B	-0.95 to -0.86
C	-0.85 to -0.76
D	-0.75 to -0.26
E	-0.25 to $+0.74$
F	$>+0.75$

(a) A decontamination factor is applicable for iodine, but not for xenon. Why is a xenon decontamination factor not needed? What other DFs should be applied at the APWR?

(b) What stability class is applicable for the meteorological conditions at the APWR?

(c) For the APWR, the ^{131}I gap activity in the fuel pin is 2.8×10^7 MBq and the ^{131}I total fuel pin activity is 2.8×10^9 MBq. What ^{131}I activity was released from the fuel pin? Why did you select the gap or total activity source term as the basis for the ^{131}I activity calculation?

(d) The activity was uniformly released through the stack with a height of 65 m over a 2-h period. Using the Pasquill–Gifford equation, calculate the maximum downwind ^{131}I concentration at the plume centerline at the site boundary. The following dispersion parameters are applicable for the conditions of part (b) at the site boundary: $\sigma_y = 100$ m and $\sigma_z = 40$ m.

02-03 You are performing an ALARA review of a maintenance task at an advanced pressurized water reactor. The task involves the repair of a primary coolant system charging pump in the vicinity of a 5-cm valve that has accumulated 3.7×10^5 MBq of radionuclide A. This valve is the dominant radiation source for the task.

Radionuclide A emits one 2.0 MeV photon per disintegration with a 100% yield. Maintenance planning estimates that it will take 4 h to repair the pump. The worker performs the task at a distance of 2 m from the valve. Assume attenuation by the valve is negligible.

(a) Calculate the unshielded effective dose rate at the worker's location. The attenuation (μ/ρ) and energy absorption (μ_{en}/ρ) coefficients are provided in the following table:

μ/ρ	0.0461 cm^2/g (lead)
μ_{en}/ρ	0.0235 cm^2/g (air)
μ_{en}/ρ	0.0258 cm^2/g (muscle)

(b) The effective dose goal for this task is 0.001 Sv. How much lead shielding is needed to achieve this goal? The density of lead is 11.35 g/cm^3 and the following point source buildup factors are applicable:

Point source buildup factors for lead	
μx	Buildup factors (2.0 MeV)
1	1.40
2	1.76
3	2.14
4	2.52
5	2.91
6	3.32
7	3.74
10	5.07
15	7.44
20	9.98

02-04 An advanced boiling water reactor completed its first 24 months of operation. Although the plant functioned well, three of the four low-power range detectors failed and must be replaced. The detectors are fabricated from stainless steel and have a mass of 10 g. As these detectors are in proximity to the core, significant activation and high dose rates are expected. The average thermal neutron flux at the detector locations is 2.0×10^{13} n/cm^2 s and the average fast neutron flux is 7.0×10^{13} n/cm^2 s.

The detector design specifications require that natural cobalt comprises no more than 0.014% by mass of stainless steel. All natural cobalt is ^{59}Co.

(a) List the production mechanisms for the following five activation products of stainless steel: ^{60}Co, ^{58}Co, ^{54}Mn, ^{56}Mn, and ^{59}Fe. Are these reactions produced by thermal or fast neutron reactions?
(b) What is the ^{60}Co activity in the detector at 30 days after reactor shutdown? The activation cross section for the ^{59}Co(n, γ)^{60}Co reaction is 37 b and its half-life is 1923 days.
(c) What is the effective dose rate from the ^{60}Co activation in the detector at 30 cm from one of the failed low-power range detectors at 30 days after shutdown? The ^{60}Co dose factor is 3.56×10^{-7} Sv m^2/MBq h.

02-05 A liquid metal fast breeder reactor produces transuranium elements including ^{239}Pu, ^{241}Pu, and ^{241}Am. What is the production mechanism for (a) ^{239}Pu, (b) ^{241}Pu, and (c) ^{241}Am?

02-06 An advanced PWR is preparing to shut down for a refueling outage. The plant manager asks you to assess some of the expected radiological conditions during the shutdown. Assume the reactor is licensed in the United States and follows Nuclear Regulatory Commission guidelines in 10CFR20, derived from ICRP-26.

(a) List five considerations when estimating the ^{131}I airborne concentration in containment 24 h after shutdown. Assume that the PWR has the following systems that reduce the iodine concentration:
(1) a primary coolant liquid clean-up system,
(2) a containment air ventilation system, and
(3) a containment air charcoal filter system.

(b) Determine the committed dose equivalent (CDE) to the worker's thyroid from a 10-h exposure to a ^{131}I air concentration of 2.96×10^{-4} MBq/m^3. The derived air concentration (DAC) for ^{131}I is 7.4×10^{-4} MBq/m^3. The worker did not use respiratory protection. Also, determine the worker's committed effective dose equivalent (CEDE). State all assumptions.

(c) List factors that should be considered in the prejob analysis for a containment entry after reactor shutdown to keep the worker's total effective dose equivalent ALARA.

(d) The plant manager is considering hydrogen peroxide treatment of the primary coolant system. Hydrogen peroxide will be added during hot, noncritical operations to increase the level of soluble Co-58 in the primary coolant system. The soluble ^{58}Co is removed from the primary coolant system by demineralizers. You expect the level of ^{58}Co in the primary coolant to increase to 3.7×10^4 MBq/m^3 as a result of the peroxide addition. State three methods for reducing the primary system ^{58}Co clean-up time.

(e) State two benefits of adding hydrogen peroxide to the primary coolant system at the onset of refueling.

(f) Given an effective dose rate of 2.5×10^{-3} mSv/h at a perpendicular distance of 2 m from the midpoint of a 2.0-m long pipe containing a uniform concentration of ^{58}Co, calculate the total activity contained in the pipe. The ^{58}Co effective dose factor is 1.66×10^{-7} Sv m^2/MBq h. State all assumptions.

02-07 You are the station health physicist at an ACR-1000 Generation III CANDU reactor during a D$_2$O moderator purification task to remove tritium oxide (HTO). In the course of the task, a spill of moderator occurs, HTO is vaporized, uniformly filling the cubicle in which the moderator purification task is being performed, and a tritium air monitor alarms. Workers at your facility are enrolled in a tritium bioassay program. The facility is equipped with workplace tritium air monitors (flow through ionization chambers). Assume the room ventilation is shut off during the moderator purification task.

The ACR-1000 is the first CANDU system operating in the United States and is licensed by the Nuclear Regulatory Commission. The health physics program follows Title 10 of the Code of Federal Regulations, Part 20 that is based on ICRP-26.

(a) Calculate the committed effective dose equivalent you would expect a worker to receive from a tritium concentration of 185 MBq/m^3 as measured by workplace air monitoring. The worker spends 1 min in the HTO atmosphere. Your procedures mandate that the HTO DAC is 0.74 MBq/m^3.

(b) The individual involved in the incident submits a postincident bioassay sample collected during the first 24 h. The results indicate tritium concentration of 1850 Bq/l in the urine. Calculate the dose received in μSv. The DCF (acute intake) $= 7.57 \times 10^{-4}$ μSv l/Bq in urine (first 24 h) and the DCF (chronic intake) $= 5.41 \times 10^{-5}$ μSv l/Bq in urine (average daily concentration).

(c) Assume that the dose equivalent calculated from the urine concentration differs from the dose equivalent that was calculated from the room air concentration. If the measurements and calculations were performed correctly, provide two likely sources of this discrepancy.

(d) Identify two techniques that can be used for tritium air monitoring. Specify one advantage and one disadvantage of each technique.

02-08 You are the senior health physicist at the Pungent Valley Nuclear Power Station, a Generation III advanced BWR licensed in the United States by the Nuclear Regulatory Commission. Plant management is considering a new demineralizer system and you are tasked with performing the radiological design review and ALARA evaluation.

The primary coolant demineralizer is a stainless steel cylinder with a height of 2 m and a diameter of 1 m. It has a processing capability of 1000 l/min and a ^{60}Co removal efficiency of 95%. The nominal operational characteristics of the system assume a routine demineralizer run time of 100 days and a routine demineralizer downtime of 60 days. The anticipated ^{60}Co activity concentration entering the demineralizer is 70.3 MBq/m^3. The ^{60}Co dose factor is 3.57×10^{-7} Sv m^2/h MBq.

(a) Name four documents (e.g., federal regulations and facility documents) that will be needed to perform this evaluation.
(b) List and briefly describe items that you should consider when evaluating the demineralizer system from an ALARA perspective.
(c) Calculate the total activity in MBq present in the demineralizer at the end of its runtime and at the end of its downtime. For the purpose of this question, ^{60}Co is the only radioisotope under consideration. The half-life of ^{60}Co is 5.27 years.
(d) For this demineralizer, what is the decontamination factor for ^{60}Co?
(e) Calculate the effective dose rate 20 m from the demineralizer at the end of its runtime and at the end of its downtime. Ignore self-shielding in the resin, water, and the shielding in the stainless steel shell in the demineralizer.

(f) List methods you could use to minimize exposure to plant personnel during maintenance of the demineralizer.

02-09 A steam generator tube rupture event has occurred at a light water cooled Generation IV fission reactor. The release rate from the boundary of the steam generator is composed of radioiodine (2 MBq/s) and noble gas (3 MBq/s).

(a) Using credible decontamination factors, what are the iodine and noble gas release rates assuming that a main steam line relief valve has lifted and not reseated?
(b) What are the release rates of these radioactive materials assuming that the release is through the condenser?
(c) From an ALARA perspective, which release pathway is preferred?

02-10 A steam generator tube leak is in progress at a Generation III PWR. If the primary to secondary leak rate is 500 l/day and the primary ^{85}Kr noble gas activity is 1000 MBq/cm^3, what is the concentration of ^{85}Kr in the air ejector? Assume the air ejector flow rate is 5000 l/min.

References

Atomic Energy of Canada Limited Report (2005) ACR-1000 Technical Summary, Mississauga, Ontario, Canada.

Bevelacqua, J.J. (1995) *Contemporary Health Physics: Problems and Solutions*, John Wiley & Sons, Inc., New York.

Bevelacqua, J.J. (1999) *Basic Health Physics: Problems and Solutions*, John Wiley & Sons, Inc., New York.

Bourdeloie, C. and Marimbeau, P. (2004) Determination of a HTGR Radioactive Material Inventory. Paper C13, 2nd International Topical Meeting on High Temperature Reactor Technology, Beijing, China, September 22–24.

Bruchi, H.J. (2004) The Westinghouse AP-1000 – final design approved. *Nuclear News*, **47** (12), 30.

Carelli, M.D. (2003) IRIS: A global approach to nuclear power renaissance. *Nuclear News*, **46** (10), 32.

Code of Federal Regulations (2007) *Standards for Protection Against Radiation*, Title 10, Part 20, U.S. Government Printing Office, Washington, DC.

DETR/RAS/99.013 (1999) A Review of the Processes Contributing to Radioactive Waste in the UK. Report prepared for the Department of the Environment, Transport, and the Regions and UK Nirex Ltd, Electrowatt-Ekono (UK) Ltd, West Sussex, UK.

Doolittle, W.W., Bredvad, R.S. and Bevelacqua, J.J. (1992) An ALARA conscious hot particle control program. *Radiation Protection Management*, **9** (1), 27.

Eichholz, G.G. (1977) *Environmental Aspects of Nuclear Power*, Ann Arbor Science Publications, Ann Arbor, MI.

Energy Information Administration Report (2003) *New Reactor Designs*, U.S. Department of Energy, Washington, DC.

Fabrikant, J.I. (1981) Guest Editorial: Health effects of the nuclear accident at Three Mile Island. *Health Physics*, **40**, 151.

Forsberg, C.W., Pickard, P.S. and Peterson, P. (2003) The advanced high-temperature reactor for production of hydrogen or electricity. *Nuclear News*, **46** (2), 30.

Foster, A.R. and Wright, R.L., Jr (1977) *Basic Nuclear Engineering*, 3rd edn, Allyn and Bacon, Boston, MA.

GIF-002-00 (2002) *A Technology Roadmap for Generation IV Nuclear Energy Systems – Ten Nations Preparing Today for Tomorrow's Energy Needs*, U.S. Department of Energy, Washington, DC.

Glasstone, S. and Sesonske, A. (1963) *Nuclear Reactor Engineering*, D. Van Nostrand Co., New York, NY.

Glasstone, S. and Walter, W.H. (1980) *Nuclear Power and Its Environmental Effects*, American Nuclear Society, La Grange Park, IL.

Hinds, D. and Maslak, C. (2006) Next-generation nuclear energy: the ESBWR. Nuclear News, **49** (1), 35.

IAEA Safety Series No. 50-SG-05 (1983) *Radiation Protection During Operation of Nuclear Power Plants*, International Atomic Energy Agency, Vienna, Austria.

IAEA Safety Series No. 110 (1993) *The Safety of Nuclear Installations: Safety Fundamentals* International Atomic Energy Agency, Vienna, Austria.

IAEA TECDOC-358 (1985) *Gas-Cooled Reactor Safety and Accident Analysis. Proceedings of a Specialists' Meeting*, Oak Ridge, May 13–15, 1985, International Atomic Energy Agency, Vienna, Austria.

IAEA TECDOC-1020 (1998) *Design Measures for Prevention and Mitigation of Severe Accidents at Advanced Water Cooled Reactors*, International Atomic Energy Agency, Vienna, Austria.

IAEA-TECDOC-1391 (2004) *Status of Advanced Light Water Reactor Designs 2004*, International Atomic Energy Agency, Vienna, Austria.

IAEA Technical Report Series No. 189 (1979) *Storage, Handling, and Movement of Fuel and Related Components at Nuclear Power Plants*, International Atomic Energy Agency, Vienna, Austria.

ICRP Publication No. 40 (1984) *Protection of the Public in the Event of Major Radiation Accidents: Principles for Planning*, Pergamon Press, New York, NY.

ICRP Publication No. 43 (1984) *Principles of Monitoring for the Radiation Protection of the Public*, Pergamon Press, New York, NY.

LaBar, M.P., Shenoy, A.S., Simon, W.A. and Campbell, E.M. (2003) The gas turbine–modular helium reactor. Nuclear News, **46** (11), 28.

Lederer, C.M. and Shirley, V.S. (1978) *Table of Isotopes*, 7th edn, John Wiley & Sons, Inc., New York.

National Research Council (2004) *The Hydrogen Economy: Opportunities, Costs, Barriers, and R&D Needs*, The National Academy Press, Washington, DC.

NCRP Report No. 55 (1977) *Protection of the Thyroid Gland in the Event of Releases of Radioiodine*, National Council on Radiation Protection and Measurements, Bethesda, MD.

NCRP Report No. 92 (1988) *Public Radiation Exposure from Nuclear Power Generation in the United States*, National Council on Radiation Protection and Measurements, Bethesda, MD.

NCRP Report No. 106 (1989) *Limit for Exposure to "Hot Particles" on the Skin*, National Council on Radiation Protection and Measurements, Bethesda, MD.

NCRP Report No. 120 (1994) *Dose Control and Nuclear Power Plants*, National Council on Radiation Protection and Measurements, Bethesda, MD.

NCRP Report No. 130 (1999) *Biological Effects and Exposure Limits for "Hot Particles,"* National Council on Radiation Protection and Measurements, Bethesda, MD.

NUREG-1560 (1996) Individual Plant Examination Program: Perspectives on Reactor Safety and Plant Performance, United States Nuclear Regulatory Commission, Washington, DC.

NUREG-1793 (2006) Final Safety Evaluation Report Related to Certification of the AP1000 Standard Design, United States Nuclear Regulatory Commission, Washington, DC.

SECY-90-016 (1990) Commission Paper, Evolutionary Light Water Reactor (LWR) Certification Issues and Their Regulatory

Relationship to Current Regulatory Requirements, January 12, 1990 and Staff Requirements Memorandum dated June 26, 1990, United States Nuclear Regulatory Commission, Washington, DC.

Shleien, B., Slaback, L.A., Jr and Birky, B. K. (1988) *Handbook of Health Physics and Radiological Health*, 3rd edn, Lippincott, Williams, and Wilkins, New York.

Stacey, W.M. (2001) *Nuclear Reactor Physics*, Wiley-VCH Verlag GmbH & Co. KGaA, Weinheim, Germany.

Torgerson, D.F. (2002) The ACR-700 – raising the bar for reactor safety, performance, economics, and constructability. *Nuclear News*, **45** (10), 24.

Twilley, R.C., Jr (2002) Framatome ANPs SWR 1000 reactor design. *Nuclear News*, **45** (9), 36.

Twilley, R.C., Jr (2004) EPR development – an evolutionary design process. *Nuclear News*, **47** (4), 26.

Uranium Information Centre (2005) UIC Briefing Paper # 77, Generation IV Nuclear Reactors, A.B.N. 30 005 503 828, GPO Box 1649N, Melbourne 3001, Australia, April.

Uranium Information Centre Nuclear Issues Briefing Paper 16(2005) Advanced Nuclear Power Reactors, A.B.N. 30 005 503 828, GPO Box 1649N, Melbourne 3001, Australia, May.

USNRC Regulatory Guide 8.8 (1979) *Information Relevant to Ensuring that Occupational Radiation Exposures at Nuclear Power Stations Will Be As Low As Reasonably Achievable (ALARA)*, United States Nuclear Regulatory Commission, Washington, DC.

USNRC Regulatory Guide 8.13 (1988) *Instruction Concerning Prenatal Radiation Exposure*, United States Nuclear Regulatory Commission, Washington, DC.

USNRC Regulatory Guide 8.29 (1981) *Instruction Concerning Risks from Occupational Radiation Exposure*, United States Nuclear Regulatory Commission, Washington, DC.

El-Wakil, M.M. (1962) *Nuclear Power Engineering*, McGraw-Hill, New York, NY.

3
Fusion Power Production

3.1
Overview

Fusion energy offers the potential for cheap, clean, and abundant energy. It also offers a number of significant advantages when compared with fission technology. In particular, fusion facilities do not suffer from many of the issues associated with fission reactors, including reactor safety, nuclear waste generation, disposition of spent reactor fuel, vulnerability to terrorist attacks, and nuclear proliferation. These factors offer considerable motivation for replacing fission reactors with fusion reactors. Unfortunately, fusion reactors are not yet a viable alternative to fission facilities.

Fusion energy is a potential source for power production in the mid- to late twenty-first century. The fusion reaction or process occurs within the plasma composed of light nuclei. A commercial fusion power facility would likely use either magnetic or inertial means to confine the plasma and facilitate the fusion process. The fusion confinement method influences the radiation types and fuel materials that must be controlled by the health physicist.

Fusion processes substantially differ from those encountered in fission reactors because they do not produce actinides or radioactive isotopes of iodine, cesium, or strontium. A fusion reactor does produce a wide variety of reaction and activation products and these products depend on the selected fusion process, the reaction energies, and the materials of construction selected for the facility. Fusion products and activation products present a challenge for the health physicist responsible for worker radiation protection at a fusion power facility. Both internal and external radiation challenges are present. In addition, tritium fuel material presents an internal hazard in its initial state prior to introduction into the fusion reactor.

In this chapter, we review the radiological hazards associated with a fusion power facility, identify the anticipated sources of radiation exposure from this facility, and identify possible as low as reasonably achievable (ALARA) measures to reduce the occupational doses. This chapter also reviews the basic physics principles and relationships that govern the fusion process. These relationships will be shown to define the basic plasma properties, govern the nuclides interacting to form the plasma, and determine their energy. The underlying physics also determines the types of radiation that are produced within the plasma.

Health Physics in the 21st Century. Joseph John Bevelacqua
Copyright © 2008 WILEY-VCH Verlag GmbH & Co. KGaA, Weinheim
ISBN: 978-3-527-40822-1

3.2
Fusion Process Candidates

Fusion involves the interaction of two light systems to form a heavier system. A variety of fusion processes are possible and could be applicable to power production. The term fusion process is used instead of fusion reaction because a fusion event that is used to produce power involves not only the reaction of individual light ions but also their density, confinement time, mode of confinement, presence of other plasma constituents that inhibit or catalyze the light ion fusion, and method to initiate, sustain, and energize the plasma. The term reaction is reserved for specific nuclear events [e.g., (n, γ), (γ, n), (n, α), (n, p), (n, 2n), and (n, 2n α)] that result from the fusion process.

Candidate fusion power processes include the following:

$$D+D \xrightarrow{50\%} T(1.01 \text{ MeV})+p(3.02 \text{ MeV}), \tag{3.1}$$

$$D+D \xrightarrow{50\%} {}^3He(0.82 \text{ MeV})+n(2.45 \text{ MeV}), \tag{3.2}$$

$$D+T \rightarrow {}^4He(3.50 \text{ MeV})+n(14.1 \text{ MeV}), \tag{3.3}$$

$$D+{}^3He \rightarrow {}^4He(3.60 \text{ MeV})+p(14.7 \text{ MeV}), \tag{3.4}$$

$$T+T \rightarrow {}^4He+n+n+11.3 \text{ MeV}, \tag{3.5}$$

$${}^3He+T \xrightarrow{51\%} {}^4He+p+n+12.1 \text{ MeV}, \tag{3.6}$$

$${}^3He+T \xrightarrow{43\%} {}^4He(4.8 \text{ MeV})+D(9.5 \text{ MeV}), \tag{3.7}$$

$${}^3He+T \xrightarrow{6\%} {}^5He(2.4 \text{ MeV})+p(11.9 \text{ MeV}), \tag{3.8}$$

$$p+{}^6Li \rightarrow {}^4He(1.7 \text{ MeV})+{}^3He(2.3 \text{ MeV}), \tag{3.9}$$

$$p+{}^7Li \xrightarrow{20\%} {}^4He(8.7 \text{ MeV})+{}^4He(8.7 \text{ MeV}), \tag{3.10}$$

$$p+{}^7Li \xrightarrow{80\%} {}^7Be+n-1.6 \text{ MeV}, \tag{3.11}$$

$$D+{}^6Li \rightarrow {}^4He(11.2 \text{ MeV})+{}^4He(11.2 \text{ MeV}), \tag{3.12}$$

$$p+{}^{11}B \rightarrow {}^4He+{}^4He+{}^4He+8.7 \text{ MeV}, \tag{3.13}$$

$$n+{}^6Li \rightarrow {}^4He(2.1 \text{ MeV})+T(2.7 \text{ MeV}), \tag{3.14}$$

where D is deuterium (^{2}H) and T is tritium (^{3}H). For consistency with the literature, D and ^{2}H and T and ^{3}H are used interchangeably.

For binary, exothermic processes, the particle energy at the reaction threshold is provided in parenthesis. A negative Q-value indicates the reaction is endothermic. Although these processes are all viable fusion candidates, the discussion focuses on the most likely near-term possibilities (i.e., D-T and D-D fusion).

If fusion becomes a practical source of energy, it will be initially realized through the D-T process. At a later stage, fusion involving only deuterium nuclei may become more important. However, the D-D process is somewhat more difficult to achieve because its inclusive cross sections are smaller in magnitude and higher densities are required to initiate and sustain the D-D fusion process.

These comments are illustrated by considering the systematics of the ^{4}He system. The binary breakup channels in increasing energy, relative to the ^{4}He ground state, are p+^{3}H, n+^{3}He, and ^{2}H+^{2}H. The D-T process receives contributions from each of these three binary channels as part of the rearrangement sequence to form ^{4}He through Equation 3.3. Moreover, the cross sections for the p+^{3}H and n+^{3}He inclusive reactions are larger than those for the ^{2}H+^{2}H inclusive reaction. These considerations provide part of the basic physics justification for the initial investigation of the D-T process.

In the D-D fusion process, the tritium nucleus formed in Equation 3.1 subsequently fuses through the D-T process (Equation 3.3). An advantage of the D-D process is that it avoids the need for a tritium fuel source, which eliminates a significant health physics hazard. However, it is unlikely that the conditions for D-D fusion will be realized on a practical scale before D-T fusion. Therefore, the subsequent discussion focuses on a D-T fusion power facility.

The D-T fusion energy output is about 94×10^6 kWh/kg of a mixture of deuterium (0.4 kg) and tritium (0.6 kg). On a per mass basis, this is more than four times the energy released from fission.

The D-T fusion process occurs within a state of the matter known as plasma. Before proceeding further, it is necessary to define the forces governing the plasma as well as the characteristics and properties of plasmas.

3.3
Physics of Plasmas

The term plasma is defined in more detail in subsequent sections of this chapter. For this section, it is sufficient to define plasma as a collection of charged particles. In the presence of electric and magnetic fields, each particle experiences a total force ($\vec{F}$). For the ith charged particle in the plasma, this force is

$$\vec{F}_i = m_i \frac{d\vec{v}_i}{dt} = q_i(\vec{E} + \vec{v}_i \times \vec{B}), \tag{3.15}$$

where m_i is the mass of the charged particle or ion, $\vec{v}_i$ is the ion velocity, $\vec{E}$ is the total electric field experienced by the ion, $\vec{B}$ is the total magnetic induction, t is the time, and q_i is the ion's charge. The magnetic induction is related to the magnetic field ($\vec{H}$) through the relationship

$$\vec{B} = \mu \vec{H}, \tag{3.16}$$

where μ is the permeability of the plasma.

The total magnetic induction is the sum of the external magnetic induction ($\vec{B}_{ext}$) used to confine the plasma and the magnetic induction generated by the ith moving charged particle ($\vec{B}_i$):

$$\vec{B} = \vec{B}_{ext} + \sum_i \vec{B}_i. \tag{3.17}$$

In a similar fashion, the total electric field is

$$\vec{E} = \vec{E}_{ext} + \sum_i \vec{E}_i, \tag{3.18}$$

where $\vec{E}_{ext}$ is the external electric field used to heat the plasma and $\vec{E}_i$ is the electric field produced by the ith charged particle.

The electric and magnetic fields are not independent quantities, but are related through the Maxwell equations:

$$\vec{\nabla} \cdot \vec{E} = \frac{\rho}{\varepsilon_0}, \tag{3.19}$$

$$\vec{\nabla} \cdot \vec{B} = 0, \tag{3.20}$$

$$\vec{\nabla} \times \vec{E} = -\frac{\partial \vec{B}}{\partial t}, \tag{3.21}$$

$$\vec{\nabla} \times \vec{H} = \vec{J} + \frac{\partial \vec{D}}{\partial t}, \tag{3.22}$$

where $\vec{\nabla}$ is the gradient operator, ρ is the plasma electric charge density, $\vec{J}$ is the plasma electric current density, ε_0 is the permittivity of free space, and $\vec{D}$ is the displacement current defined in terms of the electric field and the permittivity (ε) of the fusion plasma:

$$\vec{D} = \varepsilon \vec{E}. \tag{3.23}$$

It is readily shown that the electric field and magnetic induction can be written in terms of a scalar potential ϕ and vector potential $\vec{A}$:

$$\vec{B} = \vec{\nabla} \times \vec{A}. \tag{3.24}$$

$$\vec{E} = -\vec{\nabla}\phi - \frac{\partial \vec{A}}{\partial t}. \tag{3.25}$$

The basic physics overview of Equations 3.15–3.25 suggests that the charged particles in the plasma are subjected to forces dependent on the electric and magnetic fields. These fields are equivalent to their mathematical counterparts ϕ and $\vec{A}$ that describe potentials.

In the next section of this chapter, plasma characteristics are defined in terms of the scalar potential. In addition, the explicit plasma properties that govern the scalar potential are determined.

3.4
Plasma Properties and Characteristics

Plasma is often referred to as the fourth state of matter because of its unique properties. It consists of a collection of atoms, ions, and electrons in which a large fraction of the atoms are ionized so that the electrons and ions are essentially free. Ionization occurs when the temperature or energy of the plasma reaches a threshold value characteristic of the plasma's initial atomic constituents.

The energy (E) of an ion or electron in the plasma is related to its absolute temperature (T):

$$E = kT, \qquad (3.26)$$

where k is the Boltzmann's constant $(1.38 \times 10^{-23}$ J/K$)$. An ion with an energy of 1 eV corresponds to a temperature of

$$T = \frac{E}{k} = \frac{(1 \text{ eV})(1.6 \times 10^{-19} \text{ J/eV})}{1.38 \times 10^{-23} \text{ J/K}} = 1.16 \times 10^4 \text{ K}. \qquad (3.27)$$

Ionization is not the only process that occurs in the plasma. The ions and electrons can also recombine. However, ionization dominates recombination for practical fusion plasmas.

In general, the plasma configuration in a fusion power reactor is complex and three-dimensional effects must be considered. Moreover, the plasma is not a static system, but is quite dynamic. Plasmas are governed by a number of fundamental processes described by the Maxwell equations. These processes strongly influence the electron and ion densities and lead to an electric charge density that is related to the three-dimensional scalar potential.

The scalar potential is determined from Equation 3.25. If we impose the restriction that the vector potential is constant in time, the electric field becomes

$$\vec{E} = -\vec{\nabla}\phi. \qquad (3.28)$$

Taking the divergence of both sides of Equation 3.28 yields

$$\vec{\nabla} \cdot \vec{E} = -\nabla^2 \phi. \qquad (3.29)$$

Equations 3.19 and 3.29 are combined to obtain Poisson's equation

$$\vec{\nabla} \cdot \vec{E} = \frac{\rho}{\varepsilon_0} = -\nabla^2 \phi, \tag{3.30}$$

or

$$\nabla^2 \phi = -\frac{\rho}{\varepsilon_0}. \tag{3.31}$$

The charge density in the plasma is proportional to the difference in the ion (n_i) and electron densities (n_e):

$$\rho = (n_i - n_e)e, \tag{3.32}$$

where e is the electron charge (1.6×10^{-19} C). Combining Equations 3.31 and 3.32 yields the equation

$$\nabla^2 \phi = -\frac{(n_i - n_e)e}{\varepsilon_0}, \tag{3.33}$$

which depends on the specific plasma conditions. Equation 3.33 suggests that the electron density varies with the plasma potential. Assuming thermal equilibrium, the electron density is written in terms of a Boltzmann relationship:

$$n_e = n_\infty \exp\left(\frac{e\phi}{T_e}\right), \tag{3.34}$$

where n_∞ is the density as the electron moves to the boundary of the plasma and T_e is the electron energy (Equation 3.27).

The ions are considerably more massive than the electrons and are less affected by the potential as it attempts to overcome their inherent inertia. As the plasma tends to be electrically neutral, it is reasonable to let $n_i \approx n_\infty$, which implies that the ion density is not significantly perturbed by ϕ.

The general solution of Equation 3.33 is complex and often requires the application of numerical methods. Fortunately, a reasonable understanding of the solution can be obtained without the rigors of a three-dimensional analysis. From a health physics perspective, sufficient insight is gained by solving a corresponding one-dimensional problem.

The one-dimensional problem considers a potential $\phi(x)$. This one-dimensional potential is considered to arise from a charged grid placed at the center ($x = 0$) of the plasma. The grid produces a potential ϕ_0 as $x \to 0$, and $\phi \to 0$ as x approaches the plasma boundary ($x \to \infty$). Considering the one-dimensional limit and charge neutrality, Equation 3.33 becomes

$$\frac{d^2 \phi}{dx^2} = -\frac{e}{\varepsilon_0}(n_\infty - n_e). \tag{3.35}$$

Using Equation 3.34, Equation 3.35 is written as

$$\frac{d^2\phi}{dx^2} = \frac{en_\infty}{\varepsilon_0}\left(\exp\left(\frac{e\phi}{T_e}\right) - 1\right). \tag{3.36}$$

Although Equation 3.36 is highly nonlinear, its solution is facilitated by noting that the potential decreases as the charged particle moves away from the grid:

$$\left|\frac{e\phi}{T_e}\right| \ll 1 \quad \text{as} \quad x \to \infty. \tag{3.37}$$

Using this property of the one-dimensional charged grid potential permits the exponential to be expanded in a power series. As each successive term of the expansion rapidly decreases, only the first two terms need to be kept:

$$\exp\left(\frac{e\phi}{T_e}\right) \approx 1 + \frac{e\phi}{T_e}. \tag{3.38}$$

Equation 3.38 is used to simplify Equation 3.36:

$$\frac{d^2\phi}{dx^2} = \frac{en_\infty}{\varepsilon_0}\left(1 + \frac{e\phi}{T_e} - 1\right) = \frac{e^2 n_\infty}{\varepsilon_0 T_e}\phi. \tag{3.39}$$

Equation 3.39 has the solution

$$\phi(x) = \phi_0 \exp(-|x|/\lambda_D), \tag{3.40}$$

where λ_D is a parameter called the Debye length:

$$\lambda_D = \left(\frac{\varepsilon_0 T_e}{e^2 n_\infty}\right)^{1/2}. \tag{3.41}$$

The Debye length is an important parameter because it characterizes the behavior of interactions within the plasma. When two charged particles in a plasma are very close together they interact through their Coulomb fields as isolated, individual particles. However, as the distance between the two particles increases beyond their mean separation distance (proportional to $n_e^{-1/3}$), their interaction mechanism begins to change.

Beyond their mean separation distance, the ions interact collectively with nearby charged particles. In this collective regime, the Coulomb force from any given charged particle affects all other charged particles, which electrically polarizes the plasma. The collective interaction character reduces or shields the individual particle electric fields. The Debye length or Debye shielding length is the distance beyond which the electric field owing to any given particle is collectively shielded. Specifically, the long-range Coulomb electric field is effectively limited to a distance on the order of the Debye length in the plasma. The Debye length is also the minimum physical size needed to establish a plasma.

For scales greater than the Debye length, a plasma responds collectively to a given charge, perturbation, or field. The Debye length is also the maximum scale length over which a plasma can depart significantly from charge neutrality. Therefore, plasmas that are larger than the Debye length are quasi-neutral (or on average electrically neutral).

With this physical clarification, the basis for the approximation of Equation 3.38 may be related to the condition for establishing the plasma state. From Equation 3.40, the condition for collective motion $\phi \to \phi_0$ or $|x| > n_e^{-1/3}$ is expressed as

$$\frac{|x|}{\lambda_D} \to \frac{1}{n_e^{1/3}\lambda_D} < 1. \tag{3.42}$$

Equation 3.42 may be expressed as

$$n_e^{1/3}\lambda_D > 1, \tag{3.43}$$

or

$$n_e \lambda_D^3 \gg 1. \tag{3.44}$$

Equations 3.42–3.44 imply that many electrons (or other charged particles) exist within a cube with a side equal to the Debye length. Physically, Equation 3.44 is a necessary condition for the existence of a state of matter known as a plasma, because it represents the requirement that at distances greater than λ_D collective interactions of charged particles dominate over binary Coulomb collisions.

Meeting the conditions for a plasma does not depend on the number of charged particles, but upon the charged particle density. For example, if 10^9 ions are spread over a large volume (e.g., the volume of a Solar System), the conditions for a plasma are not met. However, if the 10^9 particles are compressed to a volume of 1 cm^3, such as in the Solar corona, then a plasma results. A quantification of the conditions necessary for a plasma to come into being is summarized in Table 3.1. Table 3.1 provides the plasma density, plasma temperature, and the Debye length for selected media.

From a health physics perspective, we have now established the requisite conditions for establishing a plasma (i.e., size $> \lambda_D$), as well as the required relationship (Equation 3.44) between density and λ_D. With the establishment of the physical size and density of the plasma, we can review specific aspects of plasma confinement in a more comprehensive manner.

Table 3.1 Characteristics of various plasmas.

Medium	Charged particle density (ions/cm^3)	T (eV)	λ_D (cm)
Interstellar gas	1	1	700
Solar corona	10^9	100	0.2
Magnetic confinement fusion plasma	10^{15}	10^4	0.002
Inertial confinement fusion plasma	10^{26}	10^4	7×10^{-9}

Derived from Gekelman (2003).

3.5
Plasma Confinement

Once the fusion process is selected, it will be necessary to confine the resulting plasma with a suitable physical mechanism. Two primary approaches are available for confining the charged particles forming the fusion plasma: magnetic confinement (MC) and inertial confinement (IC).

Magnetic confinement is based on the force a charged particle experiences in a magnetic field. The magnetic force ($\vec{F}_{mag}$) is a component of the total electromagnetic force (Equation 3.15), serves to confine the charged particles comprising the plasma, and has the form

$$\vec{F}_{mag} = q\vec{v} \times \vec{B}. \tag{3.45}$$

The cross product is rewritten in terms of the angle (θ) between $\vec{v}$ and $\vec{B}$:

$$\vec{F}_{mag} = q|v|B||\sin\theta. \tag{3.46}$$

As the fusion plasma is composed of charged particles, it is possible to confine a plasma using a suitable magnetic field. Several different magnetic field arrangements are possible including toroidal fields (TFs), poloidal fields, magnetic mirrors, Yin-Yang coils, and combinations of aforementioned configurations. These fields have been used in a number of common design types including tokamaks, Z-pitch machines, toroidally linked mirrors, and tandem mirror machines.

In magnetic confinement D-T fusion, quantities of deuterium and tritium gas are maintained to initiate and sustain the fusion process. The tritium gas (T_2) must be carefully monitored and controlled as it presents an internal intake concern. In addition, T_2 is readily converted into HTO that is considerably more radiotoxic.

Tritium gas fuel and HTO present a greater internal hazard than the solid D-T pellet used in an inertial confinement device. Therefore, ALARA concerns alone favor inertial confinement as the fusion technology of choice. However, unfused tritium fuel requires a clean-up/recovery system to minimize its health physics impact.

In inertial confinement, a small pellet of a D-T mixture is heated by a short burst of energy from either laser beams or beams of high-energy charged particles. The absorption of energy increases the pellet's temperature and also compresses it to high density. The fusion process occurs so rapidly that inertia alone prevents the pellet from disassembly while fusion is in progress. No other confinement mechanism is required.

Magnetic confinement and inertial confinement represent two extremes for producing a stable plasma condition. In magnetic confinement, the D-T plasma has a density (ρ) on the order of 10^{15} ions/cm^3 with a confinement time (τ) on the order of 0.1 s. For D-T fusion, the product of the density and confinement time must be at least 10^{14} ion s/cm^3 to satisfy the Lawson criterion for energy break-even. Break-even occurs when the energy input to establish and maintain the plasma equals the fusion energy output.

Inertial confinement of a D-T plasma requires larger densities on the order of 10^{26} ions/cm^3 with a confinement time on the order of 10^{-12} s. Given these extremes, the configuration and operating parameters for a fusion power facility significantly depend on the plasma confinement approach. It is also worth noting that for D-D fusion, the break-even value would be larger than the 10^{14} ion s/cm^3 value required for D-T fusion.

In addition to density and confinement time differences, the source geometries for inertial confinement and magnetic confinement fusion plasmas differ significantly. IC fusion can be represented as a point source as the fusion process occurs within a small D-T pellet. MC fusion occurs as a distributed source within the toroidal vacuum vessel volume.

For both IC and MC fusion, the effective dose (E) is the product of the dose equivalent rate ($\dot{E}$) and the confinement time (τ):

$$E = \dot{E}\tau. \tag{3.47}$$

For an isotropic IC point fusion source, the effective dose rate can be written as

$$\dot{E}_{IC} = \rho V g \sum_R w_R \phi_R, \tag{3.48}$$

where ρ is the plasma density, V is the D-T pellet volume, g is a geometry factor ($1/4\pi r^2$) for the point isotropic source, w_R is the radiation-weighting factor for radiation of type R, and ϕ_R is a dose coefficient. In the distributed MC fusion source, the effective dose rate is determined by integrating over the toroidal plasma volume:

$$\dot{E}_{MC} = \iiint \rho(r,\theta,\phi) g(r,\theta,\phi) \sum_R w_R \phi_R r^2 \sin\theta \, dr \, d\theta \, d\phi, \tag{3.49}$$

where (r, θ, ϕ) are standard spherical coordinates and $g(r, \theta, \phi)$ is the geometry factor including the quadrature-weighting factor for facilitating the numerical integration in Equation 3.49.

For the IC and MC fusion situations noted above, the effective dose is proportional to the product of the plasma density and confinement time:

$$E \propto \rho t. \tag{3.50}$$

For IC fusion,

$$E_{IC} \propto (10^{26} \text{ ions/cm}^3)(10^{-12} \text{ s}) = 10^{14} \text{ ions s/cm}^3, \tag{3.51}$$

and for MC fusion,

$$E_{MC} \propto (10^{15} \text{ ions/cm}^3)(10^{-1} \text{ s}) = 10^{14} \text{ ions s/cm}^3. \tag{3.52}$$

Equations 3.51 and 3.52 suggest that for break-even conditions the major radiological difference between IC and MC fusion is the radial dependence of the effective dose rates that occurs in Equations 3.48 and 3.49. The effective dose rate from the IC point fusion source geometry decreases as $1/r^2$. As the MC fusion source can be represented as a collection of point sources, each having a fraction of the total fusion power, the effective dose rate as a function of distance from the distributed MC fusion source is less than or equal to the equivalent IC point fusion source.

As the distance from the source increases to beyond three times the largest source dimension, the distributed source effective dose is essentially the same as the point source dose. From a strictly ALARA perspective, MC is favored over IC for equivalent fusion power output.

Another difference between IC and MC fusion is the temporal mode of operation. In IC fusion, pulsed operation occurs, and MC fusion results in a nearly steady state operation.

There are also radiation spectral differences that occur in IC and MC fusion. The higher densities encountered in IC fusion will reduce (soften) the average energy of neutrons and photons following the fusion event. The degree of softening of the neutron and gamma spectra depends on the selected density and source configuration. The next section of this chapter reviews the characteristics of initial D-T fusion power reactors using magnetic confinement.

3.6
Overview of an Initial Fusion Power Facility

In view of the current direction in fusion research and the selection of magnetic confinement as the design concept for the International Thermonuclear Experimental Reactor (ITER), subsequent discussion is based on the ITER tokamak design concept that utilizes D-T fusion. The tokamak principle of magnetic confinement in a torus was developed in the USSR in the 1960s. The name tokamak is derived from the initial letters of the Russian words meaning "toroidal," "chamber," and "magnetic," respectively.

The heart of the magnetic confinement system in a tokamak is the torus – a large toroidal vacuum vessel surrounded by devices to produce the confining magnetic field. Other major components of a tokamak are the superconducting toroidal and poloidal magnetic field coils that confine, shape, and control the plasma inside the vacuum vessel. The magnet system includes toroidal field coils, a central solenoid, external poloidal field coils, and correction coils. The vacuum vessel is a double-walled structure. Associated with the vacuum vessel are systems supporting plasma generation and control. These systems include the divertor system and blanket shield system.

The fusion process occurs within the vacuum vessel. Radio frequency energy and ion beams heat the plasma to reach the fusion ignition temperature. In addition to its confinement function, the magnetic field is also designed to prevent plasma from striking the inner wall of the vacuum vessel. If the plasma strikes the vacuum vessel wall, material is removed from the wall and forms a particulate, dispersed into the plasma, and is activated. As a particulate, this activated material presents both an internal and external radiation hazard. From an ALARA perspective, the quantity of this material should be minimized.

Associated with the vacuum vessel is the divertor system. The major functions of this system are plasma power exhaust, plasma particle exhaust, and impurity control. A secondary function of the divertor system is to provide vacuum vessel and field coil shielding. The plasma particle exhaust removes ^{4}He and other nuclei formed in the fusion process and through nuclear reactions.

A second system supporting vacuum vessel operations is the blanket shield system, and its major components are the first wall and blanket shield. The structure facing the plasma is collectively referred to as the first wall, and it is subdivided into a primary wall, limiters, and baffles.

The primary wall establishes the initial protection of components located behind it. Limiters provide specific protection at distributed locations around the vacuum vessel (torus). Baffles preserve the lower area of the machine close to the divertor from high thermal loads and other conditions created by the plasma.

The blanket shield supports the first wall by providing neutron shielding, a combination of stainless steel and water, for the vacuum vessel. The blanket also provides the capability for testing tritium-breeding blanket modules and for tritium production blankets. ITER has not yet specified the specific requirements for these blankets.

Production-scale facilities build upon the ITER experience, are physically larger, and have a higher power output. The ITER is a formidable structure; its main plasma parameters and dimensions are provided in Table 3.2.

There are internal, replaceable components that reside inside the vacuum vessel. The components include blanket modules, port plugs such as the heating antennae, test blanket modules, and diagnostic modules. These components absorb heat as well as most of the plasma neutrons and protect the vacuum vessel and magnet coils from excessive radiation damage. The shielding blanket design does not preclude its replacement by a tritium-breeding blanket in subsequent ITER enhancements. A decision to incorporate a tritium-breeding blanket will likely be based on the availability of tritium fuel, its cost, the results of breeding-blanket testing, and acquired experience with plasma and machine performance.

The heat deposited in the internal components and in the vacuum vessel is rejected to the environment by means of a tokamak-cooling water system designed to minimize the release of tritium and activated corrosion products into the environment. The entire vacuum vessel is enclosed in a cryostat, with thermal shields located between thermally hot components and the cryogenically cooled magnets.

The vacuum vessel fueling system is designed to inject gas and solid hydrogen pellets. During plasma start-up, low-density gaseous fuel is introduced into the

Table 3.2 ITER plasma parameters and dimensions.

Total fusion power	500–700 MW
Plasma major radius	6.2 m
Plasma minor radius	2.0 m
Plasma current	15 MA
Toroidal field @ 6.2 m	5.3 T
Plasma volume	837 m^3
Plasma surface area	678 m^2

Derived from the ITER Final Design Report (2001).

vacuum vessel chamber by the gas injection system. The plasma is generated using electron–cyclotron heating, and this phenomenon increases the plasma current. Once the operating current is reached, subsequent plasma fueling (gas or pellets) leads to a D-T process at the design power rating.

From a safety perspective, the design focuses on confinement with successive barriers provided for the control of tritium and activated material. These barriers include the vacuum vessel, the cryostat, air-conditioning systems with detritiation capability, and filtering capability of the containment building. Effluents are filtered and detritiated such that releases to the environment are minimized.

Worker radiation safety and environmental protection are enhanced by the structure housing the vacuum vessel. For worker protection, a biological shield of borated concrete surrounds the cryostat and concrete walls provide additional neutron and gamma shielding.

Accidental releases of tritium and activated material are minimized by engineered systems that maintain pressure differences to minimize any release of radioactive material. These systems are designed such that air only flows from lower to higher contamination areas. These differential pressure and airflow characteristics are maintained by the air-conditioning system.

3.7
ITER

With the aforementioned characteristics established, the design and expected development of the ITER is presented. In 2005, France was selected as the host country for the ITER with fabrication and construction costs expected to exceed $6.1 billion. The United States, China, European Union, Japan, Russia, and South Korea agreed to site the facility at Cadarache, a research facility of France's Commissariat à lEnergie Atomique. In addition to the ITER reactor, a number of support facilities are part of the project. These include the proposed Fusion Materials Irradiation Facility, a fusion simulation center, a remote experimentation center, a fusion plant technology coordination center, and a new plasma experimental device. Assuming that ITER is successful, it would be followed by a full-scale demonstration facility. A tentative timeline for the transition from ITER to a production facility is provided in Table 3.3.

Fusion reactors are typically characterized in terms of a parameter Q:

$$Q = \frac{E_{out}}{E_{in}} = \frac{P_{out}}{P_{in}}, \tag{3.53}$$

where E_{in} (P_{in}) is the energy (power) input used to initiate the fusion process and provide balance of facility loads and E_{out} (P_{out}) is the fusion energy (power) output of the device. To achieve an economic break-even, condition Q must be greater than unity and economic viability warrants a large Q-value. ITER is designed to sustain $Q > 5$ for time periods up to 300 s, and a $Q > 10$ for at least 10 s.

ITER operations include four stages. The first stage begins with a 3-year period using only hydrogen fuel (^{1}H) at D-T ignition temperatures. The hydrogen plasma

Table 3.3 Fusion reactor progression.

Date	Phase	Purpose
2008–2015	ITER construction	Test bed for subsequent operations
2015	First plasma	Initiation of testing of components
2015–2036	ITER operations	Technology feasibility Scientific demonstration Nuclear safety verification Verification of worker, public, and environmental safety and health
2036–2041	ITER decontamination	Reduction in source term using ALARA principles
~2030	Demonstration fusion power plant	Verify capability for sustained power generation
~2050	Fusion power plant	Verification of economic feasibility

Derived from Blake (2005) and McLean (2005).

permits testing of tokamak systems in a nonnuclear environment. Stage two is a 1-year period of operation with deuterium. The power output from D-D fusion is expected to be low. Nuclear operations with D-D fuel test additional systems including the heat transport, tritium processing, and particle control systems. The third stage includes D-T plasma operations with $Q \leq 10$ and $P_{out} \leq 500$ MW$_{th}$. At the end of this 3-year phase, testing of Demonstration Fusion Reactor blanket assemblies is planned. In the fourth phase, D-T operations focus upon improving D-T fusion performance.

The goal of ITER is to achieve a self-sustaining fusion reaction that relies on fusion heat without the need for external sources of heat. In Equation 3.53, $P_{in} \rightarrow 0$ and $Q \rightarrow \infty$. When this occurs, the fusion process is controlled only by the rate of fuel addition to the torus.

3.8
ITER Safety Characteristics

Although ITER is a prototype reactor, it exhibits the essential characteristics of a fusion power production facility using magnetic confinement. ITER also has favorable nuclear safety characteristics compared to fission reactors. For example, a criticality accident cannot occur through the fusion process or through the interaction of any fusion products. In addition, fissile and fertile materials are neither utilized nor produced in the D-T fusion process. For comparable power ratings, the total energy inventory in the D-T fuel and D-T plasma are several orders of magnitude lower than the total energy inventory in a commercial fission reactor core. This lower energy inventory inherently limits the extent of any off-site release of radioactive material.

Another positive benefit of ITER is the fact that the total D-T fuel inventory within the plasma containment vessel is small. If the inventory is not replenished, fusion is only sustained for about 1 min. In addition, the reaction products of D-T fusion are a neutron and ^{4}He. ^{4}He is not radioactive.

Fusion facility radioactivity inventory is minimized through the use of low-activation materials. These materials reduce the overall radiological source term and reduce the quantity of radioactive waste resulting from fusion operations.

ITER also has positive operational safety characteristics. A low-fusion power density and positive thermal characteristics facilitate a wide safety margin for response to a loss of fusion reactor cooling. The low-fusion power density and large heat transfer area permit passive cooling of plasma-facing components (PFCs) and breeder blankets if active reactor cooling is interrupted. However, the magnetic field energy associated with ITER has the potential to distort the tokamak structure.

3.9
General Radiological Characteristics

The ITER radiological hazards will be representative of those occurring in a production fusion facility. These hazards include tritium, neutron radiation, activation products, and particulates generated by plasma collisions with containment structures.

Tritium in gaseous form (T_2) and as oxides (HTO, DTO, and T_2O) will be present at ITER. The particular chemical form depends on the location within the tritium-processing system or the physical conditions encountered during a tritium-release scenario.

Neutron radiation is produced in the D-T fusion process. The 14.1 MeV neutrons pose a direct radiation hazard, have a significant potential for activation of fusion reactor components, and lead to radiation damage of reactor components. The radiation damage increases maintenance requirements and radioactive waste generation, and increases occupational radiation doses.

Activation products are the largest contributor to the radiological source term. At ITER, the most significant isotopes of stainless steel are isotopes of Mn, Fe, Co, Ni, and Mo and those of copper are Cu, Co, and Zn. During ITERs Extended Performance Phase, a reactor inventory of approximately 10^{14} MBq is anticipated. Smaller activation product inventories reside in structures outside the shield blanket or circulating as suspended corrosion products in the first wall, blanket, and divertor coolant streams.

These activation products and their activities present high-radiation fields inside the cryostat and vacuum vessel. The radiation fields are sufficiently high to require remote maintenance for systems, structures, and components within the cryostat and vacuum vessel.

Fine particles are produced as a result of ion impacts with plasma-facing components. These particles form a fine radioactive dust that could be released during maintenance inside the plasma chamber or during a severe accident.

Tritium, activation products, and toxic materials could be released during an accident or off-normal event. There are a number of energy sources that facilitate the dispersal of radioactive and toxic material, and these energy sources are summarized in Table 3.4.

Table 3.4 ITER energy inventories.

Energy source	Power or energy	Release time	Potential consequences	Control possibilities
Fusion power	1.5 GW	1000 s (pulse duration)	Melting of plasma-facing components In-vessel loss-of-coolant accidents	Normal coolant systems operations Active power shutdown Passive shutdown for large disturbances
Plasma	2.3 GJ	<1 s	Disruption/vertical displacement event Limited evaporation of plasma-facing components	Plasma control Disruption/vertical displacement event mitigation systems
Magnetic	120 GJ	s to min	Arcs Localized magnet melting Mechanical damage	Normal operation of coolant Insulation design Rapid quench detection and discharge system
Decay heat	260 GJ (in the first day) 910 GJ (in the first week)	min to yr depending on the concern	Heating near-plasma components and materials Maintenance concerns Waste management concerns	Minimization of decay heat production Defense in depth design Normal cooling systems operations Active decay heat removal systems Passive heat removal using radiative heat transfer and natural circulation
Chemical (following a reaction)	800 GJ	s to h	Overheating of plasma-facing components Hydrogen fires or explosions Overpressurization or damage of the plasma confinement structure	Using passive means, limit temperatures to about 500 °C on plasma-facing components to minimize hydrogen production Design measures to prevent ozone accumulation
Coolant	300 GJ	s to min	Pressurization of vacuum vessel, cryostat, or heat transport system vault	Overpressure suppression systems

Derived from ITER EDA, No. 7 (1996).

3.10
Accident Scenarios/Design Basis Events

The safety characteristics, radiological characteristics, and energy sources form the basis for deriving ITERs accident scenarios. Summarized in subsequent discussion, ITER accident scenarios include loss-of-coolant accidents (LOCAs), loss-of-flow accidents (LOFAs), loss-of-vacuum accidents (LOVAs), plasma transients, magnet fault transients, loss of cryogen, tritium plant events, and auxiliary system faults. These scenarios form the foundation for ITER's design basis events. Although the physical processes differ, the fusion design basis events have similarities to the fission events summarized in the previous chapter.

3.10.1
Loss-of-Coolant Accidents

LOCAs involve actively cooled components (e.g., blanket, shield, vacuum vessel, and divertor-cooling system) that remove fusion energy. Cooling media include water and helium. LOCAs are divided into two broad categories (in-vessel and ex-vessel).

An in-vessel LOCA diverts coolant into the vacuum vessel leading to pressurization or chemical reactions with hot plasma-facing components. Coolant entering the plasma chamber during plasma operations disrupts and extinguishes the plasma. However, pressure or chemical-initiated events disperse radioactive material including tritium, activation products, and fusion products. The extent of the dispersal area and quantity of radioactive material dispersed depend on the specific fusion reactor design, its operational history and operating characteristics, and details of the accident sequence. Parameters that determine the severity of the LOCA include the type and quantity of fluid leaked into the vacuum vessel; the vacuum vessel volume; the internal energy of the fluid; and for water LOCAs, the presence of condensation surfaces.

Ex-vessel LOCAs involve piping runs to heat removal systems such as steam generators or heat exchangers. As the ex-vessel piping has a larger bore than the in-vessel piping, ex-vessel LOCAs involve larger volumes of coolant than in-vessel events. Rapid detection of an ex-vessel event is required to protect the divertor and first wall from overheating when coolant is lost. The time required for detection of the ex-vessel LOCA and for shutdown of the plasma reaction depends on the plasma-facing component's heat load. For ITER, the time is on the order of seconds.

The probability of an ex-vessel LOCA is judged to be much lower than that of an in-vessel LOCA. The reduced probability is associated with the ease and regularity of scheduled inspections of heat removal systems and associated piping.

3.10.2
Loss-of-Flow Accidents

LOFAs are predominantly caused by a loss of off-site power that results in the decrease or loss-of-coolant pump output. LOFAs often lead to LOCAs because the loss of cooling flow can lead to tube overheating and subsequent tube failure if plasma shutdown is not achieved rapidly.

In-vessel LOFAs are induced by tube plugging or coolant system blockage. As in-vessel components usually involve small diameter piping, an in-vessel LOFA leads to overheating and subsequent failure of the tube or channel and results in an in-vessel LOCA, as a result of which coolant is released into the plasma chamber with disruption and termination of the plasma. Following plasma termination, cool down of components needs to be achieved to prevent further damage that could result in the release of radioactive material.

The consequences of a LOFA depend on fusion process heat loads and the design of cooling systems to manage these heat loads. Key parameters that affect a LOFA are the coolant material, divertor heat load, first wall heat load, and the heat transport system design.

3.10.3
Loss-of-Vacuum Accidents

When the vacuum established inside the plasma chamber is lost, a loss of vacuum event occurs. Vacuum disruption is realized when a gas including air leaks into the plasma chamber. Disruption follows a component failure such as a diagnostic window, port, or seal, caused by a defect, vessel erosion, component wear, radiation damage, or disruptive load or overpressurization of the plasma chamber following an in-vessel LOCA. In addition to allowing ingress into the vessel, the component failure allows radioactive material (tritium or activated material) to escape from the vessel. If air enters the vacuum vessel, it reacts chemically with the hot plasma-facing components. This interaction produces additional energy that can volatilize additional radioactive material. The severity of a LOVA depends on the specific fusion facility design.

3.10.4
Plasma Transients

Plasma transients include overpower events and plasma disruptions. Overpower conditions occur in a plasma when the balance between fusion energy generation and energy loss is disrupted. When generation exceeds loss, an increase in temperature results until the accumulation of ^{4}He and depletion of D-T fuel occurs. After about 2–10 s, a disruption and plasma shutdown occurs. Plasma disruptions include a variety of instability transients.

During a plasma disruption, confinement of the plasma is lost, the fusion process terminates, and plasma energy is rapidly transferred to the surrounding structures. This energy transfer can induce PFC ablation and possibly melting. During this energy transfer, the plasma current quenches within about a second, and magnetic forces are exerted in the vessel and support structure.

Plasma disruption can be induced by thermal plasma excursions, impurities injected into the plasma, and loss of plasma control. These conditions are expected to occur during normal operations and must be addressed through facility design. In addition, plasma disruptions generate high-energy electrons that damage PFCs and initiate failure of first wall/blanket modules or segments.

3.10.5
Magnet Fault Transients

Magnetic field transients induce forces that have the potential to damage magnet structural integrity and induce faults in other machine components. Off-normal forces yield large magnet coil displacements that impact other systems (e.g., the vacuum vessel and plasma heat transfer system piping) and produce arcs that induce localized component damage. At ITER, magnetic field transients could damage the vacuum vessel and its associated ducts, and piping and the cryostat. This damage facilitates the release of radioactive material.

Electromagnetic forces also result from equipment or operational transients that lead to electrical shorts in coils, faults in the discharge system, or power supply faults. Electrical arcs between coils, arcs to ground, and arcs at open leads facilitate localized component melting. Arcs also arise from insulation faults, gas ingress, or overvoltage conditions. The degree to which arcs or magnetic faults occur depends on the facility design and its operational characteristics. However, arcs damage structures and increase the potential for the release of radioactive material.

3.10.6
Loss of Cryogen

The loss of helium or nitrogen cryogen is a radiological safety issue because the pressures developed following the leak are possibly sufficient to breach confinement barriers. The released helium and nitrogen also displace air and present a suffocation hazard.

Releases of helium and nitrogen result from component failures or transient conditions. For superconducting magnets, quenching the superconductor without electrical discharge results in helium leakage. Cryogen plant failures also lead to the release of nitrogen gas following a liquid nitrogen release. These gas releases also provide a mode of force to mobilize radioactive material.

3.10.7
Tritium Plant Events

An accident breaching confinement barriers of the tritium processing and fueling system releases tritium in either gas or oxide form. Such events should also consider the potential for hydrogen explosions. However, tritium design standards normally require double or triple containment for systems containing hydrogen. These standards should reduce the frequency of large release and explosion events.

An explosion provides a potent mode of force to disperse radioactive material. The specific plant location of the explosion and the explosion's magnitude govern the quantity of radioactive material dispersed and the severity of the event.

3.10.8
Auxiliary System Accidents

The ITER design incorporates a number of systems associated with plasma heating and control. In general, events associated with auxiliary systems do not have major radiological consequences.

3.11
Radioactive Source Term

The aforementioned accident scenarios have the potential to release radioactive material. The extent of the release depends on the radioactive material available for release and the plant conditions that exist during the release. Table 3.5 summarizes the major at-risk inventories and the release assumptions currently recommended for use in evaluating the consequences of the postulated ITER events. A more complete listing of the assumptions used in the ITER radioactive material dispersal events is provided in subsequent discussion.

3.12
Beyond Design Basis Events

Beyond design basis events have frequencies of $<10^{-6}$/year. These types of events include vacuum vessel collapse, magnet structure collapse or movement, and building structural failure.

Collapse of vacuum vessel, collapse of magnet structural supports, or movement of magnet structural supports sever tokamak coolant lines and damage one or more of the tokamak confinement barriers. Gross building failure also damages tokamak coolant lines and structural barriers and leads to fire-related events. All of these events have the potential for a significant release of radioactive material.

3.13
Assumptions for Evaluating the Consequences of Postulated ITER Events

In its analysis of accidents, the ITER adopted a standard set of assumptions for evaluating the consequences of the postulated design basis events. This set permits common ground rules for comparing the relative severity of the postulated events. The assumption set also permits event dose limits and release limits to be calculated in a consistent manner.

Table 3.5 Radioactive material inventories in postulated ITER accident events.

Source term	Inventory available for release	Tolerable release fraction[a]	Control or mitigation strategy
In-vessel tritium as a codeposited carbon–hydrogen layer (if any)	1 kg tritium	≈30% if HTO	Administrative limit on layer buildup Dual confinement barriers against air ingress
In-vessel tritium diffusively held in beryllium and tritium in cryopumps	0.7 kg tritium	>100% if HT	Limit first wall temperatures to 500–600 °C
In-vessel tokamak dust (e.g., steel and tungsten), excluding beryllium and carbon	20 kg metal	≈30%	Administrative limit on dust Dual confinement barriers against air ingress
Oxidation-driven volatility of in-vessel steel, copper, and tungsten	Kilograms of solid near-plasma material	≈10–100% depending on temperature[b]	Limit first wall temperatures to 500–600 °C
Tritium plant circulating inventory	600 g tritium	≈75%	Administrative limit on inventories Confinement barriers Building structural integrity
Secure tritium storage	1 kg tritium	≈50%	Administrative limit on inventories Confinement barriers Building structural integrity
Hot cells, waste storage	<1 kg tritium Kilograms of activated metal	≈50% for tritium ≈10% for dust	Administrative control on tritium and dust Recycle tritium Prevent temperature increases Confinement barriers

Derived from ITER EDA, No. 7 (1996).

[a] Determined on the basis of 50 mSv dose (during the release period plus 7 days) for no evacuation, average meteorology, and ground level release conditions.

[b] Conceptual design activity analyses demonstrated that, for a tungsten surface machine at 600–700 °C, the tolerable confinement fraction was about 2% for meeting a 100 mSv dose recommendation limit.

The baseline conditions assumed in ITER accident analyses include the following:

- Design basis events use a 100 m elevated release.
- Design basis events incorporate conservative meteorology including rain.
- Beyond design basis events incorporate a ground-level release that includes building wake effects.
- Operational events and beyond design basis accidents use average meteorology conditions. Rain is not considered.
- A release duration of 1 h is assumed.
- Calculations are based on a 1 km distance from the release point to the nearest member of the public.
- The dispersion factors (χ/Q) used for accident releases, worst-case meteorology, and a ground-level release; accident releases, worst-case meteorology and an elevated release; and average annual meteorology are $2-4 \times 10^{-4}$ s/m^3, $1.4-2.7 \times 10^{-5}$ s/m^3, and 1.0×10^{-6} s/m^3, respectively. These dispersion factors do not credit ground deposition and washout effects.

Public dose criteria are needed to evaluate the acceptability of the postulated events and the need for modification of the ITER design. These criteria are summarized in Table 3.6, in which events are categorized as operational events, likely events, unlikely events, extremely unlikely events, and hypothetical events. Operational events occur during routine operations including some faults and conditions that arise because of ITER's experimental nature. Likely events are not considered to be operational events but occur one or more times during the lifetime of the facility. Unlikely events are not likely to occur during ITER's operational lifetime. Extremely unlikely events are not likely to occur by a very wide margin during ITER's operational lifetime. ITER's design basis is derived from the extremely unlikely events. Hypothetical events have an extremely low frequency (f). These events are postulated with the goal of limiting ITER's risk, and they form the basis for the beyond design basis events.

3.14
Caveats Regarding the ITER Technical Basis

The preceding discussion is based on publications developed by the ITER project. These publications summarize a success-oriented design that assumes the basic science is essentially resolved. For completeness, a summary of recent commentary that questions this assumption is presented.

Issues have been raised regarding possible gaps in the scientific foundation of the ITER. These issues arise because it has become commonplace to focus upon the political and financial aspects of new technologies such as ITER, and to

Table 3.6 Public dose criteria for postulated ITER events.

Accident consideration	Operational events	Likely events	Unlikely events	Extremely unlikely events	Hypothetical events
Annual expected frequency (f)	Expected to occur	>0.01/yr	0.01/yr $> f > 0.0001$/yr	0.0001/yr $> f > 10^{-6}$/yr	$<10^{-6}$/yr
ITER objective	Apply ALARA principles	Avoid releases	Avoid the need for any public countermeasures	Avoid the potential for public evacuation	Limit risk[d]
Dose criteria to meet the design basis[a]	0.1 mSv/yr chronic dose (all pathways integrated over all operational event categories)[b]	0.1 mSv/yr chronic dose (all pathways integrated over all likely event categories) 0.1 mSv/yr chronic dose (without ingestion)[b]	5 mSv/event chronic dose (without ingestion)	5–50 mSv/event[c]	Limit risk[d]

Derived from ITER EDA, No. 7 (1996).

[a] When dose criteria are "per year," average annual meteorology is assumed; when "per event," baseline worst-case meteorology is used for design basis events and average meteorology is used for beyond design basis events.
[b] The summation of the operational event dose and likely event doses must be ≤ 0.2 mSv/yr.
[c] The range for extremely unlikely events results from the variation among various national dose criteria. For design purposes, a value of 10 mSv during the release period plus 7-day period is utilized by ITER.
[d] The goal is to limit risk. In addition to meeting the extremely unlikely event dose criteria, the no-evacuation goal implies the need to limit doses to the local population to approximately 50 mSv/event during the release period plus 7-day period.

proceed in the development of the technology assuming that the important scientific questions have been resolved and that only engineering details remain. Some authors suggest that there are scientific gaps in the conceptual foundations of ITER and in other magnetic confinement devices that merit a thorough review.

An example of these suggested gaps is the lack of a manageable mathematical framework that reproduces the observed experimental results or is sufficiently evolved to predict planned experiments. A situation in which the theory is decoupled from predicting and focusing the experimental program is not necessarily a fatal flaw, because a number of important discoveries have occurred in the absence of a rigorous, predictive theoretical framework. However, this issue is related to two other suggested theoretical fusion issues:

- The conceptual gap between the theory and device construction does not need to be as large as currently exists with the ITER. This gap merits attention and considered investigation, but is not being aggressively pursued.
- Consensus, financial resources, publicity, and organizational structure are not sufficient to satisfy ITER's goal of achieving a Q-value on the order of 10. Given the current situation, achieving this goal is uncertain.

The aforementioned issues are also important from a health physics perspective. Any design shortcomings have a health physics impact that could lead to potentially larger releases of radioactive material and higher effective doses. However, cutting edge projects such as ITER have inherent uncertainty and any design iteration should proceed using sound ALARA principles.

3.15
Overview of Fusion Energy Radiation Protection

The D-T reaction of Equation 3.3 provides 17.6 MeV for transfer from alpha particles (3.50 MeV) and neutrons (14.1 MeV). This energy is initially transferred to other species by neutron and alpha particle collisions, which produce lower energy alpha particles, and a portion of the energy initiates other nuclear reactions including activation.

The fusion power facility has radiological hazards that are also present in contemporary facilities. For example, the tritium/HTO hazard is similar to that encountered in a Canadian-Deuterium-Uranium (CANDU) reactor that uses D_2O as the coolant and moderator. The 14.1 MeV neutrons resulting from D-T fusion are similar to the neutron hazard encountered in a low-energy accelerator facility. Therefore, health physics experience with CANDU reactors and accelerators provide insight into the radiological hazards of a fusion power facility.

A fusion power facility utilizes systems that are not found in contemporary fission reactors (e.g., the tritium fueling system, tritium clean-up system, tritium breeding and recovery system, vacuum pumping system, plasma heating system, water tritium

Table 3.7 Annual collective dose values at fission and fusion power reactors.

Facility type	Annual collective dose (Person-Sievert)
Boiling water reactor	2.21
Pressurized water reactor	1.20
Canadian deuterium	0.63
Gas cooled reactor	0.26
Future fission plant	0.7
Fusion plant (projected)	1–2

Derived from Eurajoki, Frias and Orlandi (2003).

removal system, and isotope separation system). The assessment of the occupational dose equivalent associated with each of these systems requires detailed design knowledge and related system design details such as the nature and configuration of penetrations in the vacuum vessel, activation of structural materials, water chemistry, and the leak tightness of tritium removal systems. An analysis of the radiation protection consequences of these systems is only possible once specific information regarding the occupancy factors, fusion-specific effective dose rates, frequency of operations, and number of workers involved in the operations is known. However, this information is not yet available.

Although these details should evolve as design concepts are finalized, considerable health physics information is obtained by considering the individual source terms at a fusion power facility. These source terms directly impact the facility's collective dose.

The collective dose from fusion power plants will be one of the criteria for judging their overall success. Table 3.7 summarizes the current and anticipated fission facility and anticipated fusion facilities' annual collective doses. On the basis of the information in Table 3.7, it appears that the collective effective doses at fission and fusion power facilities are comparable. It is also likely that fusion facility doses will decrease as operating experience accumulates.

3.16
D-T Systematics

The D-T fusion process involves an intermediate state in the ^{5}He system:

$$D + T : {}^2H + {}^3H \rightarrow {}^5He^* \rightarrow {}^4He + n, \tag{3.54}$$

where the "*" indicates the intermediate ^{5}He system can exist as an excited state. The formation of the intermediate ^{5}He system leads to additional reactions such as

$$^5He + {}^4He \rightarrow {}^6Li + {}^3H, \tag{3.55}$$

which produce other nuclear species. These species affect the radiation characteristics of the fusion power facility. Therefore, it is important to consider the systematics and characteristics of ^{5}He as part of the evaluation of the D-T fusion process.

Table 3.8 ^{5}He breakup channels.[a]

Channel	Threshold energy (MeV)[b]
^{4}He+n	−0.798
^{3}H+^{2}H	16.792
^{3}H+p+n	19.016
^{3}He+n+n	19.780

[a]Derived from Tilley et al. (2002).
[b]Relative to the ^{5}He ground state.

The systematics of the ^{5}He system's breakup channels are summarized in Table 3.8. Related reaction channels are provided in Table 3.9.

Table 3.8 illustrates the various low-energy rearrangement or breakup channels in the ^{5}He system. For example, without added energy, D-T fusion via Equation 3.3 occurs with the liberation of 17.6 MeV [16.792 − (−0.798 MeV)]. No other D-T reactions are likely to occur unless several MeV of excitation energy is provided. For example, ^{3}H+^{2}H→^{3}H+p+n will only occur if at least 2.2 MeV is provided (16.792 − 19.016 MeV).

Table 3.9 summarizes the variety of binary reaction channels that lead to the ^{5}He system in the exit channel of a nuclear interaction. In addition, ^{5}He is consumed in the reverse reactions with ^{5}He appearing in the entrance channel. These reactions provide an overview of the nuclides that can be formed from the basic systems of Tables 3.8 and 3.9 (e.g., ^{2}H, ^{3}H, ^{3}He, ^{4}He, and ^{5}He).

Table 3.9 Reaction channels associated with the ^{5}He system.[a]

Reaction channel	Threshold energy (MeV)[b]
^{4}He+^{4}He→^{3}He+^{5}He	−21.375
^{7}Li+γ→^{5}He+^{2}H	−9.522
^{10}B+n→^{6}Li+^{5}He	−5.258
^{6}Li+p→^{5}He+p+p	−4.497
^{7}Li+p→^{5}He+^{3}He	−4.029
^{7}Li+n→^{5}He+^{3}H	−3.265
^{4}He+^{2}H→^{5}He+p	−3.022
^{9}Be+γ→^{5}He+^{4}He	−2.371
^{6}Li+n→^{5}He+^{2}H	−2.272
^{10}B+^{2}H→^{7}Be+^{5}He	−1.877
^{9}Be+^{2}H→^{6}Li+^{5}He	−0.897
^{6}Li+^{2}H→^{5}He+^{3}He	0.997
^{3}H+^{3}H→^{5}He+n	10.534
^{3}H+^{3}He→^{5}He+p	11.298
^{7}Li+^{2}H→^{5}He+^{4}He	14.325
^{6}Li+^{3}H→^{5}He+^{4}He	15.317

[a]Derived from Tilley et al. (2002).
[b]Relative to the ^{5}He ground state.

The various reactions produce a variety of nuclides and radiation types that directly influence the radiation characteristics of the facility. The produced radiation types (e.g., n, p, and γ) and their energy determine the activation of components inside the biological shield. These various radiation types, activation processes, and radionuclides define the fusion source term. The fusion source term is discussed in subsequent sections of this chapter.

3.17
Ionizing Radiation Sources

Knowledge of the D-T fusion process and plasma characteristics permits an amplification of the previous radiation protection overview discussion. In particular, the sources of occupational radiation exposure arise from the fusion process and from nuclear materials used to support the fusion process.

The dominant types of ionizing radiation include gamma radiation from the fusion process and activation sources, beta radiation from activation sources and tritium, and neutron radiation from the fusion process. The external effective dose predominantly receives contributions from beta, gamma, and neutron radiation. Internal intakes of tritium and activation products are also a concern.

Table 3.10 provides a listing of activation products that are expected to be encountered in both fission and fusion facilities. In addition, Table 3.10 provides the half-life, fusion production mechanism, type of neutron initiating the activation reaction, component activated, and type of radiation produced. The reader should note that the fusion activation product generation mechanisms include modes that are unavailable in fission activation owing either to energy differences or interacting particle types.

In the lower energy fission neutron spectrum, the (n, γ) and (n, p) reactions predominate. The higher energy D-T fusion neutron spectrum opens additional reaction channels. In addition to (n, γ) and (n, p) reactions, more complex reactions (e.g., (n, 4n), (n, 2n, α), and (n, ^{3}He)) occur and contribute to the activation product source term. Additional discussion regarding fusion-specific activation products and their production mechanisms are discussed in subsequent sections of this chapter.

The reader should also note that Table 3.10 does not provide a complete listing of all possible reactions for a given neutron plus target system. For example, the neutron plus ^{59}Co system leads to a variety of possible reactions including the following: ^{59}Co(n, 2n)^{58}Co, ^{59}Co(n, γ)^{60}Co, ^{59}Co(n, p)^{59}Fe, ^{59}Co(n, α)^{56}Mn, ^{59}Co(n, d)^{58}Fe, ^{59}Co(n, n α)^{55}Mn, and ^{59}Co(n, ^{3}H)^{57}Fe. In general, neutrons and other particles present in the fusion plasma produce activation products. For simplicity, this chapter focuses on neutron-induced fusion activation products.

The reactions that significantly contribute to the worker's effective dose depend on the radionuclide produced and its activity. This activity is determined by the contribution from the terms comprising the activation equation. For simplicity, consider the saturation activity (A_{sat}) applicable for sustained, steady state fusion reactor operation:

$$A_{sat} = N\sigma\phi, \tag{3.56}$$

Table 3.10 Examples of activation products likely to be encountered at a fusion power facility.

Isotope produced	Half-life	Candidate production mechanisms	Activating neutron type	Component activated	Radiation types emitted
^{3}H	12.3 yr	^{2}H$(n,\gamma)^3$H Spallation ^{6}Li$(n,\alpha)^3$H Spallation	Thermal Fast Thermal Fast	Cooling water systems[a] Cooling water systems[a] Blanket assembly Soil	β^-
^{7}Be	53.3 d	Spallation	Fast	Cooling water systems[a]	γ
^{11}C	20.3 min	^{12}C$(n, 2n)^{11}$C ^{14}N$(p, \alpha)^{11}$C ^{12}C$(\gamma, n)^{11}$C	Fast NA NA	Air and cooling water systems[a] Air and cooling water systems[a] Air and cooling water systems[a]	β^+, γ
^{13}N	9.97 min	^{14}N$(n, 2n)^{13}$N ^{16}O$(p, \alpha)^{13}$N ^{14}N$(\gamma, n)^{13}$N	Fast NA NA	Air and cooling water systems[a] Air and cooling water systems[a] Air and cooling water systems[a]	β^+, γ
^{15}O	122 s	^{16}O$(n, 2n)^{15}$O ^{16}O$(\gamma, n)^{15}$O ^{14}N$(p, \gamma)^{15}$O	Fast NA NA	Air and cooling water systems[a] Air and cooling water systems[a] Air and cooling water systems[a]	β^+, γ
^{16}N	7.14 s	^{16}O$(n, p)^{16}$N ^{15}N$(n, \gamma)^{16}$N	Fast Thermal	Air and cooling water systems[a] Air and cooling water systems[a]	β^-, γ, α
^{22}Na	2.60 yr	^{23}Na$(\gamma, n)^{22}$Na ^{23}Na$(n, 2n)^{22}$Na	NA Fast	Soil Soil	β^+, γ
^{24}Na	15.0 h	^{23}Na$(n, \gamma)^{24}$Na ^{27}Al$(n, \alpha)^{24}$Na ^{24}Mg$(n, p)^{24}$Na	Thermal Fast Fast	Soil Concrete Soil and concrete	β^-, γ

(continued)

Table 3.10 (Continued)

Isotope produced	Half-life	Candidate production mechanisms	Activating neutron type	Component activated	Radiation types emitted
^{41}Ar	1.83 h	^{40}Ar(n, γ)^{41}Ar ^{41}K(n, p)^{41}Ar	Thermal Fast	Air Concrete	β^-, γ
^{54}Mn	313 d	^{55}Mn(n, 2n)^{54}Mn ^{54}Fe(n, p)^{54}Mn ^{55}Mn(γ, n)^{54}Mn	Fast Fast NA	Structural members Structural members Structural members	γ, e$^-$
^{55}Fe	2.70 yr	^{54}Fe(n, γ)^{55}Fe ^{58}Ni(n, α)^{55}Fe ^{56}Fe(γ, n)^{55}Fe	Thermal Fast NA	Structural members Structural members Structural members	γ
^{56}Mn	2.58 h	^{55}Mn(n, γ)^{56}Mn ^{56}Fe(n, p)^{56}Mn ^{59}Co(n, α)^{56}Mn	Thermal Fast Fast	Structural members Structural members Structural members	β^-, γ
^{58}Co	70.8 d	^{58}Ni(n, p)^{58}Co ^{59}Co(n, 2n)^{58}Co ^{59}Co(γ, n)^{58}Co	Fast Fast NA	Structural members Structural members Structural members	β^+, γ
^{59}Ni	75 000 yr	^{58}Ni(n, γ)^{59}Ni ^{60}Ni(γ, n)^{59}Ni	Thermal NA	Structural members Structural members	γ
^{60}Co	5.27 yr	^{59}Co(n, γ)^{60}Co ^{60}Ni(n, p)^{60}Co	Thermal Fast	Structural members Structural members	β^-, γ
^{63}Ni	100 yr	^{62}Ni(n, γ)^{63}Ni ^{64}Ni(γ, n)^{63}Ni	Thermal NA	Structural members Structural members	β^-

[a]Includes the vacuum vessel, magnetic field coil, and other heat removal cooling water systems.

where N is the number of target atoms, σ is the energy-dependent cross section for the reaction of interest, and ϕ is the energy-dependent fluence of the particle initiating the reaction of interest.

The number of atoms of a particular target depends on the materials of construction for the component being activated; the cross section depends on the specific reaction and the neutron energy; and the energy-dependent neutron fluence is governed by the fusion process, the fusion reactor configuration, and the materials of construction for the vacuum vessel and its support components. The reactor configuration and the materials of construction govern the neutron interactions, and these interactions degrade the neutron energy. Therefore, the importance of a specific reaction depends on the details of the reactor design and the fusion process utilized to produce power. Table 3.10 must be viewed within this context. Any refinement must await a more specific design for a fusion power facility.

3.18
Nuclear Materials

A variety of nuclear materials will be used in a fusion power facility. These materials include tritium and depleted uranium.

Tritium is one of the fuel components used to initiate and sustain the D-T fusion process. Depleted uranium (^{238}U) containers may be used for the storage of tritium, in fission chambers, and in various radioactive check-sources and calibration sources. The beta hazards associated with ^{3}H and ^{238}U are discussed in subsequent sections of this chapter.

3.19
External Ionizing Radiation Hazards

Fusion of the D-T system produces a variety of radiation types including alpha particles, beta particles, photons, and neutrons. Heavy ions are also produced, but they deposit the bulk of their energy within the plasma and vacuum vessel. The production of the individual radiation types depends on the selection of materials and the fusion process used in the facility. Each of these radiation types is discussed.

3.19.1
Alpha Particles

In the D-T fusion process of Equation 3.3, alpha particles are directly produced, and their energy is deposited within the plasma or in the lining of the vacuum vessel. Alpha particles are also produced by activation of vacuum vessel and plasma support components.

Alpha particles, produced through activation or nuclear reactions with materials of construction, present an internal hazard if they are dispersible. The fusion alpha hazard is not as severe as the alpha hazard associated with transuranic elements (e.g., plutonium and americium) produced in a fission power reactor or recovered in a fuel reprocessing facility.

Other alpha particle generation results from the unique materials utilized in the facility. As noted in the previous section, depleted uranium (^{238}U) containers may be used to store the reactor's tritium fuel. Alpha particles arise from the uranium series daughters that are part of the materials of construction (e.g., concrete) or dissolved in the facility's water supplies.

3.19.2
Beta Particles

Beta radiation primarily results from the decay of activation products, ^{238}U, and tritium, and it presents a skin, eye, and whole body hazard. Potential sources of beta radiation are summarized in Table 3.10, which suggests that the fusion power reactor beta hazard is similar to that encountered in a fission power reactor.

Beta radiation is also associated with the tritium fuel material and associated depleted uranium storage containers. The tritium beta particles represent an internal hazard whereas the beta radiation from the ^{238}U series is both an internal and external radiation hazard. As an equilibrium thickness of ^{238}U metal leads to an absorbed dose rate of 233 mrad/h at 7 mg/cm^2, ALARA measures are required in the vicinity of the depleted uranium storage containers to minimize the beta effective dose. The major contributor to the ^{238}U beta-absorbed dose is its daughter ^{234m}Pa ($E_\beta^{max} = 2.28$ MeV@98.6%).

Beta radiation is a health physics issue during routine operations and maintenance activities, fueling and defueling activities, and waste processing operations. Appropriate health physics measures are required to minimize the beta radiation hazard.

3.19.3
Photons

Photons are produced from the decay of fusion activation products and from nuclear reactions that occur within the fusion plasma. Examples of fusion activation products that produce photons are provided in Table 3.10. The photons emitted from activation products vary considerably in energy and half-life. As noted in subsequent discussions, shielding requirements are influenced by fusion activation gammas including the ^{16}N and ^{24}Na photons.

Photon radiation also occurs from a variety of reactions associated with the D-T fusion process. Sources of photons include bremsstrahlung and nuclear reactions such as ^{2}H(n, γ)^{3}H, ^{3}H(p, γ)^{4}He, ^{3}He(n, γ)^{4}He, and ^{2}H(^{2}H, γ)^{4}He. High-voltage equipment associated with plasma heating in MC fusion and laser support equipment in IC fusion are additional sources of photons. The primary shielding surrounding the vacuum vessel mitigates the photon radiation.

3.19.4
Neutrons

The fusion process occurring within the plasma produces fast neutrons (e.g., ≥ 14.1 MeV in D-T plasmas) and lower energy neutrons, including thermal neutrons, as the 14.1 MeV neutrons react and scatter in the various reactor components. These neutrons activate structural materials, coolant, instrumentation, and devices used to sustain the plasma (e.g., radio frequency coils and the D-T injection system). One result of activation is the creation of high dose rate components that require remote handling during maintenance operations.

After D-T fusion, some neutrons escape the vacuum vessel. The expected 14.1 MeV neutron flux, total neutron flux, and fusion power are provided in Table 3.11 for the ITER, a demonstration fusion reactor, and a fusion power reactor.

The expected neutron irradiation of inner reactor components, including the blanket and shield, dictates their required material properties (i.e., capability of withstanding operating temperatures and pressures as well as meeting the radiation damage limits). In addition, reactor components should have low activation properties to facilitate operations and maintenance activities in an ALARA manner. A limited set of structural materials has the desired activation properties including those based on ferritic martensitic steel, SiC/SiC ceramic composites, and vanadium alloys.

Neutron radiation damage impacts facility equipment lifetimes. Major components require periodic replacement because of the high-energy neutron bombardment. These components require remote handling and processing to minimize worker doses. As an example, consider the blanket assemblies surrounding the vacuum vessel that produce tritium through reactions such as $^6\text{Li}(n, {}^3\text{H})^4\text{He}$. The blanket change-out frequency ensures sufficient time to permit breeding of the required quantities of tritium to reach self-sufficiency. Radiation damage is an important consideration in determining this frequency.

Table 3.11 Comparison of selected fusion reactor parameters.

Parameter	ITER	Demonstration fusion reactor	Fusion power reactor
Fusion power (GW)	0.5	2–4	3–4
14 MeV neutron flux on the reaction chamber wall (n/cm² s)	$>10^{13}$	$>10^{14}$	$>10^{14}$
Total neutron flux on the reaction chamber wall (n/cm² s)	$>10^{14}$	$>10^{15}$	$>10^{15}$

Derived from Ehrlich, Bloom and Kondo (2000), Aymar et al. (2001), and Batistoni (2001).

Fusion neutrons also present an external radiation hazard. The 14.1 MeV neutrons are considerably more energetic than fission neutrons. Neutrons, escaping the vacuum vessel and not captured by the blanket assembly or other components, lead to occupational doses during surveillance and maintenance activities. These neutrons require shielding, and particular attention must be paid to leakage pathways, which vary with the reactor design, fusion process, and reactor-operating characteristics.

Neutrons also activate fusion reactor structures and components. Activation products are produced by the neutron fluence impinging on the various components of the fusion reactor including the vacuum vessel. Candidate component materials include stainless steel, vanadium, and ceramic materials such as Al_2O_3. Activation products include isotopes of Na, Fe, Co, Ni, Mn, and Nb that decay by beta emission, positron emission, and electron capture with associated gamma emission. A key ALARA feature is the optimization of materials that produce minimal activation products or activation products with short half-lives.

Typical neutron activation products of structural materials include ^{55}Fe, ^{58}Co, ^{60}Co, ^{54}Mn, ^{56}Mn, ^{59}Ni, and ^{63}Ni. The variety of materials used in a fusion facility and their associated trace constituents increase the diversity of activation products.

As compared to fission activation products, fusion activation products are primarily solid materials. Excluding the activation products of argon, noble gases are not dominantly produced in a fusion machine. Significant quantities of radioactive krypton and xenon are not expected.

Expected fusion activation products also include those resulting from air, water, and soil. The activation products of Table 3.10 are common to both fission and fusion processes. Subsequent discussion explores the additional complexity introduced by the higher energy D-T fusion neutron spectrum.

In addition to the expected activation products, fusion-specific activation products are produced. As the materials used in a fusion reactor are not yet completely specified, specific examples for the activation of two components (i.e., the vacuum vessel liner and vacuum vessel structural material) are provided.

Likely candidate materials for the vacuum vessel liner are vanadium, a vanadium alloy, and a vanadium composite material. Stainless steel is a likely candidate material for vacuum vessel structural material. The activation of vanadium and stainless steel are addressed in the next two sections.

3.19.4.1 Vanadium Activation – Vacuum Vessel Liner

In view of the previous discussion regarding uncertainty in the selection of materials, it is reasonable to consider natural vanadium as the vacuum vessel liner material. The dominant vanadium activation products are summarized in Table 3.12, which also summarizes the associated neutron-induced reaction products of the vanadium impurity constituents (iron, niobium, and molybdenum), dominant activation product production modes, threshold energy for the activation reaction, and activation product half-life. Some of the threshold energies are beyond those encountered in the fission process. For example, the ^{51}V(n, 4n)^{48}V reaction has a

Table 3.12 Activation of natural vanadium and associated impurity constituents in the vacuum vessel liner in a D-T fusion neutron spectrum.

Target material	Dominant production mode	Threshold energy (MeV)	Activation product	Activation product half-life
^{50}V	^{50}V(n, p^{3}He)^{47}Ca	21.5	^{47}Ca	4.5 d
	^{50}V(n, n α)^{46}Sc	10.1	^{46}Sc	84 d
	^{50}V(n, n^{3}He)^{47}Sc	20.2	^{47}Sc	3.4 d
	^{50}V(n, ^{3}He)^{48}Sc	11.8	^{48}Sc	43.7 h
	^{50}V(n, 3n)^{48}V	21.3	^{48}V	16 d
^{51}V	^{51}V(n, p α)^{47}Ca	11.7	^{47}Ca	4.5 d
	^{51}V(n, 2n α)^{46}Sc	21.3	^{46}Sc	84 d
	^{51}V(n, n α)^{47}Sc	10.5	^{47}Sc	3.4 d
	^{51}V(n, α)^{48}Sc	2.1	^{48}Sc	43.7 h
	^{51}V(n, 4n)^{48}V	32.6	^{48}V	16 d
^{54}Fe	^{54}Fe(n, α)^{51}Cr	0	^{51}Cr	27.7 d
^{56}Fe	^{56}Fe(n, 2n α)^{51}Cr	20	^{51}Cr	27.7 d
^{92}Mo	^{92}Mo(n, p)^{92m}Nb	0	^{92m}Nb	10.1 d
^{93}Nb	^{93}Nb(n, 2n)^{92m}Nb	8.9	^{92m}Nb	10.1 d

Derived from Fischer et al. (2003).

threshold energy of 32.6 MeV. In addition, the higher energy D-T fusion neutron spectrum leads to activation reactions that are more complex than the fission activation product production mechanisms.

The reactions of Tables 3.10 and 3.12 involve D-T fusion neutron production mechanisms, including the (n, γ) and (n, p) reactions, as well as more complex reactions such as (n, p ^{3}He) and (n, n α) that involve the transfer of multiple nucleons. Multiple nucleon transfer is possible because the D-T fusion neutron spectrum imparts sufficient energy to facilitate these reactions.

3.19.4.2 Activation of Stainless Steel – Vacuum Vessel Structural Material

In the ITER, the vessel structural and shielding material candidates are composed of stainless steel (SS-316). There are 12 major radionuclides produced from SS-316 activation that dominate the effective dose rate and shielding considerations after 1 day postirradiation and during the subsequent 30-day period. The activation products, their half-lives, and candidate production modes are summarized in Table 3.13. Their relative contribution to the postshutdown effective dose rate is provided in Table 3.14.

Table 3.13 Dominant ITER activation products from SS-316 irradiation spectrum.

Activation product	Half-life	Candidate production mode
^{56}Mn	2.58 h	^{56}Fe(n, p)^{56}Mn ^{55}Mn(n, γ)^{56}Mn ^{59}Co(n, α)^{56}Mn
^{57}Ni	36.1 h	^{58}Ni(n, 2n) ^{57}Ni ^{60}Ni(n, 4n) ^{57}Ni
^{58}Co	70.8 d	^{58}Ni(n, p) ^{58}Co ^{59}Co(n, 2n)^{58}Co
^{99}Mo	66.0 h	^{98}Mo(n, γ)^{99}Mo ^{100}Mo(n, 2n)^{99}Mo
^{64}Cu	12.7 h	^{63}Cu(n, γ)^{64}Cu ^{64}Zn(n, p)^{64}Cu
^{99m}Tc	6.00 h	^{99}Mo decay ^{98}Tc(n, γ)^{99m}Tc
^{54}Mn	313 d	^{54}Fe(n, p)^{54}Mn ^{55}Mn(n, 2n)^{54}Mn
^{51}Cr	27.7 d	^{50}Cr(n, γ)^{51}Cr ^{54}Fe(n, α)^{51}Cr ^{56}Fe(n, 2n, α)^{51}Cr
^{60}Co	5.27 yr	^{59}Co(n, γ)^{60}Co ^{60}Ni(n, p)^{60}Co
^{48}Sc	43.7 h	^{51}V(n, α)^{48}Sc ^{48}Ti(n, p)^{48}Sc ^{50}V(n, ^{3}He)^{48}Sc
^{59}Fe	44.6 d	^{58}Fe(n, γ) ^{59}Fe ^{59}Co(n, p)^{59}Fe ^{62}Ni(n, α)^{59}Fe ^{61}Ni(n, ^{3}He)^{59}Fe
^{92m}Nb	10.1 d	^{93}Nb(n, 2n)^{92m}Nb ^{92}Mo(n, p)^{92m}Nb

Derived from Batistoni et al. (2003).

The results of Tables 3.13 and 3.14 show similarities to fission activation products. Table 3.14 is strikingly similar to fission reactor experience because it indicates that ^{58}Co and ^{60}Co are significant activation sources in stainless steel. These isotopes will likely dominate shutdown and outage radiation fields in a manner that is similar to existing fission power facilities.

Table 3.14 Fraction of the ITER effective dose rate from an activated SS-316 shield.

Nuclide	Time postshutdown			
	1 d	7 d	15 d	1 mo
^{56}Mn	0.11	<0.0001	<0.0001	<0.0001
^{57}Ni	0.43	0.075	0.0026	<0.0001
^{58}Co	0.22	0.60	0.70	0.70
^{99}Mo	0.085	0.053	0.0089	0.0002
^{64}Cu	0.014	<0.0001	<0.0001	<0.0001
^{99m}Tc	0.024	0.015	0.0025	0.0001
^{54}Mn	0.029	0.080	0.099	0.11
^{51}Cr	0.022	0.055	0.056	0.045
^{60}Co	0.022	0.063	0.079	0.092
^{48}Sc	0.0088	0.0026	0.0002	<0.0001
^{59}Fe	0.013	0.035	0.038	0.035
^{92m}Nb	0.005	0.0095	0.0069	0.0029

Derived from Batistoni et al. (2003).

3.19.5
Heavy Ions

By definition, heavy ions have a mass greater than the proton mass. As noted in Tables 3.9 and 3.10, heavy ion species are produced in a variety of reactions. Most of the heavy ions remain confined to the vacuum vessel and deposit their energy within the plasma or in the vessel wall. Therefore, it is not likely that heavy ions will present a significant health physics concern in a D-T fusion facility. However, they contribute to vacuum vessel radiation damage and increase its maintenance and associated dose requirements.

3.20
Uncertainties in Health Physics Assessments Associated with External Ionizing Radiation

A number of uncertainties exist at the current stage of fusion power development. These uncertainties include (1) the specific design parameters of the fusion facility, (2) magnitude and angular dependence of the cross sections for some of the reactions induced by the D-T fusion neutron spectrum, (3) materials of construction and their associated lifetimes when subjected to the D-T fusion neutron spectrum, (4) specific component arrangement of the facility, (5) staffing levels including their distribution by discipline, (6) component sizes and the associated sizes of their support equipment, (7) licensing and associated regulatory requirements, and (8) business and regulatory climate that impacts the economics of an operating fusion facility. Each of these uncertainties has the potential to impact the health physics practices and requirements at the initial D-T fusion power facility.

As an illustration of the impact of these uncertainties, consider the activation of water. In particular, ^{16}O is activated via fast neutron capture to produce ^{16}N via the

^{16}O(n, p)^{16}N reaction. The ^{16}N saturation activity (A_{sat}) attributed to the ^{16}O fast neutron capture reaction is given by Equation 3.56, where N is the number of ^{16}O atoms in the target water volume, σ is the ^{16}O(n, p)^{16}N cross section, and ϕ is the activating fast neutron flux.

In a fission power reactor, the neutron spectrum has a most probable energy of about 0.7 MeV and an average energy of about 2 MeV. Only a fraction (f) of the neutron spectrum has sufficient energy to initiate the ^{16}O(n, p)^{16}N reaction with a threshold energy (E_t) of about 10 MeV. The fraction f is defined as

$$f = \frac{\int_{E_t}^{\infty} \phi(E) dE}{\int_0^{\infty} \phi(E) dE}. \tag{3.57}$$

With this definition and Equation 3.56, the ratio (r) of the ^{16}N activity in a fusion power facility and the ^{16}N activity in a fission power facility is

$$r = \frac{A_{sat}(\text{fusion})}{A_{sat}(\text{fission})} = \frac{N_{fusion} \sigma f_{fusion} \phi_{fusion}}{N_{fission} \sigma f_{fission} \phi_{fission}}, \tag{3.58}$$

where $\phi_{fission}$ and ϕ_{fusion} are the total neutron flux for fission and fusion power reactors, respectively. For equal power fission and fusion reactors

$$\phi_{fusion} \approx \phi_{fission} \tag{3.59}$$

and

$$r = \frac{N_{fusion} f_{fusion}}{N_{fission} f_{fission}}. \tag{3.60}$$

For a given fission power reactor, the values of $N_{fission}$ and $f_{fission}$ are well established. $N_{fission}$ depends on the water volume available for activation and $f_{fission}$ depends on the fuel design characteristics, moderator, reactor vessel configuration, materials of construction, and water volume configuration.

The values of N_{fusion} and f_{fusion} are uncertain. The cooling water volumes are not known and so are their specific orientations relative to the toroidal vacuum vessel. The materials of construction and their specific composition are also not established. Given these uncertainties only general observations regarding the ^{16}N activity at a fusion reactor are possible.

The first observation is the D-T fusion reaction produces 14.1 MeV neutrons (Equation 3.3). When compared with the lower energy fission neutron spectrum

$$f_{fusion} > f_{fission}. \tag{3.61}$$

It is likely that the fusion neutron spectrum has significantly more neutrons above the ^{16}O(n, p)^{16}N reaction threshold than the fission neutron spectrum.

The second observation concerns the available cooling water volumes. The water volumes available for activation in a fission reactor are on the order of 4×10^4–4×10^5 l. If the fusion reactor-cooling water volumes are comparable, then the ^{16}N activity in a fusion reactor could be a significant design consideration based on Equations 3.60 and 3.61. Given the nature of the D-T neutron spectrum, the ^{16}O(n, p)^{16}N reaction merits further investigation as fusion reactor design is more firmly established.

3.21
Internal Ionizing Radiation Hazards

The D-T fusion activation products summarized in Tables 3.10, 3.12 and 3.13 are potential internal hazards. The reactor's tritium fuel is also an internal hazard particularly when it oxidizes to the HTO form.

3.21.1
Tritium

Commercial fusion plants, using the D-T process, will maintain kilogram quantities of tritium and deuterium at the facility. These tritium inventories present an internal intake challenge. Tritium in either molecular form or other chemical forms diffuses through the vacuum vessel at high operating temperatures. In addition, tritium leakage from the vacuum vessel's coolant, through seals, valves, and piping requires health physics attention. Some tritium also diffuses into the steam system and is released into the environment.

A portion of the tritium resides in routine work areas where it presents a skin absorption, ingestion, and inhalation hazard. Tritium appears as surface contamination, which can be resuspended into the air or can directly contaminate personnel.

Tritium also resides in a variety of fusion reactor systems. For example, the tritium injection systems require careful operational control and maintenance in order to preclude leakage. In addition, systems transporting tritium or systems involved with tritium recovery merit special health physics attention. A number of health physics challenges are associated with operating and maintaining tritium transport and delivery systems including

- monitoring sealing systems having very low leakage requirements;
- performing periodic radiation and contamination surveys of large, complex surfaces;
- controlling health physics access into facility areas having a variety of radiological conditions; and
- providing methods for the temporary containment of tritium.

For HTO, the tritium activity absorbed through the skin (s) is proportional to the inhaled (i) tritium activity.

$$I_s = f\, I_i, \tag{3.62}$$

where I_s is the HTO activity absorbed through the skin, f is a skin absorption factor, and I_i is the inhaled HTO activity. Values of f range from 0.5 to 1.0.

Facility operations are complicated by the presence of tritium outside the vacuum vessel. The problem is more complex than encountered in CANDU reactors, because tritium diffuses through the vacuum vessel. As in a CANDU reactor, minimizing the leakage of systems contaminated with tritium is an essential element of the facility's contamination control program.

In CANDU reactors, 30–40% of a worker's total effective dose is because of tritium intakes. It is expected that a fraction of the effective dose in a fusion facility will also arise from tritium intakes. Therefore, the measurement of the tritium source term, sound contamination control practices, and an active bioassay program are essential elements of the radiological controls program at a fusion power reactor.

As in CANDU reactors, urinalysis is the preferred method of bioassay for HTO. However, given the various possible forms of tritium (e.g., HTO, HT, HD, DT, T_2, and HDO) that may be encountered in a fusion facility, other bioassay techniques may be required.

The activity and diversity of the tritium compounds that may be encountered in a fusion power facility require a variety of measurement techniques. These techniques are discussed in subsequent sections.

3.21.2
Particulates

Maintenance activities (e.g., cutting, grinding, and welding) generate particulate material that can become airborne. These airborne particulates enter the body through inhalation and ingestion.

Particles are also generated through the operation of systems that are in proximity to the vacuum vessel including the fuel system, coolant system, and waste extraction system. The nature of these particulate aerosols has not been fully characterized and will depend on the specific design characteristics, maintenance practices, and the neutron spectrum of the fusion power facility.

Dust is created in the inner wall of the vacuum vessel because of surface erosion. This dust is of concern because it can be activated. It is estimated that erosion accumulation is on the order of a few hundred grams, most of which will be collected by precipitators. Any dust released into accessible work areas presents an internal intake concern.

3.22
Measurement of Ionizing Radiation

The measurement of ionizing radiation in a fusion power facility is done by a variety of techniques, which facilitate the detection of alpha, beta, X-ray, gamma, and neutron radiation.

One of the dominant source terms at a fusion reactor is the external radiation derived from fusion products, activation products, and direct radiation from the fusion process. Tritium is a source of beta radiation and has the potential to significantly impact worker doses. In view of its low-energy beta spectrum, tritium is addressed separately.

3.22.1
Measurement of External Radiation

The techniques to measure various radiation types at a fusion power reactor are similar to the methods encountered at conventional fission facilities. Table 3.15

Table 3.15 General techniques for detecting ionizing radiation at a D-T fusion power facility.

Radiation type	Detector type	Energy range	Efficiency
Alpha particles	Ionization chamber	All energies for counting and spectroscopy	High
	Proportional counter	All energies including spectroscopy applications	High, but dependent on window thickness
	Geiger Mueller counter	All energies	Moderate
	Inorganic scintillation detector (ZnS)	All energies	High
	Organic scintillation detector (anthracene)	All energies	Moderate
	Semiconductor detector[a] (surface barrier and diffused junction)	All energies	Low
Beta particles	Ionization chamber	All energies	Moderate
	Proportional counter	All energies. Spectroscopy at low energies (<200 keV)	Moderate
	Geiger Mueller counter	<3 MeV	Moderate
	Inorganic scintillation detector [CsI(Tl)]	Low energies	Moderate
	Organic scintillation detector (anthracene, stilbene, and plastics)	All energies	Moderate
	Semiconductor Detector[a] (surface barrier, diffused junction, and lithium drifted silicon)	<2 MeV	Low
X-rays	Ionization chamber	All energies encountered in typical applications	Dependent on window thickness particularly at low energies
	Proportional counter	All energies encountered in typical applications	Moderate

3.22 Measurement of Ionizing Radiation

Table 3.15 (Continued)

Radiation type	Detector type	Energy range	Efficiency
X-rays (continued)	Geiger Mueller counter	All energies	Dependent on window thickness
	Inorganic scintillation detector [NaI(Tl) with thin window]	All energies	High
	Semiconductor detector[a] (surface barrier, diffused junction, and lithium drifted germanium)	All energies	High
Gamma rays	Ionization chamber	All energies	Low
	Proportional counter	All energies	Low
	Geiger Mueller counter	All energies	Low
	Inorganic scintillation detector [CsI(Tl) and NaI(Tl)]	All energies	Moderate
	Organic scintillation detector (plastics)	All energies	Low
	Semiconductor detector[a] (surface barrier, diffused junction, and lithium drifted germanium)	All energies	Moderate
Neutrons	Ionization chamber	Thermal neutron detection with BF_3 gas, boron lining, or fissionable material	Moderate
		Fast neutron detection with proton recoil from hydrogenous material	Moderate
	Proportional counter	Thermal with BF_3 gas or boron lining	Moderate
	Geiger Mueller counter	All energies via the (n, p) or (n, α) reactions	Moderate
	Inorganic scintillation detector [LiI(Eu)]	Thermal	Moderate
	Organic scintillation detector (plastics and liquids)	All energies depending on scintillation material	Low

[a] Energy resolutions are at least a factor of 10 better than scintillation detectors and this permits spectroscopic applications.

summarizes various methods that can be used in a fusion power facility to detect alpha, beta, X-ray, gamma, and neutron radiation. The applicable energy range of the technique and its efficiency are also provided. Efficiencies are given in qualitative terms (e.g., high, moderate, and low).

3.22.2
Tritium Measurement

As tritium is an important radiological consideration at a fusion power facility, having a number of methods for tritium detection is desirable. These methods include: the ion chamber tritium-in-air monitor, tritium bubbler, composition measurements, and thermal methods.

3.22.2.1 Ion Chamber Tritium-in-Air Monitors

The measurement of tritium in air presents a unique problem because its average beta particle energy is low (5.7 keV). The available beta energy is not sufficient to penetrate the walls of a detector. This difficulty is avoided if air containing tritium is pumped through an ion chamber to permit the tritium beta energy to produce ionization inside the detector. To eliminate the photon background, a practical monitor has two detectors. One is used to collect ions produced by tritium and a second chamber is sealed so it only detects the photon background. The detector circuitry is designed to measure the difference in the currents from the two detectors, and the current is calibrated to yield the tritium air concentration.

In a CANDU fission reactor, the tritium ion chamber instrument is used, but it has a number of limitations. The first limitation is that any radioactive gas is measured and provides an output that could be incorrectly interpreted as tritium. A second limitation is that the gamma compensation is adequate only in relatively low gamma fields of less than about 100 µGy/h.

Although radioactive isotopes of xenon and krypton are not produced in a fusion facility, activated air would potentially interfere with an ion chamber tritium-in-air monitor. In addition, gamma fields in excess of 100 µGy/h are likely that limits the usefulness of ion chamber tritium-in-air monitors at a fusion power facility.

3.22.2.2 Tritium Bubbler

Compared to ion chamber measurements, the tritium bubbler method is simple and accurate, though not as convenient, and is not affected by a gamma or noble gas background. This technique involves flowing tritiated air through water that traps the tritiated water vapor and then the tritium content in the water is analyzed.

A tritium bubbler consists of a pump, timer, flow gauge, and removable water container holding a volume of about 100 ml. Airflow through the water container is typically about 1 l/min with a sampling time of about 5 min. After sampling, the tritium content of the water is analyzed using liquid scintillation counting. Bubblers can be used at a fusion facility for tritium detection and evaluation.

3.22.2.3 Composition Measurements

Composition measurements determine the actual concentration of atomic or molecular species. This method is used for tritium compounds that exist in a gaseous state. The gas composition is measured using either a mass spectrometer or laser Raman spectrometer.

A high-resolution mass spectrometer measures gas species, but does not distinguish between different molecules with the same mass. For example, HT and D_2 have the same mass, and must be separated to determine the tritium concentration. All species must be measured (e.g., water as H_2O, HDO, or HTO; methane as CH_4, CD_4, or CT_4; and ammonia as NH_3, ND_3, or NT_3) to fully characterize the tritium hazard.

A Raman spectrometer is also used to measure molecular concentrations in a gas mixture. As each molecule has a unique energy spectrum and absorption characteristics, a Raman spectrometer provides a credible measurement tool for tritiated gases. For a given tritium species, the intensity of the spectrometer output is proportional to the gas concentration.

3.22.2.4 Thermal Methods

Thermal methods (calorimetry) rely on radioactive decay heat with about 0.33 W/g generated by tritium decay. The temperature increase (ΔT) resulting from the decay heat (Q) is measured and related to the mass (m) of tritium.

$$Q = mc\Delta T, \tag{3.63}$$

where c is the specific heat of the tritium species. The mass of tritium is obtained by solving Equation 3.63 for m:

$$m = \frac{Q}{c\Delta T}. \tag{3.64}$$

Calorimetry can be used for tritium in any form – solid, liquid, or gas. A limitation of the thermal method is that tritium must be the only radioactive material present. Any other radioactive material contributes to the decay heat and introduces an error into Equation 3.64.

These four measurement techniques are important because they quantify the tritium source term produced as a result of normal operational activities or off-normal events. One of the most important operational activities is facility maintenance, because these activities open contaminated systems and increase the potential for radioactive material dispersal and internal intakes.

3.23 Maintenance

Maintenance of activated structural components presents both an external as well as an internal radiation hazard. In particular, maintenance activities generate particles of a respirable size as a result of cutting, grinding, welding, and other repair activities. The health physics measures to mitigate these hazards are similar to those utilized at a commercial fission reactor.

Anticipated maintenance activities at a fusion power reactor and associated health physics concerns are summarized in Table 3.16. These activities include vacuum vessel support component maintenance during outages and power operations, vacuum vessel maintenance during outages, routine maintenance and surveillance activities, waste processing, defueling and plasma clean-up operations, and tritium addition to the vacuum vessel.

Until the design of a fusion power facility is complete, only a qualitative description of the health physics implications of maintenance operations is possible. However, general health physics considerations for vacuum vessel maintenance, vacuum vessel-cooling water system maintenance, and routine maintenance are presented here.

3.23.1
Vacuum Vessel Maintenance

Over time, the inner vacuum vessel wall suffers considerable radiation and physical damage from neutron and heavy ion interactions. It will be necessary to replace the damaged vacuum vessel surfaces every few years. Studies suggest that maintenance involves shutdown radiation fields in the vacuum vessel that are on the order of 3×10^4 Sv/h requiring mechanical or remote handling equipment. Vacuum vessel surface repair/replacement operations generate particulates that are respirable and present an internal radiation hazard. Hot particle production from activated material is also possible.

The activated vacuum vessel structure and the associated support components produce a radiation hazard that is best addressed with shielding. The majority of the structural activation products are fixed and essentially immobile. However, residual tritium contamination represents an internal concern.

3.23.2
Vacuum Vessel-Cooling Water System Maintenance

The type and design of the vacuum vessel-cooling water system impacts maintenance effective dose values associated with these systems. The vacuum vessel coolant and coolant piping will be extensively activated. The external doses from these components during maintenance are influenced by internal piping corrosion and subsequent precipitation of radioactive material in piping, valves, pumps, and heat transfer systems. Studies suggest that inspection and maintenance activities could lead to substantial occupational doses, but with appropriate chemistry control and design, collective effective dose values could be reduced to 2–3 man Sv with the prospect of further reduction to 0.5 man Sv.

3.23.3
Routine Maintenance

As noted in Table 3.16, occupational doses arise from a number of sources during routine maintenance activities. The external radiation hazard from routine maintenance activities is controlled primarily by the facility design.

Table 3.16 Health physics concerns associated with anticipated maintenance activities at a fusion power reactor.

Activity	Hazards	Concerns
Vacuum vessel support component maintenance during an outage	Activation products Fusion products Hot particles Tritium	External Internal
Vacuum vessel support component maintenance during power operations	Activation products Fusion products Hot particles Tritium Fusion neutrons Fusion gammas	External Internal
Vacuum vessel maintenance during outages	Activation products Fusion products Hot particles Tritium	External Internal
Routine maintenance and surveillance activities during power operations	Activation products Fusion products Hot particles Tritium Fusion neutrons Fusion gammas	External Internal
Waste processing[a]	Activation products Fusion products Hot particles Tritium	External Internal
Defueling and plasma cleanup operations	Activation products Fusion products Hot particles Tritium Fusion neutrons Fusion gammas	External Internal
Tritium addition to the vacuum vessel[b]	Tritium	Internal

[a] Assumes waste processing is performed at locations well separated from the vacuum vessel.
[b] Assumes tritium addition to the vacuum vessel is performed at locations well separated from the vacuum vessel, and any uranium components storing tritium are shielded.

Routine maintenance activities include component replacement and repair, instrument replacement and repair, motor refurbishment and repair, valve replacement and repair, packing adjustments, as well as support for a variety of operational activities such as filter replacement, resin sluicing and resin addition, spill clean-up, decontamination activities, fueling operations, sampling activities, and defueling operations. Both internal and external effective dose must be considered in the health physics planning to support routine maintenance.

For example, maintenance support of refueling operations has the potential to encounter a variety of hazards including tritium, activated material, hot particles, and fusion products, as well as neutron and gamma radiation. Refueling should be engineered such that it can be performed in a low dose rate area with the control of tritium receiving the major focus.

As the external radiation fields influence maintenance activities, it is important to gain an understanding of the radiation fields that may be encountered during maintenance operations. Table 3.17 summarizes two-dimensional transport calculations that predict anticipated radiation levels at the ITER. The two-dimensional model consists of a plasma region, toroidal field coils including intercoil structures, the cryostat, and the biological shield.

At the ITER, a limit of 25 μSv/h is established for hands-on maintenance. This limit assumes that maintenance personnel work for 40 h a week and 50 weeks a year. As initially evaluated, the effective dose rates between the TF coils and the cryostat (39.5 μSv/h) and between the cryostat and the biological shield (60.8 μSv/h) exceed the 25 μSv/h ITER limit. Table 3.17 suggests that the dominant effective dose rate contribution arises from ^{24}Na. The relative magnitude of each contribution is given in parenthesis in Table 3.17.

^{24}Na is produced in the concrete biological shield, and the production modes for ^{24}Na are provided in Table 3.10. The 25 μSv/h effective dose rate limit can be achieved

Table 3.17 Selected effective dose rates – 1 day following ITER shutdown[a].

ITER location	Effective dose rate (mSv/h)	Dominant nuclides[a]	Source of nuclides[b]
Between TF coils and cryostat	0.0395	^{24}Na (25%)	Biological shield (100%)
		^{60}Co (21%)	Cryostat (45%) TF intercoil structures (45%) TF coils (5%)
		^{58}Co (15%)	TF intercoil structures (69%) Cryostat (30%)
Between cryostat and biological shield	0.0608	^{24}Na (68%)	Biological shield (100%)
		^{60}Co (11%)	Cryostat (63%) Biological shield (29%)

[a]Derived from Khater and Santora (1996).
[b]The percentage contribution is provided in parenthesis.

by adding a 1-cm-thick layer of boron to the front of the concrete biological shield, which leads to the following reductions in the effective dose rate:

- A 65% reduction (owing to thermal neutron capture in the concrete) in the effective dose rate at locations between the TF coils and the cryostat. This reduction ($0.35 \times 39.5\ \mu Sv/h = 13.8\ \mu Sv/h$) meets the $25\ \mu Sv/h$ hands-on maintenance limit.

- A reduction of a factor of 3 in the effective dose rate for locations between the cryostat and biological shield. This reduction ($1/3 \times 60.8\ \mu Sv/h = 20.3\ \mu Sv/h$) also meets the $25\ \mu Sv/h$ hands-on maintenance limit.

These results indicate that the ITER design concept may require modification, but the effective dose rates can be managed through shielding modifications. The reader should note that the two-dimensional calculations are only scoping studies, and do not include the effects of streaming through ducts or penetrations that will exist in the vacuum vessel, toroidal field coils, intercoil structures, cryostat, and the biological shield of the ITER.

3.24
Accident Scenarios

The unique scenarios of postulated fusion power reactor accidents present additional radiation hazards. Some initial fusion plant designs propose to use liquid metal coolant and heat exchange systems. In a severe accident, the liquid metal coolant contacting air, water, or steam may lead to an explosive reaction that produces hydrogen gas. Such an event could lead to a loss of structural integrity with the subsequent transport and deposition of activation products, fusion products, and tritium to off-site locations.

Accident releases differ significantly from those of a fission reactor, which involve primarily noble gases and radioiodine. The final safety analysis report for a commercial fusion power reactor will address these and other fusion accident scenarios. A number of advisory groups recommend that fusion facilities be designed and operated in a manner such that no public evacuation is required even for a severe accident event.

3.25
Regulatory Requirements

In the United States, The Code of Federal Regulations Title 10, Parts 20 and 835, prescribes explicit requirements for worker protection, public protection, and ALARA. Part 20 applies to U.S. Nuclear Regulatory Commission licensees and Part 835 applies to U.S. Department of Energy licensees. In terms of regulatory requirements, attention is focused on the ALARA aspects of a fusion power facility.

In particular, 10CFR20.1101(b) states: "The licensee shall use, to the extent practical, procedures and engineering controls based upon sound radiation protection principles to achieve occupational doses and doses to members of the public that are as low as is reasonably achievable." Specific ALARA considerations are discussed in the next two sections.

3.25.1
ALARA-Confinement Methods and Fusion Process Types

Before reviewing the specific design features, ALARA aspects of the fusion confinement method are presented in Table 3.18. Table 3.18 summarizes ALARA

Table 3.18 ALARA comparison of fusion confinement methods.

Consideration	Comment	ALARA preference
Fuel type	MC fusion uses T_2 and D_2 gas and HTO production is more likely than in IC fusion. IC fusion uses a solid D-T pellet	IC fusion – the solid fuel pellet minimizes the internal intake of tritium.
Reaction geometry	MC fusion occurs within a toroidal geometry	MC fusion – near the vacuum vessel, higher dose equivalent rates occur with IC fusion for equivalent fusion powers. The MC fusion advantage disappears as the point of interest moves further from the reaction volume.
	IC fusion occurs in the small D-T pellet (point source). For equivalent fusion powers and distances from the source, the point source geometry has a higher effective dose rate value. However, the effective dose rates are within about 1% of each other when the distance from the MC source reaches three times the vacuum vessel diameter (see Problem 03-05).	
Plasma density	IC fusion operates at a higher density that softens the fusion neutron and fusion gamma spectra.	IC fusion – the vacuum vessel receives less damage because of the softer neutron spectrum. Reduced neutron damage minimizes the associated maintenance requirements.
	The MC fusion spectrum will be harder than the IC fusion spectrum.	

considerations for the selection of fuel type, reaction geometry, and plasma density for IC and MC fusion devices.

The impact of the inherent physics of the D-D and D-T fusion processes on selected facility design considerations is summarized in Table 3.19. Specifically, Table 3.19 considers vacuum vessel maintenance and change-out, production of ^{16}N, ^{3}H intakes, the Lawson criterion, and activation product generation.

Excluding all factors except radiation protection would suggest the ideal fusion facility would not be based on D-T magnetic confinement. Tables 3.18 and 3.19 suggest that ALARA considerations alone would favor a D-D inertial confinement device.

Table 3.19 ALARA Comparison of D-D and D-T Fusion Processes.

Consideration	Comment	ALARA preference
Vacuum vessel maintenance and change-out	The threshold neutron energies from D-D and D-T fusion are 2.45 and 14.1 MeV, respectively.	D-D fusion – the vacuum vessel receives less neutron damage because of the lower energy D-D neutron spectrum. This reduces maintenance requirements and the need for high dose repair activities.
^{16}N activity	The D-D fusion neutron threshold energy lies below the ^{16}O(n, p)^{16}N activation reaction threshold. The higher energy D-T fusion neutron threshold lies above the ^{16}O(n, p)^{16}N activation reaction threshold.	D-D fusion – compared to D-T fusion, the D-D fusion neutron spectrum minimizes the ^{16}N source term.
Internal intake of ^{3}H	D-T fusion uses tritium and deuterium as the fuel source. Tritium and HTO are more hazardous than deuterium. D-D fusion uses deuterium as the fuel source. Tritium is produced inside the vacuum vessel via Equation 3.1.	D-D fusion – deuterium is less hazardous than tritium.
Lawson criterion	A smaller value of $\rho\tau$ is required for D-T fusion	Uncertain – an ALARA decision will depend on specific facility design requirements and operating parameters.
Activation products	The threshold neutron energies from D-D and D-T fusion are 2.45 and 14.1 MeV, respectively.	D-D fusion – activation products with higher threshold energies are minimized by the lower energy D-D fusion neutron spectrum.

3.25.2
ALARA – Design Features

The design and operating characteristics of a fusion power reactor are not yet fully defined. In spite of this uncertainty, the ALARA design features of a fusion power reactor should be developed in a manner analogous to existing fission power facilities. Examples of these features include:

- component and structure activation are minimized through the selection of appropriate low-activation materials;
- components are designed to minimize the accumulation of radioactive material and to facilitate decontamination. This design feature is accomplished by surface preparation (e.g., electropolishing or painting) or ease of disassembly to facilitate decontamination;
- components are designed to facilitate removal and repair;
- localized ventilation is provided to minimize airborne contamination. For example, air clean-up system components are located near sources of potential airborne contamination;
- concrete surfaces are smooth and coated to facilitate decontamination;
- material substitution and purification are incorporated into the design. For example, the use of low-cobalt steel results in lower ^{60}Co activity;
- shield design considers planned power upgrade modifications, other planned modifications, maintenance activities, surveillance activities, and operational activities;
- mockups and full-scale component training aids are used to facilitate task completion;
- quick disconnects and flanged connections are utilized to facilitate the removal of components. These techniques must consider potential tritium leakage during power operations;
- containment and isolation of liquid spills are facilitated through the use of dikes, curbing, reserve tank capacity, and reserve sump capacity;
- the high-energy neutron spectrum is shielded to minimize the production of activation products and to limit radiation damage;
- modular, separable confinement structures are used as contamination control barriers;
- localized liquid transfer systems are used to isolate radioactive material and tritium-bearing fluids; and
- fully drainable systems (e.g., piping and tanks) are utilized to facilitate their decontamination and to reduce worker doses. Flush connections are also a key system design feature.

3.26
Other Radiological Considerations

Before concluding the discussion on D-T fusion, a possible process enhancement is outlined. This enhancement is the use of negative muons (μ^-) to catalyze the D-T fusion process. The radiation characteristics and properties of muons are discussed in Appendixes E–G.

A negative muon catalyzes the fusion of deuterium and tritium by forming a DTμ molecule. In a DTμ molecule, the muon binds the D-T system so tightly that fusion occurs very rapidly. After fusion of the DTμ system, the muon is released and goes on to catalyze another fusion event. This process is repeated until the muon either decays or interacts with the various species in the fusion plasma.

A muon-catalyzed fusion reaction could have a profound impact on the health physics considerations at a D-T fusion facility. Muons would not only affect the size of the fusion device but would also contribute to the facility's radiation signature.

Further discussion is deferred until the direct application of muon-catalyzed fusion in a prototypical device is achieved and sustained.

3.27
Other Hazards

Fusion power facilities have unique hazards as well as hazards common to fission power facilities. The hazards occurring at a fission power facility are outlined in Chapter 2.

The unique hazards for MC fusion include low temperatures and cooling media associated with cryogenic systems, internal intakes related to operation of tritium feed and recovery systems, and strong magnetic fields. Laser radiation and X-rays associated with their high-voltage power supplies are unique to an IC fusion power facility.

Electromagnetic fields (EMFs) are associated with magnetic confinement systems and plasma heating systems. These electromagnetic fields do not have the same frequency and the superposition of radio frequency radiation with multiple frequencies occurs. The management of EMFs having multiple frequencies is similar to managing external exposures to a variety of ionizing radiation types in that the major radiation sources are shielded as part of the facility design.

A number of hazardous materials will be utilized in a fusion power facility, including metallic components that slowly erode as a result of the fusion process, various gases, inorganic chemicals, and organic chemicals. Components in direct contact with the fusion plasma may contain beryllium, beryllium alloys, vanadium, or vanadium alloys. In the United States, limits for these materials are specified by the National Institute for Occupational Safety and Health regulations, Occupational Safety and Health Administration regulations, and industrial standards.

3.28
Other Applications

Before closing the fusion energy discussion, two additional technologies (i.e., cold fusion and sonoluminescence) are noted. It is unclear if these technologies have power-generation applications, but they are fusion-related and are presented for completeness.

3.28.1
Cold Fusion

D-D fusion has been achieved through the pyroelectric effect in which the heating of a crystal produced an electrostatic field. In the experiment, the slow heating of lithium tantalate produced a current to a tungsten electrode having an extremely small tip. The electrode was in a chamber with deuterium at low pressure, and the electrode current ionized the deuterium. The field of the electrode accelerated the deuterium ions into an erbium deuteride target. The accelerated deuterons struck the deuterium target nuclei, and produced 2.45 MeV neutrons characteristic of D-D fusion. The reaction of interest is Equation 3.2.

The pyroelectric process does not currently appear to have a practical application for the production of power. From a standpoint of neutron production, the pyroelectric process offers the potential for a simple, cost-effective, and portable neutron source. The neutron output from the initial experiment was too low for practical applications such as neutron imaging of bulk material. However, the pyroelectric effect offers the potential for enhancing the neutron output from this source. In addition, the pyroelectric source has potential applications in baggage screening, well logging, neutron radiography, and instrument calibration.

3.28.2
Sonoluminescence

Sonoluminescence is a process through which sound energy produces light. During sonoluminescence, sound waves traversing a liquid produce bubbles. These bubbles initially expand and then suddenly implode. During implosion, a flash of light may be produced.

In a series of experiments, an ultrasonic wave was directed at deuterated acetone and synchronized with pulses of neutrons. The results led to the production of additional neutrons consistent with D-D fusion. These experiments attempted to detect the products of D-D fusion (i.e., tritium and 2.45 MeV neutrons) under a number of experimental configurations including ultrasound synchronized with neutron irradiation (neutron-seeded cavitation) in deuterated acetone, neutron-seeded cavitation in normal acetone, and neutron irradiation with deuterated acetone.

Tritium and neutron emission above background only occurred in neutron-seeded cavitation. This result supports D-D fusion in a manner that is quite different from the plasma fusion process in inertial or magnetic confinement. From a health physics perspective, neutron-seeded cavitation offers significant potential as a neutron source. As a radiation source, a number of expected precautions apply. In addition,

Table 3.20 Summary of the hazards of a neutron-seeded cavitation source.

Radiation, radionuclide, or material of concern	Hazard	Control
Neutrons	The D-D fusion spectrum is an external ionizing radiation hazard	Shielding such as concrete, polyethylene, or water shields neutrons.
Tritium	Tritium produced from D-D fusion is an internal hazard. Tritium accumulates over time as the source is used.	The deuterated acetone could be periodically replaced when the tritium concentration reaches a predetermined value.
		A material that eliminates tritium migration from the source should be utilized to minimize this hazard.
Ultrasound	Nonionizing hazard produces a variety of effects in tissue.	Equipment design should maintain ultrasound fields in the vicinity of personnel below recommended levels.
Acetone	Toxic material	Protective clothing and ventilation should be utilized to minimize acetone exposures and maintain their levels below recommended levels.

the production of tritium requires additional control measures. Table 3.20 provides a summary of expected health physics controls associated with a neutron-seeded cavitation source.

3.29 Conclusions

Health physics considerations at a fusion power reactor have many elements in common with existing facilities as well as some unique features. The neutron radiation component at a fusion power reactor has similarities to neutron radiation at an accelerator facility and the tritium hazard is similar to that encountered at a CANDU reactor. When compared to a fission power reactor, a fusion power facility has unique activation products, unique materials of construction, a higher energy neutron spectrum, a broader spectrum of nonionizing radiation, and unique components and systems that support the fusion process.

Problems

03-01 Given the mass excess (Δ) noted below, (a) calculate the threshold energies for the ^{4}He binary reaction channels relative to its ground state and (b) determine the energy of each exit channel particle at threshold for D-D fusion.

Mass excess values applicable to ^{4}He	
Nuclear species	Δ (MeV)
n	8.071
p	7.289
D (^{2}H)	13.136
T (^{3}H)	14.950
^{3}He	14.931
^{4}He	2.425

03-02 In Cartesian coordinates, an alpha particle formed from D-T fusion has a velocity $\vec{v} = v_0 \hat{i}$ at a particular location in a tokamak vessel. If the magnetic field at this location is $\vec{B} = B_0 \hat{j}$ what is the instantaneous magnetic force on the alpha particle?

03-03 A technician at a D-T fusion facility is scheduled to conduct a radiation survey of a tritium recovery cubicle. The survey occurs 10 min postshutdown after a 100 day power run. The cubicle is located in the vicinity of the torus and is subjected to a thermal neutron fluence rate of 1×10^{13} n/cm^2 s and a fast neutron fluence rate of 8×10^{13} n/cm^2 s. The cubicle has a 0.1 g ^{55}Mn coupon and a 100 g ^{56}Fe coupon as part of the design verification package. These coupons are mounted within 1 cm of each other. (a) What is the ratio of the ^{56}Mn activities produced from thermal and fast capture reactions? (b) Assume the technician is 5 m from these coupons, what is the effective dose rate from these sources at the time of the survey?

^{56}Mn is produced from the ^{55}Mn(n, γ)^{56}Mn and ^{56}Fe(n, p)^{56}Mn reactions that have cross sections of 13.3 and 0.001 b/atom, respectively. The ^{56}Mn half-life is 2.58 h and its gamma constant is 2.5×10^{-4} mSv m^2/MBq h.

03-04 It is the year 2075 and the Red Cloud Fusion Facility (RCFF) has been in operation for 1 year. The D-T facility was rushed into operation with a number of engineering items to be resolved. One of these items is a confirmatory structural evaluation of the 100 MT polar crane located directly above the torus and its potential to fail during a design basis seismic event.

It is 4:00 a.m. and the RCFF is struck by a Richter Magnitude 7.5 seismic event and the polar crane falls and crushes the torus/vacuum vessel. The contents of the torus are released into the containment structure. At 6:30 a.m., an operator enters the torus containment and stays in the area for 30 min to assess the damage. Determine the radiological consequences of this event given the following information:

RCFF characteristics and radiological data

Torus Major diameter	6 m
Torus Tritium Inventory	10 g
Torus Deuterium Inventory	10 g
Torus Alpha Inventory	2.5 g
Gamma constant for the walls, floor, and ceiling	0.0005 mSv m^2/MBq h
Torus containment dimensions	50 m × 50 m × 50 m
Activity in each wall, the floor, and the ceiling	1.0×10^8 MBq
Tritium dose factor	4.27×10^{-2} mSv/MBq inhaled including absorption through the skin
Average breathing rate for the worker and public	3.5×10^{-4} m^3/s
Atmospheric dispersion factor at 1 mile	1.0×10^{-4} s/m^3
Tritium-specific activity	3.6×10^8 MBq/g
Percent of tritium released to the environment	10%

(a) Determine the maximum effective dose from the tritium intake that occurred during the worker's stay in the torus containment.
(b) What is the external gamma effective dose from the walls, floor, and ceiling? Assume the worker resides at the center of the torus containment.
(c) What is the effective dose from tritium received by a member of the public located 1 mile from the center of the torus containment?
(d) What is the potential for an alpha intake assuming the worker enters the torus containment without respiratory protection?

03-05 Show that a magnetic confinement tokamak fusion device (distributed source) has a lower effective dose rate than an equivalent inertial confinement device (point source). For simplicity approximate the torus as a thin disk source.

03-06 A review of dosimetry records at a twenty-first century 1500 MWe fusion reactor indicates the following results for operations and maintenance personnel:

Bioassay and dosimetry results by work group

Work group	Tritium bioassay	Measurable (β, γ) dose	Measurable neutron dose
Operations	Positive	Yes	Yes
Maintenance	Some positive but most are negative	Yes	Most personnel have none

Are these results credible? Why?

03-07 A fusion reactor environmental assessment is to be prepared and will compare its radiation environment to that of a fission reactor. The fusion radiation characteristics are to be compared with those of selected fission reactor types (i.e., pressurized water reactors, boiling water reactors, and Canadian deuterium reactors). In particular (a) compare the D-D and D-T fusion reactor's neutron spectrum to that of PWR, BWR, and CANDU reactors and (b) compare the likelihood of the production of tritium, noble gas, iodine, actinides, and other beta–gamma emitters for these reactor types and describe their principle means of production.

03-08 You are the radiation protection manager at the International Thermonuclear Experimental Reactor during plasma heating and dynamics testing with normal hydrogen. As these are nonnuclear tests, a number of penetrations in the torus containment are open and present an unshielded path to personnel working 200 m from the torus. Instead of loading normal hydrogen to the torus, a D-T mixture is added and an inadvertent fusion pulse occurs.

Activation analysis following the event determined that 5×10^{19} D-T fusion events occurred. Applicable flux-to-dose conversion factors and average neutron yield per fusion are noted below:

Neutron energy (MeV)	Neutron yield/fusion event	Neutron flux to dose conversion factor (Gy cm^2/n)
2.5×10^{-8}	0.1	5.1×10^{-12}
1	0.2	3.3×10^{-11}
14	0.7	7.7×10^{-11}

(a) What is the neutron-absorbed dose to the workers located at 200 m from the torus?
(b) If the event duration was 1 ms, how much ^{60}Co was produced? Assume 50 kg of ^{59}Co is available for activation. The ^{60}Co half-life is 5.27 years and the activation cross section is 37 b.
(c) The total activity produced by this event is 10^5 times the ^{60}Co activity calculated in (b). If this activity is uniformly distributed over the surface of a 10 m diameter disk, what is the effective dose rate at 1 m above the disk on its axis? Assume the total activity is represented by an effective gamma constant of 3.5×10^{-4} mSv m^2/MBq h.
(d) If the result of (c) represents the ambient radiation field in the containment building, what is its impact on vacuum vessel operations testing? Given the extent of this testing, shielding installation is not practical. Assume the testing requires 2 weeks to complete and the maximum time spent in this field is 60 h for operations personnel. Only two operators are qualified to accomplish this testing.

03-09 Following a valve repair on a defueling system designed to extract vacuum vessel dust, a maintenance worker alarms the controlled area portal monitor. A subsequent survey detects hot particle contamination on the worker's forearm, and it is removed using sticky tape. A preliminary review of the work activity indicated that the particle was in direct contact with the worker's skin for 4 h.

The particle was subsequently analyzed and its isotopic composition and activity were determined:

Radionuclide	Activity (MBq)	Dose factor[a] (Gy cm^2/MBq h)
^{47}Ca	0.15	1.51
^{58}Co	2.7	0.07
^{60}Co	0.5	1.12

[a]The dose factor is evaluated at 7 mg/cm^2.

(a) In general, what radiation types dominate hot particle absorbed doses?
(b) What is the worker's skin dose as a result of the hot particle contamination?
(c) As the station's radiation protection director, what subsequent actions would you take following this event?

03-10 The Serenity Valley Fusion Plant (SVFP) on Lake Michigan recently added tritium-breeding blanket modules to the vacuum vessel. The first blanket modules have been removed and the tritium is being recovered, using lasers to cut the blanket assemblies.

When installed, the lasers were only verified to be the "as designed" components. The manufacturer mislabeled the components received by the SVFP, and the receipt inspection only verified the component numbers. No in-service power testing was performed. Unfortunately, the installed laser power controls and components were intended for military, not civilian, applications. The lasers deliver megawatts of power to the blanket modules instead of the intended kilowatt delivery system. When the laser power strikes the blanket modules, they are vaporized and the tritium is dispersed into the atmosphere through a stack. During the event, the mean wind speed was 2 m/s and the vertical and horizontal standard deviations at the site boundary were 20 and 40 m, respectively.

The SVFP safety design calculations assume 5.0×10^{11} MBq of tritium resides in each irradiated breeding blanket module and upon worst-case laser disassembly, their contents would be released uniformly over a 10-min period.

(a) What factors affect the plume height of the released tritium?
(b) Using the Gaussian plume model, calculate the ground-level tritium activity concentration at the site boundary on the plume centerline. Assume that 10 tritium breeding blanket modules were being disassembled at the time of the event, and the effective release height was 65 m.
(c) Provide two assumptions that may contribute to the inaccuracy of the Gaussian Plume model in (b).
(d) Will the Gaussian Plume model tend to overestimate or underestimate the ground level concentration? Why?

References

10CFR20 (2007) *Standards for Protection Against Radiation*, National Archives and Records Administration, U.S. Government Printing Office, Washington, D.C.

10CFR835 (2007) *Occupational Radiation Protection*, National Archives and Records Administration, U.S. Government Printing Office, Washington, DC.

Aymar, R., Chuyanov, V., Huguet, M. and Shimomura, Y. (2001) ITER-FEAT – The Future International Burning Plasma Experimental Overview. Proceedings of the 18th IAEA Fusion Energy Conference, October 4–10, 2000, Sorrento, Italy, IAEA.

Batistoni, P. (2001) Research in the Field of Neutronics and Nuclear Data for Fusion. International Conference, Nuclear Energy in Central Europe, p. 003.1.

Batistoni, P., Rollet, S., Chen, Y., Fischer, U., Petrizzi, I. and Morimoto, Y. (2003) Analysis of dose rate experiment: comparison between FENDL, EFF/EAF, and JENDL nuclear data libraries. *Fusion Engineering and Design*, **69**, 649.

Bevelacqua, J.J. (1976) The Nuclear Four Body Problem Including Binary Breakup Channels, dissertation submitted in partial fulfillment of the requirements of the degree of Doctor of Philosophy in Physics, The Florida State University, Tallahassee, FL.

Bevelacqua, J.J. (1978) Effective three- and four-body forces in the five-nucleon system. *Canadian Journal of Physics*, **56**, 1382.

Bevelacqua, J.J. (1979) Shell-model calculations in the ^{4}He system. *Canadian Journal of Physics*, **57**, 1833.

Bevelacqua, J.J. (1981) Shell-model calculations in the A=5 system. *Nuclear Physics*, **A357**, 126.

Bevelacqua, J.J. (1986) Microscopic calculations in the A=6 system. *Physical Reviews*, **C33**, 699.

Bevelacqua, J.J. (1995) *Contemporary Health Physics: Problems and Solutions*, John Wiley & Sons, Inc., New York.

Bevelacqua, J.J. (1999) *Basic Health Physics: Problems and Solutions*, John Wiley & Sons, Inc., New York.

Bevelacqua, J.J. (2004) Muon colliders and neutrino dose equivalents: ALARA challenges for the 21st century. *Radiation Protection Management*, **21** (4), 8.

Bevelacqua, J.J. (2005) An overview of the health physics considerations at a 21st century fusion power facility. *Radiation Protection Management*, **22** (2), 10.

Bevelacqua, J.J. and Philpott, R.J. (1977) Microscopic calculations for the ^{4}He continuum. *Nuclear Physics*, **A275**, 301.

Bernham, J.U. (1992) *Radiation Protection, New Brunswick Power Corporation*, Point Lepreau Generating Station, St. John, NB.

Blake, E.M. (2005) France named ITER host country, Japan to provide Director-General. *Nuclear News*, **48** (8), 130.

Chen, F.F. (2006) *Introduction to Plasma Physics and Controlled Fusion*, 2nd edn, Springer, New York.

DOE-STD-6002-96 (1996) *DOE Standard, Safety of Magnetic Fusion Facilities: Requirements*, U.S. Department of Energy, Washington, D.C.

DOE-STD-6003-96 (1996) *DOE Standard, Safety of Magnetic Fusion Facilities: Guidance*, U.S. Department of Energy, Washington, DC.

Ehrlich, K., Bloom, E.E. and Kondo, T. (2000) International strategy for fusion materials development. *Journal of Nuclear Materials*, **283–287**, 79.

Eurajoki, T., Frias, M.P. and Orlandi, S. (2003) Trends in radiation protection: possible effects on fusion power plant design. *Fusion Engineering and Design*, **69**, 621.

Fischer, U., Simakov, S., Möllendorff, U.v., Pereslautsev, P. and Wilson, P. (2003) Validation of activation calculations using the intermediate energy activation file IEF-2001. *Fusion Engineering and Design*, **69**, 485.

Fujiwara, M.C., Adamczak, A., Bailey, J.M., Beer, G.A., Beveridge, J.L., Faifman, M.P.,

Huber, T.M., Kammel, P., Kim, S.K., Knowles, P.E., Kunselman, A.R., Maier, M., Markushin, V.E., Marshall, G.M., Martoff, C.J., Mason, G.R., Mulhauser, F., Olin, A., Petitjean, C., Porcelli, T.A., Wozniak, J. and Zmeskal, J. (2000) Resonant formation of dμt molecules in deuterium: an atomic beam measurement of muon catalyzed dt fusion. *Physical Review Letters*, **85**, 1642.

Gekelman, W. (2003) Notes on Plasma Physics-First Lecture Notes: Maxwell's Equations and Sheaths in Plasmas, University of California at Los Angeles, http://coke.physics.ucla.edu/laptag/plasma_course.dir/teacher_notes1.pdf (accessed August 11, 2006).

Glasstone, S. (1982) *Energy Deskbook*, United States Department of Energy, Technical Information Center, Oak Ridge, TN.

IAEA-TECDOC-1440 (2005) Elements of Power Plant Design for Inertial Fusion Energy: Final Report of a Coordinated Research Project 2000–2004, International Atomic Energy Agency, Vienna, Austria.

ICRP Publication 30 (1979) *Limits for Intakes of Radionuclides by Workers*, Pergamon Press, Oxford, England.

ICRP Publication 37 (1983) *Cost-Benefit Analysis in the Optimization of Radiation Protection*, International Commission on Radiological Protection, Pergamon Press, Elmsford, NY.

ICRP Publication 55 (1989) *Optimization and Decision-Making in Radiological Protection*, International Commission on Radiological Protection, Pergamon Press, Elmsford, NY.

ICRP Publication 60 (1991) *1990 Recommendations of the International Commission on Radiological Protection*, Pergamon Press, Elmsford, NY.

Information Systems on Occupational Exposure (1999) Seventh Annual Report, Occupational Exposures at Nuclear Power Plants, 1997, Nuclear Energy Agency, Organization for Economic Co-Operation and Development, Paris.

Ishida, K., Nagamine, K., Matsuzaki, T. and Kawamura, N. (2003) Muon catalyzed fusion. *Journal of Physics G: Nuclear Particle Physics*, **29**, 2043.

ITER EDA Documentation Series, No. 7 (1996) Technical Basis for the ITER Interim Design Report, Cost Review and Safety Analysis, International Atomic Energy Agency, Vienna.

Jackson, J.D. (1999) *Classical Electrodynamics*, 3rd edn, John Wiley & Sons, Inc., New York, NY.

Kamp, L.P. and Montgomery, D.C. (2003) Toroidal flows in resistive magnetohydrodynamic steady states. *Physics of Plasmas*, **10**, 157.

Kamp, L.P.J. and Montgomery, D.C. (2004) Toroidal steady states in visco-resistive magnetohydrodynamics. *Journal of Plasma Physics*, **70**, 113.

Khater, H.Y. and Santora, R.T. (1996) *Radiological Dose Rate Calculations for the International Thermonuclear Experimental Reactor (ITER)*, UWFDM-1024, Fusion Technology Institute, University of Wisconsin, Madison, WI.

Knoll, G.F. (2000) *Radiation Detection and Measurement*, 3rd edn, John Wiley & Sons, Inc., New York, NY.

McLean, A. (2005) The ITER fusion reactor and its role in the development of a fusion power plant. *Radiation Protection Management*, **22** (5), 27.

Montgomery, D. (2006) Possible gaps in ITER's foundations. *Physics Today*, **59** (2), 10.

Mustoe, J., Ali, S.M., Pace, L.D., Forty, C.B.A., Friedrich, B.C., Sandri, S. and Thompson, H.M. (1996) Occupational exposure: the role of circuit chemistry and tube coatings in reducing ORE of advanced water cooled fusion plants and the potential effects of magnetic fields, in Sixth IAEA Technical Committees on Development in Fusion Safety, Naka, Japan.

Naranjo, B., Gimzewski, J.K. and Putterman, S. (2005) Observation of nuclear fusion driven by a pyroelectric crystal, *Nature*, **434**, 1115.

NCRP Report No. 17 (1954) *Permissible Dose from External Sources of Ionizing Radiation*, National Committee on Radiation Protection, Bethesda, MD.

NCRP Report No. 107 (1990) *Implementation of the Principle of as Low as Reasonably Achievable (ALARA) for Medical and Dental Personnel*, National Council on Radiation Protection and Measurements, Bethesda, MD.

NCRP Report No. 120 (1994) *Dose Control at Nuclear Power Plants* National Council on Radiation Protection and Measurements, Bethesda, MD.

NCRP Report No. 130 (1999) *Biological Effects and Exposure Limits for "Hot Particles"*, National Council on Radiation Protection and Measurements, Bethesda, MD.

NEA-6164-ISOE (2006) Occupational Exposures at Nuclear Power Plants, Fourteenth Annual Report of the ISOE Program, 2004, Nuclear Energy Agency, Organization for Economic Co-Operation and Development, Paris.

Pfalzner, S. (2006) *An Introduction to Inertial Confinement Fusion*, Taylor and Francis, Boca Raton, FL.

Raeder, J., Cook, I., Morgenstern, G.H., Salpietro, E., Bunde, R. and Ebert, E. (1995) Safety and Environmental Assessment of Fusion Power (SEAFP), Report of the SEAFP Project, EURFUBRU XII-217/95, Brussels.

Shleien, B., Slaback, L.A., Jr and Birky, B.K. (eds), (1998) *Handbook of Health Physics and Radiological Health*, 3rd edn, Lippincott Williams & Wilkins, New York, NY.

Spitzer, L. (1962) *Physics of Fully Ionized Gases*, 2nd edn, Interscience, New York, NY.

Summary of the ITER Final Design Report, G-A0 FDR 4 01-07-21R.04, International Thermonuclear Fusion Experimental Reactor Project (2001), http://www.iter.org (accessed on August 10, 2006).

Thompson, H.M. (1992) Occupational exposure control in the design of a major tritium handling facility, in *Occupational Radiation Protection*, BNES, London.

Thompson, H.M. (1994) Occupational problems with tritium and approaches for minimizing them, in Proceedings of the IRPA Congress: Portsmouth 94, Nuclear Technology Publishing.

Thompson, H.M. The Health Physics Challenges of Fusion Power, http://www.srp-uk.org/srpcdrom/p6-16.doc (December 26, 2004).

Thompson, H.M. and Mustoe, J. (1998) Field Hazard Management: An Approach to the Control of Magnetic Field Hazards in Fusion Plant Design, in Fusion Technology 1998: Proceedings of SOFT'98, EURATOM-CEA, Cadarache, France.

Tilley, D.R., Cheves, C.M., Godwin, J.L., Hale, G.M., Hoffman, H.M., Kelley, J.H., Shaw, G.G. and Weller, H.R. (2002) Energy levels of light nuclei: A = 5, 6, 7. *Nuclear Physics*, **A708**, 1.

Unnumbered UCLA Report (1994) A Synopsis of Major Issues Discussed at the Third Meeting of the Utility Advisory Committee, Fusion Power Plant Studies Program, University of California Los Angeles, February 10, 1994.

Unnumbered UCSD Report (1995) Report of the Sixth Joint Meeting of the Fusion Power Plant Studies Utilities Advisory Committee and EPRI Fusion Working Group, University of California San Diego, February 16–17, 1995.

Xu, Y. and Butt, A. (2005) Confirmatory experiments for nuclear emissions during acoustic cavitation. *Nuclear Engineering and Design*, **235**, 1317.

III
Accelerators

Accelerators and their health physics aspects are the focus of Part Three. Lepton and hadron colliders including electron–positron, proton–antiproton, heavy ion, and muon colliders are included in Chapter 4. A discussion of the Large Hadron Collider and the planned International Linear Collider is also provided.

Chapter 5 reviews photon light sources and their unique health physics challenges. Free-electron lasers with an output in the X-ray and γ-ray regions are included in the discussion and unique high-intensity γ-ray sources are also addressed.

4
Colliders and Charged Particle Accelerators

4.1
Introduction

In the twenty-first century, high-energy accelerators will produce more intense radiation than their twentieth century counterparts. New, exotic particle production is also expected. These higher energies and new radiation types will challenge the twenty-first century health physicist and require the integration of additional technical knowledge into the training and professional development of a health physicist.

Accelerator types utilized in the twenty-first century will include fixed target as well as colliding-beam machines. These machines accelerate ions of various types with energies that will likely span the TeV–PeV range. Colliders utilize the collision of beams of particles to achieve higher energies than fixed target accelerators and also offer the opportunity to explore areas (e.g., particle–antiparticle collisions) that are more difficult to be achieved in fixed target experiments.

The twenty-first century colliders will include both lepton and hardon machines. Electron–positron colliders and hadron–hadron colliders have been successful research tools and their use and viability will continue well into the twenty-first century. Other emerging accelerator types, including muon colliders, are also possible, and these machines will have a dramatic impact on the health physics profession. Each of these broad classes of accelerators is addressed in this chapter.

4.2
Candidate Twenty-First Century Accelerator Facilities

Accelerator facilities represent a balance of technology, the desires of the high-energy physics community, and government/public support. New, higher energy linear and circular colliders will emerge in the twenty-first century, and experiments will be conducted using both fixed target and colliding-beam configurations. New accelerator applications will rely on superconducting technology that has grown dramatically in importance during the last two decades of the twentieth century and will become

Health Physics in the 21st Century. Joseph John Bevelacqua
Copyright © 2008 WILEY-VCH Verlag GmbH & Co. KGaA, Weinheim
ISBN: 978-3-527-40822-1

increasingly important in the twenty-first century. Table 4.1 provides a summary of the existing and planned accelerator facilities that will operate in the twenty-first century.

As noted in Table 4.1, particle physics will undergo a major transition in the early part of the twenty-first century. The next generation of facilities include the Large Hadron Collider (LHC) at the *Centre (Organisation) European pour la Recherche Nucleaire* (CERN). The LHC represents a major step forward in the application of superconducting magnet technology and is the world's highest energy accelerator.

Discussions are currently going on to set the future direction for accelerators following the LHC. Possibilities include the International Linear Collider (ILC); a muon–muon collider; the Very Large Hadron Collider; free-electron lasers producing gamma- and X-rays; machines that support neutrino physics, dark matter, and proton decay; accelerator and detector research and development; and new emerging technologies to accelerate particles.

Two of these machine types, the muon–muon collider and free-electron laser, capable of producing high-intensity X-rays and γ-rays, have unique challenges in the field of health physics. The free-electron laser and its health physics challenges are addressed in the next chapter of this book.

The Large Hadron Collider began operations with proton on proton experiments at CERN in 2008. At the same time, there is a growing scientific consensus that the next major facility in high-energy physics should be an international electron–positron collider that is best operated as a linear collider. This machine is currently referred to as the International Linear Collider.

Given this background, the focus of this chapter is accelerators that will emerge, reach prominence, or have the potential to significantly affect the health physics profession in the twenty-first century. Before reviewing the impacts of health physics and the unique characteristics of the twenty-first century accelerator types, it is important that the reader reviews the background physics associated with these high-energy machines. This physics background includes the Standard Model of Particle Physics, Special Theory of Relativity, knowledge of muon characteristics, and the understanding of luminosity. Readers that need a review of the requisite concepts should consult Appendices E–H.

The physics and particle properties establish a basis to further address the twenty-first century accelerators, the interactions that govern their health physics impact, and their associated health physics characteristics. As a prelude to this presentation, a summary of low-energy accelerator radiation characteristics and health physics concerns is presented.

4.2.1
Radiation Characteristics of Low-Energy Accelerators

Before reviewing the twenty-first century accelerator characteristics, it is worth reviewing the radiation characteristics of low-energy accelerators. Typical low-energy accelerators are found at universities and industrial facilities, and these machines accelerate a variety of particles including protons and electrons. The particles produced by low-energy proton and electron accelerators and their associated health physics concerns are

Table 4.1 Selected properties of existing and planned twenty-first century accelerator facilities.

Facility[a]	Start date	Maximum beam energy (GeV)	Luminosity ($10^{30}/cm^2 s$)	Average beam current per species (mA)	Circumference or length (km)	Peak magnetic field (T)
BEPC-II	2007	2.1 (e^+, e^-)	1000	910	0.238	0.677–0.766
CESR-C	2002	6 (e^+, e^-)	35 @1.9 eV/beam	55	0.768	2.1
FEL	[b]	X-rays[b]	[b]	[b]	[b]	[b]
	[b]	γ-rays[b]				
HERA	1992	30 e	75	40 e	6.336	0.274 e
		920 p		90 p		5 p
ILC	[b]	[b]	[b]	[b]	[b]	[b]
KEKB	1999	8 e^-	11 305	1130 e^-	3.016	0.25 e^-
		3.5 e^+		1500 e^+		0.72 e^+
LHC	2008	7000 p–p	10 000	584	26.659	8.3
	~2008	2760/n[c] Pb–Pb	0.001	6.12	26.659	8.3
Muon–muon collider		[b]	[b]	[b]	[b]	[b]
PEP-II	1999	7–12 e^-	6777	1200 e^-	2.2	0.18 e^-
		2.5–4 e^+		1800 e^+		0.75 e^+
RHIC	2000	100 p–p	6 p–p	48	3.834	3.5
		100/n[c] Au–Au	0.0004 Au–Au	33		
		100/n[c] d–Au	0.07 d–Au	7.7 d		
				38 Au		
Tevatron	1987	980 p	50	66 p	6.28	4.4
		980 $\bar{p}$		8.2 $\bar{p}$		
VEPP-2000	2005	1.0 (e^+, e^-)	100	300	0.024	2.4
VLHC	[b]	[b]	[b]	[b]	[b]	[b]

Derived from Particle Data Group (2004, 2006).

[a] Facility names: BEPC: Beijing Electron–Positron Collider (China); CESR: Cornell Electron–Positron Storage Ring, Cornell University (USA); FEL: Free-electron laser; HERA: Hadronen–Elektronen Ring Anlage (Hadron Electron Ring Accelerator, Deutsches Elektronen Synchrotron (DESY)); ILC: International Linear Collider; KEK-Kou Enerugi Kenkyujo (The High Energy Research Accelerator Organization) (Tsukuba, Japan); LHC: Large Hadron Collider (Centre (Organisation) European pour la Recherche Nucleaire); PEP: Positron Electron Project (Stanford Linear Accelerator Center (Stanford University, USA)); RHIC: Relativistic Heavy-Ion Collider (Brookhaven National Laboratory, USA); Tevatron: Fermi National Accelerator Laboratory Batavia, IL, USA; VEEP-2000: Electron–Positron Accelerator Facility (Novosibrisk, Russia);VLHC: Very Large Hadron Collider;

[b] Machine characteristics, operating parameters, and operational date to be determined.

[c] Energy per nucleon (n).

summarized in Tables 4.2 and 4.3. These tables provide a qualitative description of the characteristics of low-energy accelerators in terms of the interactions that are rough averages over target nuclei that encompass the elements in the periodic table.

Tables 4.2 and 4.3 are predictable in terms of the excitation properties of nuclei. Neutron production is expected on the basis of (p, n) and (γ, n) reaction characteristics. These reaction characteristics are purely kinematic and are the consequences of the energy levels in nuclei and their relation to the neutron reaction channel. Pions are produced whenever the nuclear excitation energy exceeds the pion rest mass. Muons are a direct consequence of pion decays. The properties of pions and muons are summarized in Appendix E.

Muons become more important from a health physics perspective as the energy increases. For this reason, muons are examined in greater detail in Appendix G.

Heavy-ion accelerators also operate at the energies spanned by Tables 4.2 and 4.3. Once the heavy-ion energy exceeds the Coulomb barrier, neutrons dominate the radiation characteristics of low-energy heavy-ion accelerators. Adler's relationship provides the energy of the Coulomb barrier (E_C) in MeV:

$$E_C = \frac{Z_1 Z_2 (1 + A_1/A_2)}{A_1^{1/3} + A_2^{1/3} + 2}, \tag{4.1}$$

where Z_1 and Z_2 are the charge of the heavy ion and target nucleus, respectively, and A_1 and A_2 are their respective mass numbers.

With the low energy health physics concerns established, higher energy impacts are addressed during discussion of the various accelerator types. These health physics impacts are addressed in subsequent discussion.

Table 4.2 Low-energy proton accelerators.

Beam energy (MeV)	Region	Radiation/particles produced	Health physics concerns
<6–8	Elastic scattering	Protons with a range of less than 1 mm in most solids and less than a meter in air	Direct exposure to the beam or scattered radiation
6–100	Inelastic scattering	Neutrons Nuclear fragments	Neutrons dominate the shielding requirements
>100	Particle production	Pions (beam energies >140 MeV) Muons from pion decay Neutrons Protons	Most particles are produced in the beam direction. As accelerator energies increase, muon production increases

Source: Bevelacqua (1995).

Table 4.3 Low-energy electron accelerators.

Beam energy (MeV)	Region	Radiation/particles produced	Health physics concerns
<6	Very low energy	Ionization and bremsstrahlung (X-rays)	Primary electron beam and photons it produces must be shielded Backscattered electrons may be important within shielded enclosures
6–50	Giant resonance (GR)	Electrons Photons produced by electrons Neutrons produced by photons excite the nucleus	Bremsstrahlung is the dominant source of radiation, but neutron radiation requires hydrogeneous shielding such as concrete As the energy increases, the bremsstrahlung is increasingly forward peaked Giant resonance neutrons (at threshold) are produced isotropically For exposures inside the shield, electrons are a major concern
30–150	Low energy	Neutrons Bremsstrahlung	The neutron production cross section is lower than that in the GR region Bremsstrahlung dominates the shielding considerations For exposures inside the shield, electrons and bremsstrahlung are the major contributors to dose
>140	Particle production	Pions Muons Neutrons	For shields thicker than about 120 cm, neutrons become the dominant design concern Bremsstrahlung is the major source of radiation inside the shield

Source: Bevelacqua (1995).

4.3
Types of Twenty-First Century Accelerators

The likely candidates for twenty-first century particle accelerators include energy and luminosity upgrades of the twentieth century accelerator types and the introduction of new accelerator types. In this chapter, the following types of machines are considered as examples of these twenty-first century accelerator concepts: spallation neutron source (SNS), high-energy electron–positron accelerator, Large Hadron Collider/colliding-beam hadron collider, Relativistic Heavy-Ion Collider (RHIC), International Linear Collider, muon–muon collider, and the Very Large Hadron Collider (VLHC). Each of these accelerator concepts present unique health physics considerations.

With this background, the health physics of twenty-first century accelerators begins with a machine that overlaps the twentieth and twenty-first centuries. This accelerator is the spallation neutron source of the Oak Ridge National Laboratory (ORNL).

4.3.1
Spallation Neutron Source

Spallation neutron sources operating in the twentieth century had proton energies and currents that did not exceed 800 MeV and 500 µA. The twenty-first century SNS sources will exceed these values. In the first decade of the twenty-first century, the Oak Ridge National Laboratory's SNS and Japan Proton Accelerator Research Complex (J-PARC) machine will have proton energies (currents) of 1 GeV (1.4 mA) and 3 GeV (333 µA), respectively. As the ORNL machine is operational and J-PARC is still under construction, the SNS concept is illustrated by reviewing the ORNL facility.

4.3.1.1 Machine Overview
Neutron beams are produced in the SNS facility by bombarding a mercury target with energetic protons. The protons excite mercury nuclei through spallation, releasing neutrons that are assembled into beams and guided to neutron instruments. Up to 24 beamlines will exist when the SNS is fully operational. The SNS is designed with the flexibility to be upgraded to higher powers to meet the needs of the twenty-first century research.

The SNS will explore the most intimate structural details of a vast array of novel materials. To accomplish its mission, the facility will consist of specialized components including an ion source, a linear accelerator (LINAC), an accumulator ring, and a mercury target.

4.3.1.2 Ion Source
The SNS's ion source system includes an ion generator, beam formation and control hardware, and low-energy beam transport and acceleration systems. The ion source produces negative hydrogen (H^-) ions having a pulsed format and an energy of 2.5 MeV. This beam is delivered to a LINAC.

4.3.1.3 LINAC
The LINAC accelerates the H^- beam from 2.5 to 1000 MeV (1 GeV). It is a combination of normal conducting and superconducting radio frequency cavities

that accelerate the beam. A magnetic field package provides beam focusing and trajectory control. Following exit from the LINAC, the beam is injected into the accumulator ring to allow the high-power beam to be safely controlled.

4.3.1.4 Accumulator Ring

The accumulator ring structure bunches and intensifies the H^- beam for delivery to a graphite foil. The H^- beam traverses the foil that strips electrons from the H^- beam to produce protons (H^+) that circulate in the accumulator ring. After approximately 1200 trips through the accumulator ring, these protons are directed with a frequency of 60 Hz and a pulse width of about 1 µs into the mercury target.

4.3.1.5 Hg Target

The short, powerful pulses of the incoming 1-GeV proton beam deposit considerable energy in the liquid mercury spallation target. For each proton striking a mercury nucleus, 20–30 neutrons are expelled. Neutrons emitted from the target have energies that are reduced for research applications by passing them through cells filled with water to produce thermal neutrons or through containers of liquid hydrogen at a temperature of 20 K to produce cold neutrons. Cold neutrons are useful for polymer and protein research.

4.3.1.6 Applications

SNS neutrons have numerous applications in determining the microscopic properties of materials including their physical, electrical, magnetic, biological, and chemical properties. Neutrons also have advantages when compared to other radiation types (e.g., X-rays and electrons) to determine material properties. Beams of X-rays and electrons interact with the electrons in the materials through electromagnetic and electrostatic interactions, respectively. Neutrons interact via the strong interaction and probe materials more deeply than X-rays and electrons.

4.3.1.7 SNS Design Decisions

A number of SNS design features are specifically included to enhance the radiological characteristics of the machine. The SNS accelerator systems are designed to be accessible to the practicable extent. In general, acceptable uncontrolled beam losses are limited to 1 W/m to limit activation and residual dose rates.

Shielding is incorporated into the design to limit dose rates of the accessible area. For example, about 5 m of soil covers the accelerator and ring tunnels to limit effective dose rates to below 2.5 µSv/h in the open air for normal operational beam losses of 1 W/m and below 250 mSv/h for full beam loss events. Any localized hot spots are addressed on an individual basis.

4.3.1.8 Radiation Protection Regulations

At SNS, the control of occupational radiation is in accordance with the US Department of Energy (DOE) Regulation 10 CFR Part 835, *Occupational Radiation Protec-*

tion, and supplemented by the guidance provided in the US DOE Order 420.2A, *Safety of Accelerator Facilities*. SNS guidelines state that the radiation should be 2.5 µSv/h for uncontrolled and occupied areas and 250 mSv/h for accident conditions.

To enhance worker radiation protection, defense-in-depth systems are provided. These systems supplement accelerator shielding and other primary control systems and include a personnel protection system (PPS) and a machine protection system.

The personnel protection system includes an automatic beam cutoff system. This system is based on inputs from fission chambers located at specific locations outside the shielding and an interlock system to enter into radiation-controlled areas. PPS acts within about 2 s for the worst-case accidents.

The machine protection system includes beam current and radiation monitors. This system designed to terminate machine operates within one to two pulses.

4.3.1.9 Health Physics Considerations

Even with its unique aspects, the SNS ionizing radiation hazards are dominated by neutron and gamma radiation. To illustrate the relative magnitude of these radiation components, the neutron (gamma) effective dose rate at the top of the beam stop shielding is 8.32 mSv/h (1.86 mSv/h).

The SNS radiological characteristics are similar to those at a conventional accelerator and vary considerably from the high-energy accelerators of the twenty-first century. As an initial example of a twenty-first century accelerator, electron–positron colliders are considered next.

4.3.2
Electron–Positron Colliders – Existing Machines

4.3.2.1 Overview

An electron–positron collider accelerates electrons and positrons in circular rings before colliding the individual beams. There are a number of electron–positron colliders that have operated, are currently operating, or are being upgraded. These include the Large Electron–Positron (LEP) collider and machines noted in Table 4.1. A new machine, the International Linear Collider, is under design and is addressed in subsequent discussions.

From an experimental physics perspective, electron–positron colliders have a number of advantages when compared to hadron colliders. First, the collision results are less complex in terms of the particles produced, because electrons and positrons are fundamental particles without any underlying structure or features. Hadrons are composed of quarks, but the electron and positrons have no such substructures. Therefore, the final-state interactions of the lepton are less complex than the structures that are produced from the interaction of the hadron's quarks. Particle interaction complexity is not the only advantage of electron–positron colliders. Additional lepton and hadron characteristics are outlined in Appendix E.

The lepton colliders are also capable of achieving larger luminosities (see Appendix H) than hadron colliders. In addition, an order of magnitude less energy is required in electron–positron machines than in hadron colliders to achieve similar experimental results. For example, an electron–positron collider with a center-of-mass energy of 2 TeV is roughly equivalent to a 20-TeV center-of-mass energy hadron collider. In spite of these advantages, electron–positron collider health physics concerns exist.

Electron–positron colliders produce more bremsstrahlung than hadron colliders. This bremsstrahlung production serves to limit the upper energies achieved by circular electron–positron colliders. In addition, electric power requirements rapidly increase with increasing energy unless beam power recovery mechanisms are developed and implemented.

The bremsstrahlung produced in a circular electron–positron collider is a fundamental concern that can only be decreased by increasing the circumference of the machine. The logical conclusion is to use an accelerator with an infinite radius (i.e., a linear collider). This is most easily achieved by replacing the dual beams in a circular collider with colliding beams from two linear colliders.

A promising option for boosting the energies in electron–positron colliders is the use of superconducting magnets in the accelerating structure. TeV-scale electron–positron colliders based on superconducting magnets are possible on the basis of the development of new superconducting materials. Ease of fabrication of superconducting magnets also affects the ability to construct a TeV-scale electron–positron linear collider.

The electron and positron beams produce a variety of radiation types that are derived from the direct beam and its interactions. Secondary radiation is produced from bremsstrahlung when beam particles strike accelerator components and from synchrotron radiation when beam particles are deflected by magnetic fields. Table 4.4 illustrates the consequences of these two principal means of producing secondary radiation. These secondary radiation categories and their health physics consequences are addressed in more detail in subsequent discussions.

The effects noted in Table 4.4 also occur in proton accelerators. The magnitude of the effect depends on the beam luminosity, beam energy, and particular accelerator configuration.

The primary electron (positron) beams are contained within beam tubes, and secondary radiation is produced when the primary particles exit the beam tube either by design or by accident. When electrons (positrons) exit the beam tube, they strike accelerator components such as the beam tube structure, vacuum components, collimators, or structural members. When this occurs, the beam particle decelerates and radiates photons through the process of bremsstrahlung. The high-energy bremsstrahlung photons produce electron–positron pairs that lead to additional bremsstrahlung. This process repeats itself and produces an electromagnetic shower or cascade that contains numerous particles and a spectrum of photons having energies up to the kinetic energy of the initial beam particles.

A second category of secondary radiation occurs when the beam particles traverse the magnetic fields of the accelerator, which produces a force that alters the particle's

Table 4.4 Qualitative description of secondary radiations emitted from an electron–positron collider.

Secondary radiation	Health physics consequences
Bremsstrahlung when the electron–positron beams strike accelerator components	Electromagnetic cascade radiation containing high-energy photons, electrons, and positrons High-energy radiation including neutrons, pions, muons, and other hadrons Activation of accelerator structures and components Activation of air, cooling water, and soil Ozone and oxides of nitrogen produced in the air
Synchrotron radiation when electrons–positrons are deflected by magnetic fields	Electromagnetic cascade Photons Neutrons Activation of accelerator structures and components Activation of air, cooling water, and soil Ozone and oxides of nitrogen produced in the air

Derived from CERN 84-02 (1984).

trajectory. It also changes the particle's velocity and leads to the emission of photon radiation. This process is known as synchrotron radiation. Synchrotron radiation is related to bremsstrahlung because a change in velocity or acceleration is involved in both processes. However, the synchrotron radiation differs from the bremsstrahlung spectrum.

With bremsstrahlung, the photon energy extends from zero up to the energy of the beam particle. However, synchrotron radiation is governed by the configuration and strength of the magnetic field. Therefore, the synchrotron spectrum is machine specific. For example, the decommissioned large electron–positron collider of CERN had a synchrotron spectrum that extended from the range of visible light to a maximum intensity that occurred in the range of a few hundred keV. The synchrotron radiation intensity rapidly decreases from its peak value as the photon energy increases above a few MeV. Both bremsstrahlung and synchrotron radiation induce an electromagnetic cascade.

The net result of the electromagnetic cascade is the deposition of energy in materials that are penetrated. This energy includes both particles stopped in the material and photon absorption. The photons produce additional secondary radiation and particles (e.g., photoneutrons) that activate accelerator materials. These same mechanisms lead to effective doses when personnel are in the presence of this radiation. These secondary radiations should be attenuated to insignificant levels by the concrete and Earth shielding outside the accelerator tunnels containing the beam tubes. As noted in Table 4.4, activation products of air, water, and soil, as well as the generation of toxic gases, are produced.

From a health physics perspective, the energy loss of the circulating, accelerating electrons and positrons produces synchrotron radiation (photons). Given the mass of the electrons and positrons, their trajectories are easily altered. Therefore, synchrotron radiation is expected to be a large fraction of the available beam power. The synchrotron radiation requires shielding, and the extent of the shielding depends on the specific location within the accelerator facility.

The amount of synchrotron radiation depends on the specific design characteristics of the electron–positron collider. Dominant factors governing the production of synchrotron radiation are the beam power and radius of curvature of the accelerator ring. From a practical standpoint, radiation generated from the circulating electron and positron beams occurs within the unoccupied shielded ring and is not normally a health physics issue.

The dominant contributors to the radiation environment at an electron–positron facility include electromagnetic cascade showers, external bremsstrahlung, photoneutrons, muons, and synchrotron radiation.

4.3.2.2 Electromagnetic Cascade Showers

An electromagnetic cascade shower induced by an electron or positron is a sequence of bremsstrahlung and pair production processes. The cascade originates when bremsstrahlung is produced from the deceleration of an electron or positron. The bremsstrahlung photons lead to pair production followed by additional bremsstrahlung that rapidly disperses the kinetic energy of the incident electron into an array of photons, electrons, and positrons. Photonuclear reactions are also initiated by the cascade photons and produce secondary neutrons through (γ, n) reactions.

At high energies compared to a particle's rest mass, energy loss by radiation greatly exceeds the energy loss by ionization. The average distance an electron travels for reducing its energy by a factor of $1/e$ approaches a constant value known as the radiation length (X_o), which depends on the atomic number (Z) and atomic weight (A) of the medium it traverses. For electrons,

$$X_o \approx 716 A \left[Z \left((Z+1) \ln \left(\frac{183}{Z^{1/3}} \right) \right) \right]^{-1} \text{ (g/cm}^2\text{).} \tag{4.2}$$

For photons traversing a medium, the average distance traveled (S) to produce an electron–positron pair is approximately given by

$$S \approx \frac{9}{7} X_o. \tag{4.3}$$

The average energy loss through radiation $(dE/dX|_{rad})$ is roughly proportional to the particle's energy. With this approximation, the radiated energy decreases exponentially as a function of the number of radiation lengths traversed. However, the energy loss by ionization $(dE/dX|_{ion})$ varies slowly with energy. Below a kinetic energy value known as the critical energy (E_{crit}), radiation is no longer the dominant mechanism for energy loss. Therefore, once an electron's energy falls below E_{crit}, it no longer plays an important role in the electromagnetic cascade shower.

Values of E_{crit} are characteristic of the medium traversed by the particle and are given by the approximation,

$$E_{crit} \approx \frac{800}{Z+1.2}, \quad (4.4)$$

where the critical energy is expressed in MeV and Z is the atomic number of the attenuating medium. Representative values of the critical energy for materials used in the construction of an electron–positron collider are summarized in Table 4.5.

The cascade shower builds rapidly in the first layers of the medium, and the approximate number of electrons and positrons doubles for each relaxation length until a broad maximum is achieved. The maximum number of particles created in the cascade and the depth at which the maximum occurs (X_{max}) depend on the incident electron or positron energy (E_o) and the traversed medium through the parameter E_{crit}. The location of this maximum is approximated by the relationship

$$X_{max} \approx 1.01 X_o \left[\ln\left(\frac{E_o}{E_{crit}}\right) - 1 \right]. \quad (4.5)$$

For energies between 50 and 100 GeV, the maximum occurs near eight X_o if the shower develops in high-Z material. The average number of particles in the cascade shower for this energy range is on the order of 500–1000 for electrons/positrons incident on high-Z material.

Beyond the maximum location, the shower is gradually absorbed as its energy is deposited in the medium. At such depths, there are few electrons remaining beyond the critical energy and the photon spectrum is predominantly populated by low-energy quanta. The character of the photon spectrum is influenced by the behavior of the photon attenuation coefficient as a function of energy that provides a natural enhancement near the minimum of the Compton scattering coefficient

Table 4.5 Radiation parameters used in construction of electron–positron colliders.

Parameter	Units	Material					
		Air	Water	Concrete	Al	Cu	Pb
Z or Z_{eff}	e	7.2	6.6	11.6	13	29	82
X_o	g/cm^2	36.61	36.08	25.71	24.01	12.86	6.37
E_{crit}	MeV	102	92	51	51	24.8	9.51
E_{Comp}	MeV	45	55	25	21.6	8.4	3.6
μ_{Comp}	cm^2/g	0.0160	0.0166	0.0209	0.0215	0.0304	0.0419

Derived from CERN 84-02 (1984).

(μ_{Comp}). The photon energy for which μ_{Comp} has its minimum value is E_{Comp} (see Table 4.5).

Electromagnetic cascade showers are an important health physics consideration in high-energy electron accelerators. Showers ensure that the electron/positron energy is dissipated within a reasonable distance. Electromagnetic cascade showers affect radiation protection considerations in a variety of ways including (1) shower-induced radiation in areas occupied by personnel, (2) shower-induced radiation affecting sensitive equipment, (3) production of secondary radiation, (4) production of induced activity, and (5) production of radioactive and toxic gases. Specific details of electron-/positron-induced cascade showers are investigated using Monte Carlo techniques with codes such as EGS4 and MORSE (see Appendix J).

4.3.2.3 External Bremsstrahlung

In addition to the initiation of a cascade, an electron beam produces bremsstrahlung. If the electron beam strikes an object external to the evacuated beam tube, external bremsstrahlung is produced. External bremsstrahlung should be evaluated for required shielding and other controls.

An electromagnetic cascade shower, originating in the initial struck object, generates the external bremsstrahlung field. The electromagnetic cascade then proceeds well beyond the initial struck object and propagates through a variety of materials including air, concrete, soil, or other shielding present.

It is also possible that the shower occurs in the vicinity of the initial struck object. This is the limiting case for estimating the required shielding of components.

It was previously noted that the shielding properties of materials is approximated in terms of the Compton minimum for the shower. This approximation can be applied to the limiting case at high energies when the shower occurs along the beam direction. As a first approximation, the following effective dose rate ($\dot{H}$) relationships at 0° and 90° relative to the beam direction may be used for shielding scoping studies for the external bremsstrahlung component,

$$\dot{H}(0°) = 300 E_o, \tag{4.6}$$

and

$$\dot{H}(90°) = 100, \tag{4.7}$$

where these dose rate equations are in units of Sv/h at 1 m, per incident kW of electron beam power, and E_o is in MeV.

As expressed in Equation 4.7, the 90° effective dose rate is approximately proportional to beam power and independent of the beam energy. This relationship is based on the most penetrating radiation components considered in the shielding design. However, effective dose rates at 90° near an unshielded target may be significantly higher because of the contribution of lower energy components that are quickly attenuated in the presence of shielding. Specific geometry effects are not considered in Equation 4.7.

4.3.2.4 Photoneutron Production

Neutrons merit attention because of their potential to produce biological damage in tissue. Above a threshold of about 6 MeV for heavy nuclei and 12 MeV for most light nuclei, neutrons are produced in the interaction of a photon with a nucleus.

Above photon energies of about 25 MeV, the photoneutron cross section is reasonably well described by a quasi-deuteron (qd) model in which the photon is absorbed by a neutron–proton pair within the nucleus. The cross section for the qd process is,

$$\sigma_{qd} = \sigma_d D(A-Z)\frac{Z}{A}, \tag{4.8}$$

where $(A-Z)Z$ is the number of quasi-deuteron pairs, σ_{qd} is the photodeuteron cross section, and D is the quasi-deuteron constant that is a quantification of the likelihood that a quasi-deuteron pair is within a suitable interaction distance of the photon. Values of D are extracted from a data range from 7 to 12. Following the photon interaction, the emitted proton or neutron produces an intranuclear cascade and the emitted neutron spectrum is governed by the characteristics of this cascade.

Above 140 MeV, the photoneutron cross section increases because the photopion production threshold is reached. There are a number of resonances in the (γ, n) cross section below 1.1 GeV. The first peak is the largest and is located at about 300 MeV with a width of about 110 MeV. Above 1.1 GeV, the photoproton and photoneutron total cross sections decrease slowly to an asymptotic value of about 100 μb. Neutrons are emitted from the excited nucleus through the intranuclear cascade, evaporation, and from the extranuclear cascade generated by photopions. The extranuclear cascade is the hadronic analogue of the leptonic electromagnetic cascade outlined previously.

4.3.2.5 Muons

Muon pair production in the Coulomb field of a nucleus is possible above a photon energy of about 211 MeV. This process is analogous to electron–positron pair production, but the muon pair production cross sections are smaller by a factor of about 40 000 owing to the differences in electron (0.511 MeV) and muon (105.7 MeV) masses:

$$\frac{\sigma_{\mu\text{-pair production}}}{\sigma_{e\text{-pair production}}} = \left(\frac{m_e}{m_\mu}\right)^2 = \left(\frac{0.511\,\text{MeV}}{105.7\,\text{MeV}}\right)^2 = \frac{1}{4.28\times10^4} = 2.34\times10^{-5}, \tag{4.9}$$

The dominant muon pair production process is coherent muon production. In coherent production, the target nucleus remains intact as it recoils from the photon interaction. In a small percentage of the time, the nucleus breaks up with the resultant emission of muons. Muons also result from the decay of photopions and photokaons:

$$\pi^+ \rightarrow \mu^+ + \nu_\mu, \tag{4.10}$$

$$\pi^- \to \mu^- + \bar{\nu}_\mu, \tag{4.11}$$

$$K^+ \to \mu^+ + \nu_\mu, \tag{4.12}$$

$$K^- \to \mu^- + \bar{\nu}_\mu. \tag{4.13}$$

In Equations 4.10 – 4.13, the muon contribution to the absorbed dose is usually small compared to the direct production of muon pairs if the meson decay path is not too long.

The muon fluence is highly peaked along the beam direction, and the muon flux density per kilowatt of beam power is roughly proportional to E_o. The total number of muons per kilowatt of beam power does not increase rapidly with incident electron energy. As the electron energy increases, the muon beam becomes more tightly collimated (peaks along the beam direction) and the fluence increases at 0°.

Although muons are similar to electrons, their larger mass does not permit them to efficiently radiate energy by bremsstrahlung. Therefore, they do not efficiently participate in the electromagnetic cascade shower. Leptons, including muons, do not interact with nuclei via the strong interaction. The remaining significant muon-stopping mechanism is through energy loss by ionization. Appendix G describes the muon range in soil and other materials as a function of energy. The soil overburden is commonly used as shielding to attenuate the muon radiation. To minimize the muon radiation concern, low beam power, shielding, and geometry are used.

4.3.2.6 Synchrotron Radiation

When electrons and positrons traverse a magnetic field, such as a dipole magnet in an injector/accelerator main ring or through the fields of the quadrupole/focusing magnet, they experience a centripetal acceleration in maintaining the desired orbit. This acceleration results in the emission of synchrotron radiation. The synchrotron radiation spectrum is typically of lower energy than the bremsstrahlung spectrum and occurs where the beam particles are deflected. This occurs primarily in the curved section of the accelerator.

Two parameters determine the effects of the synchrotron radiation. These are the radiated power per unit beam length and the critical energy. The critical energy is defined further in the text.

The emitted synchrotron radiation depends on a number of parameters including the beam energy and strength and shape of the magnetic field. Associated with the synchrotron radiation output is the electron energy loss.

The total electron energy loss per revolution in a circular orbit (dE) is

$$dE = \left(\frac{4\pi}{3}\right) m_e r_e \frac{\gamma^4}{\rho}, \tag{4.14}$$

where m_e is the rest mass of the electron, r_e is the classical radius of the electron, ρ is the radius of the electron's orbit, and γ is the ratio of total energy E to the rest energy of the electron or positron. If E is expressed in GeV and L and ρ in m, the total energy loss dE/dL in keV/m becomes

$$\frac{dE}{dL} = \frac{14.08 E^4}{\rho^2}. \tag{4.15}$$

The total power emitted by the electron is determined by using Equation 4.15 and the definition of power as the energy delivered per unit time. The total power (P) emitted in watts per revolution for a circulating current I (in mA) is then

$$P = \frac{88.46 E^4 I}{\rho}. \tag{4.16}$$

The total power expression of Equation 4.16 represents an integral over all photon energies, but provides no information regarding the photon spectrum. It is desirable to know the distribution of the number (N) of photons emitted per unit energy interval dE per unit time (t). The photon spectrum for a single radiating electron is given by the relationship

$$\frac{d^2 N}{dE dt} = \left(\frac{\alpha}{\pi\sqrt{3}\hbar}\right) \frac{1}{\gamma^2} \int_r^\infty K_{5/3}(\eta) d\eta, \tag{4.17}$$

where E is the photon energy, α is the fine structure constant ($\alpha = e^2/\hbar c \approx 1/137$), $\hbar$ is Planck's constant divided by 2π, the integrand is a modified Bessel function of order 5/3, and the lower limit of integration r is

$$r = \frac{E}{E_c}, \tag{4.18}$$

where E_c is the primary critical energy. If E is in GeV and ρ is in m, the primary critical energy in keV is given by

$$E_c = \frac{2.218 \, E^3}{\rho}. \tag{4.19}$$

For relativistic electrons, the integral of Equation 4.17 is simplified through a change in variables using the relationship between distance (s) and time (t)

$$s = ct. \tag{4.20}$$

Using $ds = c dt$ and integrating over all photon energies leads to an expression for the number of photons emitted per meter of electron orbit (dN/ds):

$$\frac{dN}{ds} = \frac{19.4 \, E}{\rho}. \tag{4.21}$$

An application of the synchrotron radiation relations is provided in Table 4.6. Table 4.6 characterizes the synchrotron radiation emitted from the decommissioned

Table 4.6 Parameters determining synchrotron radiation at the large electron–positron collider.[a]

Parameter	LEP energy (GeV)		
	51.5	86	100
Current per beam (mA)	3.0	3.3	5.5
Beam charge for both beams (C)[b]	6.5×10^4	7.1×10^4	11.9×10^4
Critical energy (keV)	97.8	455.2	715.7
Radiated power for both beams (W/m)	62	529	1613

[a] Derived from CERN 84-02 (1984).
[b] Based on 3000 h operation.

LEP collider at CERN. This table provides the beam current, beam charge, and radiated power for the electron and positron beams. These parameters and the critical energy are provided as a function of the LEP energy.

4.3.2.7 Radiation Levels at the Large Electron–Positron Collider

As noted previously, LEP has been deactivated and its infrastructure is now incorporated into the LHC. Accordingly, the following discussion provides an indication of the radiation levels at an electron–positron collider.

The historical LEP radiation levels are considered in a broad manner. The expected radiation levels outside the shielding and inside the machine tunnel are reviewed.

4.3.2.8 LEP Radiation Levels Outside the Shielding

At LEP, shielding was adequate to ensure that radiation dose rates were less than 2.5 µSv/h at all freely accessible locations outside the shielding. Dose rates from a klystron gallery on top of the shield reached levels of 25 µSv/h because of radiation penetrating the shield, radiation scattering through the waveguide, and small contributions from X-rays resulting from the operation of klystrons. Areas exceeding 25 µSv/h (e.g., on the roof of the accumulator tunnel) were radiologically controlled and made inaccessible during accelerator operation.

4.3.2.9 Radiation Levels Inside the LEP Machine Tunnel

Within the accelerator tunnel, the dominant radiation source term was attributed to bremsstrahlung X-rays. X-ray emission depends on a number of factors including beam power loss, electron energy, target material, target thickness, and emission angle. The dose rate to an object in the tunnel is a function of its position relative to the point of beam loss and on the influence of local shielding.

The dose rates for high-energy machines are often obtained using computer models (see Appendix J), and simple empirical relationships are limited in applicability. An empirical relationship based on measured absorbed dose rates for electrons striking thick targets at 33 MeV, 100 MeV, and 5 GeV is

$$D = \frac{2.7 P E^{1/2}}{R^2 \theta^{3/2}}, \tag{4.22}$$

Table 4.7 Dose rates at LEP inside the preinjector tunnel during machine operation.

Location	Dose rate	
	At 1 m (Sv/h)	At 20 cm (Sv/yr)
Gun-Buncher	0.05	5×10^3
First LINAC	0.30	3×10^4
Converter	60	6×10^6
Second LINAC	0.04	4×10^3
Electron–positron accumulator ring	0.15	1.5×10^4
Proton synchrotron ring	0.08	8×10^3

Derived from CERN 84-02 (1984).

where D is the absorbed dose rate in Gy/h, E is the electron energy in MeV, P is the electron beam power loss in W, and R is the distance in m at an angle θ in ° to the beam direction. Equation 4.22 is based on a fixed target, and applying it to twenty-first century colliding-beam machines is not a straightforward venture.

As an example of colliding-beam machine radiation levels, the dose rates at LEP inside the preinjector tunnel during machine operation are summarized in Table 4.7. Table 4.7 provides both short-term and annual dose rates for a variety of locations at LEP.

The dose rates mentioned in Table 4.7 demand sound radiological practices including access control. They also provide an indication of the International Linear Collider radiation levels. Before reviewing the ILC, hadron colliders are addressed with an emphasis on the Large Hadron Collider.

4.3.3
Hadron Colliders

Hadron colliders include the Tevatron, RHIC, and the LHC (see Table 4.1). These colliders utilize collisions of protons, protons and antiprotons, and heavy ions.

The LHC is the world's highest energy hadron collider. Other operating hadron colliders include the Tevatron, a proton–antiproton machine, and RHIC. The health physics aspects of the various hadron colliders are addressed in this section.

4.3.3.1 Large Hadron Collider
The Large Hadron Collider began its initial operation in the year 2008 and collides protons in the 27-km circumference tunnel previously used by the Large Electron–Positron collider. Each proton beam has an energy of 7 TeV, leading to a total collision energy of 14 TeV. The LHC also collides heavy ions such as lead with a collision energy of about 2.76 TeV/n. As currently configured, the LHC is able to produce particles that are about 10 times more massive than the heaviest known particle. As such, the LHC is capable of testing the predictions of the standard model and models beyond it, including supersymmetry. It is possible that the Higgs

boson and the lightest supersymmetry particles will be detected when the LHC operates.

The LHC tunnel is located 100 m underground, in the region between the Geneva, Switzerland, airport and the nearby Jura mountains. Inside the LHC tunnel, protons travel around the 27-km ring at nearly the speed of light. These protons collide with a second proton beam traveling in the opposite direction. As the protons collide, a portion of their quarks/gluons transform into a new sequence of particles that are detected by a series of detectors or experiments known by their designations ALICE (a large ion collider experiment), ATLAS (a toroidal LHC apparatus), CMS (compact muon solenoid), LHCb (Large Hadron Collider beauty), TOTEM (Total cross section, elastic scattering and diffraction dissociation), and LHCf (Large Hadron Collider forward). Each detector investigates particle collisions from a different perspective and with different technologies.

To date, the LHC has six approved experiments/detectors. Two large, general-purpose detectors (ATLAS and CMS) examine the collection of particles created in LHC collisions. Two medium detectors (ALICE and LHCb) are designed to study collisions in a more specific manner. Additional smaller experiments (TOTEM and LHCf) are approved, and other proposed detectors will search for exotic particles such as magnetic monopoles. A more detailed discussion of these detectors follows.

4.3.3.1.1 CMS The CMS experiment is a large detector located in an underground chamber at Cessy, France, just across the border from Geneva. It is a cylindrical detector, 21 m in length and 16 m in diameter, that weighs approximately 12 500 MT. CMS is optimized for the detection of muons. The main goals of the CMS experiment are the discovery of the Higgs boson, detection of particles that provide support for supersymmetry, and study of heavy-ion collisions at extreme energies.

4.3.3.1.2 ATLAS ATLAS is a general-purpose detector that is 45 m in length and 25 m in diameter, and weighs about 7000 MT. Its design facilitates the detection of a wide variety of particle types and energies, including particles produced through new physical processes.

The ATLAS detector consists of a series of concentric cylinders that encompass the interaction region of the colliding proton beams. ATLAS is composed of four complementary detectors: the inner detector, the calorimeters, the muon spectrometer, and the magnet systems. The inner detector provides a precise determination of the particle trajectory. Calorimeters measure the energy of readily stopped particles. The muon system yields additional data for the penetrating muons. Finally, the magnet systems alter the trajectories of charged particles in the inner detector, and the muon spectrometer facilitates the determination of their kinematic properties.

4.3.3.1.3 LHCb The LHCb detector is designed to specifically investigate the interactions of heavy particles containing a bottom or beauty quark. In addition to determining particle properties, it has the capability to detect violations of the charge conjugation and parity symmetries.

4.3.3.1.4 TOTEM
The TOTEM detector is dedicated to the measurement of the total proton-proton cross section and study of elastic scattering and diffractive dissociation at the LHC. More specifically, TOTEM measures the total cross section with an absolute error of about 1 mb by using a luminosity independent method.

4.3.3.1.5 ALICE
ALICE detector is dedicated to the investigation of nucleus–nucleus interactions at LHC energies. Its purpose is the study of the physics of strongly interacting matter at extreme energy densities and to investigate the quark–gluon phase of nuclear mater. The existence of quark–gluon matter and its properties are the key issues in understanding the strong interaction, its symmetry properties, and its quark confinement characteristics. Using ALICE, the LHC performs a comprehensive study of the hadrons, electrons, muons, and photons produced in the collision of heavy nuclei.

These experiments detect the results of a wide variety of interactions that occur in a hadron collider. A discussion of the particles involved in these various interactions follows. This discussion specifically addresses antiprotons, protons, neutrons, muons, hadronic cascade particles, heavy ions, and synchrotron radiation. High-power beam loss events are also discussed

4.3.3.1.6 Antiprotons
Although the LHC is not currently configured for antiprotons, these particles are included in the discussion for completeness. However, only those aspects of the radiation field emitted by proton–antiproton or nucleon–antiproton interactions that are important for radiation protection purposes are emphasized.

At low energies, the dominant process in proton–antiproton collisions is annihilation. Calculations of the proton–antiproton annihilation process are summarized in Table 4.8. In Table 4.8, the annihilation events are presented for interactions in air and steel for energies near the rest energy. The results of Table 4.8 utilize the intranuclear cascade model with other models for various stages of nuclear interactions including the quark–gluon model, Fermi breakup model, preequilibrium model, and evaporation model.

Momentum conservation requires the emission of at least two particles from each annihilation event. For antiproton–proton annihilations occurring at rest, the total energy available is twice the rest energy of the proton or 1876.5 MeV. The results of Table 4.8 are consistent with this available energy. From a health physics perspective, Table 4.8 suggests that the dominant species that escapes from the shield or scatters in air are protons, neutrons, pions, and photons. Muons also result from the decay of charged pions.

The dominant mesons are the neutral and charged pions. From Table 4.8, neutral pions have an average kinetic energy of about 220 MeV in air. As the mean lifetime of the π° is 8.4×10^{-17} s, these particles travel only a short distance before decaying. The

Table 4.8 Average multiplicities (<N>) and energies (<E> (MeV)) for protons, neutrons, pions, kaons, photons, electrons, and antiprotons emitted from antiproton annihilations in air and stainless steel.

Particle	Mass (MeV)	Air		Steel	
		<N>	<E> (MeV)	<N>	<E> (MeV)
p	938.3	1.761	84.92	4.290	50.18
n	939.6	3.538	180.26	6.011	92.70
π^+	139.6	0.556	237.95	0.595	216.61
π^-	139.6	0.851	243.60	0.871	225.26
π°	135.0	0.906	221.40	0.931	206.47
K^+	493.7	3.4×10^{-4}	72.04	1.2×10^{-3}	75.80
K^-	493.7	1.4×10^{-4}	49.41	3.3×10^{-4}	45.10
γ	0	0.137	307.96	0.126	336.14
e^-	0.511	9.4×10^{-4}	149.00	8.7×10^{-4}	165.82
e^+	0.511	7.9×10^{-4}	181.13	8.7×10^{-4}	210.50
$\bar{p}$	938.3	5.7×10^{-4}	0.27	1.4×10^{-4}	0.21

Derived from Fermilab-Pub-02/043-E-REV (2002) and Particle Data Group (2006).

dominant neutral pion decay mode is photon emission with a branching ratio of 98.8%.

Following antiproton annihilation, the charged pions have an average energy of about 240 MeV in air. With a mean life of 2.60×10^{-8} s, their range is larger than the neutral pion and their decay mode has a different character:

$$\pi^+ \rightarrow \mu^+ + \nu_\mu, \tag{4.23}$$

$$\pi^- \rightarrow \mu^- + \bar{\nu}_\mu. \tag{4.24}$$

Therefore, the radiation field from pion decay is composed of two components. There is a photon component from the neutral pions, and a charged particle component owing to the charged pions. In addition to the radiation components derived from pion decay, there is also proton and neutron radiation. Neutrons dominate the radiological and shielding considerations.

The results of Table 4.8 suggest that more energetic nucleons (neutrons and protons) are produced from proton–antiproton annihilations in air. Therefore, they are more important from a shielding perspective. The results in air are likely to be more representative of annihilations occurring in materials such as plastics.

The radiation dose is directly related to the number of annihilation events. As antiprotons are more difficult to create and store, the number of antiprotons governs the annihilation rate and, hence, the radiological considerations.

4.3.3.1.7 Proton Reactions Low-energy proton interactions were addressed in Table 4.2. Above about 500 MeV, proton reactions produce secondary radiation types including pions, kaons, muons, positrons, and electrons. At facilities with multiple beam lines, care must be taken to ensure that the beam only enters active areas configured for beam acceptance. Therefore, safety interlocks and beam path control are needed to ensure that the proton beam does not enter areas occupied by personnel.

4.3.3.1.8 Neutrons Normally, neutrons are the dominant prompt radiation hazard at proton accelerators above 10 MeV (see Table 4.2). In view of the interaction mechanisms of the electromagnetic and hadronic cascades, the shield preferentially removes photons and charged particles, so they usually account for only a small contribution to the total effective dose rate outside the shield. Neutrons normally dominate the effective dose.

For proton energies between 200 MeV and 1 GeV, an increased number of nuclear reactions occur. Highly excited compound nuclear states decay by the emission or evaporation of neutrons that lead to the development of hadronic cascades. Other particles and light nuclei are also emitted through evaporation, but these radiation types are readily absorbed in accelerator shielding. Protons and neutrons are produced in roughly equal numbers. Because their energy increases, protons are of increasing importance from a radiation protection standpoint.

At energies above of 1 GeV, neutrons production still occurs. In addition, enhanced numbers of secondary particles are produced. Both neutrons and other particles initiate cascades in shielding that produce radiation sources extending to a larger spatial volume than the trajectory volumes of either the primary or the secondary particles. In the energy region of a few tens of GeV, measurements of the angular distribution of hadrons (principally neutrons, protons, and pions) confirm that as the energy of the incident particle increases, particle production in the beam direction becomes more pronounced.

4.3.3.1.9 Muons At proton energies above about 300 MeV, the production of charged pions becomes important. Above 1 GeV, charged kaons are produced. As the proton energy increases beyond these thresholds, pions and kaons are produced by both the primary particles and energetic secondary particles present in the hadronic cascade.

Both pions and kaons have short half-lives and rapidly decay into muons and neutrinos. Muons and other leptons do not interact via the strong interaction. To first order, charged leptons are attenuated through ionization energy loss mechanisms. For high energies, the muon range becomes quite large, and this range precludes the construction of reasonably sized shields unless they are constructed using soil. Muon ranges in water, polyethylene, air, concrete, standard rock, and soil as a function of energy are summarized in Appendix G.

4.3.3.1.10 Hadronic (Nuclear) Cascade The hadronic cascade is an important consideration in determining the shielding of high-energy nucleon and high-energy electron machines. In both cases, the nuclear cascade is the most important means of transporting radiation through the shield.

At a proton accelerator, the hadronic cascade is initiated when the beam interacts with accelerator or extraction system components. At electron accelerators, high-energy electrons produce hadrons from photon-induced reactions. These reactions include the photodisintegration of (n–p) pairs within the nucleus and photoproduction of pions that are then reabsorbed within the nucleus. The emitted neutrons and protons initiate a hadronic cascade. The six processes of the hadronic cascade are illustrated in a tabular form in Table 4.9 and include muon production, the electromagnetic cascade, the intranuclear cascade, the extranuclear cascade, evaporation of nucleons and nuclear fragments, and induced activity.

In Table 4.9, a cascade or sequence of reactions is initiated by an incident hadron. Energetic protons that strike a target make multiple collisions with nucleons in a

Table 4.9 Processes of the hadronic cascade.

Process	Initial particles produced as a result of the incident hadron collision	Timescale (s)	Typical energy per particle (MeV)	Percent of energy deposition
Muon production	$\pi^+ \rightarrow \mu^+ + \nu_\mu$ $\pi^- \rightarrow \mu^- + \bar{\nu}_\mu$ $K^+ \rightarrow \mu^+ + \nu_\mu$ $K^- \rightarrow \mu^- + \bar{\nu}_\mu$	10^{-8}	Any	10
Electromagnetic cascade	$\pi^\circ \rightarrow \gamma + \gamma$ $\gamma \rightarrow e^+ + e^-$ $\gamma \rightarrow \mu^+ + \mu^-$ (γ, n) reactions	10^{-16}	Any	20
Intranuclear cascade	Spallation products dominated by protons, neutrons, pions, and kaons	10^{-22}	<200	30
Extranuclear cascade	Nuclear collisions produce a variety of radiation types including protons, neutrons, pions, and kaons	10^{-23}	>200	30
Evaporation of Nucleons and Nuclear fragments	Evaporation products are dominated by protons, neutrons, deuterons, and alpha particles	10^{-19}	<30	10
Induced activity	Activation reactions lead to the emission of radiation types dominated by photons, beta particles, and alpha particles	Seconds to years	<10	<1

Derived from NCRP 144 (2003) and ICRU (1978).

nucleus causing spallation. This process is referred to as an intranuclear cascade and dominantly produces protons, neutrons, pions, and kaons.

High-energy particles such as neutrons and protons emitted in the course of the hadronic cascade process collide with other nuclei, causing additional reactions. This process is referred to as an extranuclear cascade. The residual nuclei of the cascade are in an exited state, and yield additional neutrons (evaporated neutrons). In these processes, about 10 fast neutrons are emitted for each proton initiating the cascade.

4.3.3.1.11 Heavy Ions Heavy ions are often used to produce high neutron fluences. The distributions of reaction products become more forward peaked as the incident ion energy increases. As neutrons are a primary result of heavy-ion bombardment, a summary of available data regarding neutron yields is provided.

Experimental results for neutron yields for various incident heavy ions for the specific energy region from 3 to 86 MeV/n is parameterized in terms of relatively simple functions. Although this energy region is much less than expected in twenty-first century machines, the results do provide insight into neutron yield systematics.

The total neutron yield Y (neutrons/ion) is approximately fit as a function of the target atomic number (Z) and the specific ion energy W (MeV/n):

$$Y(W, Z) = C(Z) W \eta(Z), \tag{4.25}$$

where

$$\eta(Z) = 1.22\sqrt{Z}, \tag{4.26}$$

$$C(Z) = \frac{1.95 \times 10^{-4}}{Z^{2.75}} \exp\left[-0.475 (\ln Z)^2\right]. \tag{4.27}$$

The actual values of the functions $\eta(Z)$ and $C(Z)$ used to obtain the parameterization of Equations 4.25–4.27 are provided in Table 4.10. The parameterization of Equations 4.25–4.27 are only approximate, but are sufficiently accurate for most purposes.

4.3.3.1.12 Synchrotron Radiation The LHC, using high-field-strength magnets, produces on the order of 0.1 W/m per beam of synchrotron radiation power. This power is to be extracted from the cryogenic equipment of the superconducting magnets, and this places a significant demand on the cryogenic requirements. In addition to engineering issues, the production of synchrotron radiation presents a health physics hazard, and the associated photon radiation merits attention. Chapter 5 provides additional discussion regarding synchrotron radiation and its health physics consequences within the context of photon light sources.

Table 4.10 Values of the functions $\eta(Z)$ and $C(Z)$.

Element	Z	$\eta(Z)$	$C(Z)$
H	1	1.5	1.7×10^{-4}
He	2	2.6	3.9×10^{-6}
C	6	2.7	2.5×10^{-6}
O	8	3.6	3.6×10^{-7}
Ne	10	7.0	2.7×10^{-10}
Ar	18	7.0	5.1×10^{-11}
Kr	36	7.9	6.0×10^{-12}
Pb	82	11.0	1.7×10^{-13}

Derived from NCRP 144 (2003).

4.3.3.1.13 High-Power Beam Loss Events In addition to normal operational conditions, off-normal events contribute to the radiation environment of the LHC. Of these events, those involving beam loss have a significant health physics impact.

Each of the two circulating 7 TeV proton beams of the LHC contains nominally 350 MJ of energy. An accidental beam loss event causes severe damage to the collider equipment. Such a malfunction is initiated through electronic or mechanical malfunctions or through more exotic effects such as a high-energy cosmic particle impinging upon a sensitive element of the trigger. The LHC conceptual design report (CRD) addresses this type of an event. The CDR outlines a protection system that protects the machine components from a beam loss event. This report suggests that with this system, the peak temperature rise in all the affected components is acceptable. All the LHC dipole and quadrupole magnets are protected against damage in such an accidental event.

In addition to hardware and engineering issues, a beam loss event presents a significant radiation hazard. The magnitude of the hazard depends on the location of the event, and the beam energy and luminosity. Protons, neutrons, muons, and photons are part of the beam loss source term.

Interlocks that terminate the beam mitigate a beam loss event. Beam termination follows the sensing of predetermined radiation levels or beam line vacuum conditions. Installed shielding also mitigates the consequences of an LHC beam loss event.

4.3.4
Heavy-Ion Colliders

A heavy ion collider accelerates nuclear cores for impact with other beams or fixed targets. Future heavy-ion machines will likely exceed the energy of the RHIC at The Brookhaven National Laboratory (BNL) and the LHC's heavy-ion package. This increase in energy will lead to radiation levels that exceed those at the RHIC and the LHC. The discussion focuses on RHIC as this facility has several years of operational history and is representative of a future heavy-ion accelerator.

RHIC is a unique facility in that heavy ions are collided at relativistic energies. The severity of the RHIC radiation hazard is illustrated by noting that the in-beam effective dose rate for a gold ion beam exceeds MSv/h radiation levels.

The configuration at RHIC provides insight into the radiation characteristics of subsequent heavy-ion colliders. At RHIC, the facility consists of a number of interfacing facilities that generate, transfer, and collide the heavy-ion beams and then detect the products of the reaction. These facilities include van de Graaff accelerators, beam transfer lines, synchrotrons, and a storage ring.

RHICs tandem van de Graaff facility (TVDGF) consists of two 15 MV electrostatic accelerators, each about 24 m long, aligned end-to-end. The TVDGF supplies the initial ions and provides them with their initial energy and beam configuration. In addition to heavy ions, some experiments at RHIC use colliding beams of protons. For these experiments, energetic protons are supplied by a 200-MeV LINAC. Protons from the LINAC are transferred to a Booster facility.

Heavy ions and protons are transported using a tandem to Booster facility (TBF). The TBF is a 700-m-long tunnel and beam transport system that increases the ion's velocity to about 0.05 c. Following exit from the TBF, the ions are delivered to a Booster synchrotron.

The Booster synchrotron is a circular accelerator that further increases the ion's energy. This synchrotron feeds the beam into the Alternating Gradient Synchrotron (AGS). Upon entering the AGS from the Booster synchrotron, ions travel at about 0.37 c. The AGS further accelerates the ions until their velocities reach about 0.997 c.

Upon exiting the AGS, the beam enters another beam line called the AGS-to-RHIC (ATR) transfer line. At the end of the ATR, there is a switching magnet that sends the ion bunches down one of two beam lines. Bunches are directed to either a clockwise ring or a counterclockwise ring. Within the heavy-ion collider ring, the counter-rotating beams are accelerated and then collided. RHICs 4-km ring has six intersection points where its two rings of accelerating magnets cross, allowing the particle beams to collide. Specialized detectors are designed for specific interaction characteristics to detect the collision products.

There are currently four detectors supporting RHIC operations, and space is available for two additional detectors. If the RHICs ring is viewed as an analogue clock face, the four current detectors are at 2 o'clock (BRAHMS [Broad Range Hadron Magnetic Spectrometer]), 6 o'clock (STAR [Solenoidal Tracker]), 8 o'clock (PHENIX [Pioneering High Energy Nuclear Interaction Experiment]), and 10 o'clock (PHOBOS). There are two additional locations at 12 and 4 o'clock where future detectors could be placed.

The BRAHMS detects charged hadrons as they traverse the detector. BRAHMS measures the hadron's momentum, energy, and other characteristics as a function of angle.

The STAR at RHIC tracks the particles produced in each heavy-ion collision. It searches for signatures of the quark–gluon plasma.

The PHENIX detector measures the properties of photons, electrons, muons, and hadrons following heavy-ion collisions. By studying the reaction products of the heavy-ion collision, details of the event are determined.

The PHOBOS detector measures the temperature, size, and density of the fireball produced in gold ion collisions. These measurements provide a vehicle to investigate the quark–gluon plasma.

4.3.4.1 Examples of RHIC Radiological Hazards

One of the steps involved in RHIC operations is the transport of ions from the TVDGF to the booster. From the experience of the university tandem van de Graaff, it is well known that deuteron beams generate high effective dose rates. Accordingly, high dose rates are associated with the TVDGF and along the transport line when deuterons are accelerated and transported. Therefore, a consideration of deuterons imposes stringent constraints on the radiation protection system.

The deuteron effective dose is energy dependent. Limiting the deuteron beam's energy can be used to control the dose. Measurements provide the energy dependence of the effective dose for stopping a deuteron beam. For example, the effective dose rate at 30.48 cm from a stopped 12 MeV deuteron beam at an angle of 30° is 25 Sv/h. This value scales experimental results to the RHIC output of 100 µA (DC). At RHIC, the TVDGF can produce deuterons beam energies up to 30 MeV.

For a beam loss event involving 2.2×10^6 deuterons/h, 0.45 mSv/h results after these deuterons traverse about 1 m of soil. These results suggest that controls are needed to limit the effective dose during a beam loss event. Methods to limit this dose are addressed in the following section of this chapter.

4.3.4.2 Radiation Protection Philosophy

The RHIC radiation protection philosophy is based on the US Department of Energy's Hazard Analysis and Hazard Control Process. The result of this process is the establishment of a safety analysis that (1) provides operational limits to manage personnel radiation dose, (2) provides safe collider operations, and (3) manages anticipated events such as a loss of beam event.

4.3.4.3 Personnel Safety Envelope

RHIC established a safety envelope for collider beam operations. To limit personnel effective dose, limits are established for facility personnel and other individuals. The safety limits for collider operations include (1) less than 0.25 mSv in 1 year to individuals in facilities adjacent to RHIC, (2) less than 0.05 mSv in 1 year to a person located at the site boundary, (3) off-site drinking water concentrations and on-site potable well water concentrations not resulting in 0.04 mSv or greater to an individual in 1 year, and (4) less than 12.5 mSv in 1 year to a RHIC staff member.

These dose limits are tied to specific facility operating parameters. The parameters are related to collider parameters and beam loss events.

4.3.4.4 Collider Safety Envelope Parameters

To achieve the RHIC personnel safety envelope dose limits, collider safety envelope parameters have been derived. Limits have been established for a number of collider particles that could be involved in a beam loss event. RHIC established that the maximum number of heavy ions in each ring should not exceed the equivalent of 2.4×10^{11} Au ions at 100 GeV/n. In a similar manner, the maximum number of protons in each ring shall not exceed 2.4×10^{13} at 250 GeV.

4.3.4.5 Beam Loss Control

As noted in a previous section, beam loss events are significant radiological hazards. Radiation monitoring of a beam loss event is accomplished in an ALARA manner. To maintain radiological control of a beam loss event, specific dose limits are established. For uncontrolled areas, beam loss induced radiation is limited to less than 0.005 mSv in an hour. For repeated events, the limit is less than 0.25 mSv in a year. In controlled areas, the limits are 0.05 mSv in an hour and less than 1 mSv in a year for repeated events.

4.3.4.6 Particle Accelerator Safety System

The RHIC safety envelope is supplemented by a particle accelerator safety system. An access control system must be functional during beam operations. During beam operations, area radiation monitors, interlocked to the access control system, are operational and under configuration control to ensure that these systems provide the required beam shutdown capability.

4.4
Planned Accelerator Facilities

There are a number of accelerator facilities in various stages of planning that are candidates for a replacement for the LHC, a new electron–positron collider, and a muon collider. Many of the candidate machines present new accelerator concepts that are quite different from the existing accelerators. Each of these facilities and their health physics implications are addressed.

The planned twenty-first century accelerators are research tools that investigate fundamental issues associated with the nature of matter and energy as well as the nature of the universe. These issues include, but are not be limited to, (1) the emergence of new physical laws and symmetries, (2) the nature of dark energy and dark matter and how they are produced, (3) the emergence of extra spatial or temporal dimensions, (4) the unification of the fundamental interactions, (5) the emergence of new flavors or particle generations, (6) the quantification of neutrino masses and existence of additional neutrino generations, (7) the stability of the universe, (8) the paucity of antimatter relative to matter in the universe, (9) the existence and properties of the Higgs boson, and (10) the existence and properties of supersymmetry particles.

The investigation of these issues will be accomplished using hadron and lepton colliders and accelerator designs using emerging technologies. The preeminent tools for the next few decades will be the LHC located at CERN and an International Linear Collider (ILC) that is in the process of being designed and sited. The LHC is designed to resolve a portion of the issues noted above, and the ILC will be a complementary platform to the LHC. A next-generation hadron machine, the Very Large Hadron Collider and a muon collider are expected to emerge as the twenty-first century progresses and will provide additional capability to resolve fundamental issues.

4.4.1
International Linear Collider

Current design philosophy proposes that the International Linear Collider begin operations at 500 GeV, with the capability to upgrade to about 1 TeV. A TeV-scale electron–positron linear collider has the capability to provide new insights into the structure of space, time, matter, and energy.

The design, research, and development of the ILC involve a number of high-energy physics laboratories. These laboratories include DESY, Fermilab, KEK, and the Stanford Linear Accelerator Center (SLAC). CERN is currently focusing on the LHC, but will likely have a greater involvement in the ILC in the future. In the United States, the Lawrence Berkeley Laboratory (LBL), Lawrence Livermore National Laboratory (LLNL), Brookhaven National Laboratory, Argonne National Laboratory (ANL), the Jefferson Laboratory, and several universities have also contributed to the ILC effort. Considerable support has also been derived from a number of international universities and laboratories.

Although the location of the ILC is yet to be determined, the basic structures are established and include injectors, damping rings, and linear accelerators. The ILC's major components are an electron source/LINAC, positron source/LINAC, electron-damping ring, positron-damping ring, main LINACs, and an interaction area. These components are described in subsequent discussions.

4.4.1.1 Electron Source/LINAC
Electrons are initially generated from a laser initiator. High-intensity light pulses from a titanium–sapphire laser strike a target and produce electrons. Each 2-ns laser pulse creates about 10^9 electrons. An electric field directs each bunch of electrons into a 250-m linear accelerator that increases the electron energy to 5 GeV. The 5-GeV electrons are then transferred to a damping ring.

4.4.1.2 Positron Source/LINAC
Positrons are produced through a series of steps that initially involve transporting an electron beam through an undulator or photon-generating device. Undulators and other photon sources are addressed in Chapter 5.

Magnets within the undulator generate bremsstrahlung photons in the forward (beam) direction. Just beyond the undulator, electrons return to the main accelerator while photons strike a titanium alloy target, producing positrons. A 5-GeV accelerator directs the positrons to the first of the two positron-damping rings.

4.4.1.3 Electron-Damping Ring

After exiting the source, the electron bunches enter a 6-km circumference, damping ring. The electron bunches traverse a wiggler, which is an electromagnetic device that alters their trajectory, and leads to a more uniform, compact spatial distribution of electrons. Each electron bunch circles the damping ring 10 000 times in about 0.2 s. A series of dipole and quadrupole magnets ensure that the bunches remain within the damping ring. Dipole magnets maintain the desired trajectory and bend the electron bunches around the ring. Quadrupole magnets focus and shape the electron bunches. Following exit from the damping ring, the electron bunches enter the main electron LINAC.

4.4.1.4 Positron-Damping Ring

The ILC has two identical positron-damping rings, located in one 6-km circumference tunnel. Two positron-damping rings are required, because positrons produce bremsstrahlung while traversing the ring. The bremsstrahlung photons strike the interior of the positron beam tube and produce electrons. These electrons strike primary beam positrons, annihilation reactions occur, and these reactions deplete the positron density in the damping ring. To limit the generation of photoelectrons, the damping ring positron density is reduced by about a factor of two. Accordingly, two positron-damping rings are required to ensure the same number of positrons as electrons.

4.4.1.5 Main LINACs

Two main linear accelerators, one for electrons and one for positrons, accelerate the lepton bunches toward a collision area. Electric fields accelerate the electrons and positrons to 250 GeV. The two 12-km-long tunnel segments of each LINAC are located about 100 m below the ground and house the two accelerators. The 100 m of Earth provides shielding for the primary beam and secondary particles produced from primary beam interactions.

4.4.1.6 Interaction Area

In the interaction area, the 250-GeV electron beam collides with a 250-GeV positron beam. The collision occurs at 500 GeV that is well beyond the accessible energy region in current electron–positron colliders. In its baseline configuration, the ILC includes two collision points and space for two detectors. The ILC energy and detection capability offers the possibility of resolving a portion of the fundamental physics issues noted previously.

4.4.1.7 Evolving ILC Design

The ILC design is evolving and changes in its configuration should be expected. For example, damping ring configurations have included separate electron and positron damping rings and a dual-use ring. The dual-use ring is the baseline configuration for this book.

The dual-use damping ring reconfigured the ILC footprint by combining the electron and positron damping rings in one tunnel and relocates them to the center of the complex. This option reduces the facility construction cost and consolidates a number of systems into the central damping ring complex.

4.4.1.8 ILC Health Physics

The health physics issues of lepton colliders such as the ILC are similar to those encountered at twentieth century electron–positron colliders. These concerns include electromagnetic cascade radiation containing high-energy photons, electrons, and positrons; high-energy radiation including neutrons, pions, muons, and other hadrons; activation of accelerator structures and components; activation of air, soil, and cooling water; and production of ozone and oxides of nitrogen in the air. The magnitude of these concerns depends on the ultimate energy and luminosity achieved by the ILC and its successors.

Cascade and beam loss events are the expected ILC radiation hazards that lead to elevated radiation doses. Engineering controls are then provided to either manage or reduce these doses. These controls include conventional shielding and beam interlocks using ionization chambers. The expected areas of relatively high radiation dose include the collimators in the beam delivery system, damping ring injection regions, and beam dump areas downstream the interaction zone.

In addition to conventional radiation issues, linear colliders have unique hazards. For example, a beam loss event involving an electron (positron) bunch from a single train has enough power to melt or vaporize accelerator materials that it strikes at normal incidence. These severe beam loss events alter the radiation characteristics of the facility and must be mitigated by radiation or vacuum interlocks.

Additional radiation types may also emerge as the electron energies increase. The radiation properties of quanta characterizing new physics (e.g., dark energy, dark matter, extra spatial or temporal dimensions, new quark flavors, additional particle generations, Higgs bosons, and supersymmetry particles) are impossible to predict.

An indication of the particle types and energies encountered at the ILC is provided by the design and configuration of its planned detectors. These detectors include a calorimeter and muon counter that function for hadron energies from 1 to at least 80 GeV, and electron and positron energies spanning the maximum ILC accelerator energy. Radiation produced from the beam interactions include photons, pions, protons, antiprotons, muons, neutrons, antineutrons, kaons, deuterons, alpha particles, and other radiation types included in the Standard Model of Particle Physics.

The ILC is not the only planned accelerator that incorporates leptons as its primary beam. Plans also exist for muon colliders.

4.4.2
Muon Colliders

The presentation of the health physics aspects of muon colliders is more detailed than the Large Hadron Collider, the International Linear Collider, and the Very Large

Hadron Collider discussion because it is a unique accelerator type. Muon colliders cannot rely on previous operational experience derived from electron–positron colliders or hadron colliders. These existing machines form the operational template for the LHC, the ILC, and the VLHC. Muon colliders represent virgin territory and merit an expanded presentation.

TeV energy muon colliders have the potential to produce annual neutrino effective doses that exceed hundreds of mSv within and outside the facility. At this energy, the effective dose profile resides in a narrow conical plume that spreads over distances greater than tens of kilometers. Controlling and measuring these neutrino effective doses represent a significant health physics challenge. Knowledge of neutrino characteristics and interaction properties facilitates their detection and the calculation of effective dose values.

4.4.2.1 Neutrino Characteristics

Neutrinos were initially postulated as a means to preserve the conservation of momentum during beta decay. Following over 50 years of research, neutrinos remain elusive and some of their properties are still not well understood.

Neutrinos are electrically neutral particles, interact solely through the weak interaction, and have very small interaction cross sections. They are present in the natural radiation environment because of cosmic rays and Solar and terrestrial sources, and are also produced at contemporary reactors and accelerators. From a health physics perspective, these sources produce neutrino effective doses that are inconsequential. Although this will remain true for a number of years, twenty-first century muon accelerators or colliders will produce copious quantities of TeV energy neutrinos. In the TeV energy region, the health physics consequences of neutrinos can no longer be ignored. Upon operation of these accelerators, neutrino detection and the determination of neutrino effective doses will no longer be academic exercises but will become practical health physics concerns.

In a muon collider, neutrinos are produced when muons decay. The neutrino effective dose arises from neutrino interactions that produce showers or cascades of particles (e.g., neutrons, protons, pions, and muons described in Appendix E). It is the particle showers that produce the dominant contribution to the neutrino effective dose.

Concerns for consequential neutrino effective doses have been postulated. For example, it has been suggested that the final stages of stellar collapse produce neutrino effective doses that are sufficiently large to lead to the extinction of some species on Earth. This concern has been challenged, but the concern for neutrino effective doses on the order of hundreds of mSv/year or greater remains. This concern also exists for the muon colliders that could become operational in the next few decades of the twenty-first century.

In a muon collider, both μ^- and μ^+ beams are accelerated and then collided. The acceleration is accomplished using either a linear muon accelerator or a circular muon accelerator or storage ring.

A portion of the muons decay (see Appendix E) and produce neutrinos and antineutrinos. The neutrinos interact through a variety of processes that are complex.

To simplify the discussion, four processes (A, B, C, and D) are defined to describe neutrino interactions with matter.

Process A involves neutrino scattering from atomic electrons. Electrons that recoil from elastic neutrino scattering deposit energy in tissue to produce a neutrino effective dose. Process A occurs over a wide range of energy and the electron tissue interaction involves the multiple scattering of electrons.

In Process B, neutrinos interact coherently with nuclei. This process is only effective for low neutrino energies where the neutrino wavelength is too long to resolve the individual nucleons within the nucleus. At higher energies, Processes C and D become more important.

Process B leads to low-energy ions having large linear energy transfer values. These ions deposit their energy into tissue according to their ranges that are typically $\ll 1$ cm. Although Process B is independent of the neutrino generation, the cross section for neutrinos is about twice the antineutrino cross section.

Process C involves neutrino scattering from nucleons without shielding between the neutrinos and tissue. At energies below about 0.5 GeV, tissue dose is because of recoiling nucleons. As the neutrino energy increases above about 0.5 GeV, secondary particle production increases. Eventually, these secondary particles produce particle showers or cascades in tissue. Process C is independent of the neutrino generation, affecting all three generations in the same manner.

Process D is similar to Process C with the exception that the neutrinos are shielded before striking tissue. Neutrinos with energy greater than about 0.5 GeV, emerging from a layer of material (e.g., Earth shielding), result in a larger effective dose than unshielded neutrinos. The increase in effective dose arises from the fact that the tissue is exposed to the secondary particles produced by neutrino interactions in the shielding material as well as the neutrino beam. Process D is also independent of the neutrino generation.

4.4.2.2 Neutrino Beam Characteristics at a Muon Collider

Neutrinos are produced when the accelerator muon beam particles decay. Weak interactions of muon neutrinos are described in terms of two broad categories: charged current and neutral current interactions. Charged current interactions involve the exchange of W^+ and W^- bosons to form secondary muons. Neutral current interactions produce uncharged particles through the exchange of Z^o bosons. Both types of interactions produce hadronic particle showers. Therefore, the neutrino-induced radiation hazard includes secondary muons and hadronic cascade showers. The hadronic showers have a much shorter range than the muons, and the number of particles in a hadronic shower is quite large. The neutrino radiation hazard arises from these penetrating charged particle showers. For TeV energy neutrinos, direct neutrino interactions in human account for less than 1% of the total effective dose because hadrons from the neutrino interactions typically exit a person before producing a charged particle shower.

The muon beam and subsequent neutrino beam are assumed to be well collimated and to have a minimum divergence angle. The beam divergence is analogous to the divergence of a laser beam as it exits its aperture.

For practical situations, the muons in the accelerator beam have a small divergence angle and are periodically focused using electromagnetic fields to ensure their collimation. No beam divergence is assumed in the subsequent calculations. Therefore, the actual beam is somewhat more diffuse than assumed in the neutrino effective dose calculations. The neutrino beam still produces particle showers, but they are somewhat broader and less intense than the assumed well collimated result.

The magnitude of the effective dose from a particle shower is dependent on the material in the interaction region lying directly upstream of the individual being irradiated. Calculations of the neutrino effective dose consider configurations where a person is (1) completely bathed in the neutrino beam and (2) is surrounded by material that produces particle showers from neutrino interactions. These requirements lead to a bounding set of effective dose predictions.

These assumptions are too pessimistic for the TeV-scale energies that will be encountered in mature twenty-first century muon colliders, because they suggest that the person be uniformly bathed in the neutrino beam and the resultant particle showers. However, considering the current level of muon collider design knowledge, they provide a bounding neutrino effective dose result.

Basic physics principles suggest that the neutrino interactions are more peaked in the beam direction as the muon energies increase. In addition, the neutrino beam radius (r) is relatively small and is given by

$$r = \theta L, \tag{4.28}$$

where θ is called the characteristic angle, also known as the opening half-angle or half-divergence angle of the muon decay cone. This angle is given by the relationship

$$\theta = \frac{mc^2}{E}. \tag{4.29}$$

In Equations 4.28 and 4.29, L is the distance to the point of interest such as the distance from the muon decay location to the Earth's surface, E is the muon beam energy, and mc^2 is the rest mass of the muon (105.7 MeV). As the muon energy increases, the neutrino beam radius and extent of the resultant hadronic showers are smaller than the size of a person (see Table 4.11).

Table 4.11 Candidate muon collider parameters.

E (TeV)	2	5	50
Distance to the point of interest (km)	62	36	36
Neutrino beam radius (m)	3.3	0.8	0.08
Collider depth (m)	300	100	100

Derived from King (1999).

The characteristic angle varies inversely with energy. If E is expressed in TeV

$$\theta \approx \frac{10^{-4}}{E(\text{TeV})}. \qquad (4.30)$$

Therefore, the emergent neutrino beam consists of a narrow diverging cone.

Table 4.11 summarizes candidate muon collider parameters. The muon colliders are constructed below the Earth's surface to provide muon shielding. However, the neutrino attenuation length is too long for the beam to be appreciably attenuated by any practical amount of shielding, including the expanse of the ground between the collider and its exit from the surface of the Earth. Therefore, the ALARA principle as applied to neutrinos no longer includes shielding as an element. In fact, shielding the neutrino beam produces hadronic showers and is anti-ALARA. This peculiar behavior is because of the weak interaction, the uncharged nature of the neutrino, and the TeV energies of the twenty-first century muon colliders.

The neutrinos exiting a muon collider not only have a narrow conical shape but also have a range that is quite long. The long, narrow plume of neutrinos produces secondary muons and hadronic showers at a significant distance from the collider. This distance will be greater than tens of kilometers for TeV muon energies.

4.4.2.3 Neutrino Interaction Model

Neutrinos interact directly with tissue or with intervening matter to produce charged particles that result in a biological detriment. The radiation environment is complex and simulations (e.g., Monte Carlo methods of Appendix J) are used to model the dynamics of the neutrino interaction, including the energy and angular dependence of each particle (e.g., ν_e, $\bar{\nu}_e$, ν_μ, $\bar{\nu}_\mu$, e, μ, and hadrons) involved in the interaction. Performing a neutrino simulation is dependent on specific accelerator characteristics and does not significantly add to the health physics presentation. Rather than performing a Monte Carlo simulation, an analytical approach is used to quantify the neutrino effective dose. This approach is acceptable in view of the current uncertainties in muon collider technology and the nature of the neutrino interaction for both charged current and neutral current weak processes.

4.4.2.4 Neutrino Effective Dose

A muon collider provides a platform for colliding beams of muons (μ^-) and antimuons (μ^+). The collider incorporates a pair of linear accelerators with intersecting beams or a storage ring that circulates the muons and antimuons in opposite directions prior to colliding the two beams. The accelerator facility energy is usually expressed as the sum of the muon and antimuon energies. For example, a 100-TeV accelerator consists of a 50-TeV muon beam and a 50-TeV antimuon beam.

In a muon collider, neutrinos are produced from the decay of muons:

$$\mu^- \rightarrow e^- + \nu_\mu + \bar{\nu}_e, \qquad (4.31)$$

$$\mu^+ \rightarrow e^+ + \bar{\nu}_\mu + \nu_e. \qquad (4.32)$$

Because muon colliders produce large muon currents, neutrinos are copiously produced from the decay of both muons and antimuons.

Neutrino effective dose calculations are performed for two potential muon collider configurations. The first configuration utilizes the intersection of the beams of two muon linear colliders. The linear collider effective dose model incorporates an explicit representation of the neutrino cross section and evaluates the dose-assuming specific values for the muon energy, number of muon decays per year, and accelerator operational characteristics (e.g., accelerator gradient or the increase in muon energy per unit accelerator length). The operational parameter approach is more familiar to high-energy physicists, but it serves to illustrate the sensitivity of the neutrino effective dose to the key muon collider's operating parameters.

The second configuration is a circular muon collider. The neutrino effective dose for the circular muon collider involves an integral over the energy of the differential fluence and fluence-to-dose conversion factor. This approach is more familiar to health physicists, but much of the muon collider's operating parameters are absorbed into other parameters and are not explicitly apparent. Using both approaches yields not only the desired neutrino effective dose but also illustrates the sensitivity of the dose to a number of accelerator parameters and operational assumptions.

4.4.2.5 Bounding Neutrino Effective Dose – Linear Muon Collider

The bounding neutrino effective dose from a linear muon collider is derived from a straight section (ss) of a circular muon collider. This derivation is based on a limiting condition from a circular accelerator with a number of straight sections as a part of the facility. Parameters unique to the circular collider such as the ring circumference and the straight section length appear in intermediate equations but cancel in the final effective dose result. In the linear muon collider, the muon beam is assumed to be well collimated.

In a linear muon collider, the total neutrino effective dose (H) is defined in terms of a dose contribution $\delta H(E)$ received in each energy interval E to $E + dE$, as the muons accelerate to the beam energy E_o:

$$H = \int_0^{E_o} dE \delta H(E). \tag{4.33}$$

Effective dose is usually represented by the symbol E, but H is also used for effective dose to avoid confusion when energy explicitly appears in an expression. The effective dose contribution $\delta H(E)$ is

$$\delta H(E) = H' \frac{1}{f_{ss}} \frac{df(E)}{dE}, \tag{4.34}$$

where $(df(E)/dE)dE$ is the fraction of muons that decay via Equation 4.31 in the energy interval from E to $E + dE$

$$\frac{df(E)}{dE} = \frac{1}{\gamma \beta c \tau g}, \tag{4.35}$$

where γ is the Lorentz factor, τ is the muon mean lifetime, and g is the accelerator gradient.

$$\gamma = \frac{E_o}{mc^2}, \tag{4.36}$$

$$\beta = \frac{v}{c}, \tag{4.37}$$

$$\tau = 2.2 \times 10^{-6} \text{ s}, \tag{4.38}$$

$$g = \frac{dE}{dl}. \tag{4.39}$$

The other parameters appearing in Equations 4.34 – 4.39 include f_{ss} (the ratio of the straight section length to the ring circumference), the ring circumference C, and H' (the effective dose that is applicable as the muon energy reaches the TeV energy range), where

$$f_{ss} = \frac{l_{ss}}{C}, \tag{4.40}$$

$$C = \frac{2\pi E_o}{0.3 \bar{B}}, \tag{4.41}$$

v is the muon velocity, l_{ss} is the straight section length, E_o is the muon energy, $\bar{B}$ is the ring's average magnetic induction, and N is the number of muon decays in a year.

In the narrow beam approximation, the effective dose is independent of distance (L) for $L < 5\, E_o$, where L is expressed in km and E_o in TeV. Using this approximation,

$$H' = K' N l_{ss} \bar{B} E X, \tag{4.42}$$

where K' is a constant that depends on the units used to express the various quantities appearing in Equation 4.42, and $X = X(E)$ is the energy-dependent cross-section factor defined in subsequent discussions.

Combining these results leads to the annual neutrino effective dose (H) in mSv/year:

$$H = K \frac{N}{g} \int_0^{E_o} EX(E) dE, \tag{4.43}$$

where $K = 6.7 \times 10^{-21}$ mSv GeV/m TeV^2, if g is expressed in GeV/m, N is expressed in muon decays per year, E is the muon energy in TeV, and the cross-section factor is dimensionless.

In deriving the linear muon collider effective dose relationship, a number of assumptions were made. These assumptions are explicitly listed to ensure that the reader clearly understands the basis of Equation 4.43:

- The narrow beam approximation is applicable.
- The irradiated individual is assumed to be within the footprint of the neutrino beam.
- The irradiated individual is assumed to be within the footprint of the hadronic cascade that results from the neutrino interactions.
- The individual is irradiated by only one of the linear muon accelerators whose energy is one-half the total linear muon collider energy.
- Given the TeV muon energies and the Earth shielding present, charged particle equilibrium exists.
- Given the TeV muon energies, Process D dominates the neutrino effective dose.
- The neutrino and hadronic radiation uniformly irradiate the individual.
- The muon beam is well collimated.
- The neutrino effective dose calculation assumes a 100% occupancy factor.
- The neutrino effective dose is an annual average based on the number of muon decays in a year.

The cross-section factor is a parameterization of the neutrino cross section (see Table 4.12) in terms of a logarithmic energy interpolation. The numerical factors in the Table 4.12 expressions (1.453, 1.323, 1.029, 0.512, and 0.175) are the total summed neutrino–nucleon and antineutrino–nucleon cross sections divided by the energy at neutrino energies of 0.1, 1, 10, 100, and 1000 TeV, respectively, given in units of 10^{-38} cm^2/GeV. As an approximation, the muon energies in Table 4.12 are equal to the corresponding neutrino energies. The cross-section factor is a dimensionless number and is normalized such that $X(E = 0.1 \text{ TeV}) = 1.0$.

Equation 4.43 is approximated by replacing the energy-weighted integral of $X(E)$ by its value at $E = E_o/2$. This choice is acceptable given the energy dependence of the cross section and the associated uncertainties in the collider design parameters. With this selection, the annual neutrino effective dose (mSv/year) becomes

$$H = \frac{K}{2} \frac{N}{g} X\left(\frac{E_o}{2}\right) E_o^2. \tag{4.44}$$

Table 4.12 Cross-section factor X(E) as a function of muon energy.[a,b]

Muon energy range (TeV)	X(E)
$E < 1$	$(-1.453\, \alpha + 1.323\, (\alpha + 1))/1.453$
$1 < E < 10$	$(1.323\, (1 - \alpha) + 1.029\, \alpha)/1.453$
$10 < E < 100$	$(1.029\, (2 - \alpha) + 0.512\, (\alpha - 1))/1.453$
$100 < E < 1000$	$(0.512\, (3 - \alpha) + 0.175\, (\alpha - 2))/1.453$
$E > 1000$	$(0.175/1.453)\, 3^{3-\alpha}$

[a] $\alpha = \log_{10}(E)$ where E is the muon energy expressed in TeV.
[b] Derived from Quigg (1997).

Table 4.13 Annual neutrino effective dose for a linear muon collider using the narrow beam approximation.

Accelerator facility energy (TeV)	Muon energy (TeV)	Annual neutrino effective dose (mSv/yr)
0.1	0.05	5.7×10^{-5}
1	0.5	5.2×10^{-3}
10	5	0.45
100	50	30
500	250	440
1000	500	1400

As a practical example, consider a 1000-TeV muon linear accelerator assuming $E_o = 500$ TeV (i.e., two 500 TeV linear muon accelerators) and $N = 6.4 \times 10^{18}$ muon decays per year. Using these values in Equation 4.44 with a $g = 1$ GeV/m value leads to

$$H(2 \times 500\,\text{TeV}) = 1.4\,\text{Sv/year}, \tag{4.45}$$

which is a significant effective dose value that cannot be ignored. Health physicists at a twenty-first century linear muon collider must contend with a large neutrino effective dose within and outside the facility. Table 4.13 provides the expected annual neutrino effective dose for a variety of accelerator energies using the same N and g values noted above and the narrow beam approximation.

The values of Table 4.13 suggest that the annual effective dose limit for a US radiation worker (50 mSv/year) and the annual effective dose limit to the public (1 mSv/year) can be exceeded at TeV energy muon accelerators. Selecting an accelerator location is an issue for TeV energy muon linear colliders because of the public radiation concerns arising from neutrino interactions. Locating an accelerator may be restricted to low population or geographically isolated areas to minimize the public neutrino effective dose.

4.4.2.6 Bounding Neutrino Effective Dose – Circular Muon Collider

The bounding effective dose for a circular muon collider could be obtained using the methodology of the previous section. However, a number of operational assumptions including the ring circumference and average magnetic induction would be required. Instead, an approach more familiar to health physicists is used. This approach also permits the determination of the neutrino effective dose as a function of distance.

Effective doses of the circular muon collider are determined by considering the energy distribution or differential fluence, which is $dN_i(E_i)/dE_i$ where N_i is the number of neutrinos of generation i per unit energy, E_i is the neutrino energy, and $i = 1$–3 for the three neutrino generations. The neutrino effective dose H is determined once the neutrino fluence to effective dose conversion factor $C(E_i)$ is known. The dose conversion factors used in this chapter provide an approach for treating the neutrinos and their antiparticles in the first two generations. In view of the limited data, third-generation neutrinos are not considered, but these neutrinos may become more important as the accelerator energy increases.

One of the initial goals of a muon accelerator is the development of a pure muon neutrino beam to investigate the magnitude of the neutrino mass because a massless neutrino is the key tenant of the standard model. The subsequent discussion is limited to muon neutrinos that result from muon decays. With this background, the muon neutrino effective dose in a circular muon collider is

$$H = \int_0^{E_o} \frac{dN(E)}{dE} C(E) \, dE, \tag{4.46}$$

where E_o is the energy of the primary muons before decay.

The differential fluence value in the laboratory system is averaged over all neutrino production angles and assumes that the accelerator's shielding is thick enough to attenuate the primary muon beam, and that it is thicker than the range of all secondary radiation. Accordingly, the neutrino radiation is in equilibrium with its secondary radiation.

Using the equilibrium condition and averaging over all production angles provides a differential fluence relationship for the neutrino radiation from a circular muon collider,

$$\frac{dN(E)}{dE} = \frac{2}{E_o}\left(1 - \frac{E}{E_o}\right)\Phi, \tag{4.47}$$

where $N(E)$ is the number of neutrinos per unit energy, E is the neutrino energy, E_o is the energy of the primary muons before decay, and Φ is the integral neutrino fluence (total number of neutrinos per unit area) following the muon decays.

For secondary particle equilibrium, the normally assumed neutrino fluence to effective dose conversion factor is used:

$$C(E) = KE^2, \tag{4.48}$$

Equation 4.48 was derived for the neutrino energy range of 0.5 GeV–10 TeV. In Equation 4.48, $K = 10^{-15}\,\mu\text{Sv}\,\text{cm}^2/\text{GeV}^2$. In view of the trend in the neutrino data, Eq. 4.48 is used for energies beyond those usually considered. Without relevant data, this is a reasonable first approximation because increasing energy and increasing number of secondary shower particles (hadrons) is the main reason for the rising fluence to effective dose conversion factor with increasing neutrino energy for the equilibrium (shielded neutrino) case or process D described earlier. It is also reasonable because the neutrino attenuation length (λ) decreases with the increase in the energy of the primary neutrinos. Although TeV energy units are used in the final result, GeV units are used in the derivation of the neutrino

4.4 Planned Accelerator Facilities

effective dose to facilitate comparison with the literature. Before developing the neutrino effective dose relationship for a circular muon collider, the neutrino attenuation length is briefly examined.

The neutrino attenuation length is written in terms of the neutrino interaction cross section σ_v:

$$\lambda = \frac{A}{\rho N_A \sigma_v} = \frac{1}{N \sigma_v}, \tag{4.49}$$

where A and ρ are the atomic number and density of the shielding medium, N_A is Avogadro's number, N is the number density of atoms of the shielding medium per unit volume, and σ_v is on the order of

$$\sigma_v \rightarrow O[10^{-35} \text{ cm}^2 (E/1 \text{ TeV})], \tag{4.50}$$

where the neutrino energy (E) is expressed in TeV.

Using Equations 4.49 and 4.50 permits the neutrino attenuation length to be written as

$$\lambda = 0.5 \times 10^6 \text{ km} \left(\frac{1 \text{ TeV}}{E}\right) \left(\frac{3 \text{ g}}{\text{cm}^3 \rho}\right). \tag{4.51}$$

As the neutrino attenuation length is very long, the neutrino fluence is very weakly attenuated while traversing a shield. Therefore, shielding is not an effective ALARA tool for neutrinos.

The effective dose arising from an energy independent neutrino fluence spectrum is accomplished by performing the integration of Equation 4.46 using Equations 4.47 and 4.48:

$$H = \int_0^{E_o} \frac{2}{E_o} \left(1 - \frac{E}{E_o}\right) \Phi(KE^2) \, dE, \tag{4.52}$$

$$H = \frac{2K}{E_o} \int_0^{E_o} \left(E^2 - \frac{E^3}{E_o}\right) \Phi \, dE, \tag{4.53}$$

$$H = \frac{K}{6} E_o^2 \Phi, \tag{4.54}$$

where H is the annual neutrino effective dose in μSv and Φ is the total number of neutrinos per unit area that is assumed to be independent of energy.

The neutrino fluence Φ is the total number of neutrinos traversing a surface behind the shielding. The surface is governed by the divergence of the neutrino beam and the distance r from the neutrino source. The neutrino's half-divergence angle (θ) is

$$\theta = \frac{mc^2}{E} = \frac{1}{\gamma} \approx \frac{1}{10 E_o}, \tag{4.55}$$

where mc^2 is the rest mass of the muon in MeV, E is the muon energy in GeV, and E_o is the energy of the primary muon beam in GeV.

The neutrino fluence Φ at a given distance r from the muon decay point is just the number of neutrinos N per unit area:

$$\Phi = \frac{N}{\pi(\theta r)^2}. \tag{4.56}$$

Combining Equations 4.54 – 4.56 and using the numerical value for K in Equation 4.48 yields a compact form for the annual neutrino effective dose from a circular muon collider:

$$H = \frac{10^{-15} E_o^2}{6} \frac{N}{\pi(\theta r)^2} = \frac{10^{-15} E_o^2}{6} \frac{N(10E_o)^2}{\pi} \frac{1}{r^2} = \frac{10^{-13} E_o^4 N}{6\pi r^2} \left(\frac{\mu \text{Sv cm}^2}{\text{GeV}^4}\right). \tag{4.57}$$

The circular muon collider neutrino effective dose of Equation 4.57 has a very strong dependence on the energy of the primary muon beam.

Equation 4.57 provides the neutrino effective dose, assuming all muons decay at the same point. Recognizing that the muons decay at all storage ring locations with equal probability provides a more physical description of the effective dose. For facilities such as the European Laboratory for Particle Physics (CERN), the neutrino effective dose is calculated as an integral over the length of the return arm (l) of the storage ring pointing toward the surface from d to $d+l$, where d is the thickness of material traversed by the neutrino beam between the end of the return arm and the surface of the Earth along the direction of the return arm. The quantity d may also be described as the approximate minimum thickness of Earth needed to absorb the circulating muons if beam misdirection or total beam loss occurs (i.e., the beam exits the facility). Recognizing that the muons decay at any location along the return arm leads to the neutrino effective dose:

$$H = \frac{10^{-13} E_o^4}{6\pi} \int_d^{d+l} \frac{N \, dr}{l \, r^2} = \frac{10^{-13} E_o^4 N}{6\pi l} \left(\frac{1}{d} - \frac{1}{d+l}\right)\left(\frac{\mu \text{Sv cm}^2}{\text{GeV}^4}\right). \tag{4.58}$$

The parameters for a possible muon facility at CERN are used to illustrate an application of Equation 4.58. For a 50-GeV (0.05-TeV) muon energy in the storage ring, $N = 10^{21}$ muons per year decaying in the ring, a return arm length pointing toward the surface ($l = 6.0 \times 10^4$ cm), and a 100-m thickness of material (d) traversed by the neutrino beam between the end of the return arm and the surface, a surface neutrino effective dose of 47 mSv/year is predicted for the CERN muon collider. As the planned CERN design has 3 return arms, the effective dose rate at the end of one of the arms is about 16 mSv/year (47 mSv/3). Increasing energy leads to higher muon effective dose rates, and muon-shielding requirements force the collider deeper underground (see Table 4.14).

Table 4.14 Geometrical parameters for representative cases of circular muon colliders.[a]

Muon energy (TeV)	Collider's depth below Earth's surface (m)	Horizontal distance for the beam exiting the Earth (km)	Half-angle subtended by a horizontal beam with respect to Earth's center (mrad)	Neutrino beam half-divergence angle (μrad)
1	100	36	5.6	106
2	100	36	5.6	53
5	200	51	8	21
10	500	80.5	12.5	11

[a] Derived from Johnson, Rolandi and Silari (1998).

These results suggest that the circular muon collider be installed underground to shield the muon beam in the event that the beam becomes misdirected. The required shielding is determined by the muon energy loss:

$$\frac{dE}{dx} = 0.6 \frac{\text{TeV}}{\text{km}} \left(\frac{\rho}{3 \text{ g/cm}^3} \right). \tag{4.59}$$

When compared with muons, neutrinos have a much smaller interaction cross section. The Earth shielding that completely attenuates the muons has a negligible effect on the neutrinos. Accordingly, the neutrinos produce a nontrivial annual effective dose at the Earth's surface where the beam emerges. To evaluate the magnitude of this neutrino effective dose, assume that the Earth is a sphere, and a horizontal, circular muon collider is situated at a depth d below the Earth's surface. The exit point of the neutrino beam from the Earth is at a horizontal distance L given by

$$L = \sqrt{2dR - d^2} \approx \sqrt{2dR} \approx 36 \text{ km} \sqrt{\frac{d}{100 \text{ m}}}, \tag{4.60}$$

where $R = 6400$ km is the Earth's radius. Table 4.14 provides representative values of d and L.

In addition to d and L, a number of other relevant parameters associated with the circular collider of Equation 4.60 are summarized in Table 4.14. In Table 4.14, φ is the half-angle subtended by the horizontal accelerator beam with respect to the Earth's center before it exits the surface:

$$\sin \varphi = \frac{L}{R}. \tag{4.61}$$

The functional form of Equation 4.57 suggests that the calculation of neutrino effective doses from a circular muon collider is dependent on the assumed physical configuration (r) and beam characteristics (E_o and N). An estimate of the neutrino effective dose for a circular muon collider is made using Equation 4.57. For comparison with Equation 4.44, Equation 4.47 is rewritten in terms of TeV and mSv units:

$$H = \frac{10^{-4} E_o^4 N}{6\pi r^2} \left(\frac{\text{mSv cm}^2}{\text{TeV}^4} \right), \qquad (4.62)$$

where N is the number of muon decays per year, E_o is the muon energy in TeV, r is the distance from the point of muon decay, and H is the annual neutrino effective dose in mSv. For consistency with the linear muon collider assumptions, 6.4×10^{18} muon decays per year have been assumed.

In deriving the circular muon collider effective dose relationship, a number of assumptions were made. These are explicitly listed to ensure that the reader clearly understands the basis for Equation 4.62:

- The neutrino beam is limited to muon neutrinos only.
- Given the TeV muon energies and the Earth shielding present, charged particle equilibrium exists.
- In deriving the muon neutrino effective dose to fluence conversion factor, the effects of the third lepton generation are not considered.
- Given the TeV muon energies, Process D dominates the neutrino effective dose.
- The muon neutrino effective dose to fluence conversion factor is assumed to be valid for TeV–PeV-scale energies.
- The irradiated individual is assumed to be within the footprint of the neutrino beam.
- The irradiated individual is assumed to be within the footprint of the hadronic particle shower that results from the neutrino interactions.
- The individual is irradiated by only one of the muon beam's decay neutrinos whose energy is one-half the total circular muon collider energy.
- The neutrino and hadronic radiation uniformly irradiate the individual.
- The muon beam is well collimated.
- The neutrino effective dose calculation assumes a 100% occupancy factor.
- The neutrino effective dose is an annual average based on the number of muon decays in a year.

Table 4.15 summarizes the results of neutrino effective dose calculations as a function of distance from the muon decay location (r) for a circular muon collider. As the facility energy is the sum of the muon and antimuon beam energies, a 100-TeV accelerator consists of a 50-TeV muon beam and a 50-TeV antimuon beam. Table 4.15 is truncated at 100 TeV because accelerator construction costs and the physical size of the facility limit the circular muon collider energy. In addition, the long, thin conical radiation plumes present a radiation challenge well beyond the facility boundary. For example, a 25-TeV circular muon collider produces a neutrino effective dose of 37 mSv/year at a distance of 1500 km from the facility. Although the neutrino

Table 4.15 Annual neutrino effective dose for a circular muon collider.

Accelerator energy (TeV)[a]	H (mSv/yr)			
	$r = 5$ km	$r = 25$ km	$r = 100$ km	$r = 1500$ km
0.1	8.5×10^{-4}	3.4×10^{-5}	2.1×10^{-6}	9.4×10^{-9}
2	140	5.4	0.34	1.5×10^{-3}
25	3.3×10^{6}	1.3×10^{5}	8.3×10^{3}	37
100	8.5×10^{8}	3.4×10^{7}	2.1×10^{6}	9.4×10^{3}

[a]The muon beam energy is half the accelerator energy.

Table 4.16 Circular muon collider physics and cost parameters.

Accelerator energy (TeV)	0.1	3	10	100
Circumference (km)	0.35	6	15	100
Average magnetic field (T)	3.0	5.2	7.0	10.5
Cost	Feasible	Challenging	Challenging	Problematic

effective dose plume has a diameter of 12 m only at 1500 km, it presents a radiation challenge for muon collider health physicists and management.

Physics and cost parameters associated with 0.1-, 3-, 10-, and 100-TeV circular muon colliders are summarized in Table 4.16. The collider cost presents a funding challenge as the TeV muon energies are reached. In addition to funding issues, the control of radiation from the muon beams and neutrino plumes must be addressed.

As the collider energy increases, muon-shielding requirements dictate a subsurface facility. The impact of locating the muon collider deeper underground with increasing accelerator energy is also investigated. Using Equation 4.62 and the data summarized in Table 4.14, permit the calculation of the neutrino effective dose upon its exit from the Earth's surface. If the same beam properties are assumed as for the linear muon collider (e.g., $N = 6.4 \times 10^{18}$ muon decays per year) and it is assumed that $r = L$ (Table 4.14), then the magnitude and size of the resultant radiation plumes derived from Equation 4.62 are summarized in Table 4.17.

Although the effective dose results at the Earth's surface are significant, they occur over a relatively small area. The results also assume a 100% occupancy factor for this small area. The magnitude of the neutrino effective dose merits attention and warrants emphasis on the ALARA principle.

4.4.2.7 ALARA Impacts of Muon Colliders

As noted previously, neutrinos are electrically uncharged and only interact through the weak interaction. Their small, but nonzero, interaction cross section creates a unique situation in terms of the behavior of the neutrino effective dose, particularly in terms of the shape and energy dependence of their radiation profile. These properties

Table 4.17 Neutrino plume characteristics for an underground circular muon collider.

Muon energy (TeV)[a]	d (m)[b]	L (horizontal distance at the Earth's surface) (km)[c]	Beam radius at the Earth's surface (m)[d]	H at the Earth's surface (mSv/yr)
1	100	36	3.6	2.6
2	100	36	1.8	42
5	200	51	1.0	820
10	500	80.5	0.8	5200

[a] The accelerator energy is twice the muon energy.
[b] Accelerator depth below the surface of the Earth.
[c] Horizontal exit point distance from the surface of the Earth.
[d] The half-divergence angle is determined from Equation 4.30.

lead to a modification of the ALARA concepts when they are applied to the muon colliders.

The basic ALARA principle suggests that the effective dose at a given location be reduced if the exposure time is reduced, the distance from the source is increased, or shielding is added between the source and the point of interest. In the twenty-first century, the ALARA principle must be modified at a TeV energy muon collider. The ALARA principle of time is still valid for muons and neutrinos. Decreasing the exposure time reduces the neutrino and muon effective doses.

The ALARA principle of distance is ineffective when neutrinos are involved. As neutrinos interact very weakly, relatively long distances are not effective in significantly reducing the neutrino effective dose. In fact, the neutrino beam remains a hazard for hundreds of kilometers. However, distance is still effective for reducing the muon effective dose.

Unlike other radiation types, shielding neutrinos is anti-ALARA. The magnitude of the particle showers produced by neutrino interactions is governed by the quantity of shielding material between the neutrino beam and the point of interest. However, shielding muons is an effective ALARA measure.

From the standpoint of TeV energy neutrino radiation, a linear muon collider has a number of advantages over circular muon colliders. First, the radiation is confined to two narrow beams that can be oriented to minimize the interaction of the neutrinos. A simple ALARA technique would be to orient the linear accelerators at an angle such that the neutrino beams would exit the accelerator above the ground. This would minimize the residual neutrino interactions with the Earth or man-made structures. Second, the spent muons can be removed from the beam following collisions or interactions before they decay into high-energy neutrinos.

4.4.2.8 Other Radiation Protection Issues

A number of radiation protection issues associated with TeV energy muon colliders will challenge twenty-first century health physicists. The issues related to large neutrino effective dose values and effective neutrino dosimetry have been previously noted. Before the construction of a muon collider, thorough studies need to be performed to define the accelerator's radiation footprint; define muon collider shielding requirements; assess induced activity within the facility and the environment (e.g., air, water, and soil), including the extent of groundwater activation; assess radiation streaming through facility penetrations (e.g., ventilation ducts and access points); assess various accident scenarios such as loss of power or beam misdirection; and assess the various pathways for liquid and airborne release of radioactive material. Facility waste generation and decommissioning are other areas that require evaluation.

In addition to the aforementioned radiation protection issues, the TeV energy neutrino beam creates unique radiation protection concerns at muon colliders. Above about 1.5 TeV, the neutrino-induced secondary radiation poses a significant hazard even at distances on the order of tens to hundreds of kilometers. The neutrino radiation hazard presents both a physical as well as political challenge.

These issues also complicate the process for locating a suitable site for a TeV energy muon collider. There are a number of ways to minimize the radiation hazard at a muon collider. Potential solutions to reduce the neutrino effective dose associated with twenty-first century muon colliders include the following:

(1) Using radiation boundaries or fenced-off areas to denote locations with elevated effective dose values.
(2) Building the collider where human exposure is minimized. This would include building the collider on an elevated ground or at an isolated location.
(3) Using linear muon colliders at the higher TeV energies such that their interaction region is above the Earth's surface.
(4) Minimizing the straight sections in the ring of circular muon colliders, burying the collider deep underground to increase the distance before the neutrino beam exits the ground, and orienting the collider ring to take advantage of natural topographical features.

Orders of magnitude reductions in the neutrino effective dose are required for the muon colliders noted herein (see Tables 4.13, 4.15 and 4.17) to meet current regulations for public exposures. Many of the possible solutions noted above will be inadequate for the TeV energy muon colliders. The most feasible options for locating and operating the highest TeV energy muon collider are to use either

(1) an isolated site where no one is exposed to the neutrino radiation before it exits the Earth's surface or
(2) A linear muon collider constructed such that the individual muon beams collide in air well above the Earth's surface.

For Option 1, the accelerator could either be constructed at an elevated location or at an isolated area. The area needs to be large, perhaps having a site boundary with a diameter greater than 100 km. This requirement restricts the available locations.

Option 2 would be technically feasible, but the resultant skyshine needs to be evaluated.

In addition to concerns for locating a suitable site for a muon collider facility, routine operational radiation protection issues need to be addressed. Some of these operational issues lead to significant, unanticipated radiation levels in controlled as well as uncontrolled areas and include the following:

(1) *Beam alignment errors* – Beam alignment errors are caused by a variety of factors including power failures, maintenance errors, and magnet faults. Both human errors and mechanical failures lead to beam alignment issues.
(2) *Design errors* – Design errors lead to inadequate shielding, beam misalignment, beam confinement failures, or beam stop inadequacies.
(3) *Unauthorized changes* – Changes in the beam energy or beam current that exceed the authorized operating envelope lead to elevated fluence rates, the creation of unanticipated radiation types, or the creation of particles with higher energy than anticipated.
(4) *Activation* – The activation of air, water, and soil and facility structures present facility and site radiation concerns. The magnitude of toxic and radioactive gases requires evaluation.
(5) *Control of miscellaneous radiation sources* – The control of secondary radiation sources, radio frequency equipment, high-voltage power supplies, and other experimental equipment merits special attention. These radiation sources are more difficult to control than primary or scattered accelerator radiation because health physicists may not be aware of their existence, the experimenters may not be aware of the hazard, or the miscellaneous radiation sources are at least partially masked by the accelerator's radiation output. These miscellaneous radiation sources produce X-rays as well as other types of radiation.

These operational radiation protection issues require close control because the energies involved have the potential to produce large and unanticipated effective dose values. In addition, the neutrino beam influences radiation levels well beyond the facility boundary. A neutrino radiation plume extending well beyond the facility boundary presents a unique challenge for twenty-first century health physicists at a TeV energy muon collider.

Neutrino radiation will be an issue for health physics and a design constraint for muon colliders, particularly at TeV energies. TeV energy muon colliders require careful site selection and the neutrino effective dose dictates that these machines are constructed in isolated or elevated areas. With the operation of TeV energy muon colliders, the neutrino effective dose can no longer be neglected. Neutrino detection, neutrino dosimetry, and the determination of the neutrino effective dose will no longer be academic exercises but will become operational health physics concerns. Keeping public and occupational effective neutrino doses below regulatory limits will require careful and consistent application of the ALARA principle.

4.4.3
Very Large Hadron Collider

As science pushes for increasing accelerator energies, initial design considerations have been proposed for a next-generation hadron collider that follows the LHC. This machine is designated as the Very Large Hadron Collider. The VLHC could be sited at an existing accelerator such as Fermilab, CERN, or DESY and would be developed in a phased manner.

In a Fermilab study, the existing Tevatron accelerator complex serves as the VLHC injector. The Stage-1 VLHC would reach a collision energy (center of mass) of 40 TeV and be housed in a 233-km circumference tunnel. The Stage-2 VLHC, constructed after the scientific potential of the first stage has been fully realized, would reach a collision energy of at least 175 TeV with the installation of high-field magnets. Other studies are proposing a higher energy than the cancelled Superconducting Super Collider (SSC) project and are exploring low-cost magnets and tunnels for a facility on the order of 100 TeV in the center of mass.

The health physics issues at the VLHC are expected to be similar to those currently encountered at existing facilities such as those at the Fermi National Laboratory (Tevatron), the Large Hadron Collider at CERN, and the Relativistic Heavy-Ion Collider at the Brookhaven National Laboratory. Experience gained during operation of the LHC may indicate new challenges. However, it is difficult to assess any new radiation types, as the operational period of the LHC has just begun and the physics predictions are uncertain.

The proposed energy of 100–175 TeV for the VLHC has the potential to open pathways for the creation of new radiation types that are unavailable to existing machines. These new and exotic particles could include magnetic monopoles, new generations of quarks and leptons, dark matter/dark energy, the Higgs boson, and supersymmetry particles (e.g., squarks, sneutrons, sleptons, neutralinos, charginos, and smuons). Supersymmetry is a theory that directly associates half-integer fermion matter fields and integer boson gauge fields to relate a conventional particle (e.g., a proton) to its supersymmetric partner (e.g., the sproton). Any new and exotic particles may have a unique interaction and radiation characteristics that could challenge twenty-first century health physicists.

As an example of the potential health physics challenges, consider the synchrotron radiation power produced at the VLHC compared to that produced in the LHC. The VLHC in its second stage would produce 5 W/m per beam. This power output is about 50 times greater than the corresponding LHC synchrotron radiation. Any increase in synchrotron radiation output must be shielded and managed in an ALARA manner.

4.5
Common Health Physics Issues in Twenty-First Century Accelerators

In spite of the uncertainties associated with twenty-first century accelerator operations and emerging radiation types, there are a number of common health physics issues that should be encountered in subsequent generations of accelerators. These issues are summarized in subsequent discussions and include the radiation hazards associated with activation reactions and cascade sequences.

4.5.1
Sources of Radiation

Substantial shielding is often required around the ion source, accelerating sections, user facilities, target area, and beam stops of an accelerator. A qualitative description of the radiation dose at these locations for electron, proton, and heavy-ion accelerators is summarized in Table 4.18. This table supports the conclusions that radiation challenges intensify as the accelerator energies increase. The increase may be more dramatic than anticipated if new physics and new radiation types emerge at higher energies, which will occur as the twenty-first century proceeds.

The sources of radiation contributing to the radiation exposures summarized in Table 4.18 are varied. Activation is a source of radiation in all accelerator types likely to be constructed. A summary of radionuclides produced in various materials as a result of activation is provided below.

The various accelerator types also have radiation components that include nuclear and electromagnetic cascades, bremsstrahlung, neutrons, muons, and electrons. At energies above about 10 MeV, neutrons usually present the dominant source of occupational radiation dose at proton accelerators. Proton accelerators also result in other types of particle production through direct interactions including (p, p), (p, γ), (p, d), (p, α), (p, π^+), and (p, π^-) reactions. For proton energies above about 10 GeV, muon production becomes important.

At high-energy proton accelerators, the extranuclear hadron cascade produces the dominant source term for induced activity. In this process, nuclei are produced in excited states that decay in a variety of modes including neutron emission or nuclear evaporation. The hadron cascade continues to produce radionuclides until their energies are reduced below the thresholds for subsequent nuclear reactions. Meson and muon production also occurs.

Table 4.18 Potential for radiation doses from particle accelerator component systems.

System or component	Particles accelerated					
	Electron energy (MeV)			Proton or heavy-ion energy (MeV)		
	10–100	100–1000	>1000	<10	10–100	>100
Ion source	Low	Moderate	Moderate	Low	Low	Moderate
Accelerator	Moderate	Moderate	High	Low	Moderate	High
Beam delivery	High	High	High	Moderate	Moderate	High
Target/user	High	High	High	Moderate	High	High
Beam stop	Moderate	High	High	Moderate	High	High

Derived from NCRP 144 (2003).

In a hadron cascade, a proton produces about four interactions for each GeV of energy. Therefore, the number of interactions rapidly increases as accelerator energies increase from the GeV → TeV → PeV energy regions. At 1 PeV (10^3 TeV or 10^6 GeV), the hadron cascade undergoes greater than 6 orders of magnitude more interactions than at 1 GeV. The TeV → PeV energy transition will be a challenge in terms of the number of interactions, the possible emergence of new radiation types, the magnitude of radiation levels, and the complex behavior of known radiation types. Examples of this complexity were illustrated in the previous discussion of TeV-scale muon colliders.

At electron accelerators, activation occurs through the production of secondary particles (e.g., neutrons and muons) produced through electron interactions. For example, electrons produce photons and these photons produce neutrons through (γ, n) reactions. Neutrons and other particles are also produced as the electromagnetic cascade advances in a medium. As the energy increases, electron accelerators produce a variety of radiation types of health physics concern including photons, neutrons, heavy ions, pions, kaons, and muons.

In the twenty-first century, accelerators will increase in energy to the TeV and PeV energy range. As the energy increases, the number of possible reaction channels also increases. This is illustrated by recalling the reactions involved in the DD fusion process. At the D + D threshold, the following reactions occur:

$$D+D \rightarrow {}^4He+\gamma, \tag{4.63}$$

$$D+D \rightarrow D+D, \tag{4.64}$$

$$D+D \rightarrow {}^3He+n, \tag{4.65}$$

$$D+D \rightarrow {}^3H+p. \tag{4.66}$$

As the deuteron energy increases, multiparticle breakup channels occurs:

$$D+D \rightarrow D+n+p, \tag{4.67}$$

$$D+D \rightarrow n+p+n+p. \tag{4.68}$$

Above the pion threshold, pion production occurs:

$$D+D \rightarrow {}^4He+\pi^o, \tag{4.69}$$

$$D+D \rightarrow {}^4He+\pi^+ +\pi^-. \tag{4.70}$$

More complex reactions involving hadrons, mesons, and leptons occur as the deuteron energy increases through the GeV, TeV, and PeV energy ranges.

In general, as the accelerated particle's energy and the number of reaction channels increase, there is also an associated increase in the type and diversity of activation products. For example, low-energy proton reactions for ^{40}Ca targets primarily produce elastic scattering. As the energy increases, the neutron, deuteron, and alpha channels open. At higher energies, meson production reactions and spallation reactions occur and lead to the fragmentation of the ^{40}Ca nucleus. Meson and lepton production also occurs.

4.5.2 Activation

The accelerator beam or the reaction products of the beam and target induce nuclear reactions that lead to activation in the target, beam stop, accelerator structures, and in air, water, and soil in the vicinity of the target area. Activation occurs by nuclear reactions that either produces radionuclides directly or through secondary interactions. Air, water, and soil activation products are illustrated in Table 4.19 and activation production in accelerator materials including plastic, oil, aluminum, steel, stainless steel, and copper are summarized in Table 4.20.

4.5.3 Radiation Shielding

Radiation shielding provides a means of mitigating the radiation hazard from lepton and hadron accelerators. A variety of factors are considered in accelerator shielding design. These factors include the accelerator type, beam particle, beam profile, beam energy, target material and configuration, facility configuration, future facility upgrades, beam stop material and configuration, ALARA considerations, available space, comparison with other facilities, construction techniques, environmental radiation levels, induced radioactivity, radiation-weighting factors, radiation exposure history at the institution, regulatory limits, available shielding materials, source terms, and trends in regulatory limits with time. Once these factors are known, the shield design proceeds using the analytical techniques of Appendix C or the numerical approaches of Appendix J.

4.5.4 Radiation Measurements

Radiation detectors frequently utilized in accelerator facilities include active, real-time techniques (e.g., ionization chambers, Geiger–Mueller (GM) counters, proportional counters, fission chambers, and counter telescopes) and passive techniques (e.g., thermoluminescent dosimeters (TLD), nuclear emulsions, track-etch techniques, bubble dosimeters, and activation measurements). Special problems are encountered when operating active detectors in the pulsed fields of an accelerator.

Table 4.19 Primary activation products of air, water, and soil.

Air	^{11}C, ^{13}N, and ^{15}O
Water	^{3}H, ^{7}Be, ^{11}C, ^{13}N, and ^{15}O
Soil	^{3}H and ^{22}Na

Table 4.20 A summary of activation products for materials commonly utilized in accelerator environments.

Irradiated material	Radionuclides produced	Half-life
Plastic and oils	^{7}Be	53.3 d
	^{11}C	20.3 min
Aluminum	All of those above plus:	
	^{18}F	110 min
	^{22}Na	2.60 yr
	^{24}Na	15.0 h
Steel	All of those above plus:	
	^{42}K	12.4 h
	^{43}K	22.3 h
	^{44}Sc	3.93 h
	^{44m}Sc	2.44 d
	^{46}Sc	83.8 d
	^{47}Sc	3.35 d
	^{48}Sc	1.82 d
	^{48}V	16.0 d
	^{51}Cr	27.7 d
	^{52}Mn	5.59 d
	^{52m}Mn	21.1 min
	^{54}Mn	312 d
	^{56}Co	77.3 d
	^{57}Co	272 d
	^{58}Co	70.9 d
	^{55}Fe	2.73 yr
	^{59}Fe	44.5 d
Stainless steel	All of those above plus:	
	^{60}Co	5.27 yr
	^{57}Ni	35.6 h
	^{60}Cu	23.7 min
Copper	All of those above plus:	
	^{65}Ni	2.52 h
	^{61}Cu	3.35 h
	^{62}Cu	9.74 min
	^{64}Cu	12.7 h
	^{63}Zn	38.5 min
	^{65}Zn	244 d

Derived from NCRP 144 (2003).

As noted previously, the radiation environment at a particle accelerator results from a wide range of physical phenomena and consists of several radiation types distributed over a broad range of energies. The radiation fields are also time dependent and this dependence is affected by the accelerator duty cycle, details of the accelerator system, and beam extraction system. Accelerator radiation fields are complex, but their description can be simplified in terms of three basic rules:

(1) If muons are produced, neutrons are also produced. However, the location of muon production and neutron production do not necessarily coincide.
(2) Very high-energy neutrons are accompanied by high-energy (fast), intermediate-energy, and thermal neutrons.
(3) Neutrons are accompanied by photons.

The measurement of various radiation types including heavy ions, photons, and neutrons was summarized in Table 3.15. However, there are unique aspects that should be considered when measuring the various radiation types in an accelerator environment.

The response of active detectors around short-pulse accelerators is valid only if the count rate is a small fraction of the machine pulse rate. This restriction applies to scintillation detectors, gas-filled detectors in pulse mode, and semiconductor detectors. For this situation, the true count rate (n) is obtained from the observed count rate (m) using the relationship

$$n = (f)\ln\left(\frac{f}{f-m}\right), \quad (4.71)$$

where f is the accelerator pulse repetition frequency.

As noted previously, accelerator radiation is detected by a number of instrument types. All of these instruments detect a variety of radiation types, and these measurements must be interpreted with care. For example, issues arise with the use of ionization chambers in accelerator fields including the effects of radio frequency interference, pulsed radiation fields, the small cross-sectional area of the accelerator beam, and volume recombination in the detector.

Neutron fields are complex because neutrons are produced in a variety of reactions and span a wide energy range. Because of this complexity, additional information is provided in Tables 4.21 – 4.23 to supplement the measurement techniques of Table 3.15

Tables 4.21 – 4.23 provide a summary of reactions used in neutron detectors in accelerators and the characteristics of those detectors. Table 4.21 reviews selected activation reactions used for the determination of thermal neutron fluence rates. Neutron activation reactions capable of detecting a variety of neutron energies and their characteristics are provided in Table 4.22. Table 4.23 summarizes a selected set of active thermal neutron detectors.

Table 4.21 provides methods for thermal neutron detection using (n, γ) reactions. These are passive detectors that utilize activation reactions.

Active thermal neutron detectors include methods based on the ^{10}B(n, 2α)^{3}H, ^{3}He(n, p)^{3}H, and ^{6}Li(n, α)^{3}H reactions. These (n, α) and (n, p) methods are summarized in Table 4.23.

Table 4.22 summarizes methods that utilize threshold reactions to detect neutrons. By utilizing a number of these reactions, the accelerator's neutron spectrum

Table 4.21 Activation reactions used in determining thermal neutron fluence rates of particle accelerators.[a]

Reaction	Decay products	Half-life	Detector	Sensitivity[b]
^{115}In(n, γ)^{116}In	$β^-$ γ 0.42 MeV @29% 1.1 MeV @58% 1.3 MeV @84%	54.2 min	β particle detector NaI γ spectrometer	Four foils, 6.6 cm × 15.2 cm, total mass = 46 g, sensitivity 300 cpm
^{197}Au(n, γ)^{198}Au	$β^-$ γ 0.42 MeV @95%	2.7 d	β particle detector NaI γ spectrometer	2.54 cm diameter foil, mass 0.5 g: sensitivity 1.8 cpm 5.08 cm diameter foil, mass 1 g: sensitivity 13.4 cpm
^{23}Na(n, γ)^{24}Na	$β^-$ γ 1.39 MeV @100% 2.75 MeV @100%	15 h	γ spectrometer	Na$_2$CO$_3$ cylinder 4.5 cm × 2 cm, mass 12 g Na: sensitivity 3 cpm

[a] Derived from NCRP 144 (2003).
[b] Sensitivity at saturation and zero decay time for unit neutron fluence rate ≈ 1 n/cm² s.

can be determined. Knowledge of the neutron spectrum outside the various accelerator components is an important consideration in shield design as the cross sections that govern the attenuation of the various radiation types are energy dependent.

Neutron detection is important because neutrons induce many activation reactions. Table 2.2 provided a summary of thermal and fast neutron reactions that occur in a power reactor. Many of these reactions also occur in accelerators because their steel structural components have Mn, Fe, and Co constituents.

In addition to the photon and neutron source terms, muons also contribute to the effective dose. Muons are leptons and with the exception of their mass behave in a manner that is similar to electrons when interacting with matter. Ionization chambers, counter telescopes, nuclear emulsions, silicon detectors, thermoluminescent detectors, and scintillation detectors can be used to detect muons.

4.5.5
Environment

Experience to date at high-energy accelerators indicates that the dominant radiological impact on the environment is in the form of prompt radiation. Muons, neutrons, and photons dominate the prompt radiation field affecting the environment in the vicinity of the accelerator. Of these three, neutrons are usually the most

Table 4.22 Selected characteristics of neutron activation detectors.[a]

Detector	Reaction[b]	Energy range (MeV)	Half-life	Typical detector size	Cross section Peak (mb)	Cross section High-energy (mb)	Detected particle
Sulfur	^{32}S(n, p)^{32}P	>3	14.3 d	2.54 cm diameter, 4 g disk	500	10	β^-
Aluminum	^{27}Al(n, α)^{24}Na	>6	15 h	16.9–6600 g	11	9	γ
Aluminum	^{27}Al(n, 2p 4n)^{22}Na	>25	2.6 yr	16.9 g	30	10	γ
Plastic scintillator	^{12}C(n, 2n)^{11}C	>20	20.4 min	13–2700 g	90	30	β^+, γ
Plastic scintillator	^{12}C(n, X)^{7}Be	>30	53 d	16.9 g, 2.54 cm high	18	10	γ
Mercury	^{198}Hg(n, X)^{149}Tb	>600	4.1 h	≤500 g	2	1	α, γ
Gold foils	^{197}Au(n, X)^{149}Tb	>600	4.1 h	2.54 cm diameter, 0.5 g	1.6	0.7	α, γ
Copper foils	Cu(p, X)^{24}Na	>600	14.7 h	5.6 cm diameter, 9 g	4	3.6	γ

[a] Derived from NCRP 144 (2003).
[b] X indicates a spallation reaction.

Table 4.23 Examples of active thermal neutron detectors.

Type	Advantages	Disadvantages
BF$_3$	Excellent photon rejection Low cost	Energy resolution suffers beyond 67–80 kPa filling pressure
^{3}He	Filling pressure up to 1 MPa More sensitive and more stable than BF$_3$ Good photon rejection	Expensive
^{6}LiI(Eu)	High sensitivity (solid) Compact size (typically 4 mm× 4 mm× 1 mm) helps to reduce response anisotropy	Photon rejection is weaker than gas counters Light-guide and photomultiplier tube partially reduce the advantage of compact size

Derived from NCRP 144 (2003).

important, but all of these radiation types contribute to both the direct as well as scattered radiation components of environmental radiation. In addition to direct radiation, skyshine contributes to the radiation environment outside an accelerator.

Skyshine is the radiation emitted from an accelerator that is reflected by the atmosphere back to Earth. By penetrating the thinner shielding of roof structures, scattered neutron and photon radiation contributes to the off-site radiation environment.

Skyshine and direct radiation are the dominant contributors to environmental radiation from an accelerator. Compared to direct radiation and skyshine, and several times smaller in magnitude, are accelerator produced activation products of air, groundwater, soil, and activated accelerator component pathways. However, activated air, water, and soil merit periodic monitoring. Nonradioactive gases such as ozone and oxides of nitrogen are also produced.

4.5.6
Operational Radiation Safety

Many of the elements of an operational radiation safety program for accelerators are common to other radiological facilities. Of particular interest are the high dose rates that can be encountered. These dose rates dictate that access control and beam interlocks are significant components of operational radiation safety. Ionizing radiation, nonionizing radiation, and toxic gases are hazards that must be the integrated aspects of operational accelerator safety.

4.5.7
Safety Systems

During accelerator operations, radiation is produced when the beam interacts with the target and other accelerator materials. Secondary radiation (e.g., neutrons, muons, and photons) generates additional radiation through atomic or nuclear interactions, including hadronic and electromagnetic cascades. Radiation safety systems (RSS) are used to protect personnel from accelerator radiation. The primary components of RSS include shielding, personnel protection systems, and beam containment systems (BCS).

PPS is an access control system that prevents personnel from entering areas with dangerous radiation levels. BCS prevent dangerous levels of radiation outside the shielded enclosure. Other safety systems such as burn-through monitors and interlocked radiation detectors can also be integrated into the RSS to terminate beam operations. These systems are facility specific and depend on the accelerator type, energy, and experimental arrangement.

4.6
Other Applications

Accelerators also have applications in other health physics areas including the transmutation of nuclear waste and in cancer therapy. These applications are reviewed in a subsequent volume of *Health Physics in the 21st Century*.

Problems

04-01 The nuclear theory group at the Large Hadron Collider is planning to perform an experiment involving a ^{208}Pb beam striking a ^{238}U target. (a) What is the Coulomb barrier for this interaction? (b) If the beam energy is 1200 GeV, what radiation types are produced and which of them dominate the health physics considerations? (c) If the beam current is 1 mA and the accelerated ions have a charge of +20 e, how many lead ions strike the target per second? (d) For the conditions of part (c), 10 neutrons are produced for every lead-ion striking the target. Assuming that the neutrons are produced in an isotropic manner, what is the neutron fluence at 5 m from the target?

04-02 In support of an International Thermonuclear Experimental Reactor shielding design experiment, an accelerator produces 14 MeV neutrons using the ^{3}H(d, n)^{4}He reaction. For this experiment, the deuteron fluence rate striking the target is 6.25×10^{13} d/cm^2 s, the tritium target activity is 3.7×10^5 MBq in 1 cm^2 of active target area, the beam diameter is 1 cm^2, and the tritium half-life is 12.3 years.

(a) If the 14 MeV neutron dose factor is 5×10^{-10} Sv cm^2/n, what is the neutron effective dose rate at a point 1.2 m from the target? The total reaction cross section for the ^{3}H(d, n)^{4}He reaction is 5 b. For this problem assume that the neutrons are produced isotropically.
(b) List and justify the major elements of an accelerator radiation protection program for this facility.

(c) Lead and normal polyethylene are used to construct a temporary shield around the target. In what order should one place these materials? Why?

0403 You are the health physicist at the University of Eastern Wyoming Muon Factory, a high-energy accelerator facility. An open-air cylindrical ionization chamber detects muons in this facility. The chamber is bombarded by a uniform flux density of high-energy (minimum ionizing) muons incident normal to the long axis of the chamber. The radiation field is constant in time, and there is no pulse structure of significance. In this radiation field, the current collected from the anode of the ionization chamber has a value of 10^{-12} A.

Assume that the passage of the muons through the entire length of the chamber represents insignificant degradation of the muon energy and does not significantly alter their direction. The effective dose per unit fluence for the muons is 4×10^{-4} µSv cm²/µ.

The radius of the ion chamber is 5 cm and its length is 20 cm. For purposes of this problem, the chamber walls are taken to be approximately "tissue equivalent." The density of air at STP is 1.293 g/l. Measurements are performed at 20 °C and 1 atm pressure.

The mass stopping power of high-energy muons in air is 2.0 MeV cm²/g, and the mass stopping power of the ionization chamber walls is equal to that of the gas.

(a) Calculate the effective dose rate from the measured chamber current, assuming that the anode is 100% efficient in collecting this current.
(b) List five conditions that could affect the accuracy of the ionization chamber measurements.
(c) How are muons created in an accelerator?
(d) What is the charge of a muon?
(e) Compare the muon's mass, lifetime, and decay mode with that of other light elementary particles.
(f) List the factors that should be considered when determining where an ionization chamber should be located to measure the radiation field from a misdirected particle beam that might create a temporary muon radiation field.
(g) List the hazards (other than ionizing radiation) associated with high-energy accelerator facilities.

0404 During start-up of a high-luminosity 100-TeV colliding-beam proton accelerator, a magnet failure causes one of the proton beams to exit the detector area, strike a steel structural member, and produce a sustained burst of neutrons. The condition was undetected for several minutes and automatic beam shutdown interlocks failed to function.

The neutrons entered an area occupied by several technicians who receive an unanticipated acute exposure. The event produced a total of 10^{18} neutrons.

(a) The accelerator supervisor recognizes that the event has significant radiological consequences and organizes a search and rescue team. As the accelerator health

physicist, you are asked if the team can enter the irradiation area. What are your primary considerations in developing your recommendation?
(b) Describe a method that could be used to quickly screen persons who may have been irradiated during the beam misdirection event.
(c) Assume that the workers received 8 Gy (deep dose). Describe medical interventions that could positively affect the workers' health consequences if administered within the first month following the accident.

0405 An experimenter is performing high-energy proton bombardment of a target to assess the long-term buildup of activation products. A control failure leads to a dramatic increase in beam current and failure of the target confinement system. The target failure allows ^{15}O to be released at a constant rate into the target room air space. Assume instant and complete mixing of ^{15}O with room air. The target room dimensions are 6 m × 6 m × 3 m and the target area exhaust rate is 30 m^3/min. The ^{15}O release rate is 2.6×10^9 atoms/s, and the ^{15}O half-life is 122 s. Assume the accelerator is licensed in the United States under regulations based on ICRP-26.

(a) Will room ventilation or radioactive decay be the dominant removal mechanism?
(b) What is the room activity concentration of ^{15}O (in Bq/m^3) after 4 min of release?
(c) Assume that a second experimenter is in the target area air space during a subsequent event, and remained there for 6 min. The experimenter is concerned because she calculated an average ^{15}O air concentration of 12.9 MBq/m^3 that greatly exceeded the ^{15}O Derived Air Concentration (DAC) value of 4000 Bq/m^3 for submersion. Give two reasons why exceeding this DAC does not necessarily mean that a dose limit has been exceeded.

0406 An accident occurred at a 1000 GeV proton–antiproton collider and resulted in the irradiation of an experimenter's left hand. The dose reconstruction following the event indicates that 1×10^8 negative pions were captured by ^{16}O nuclei/cm^3 of the affected body tissue. The reaction products and energy emitted per capture are summarized below. The laboratory director will speak to the press in 15 min, and directs you to estimate the dose delivered to the hand. Given the time constraints, what is the absorbed dose delivered to the experimenter's hand? Assume that the tissue is reasonably approximated by water.

Reaction products and average energy per capture of a negative pion by a ^{16}O nucleus

Emitted particle	Average energy per capture (MeV)
Neutrons	100
Protons ($E < 2$ MeV)	150
^{15}O, ^{14}O, ^{13}O, and ^{12}O fragments	100
High-energy photons	200
Neutrinos	150

04-07 Verify Equation 4.45.

04-08 On March 17, 2099, the Super-Duper Hadron Collider (SDHC) initiated operations with the collision of two 1.5 PeV proton beams. The collision produces the first supersymmetry particle, the sproton. High-energy theorists are puzzled because sprotons have a much longer half-life than predicted. A total of 10^5 sprotons and 10^5 antiprotons are produced and stored in separate storage rings prior to their collision. The collisions occur uniformly along a 100-m linear collision zone and lead to the result

$$p_s + \bar{p}_s \rightarrow \gamma + \gamma.$$

(a) What is the effective dose rate at a perpendicular distance of 10 m from the end of the collision zone? The photons have an average energy of 0.1 TeV and all annihilation events occur within 1 µs. The dose factor for the annihilation photons is 1.35×10^{-7} (Sv m²/MBq h) E Y, where E is the photon energy in MeV and Y is the photon yield.

(b) If the SDHC's design effective dose goal is 0.2 mSv/h, how much shielding is required to achieve this effective dose rate? A carbon–thorium–concrete composite is the only viable material to shield sprotons. The HVL for the composite is 5 cm.

(c) If the sproton decays into heavy charged hadrons, how could these decay products be detected?

04-09 A heavy-ion collision at the Big Bang National Laboratory (BBNL) inadvertently produces superheavy element $^{472m}_{164}X$ with a half-life of 37 years. Scientists at the BBNL immediately realize the energy potential of the metastable state and begin a series of experiments to determine its properties. As a part of this determination, the crystal lattice structure of a 100-g sample of $^{472m}_{164}X$ is investigated using X-ray diffraction techniques. When the X-ray device is energized, all $^{472m}_{164}X$ nuclei simultaneously deexcite with each metastable nucleus emitting 10 gamma-rays having an average energy of 1 MeV.

(a) What is the unshielded γ-ray absorbed dose received by an individual located 10 km from the $^{472m}_{164}X$ source? Assume the dose conversion factor for the photons is 1 Gy/h = 5.5×10^7 γ/cm² s.

(b) To verify the $^{472m}_{164}X$ energy output, a calorimeter was placed 100 m from the source. If the calorimeter mass is 100 g of water equivalent, what is its temperature increase following the deexcitation of the metastable nuclei in the 100 g sample?

(c) If the testing is performed on a flat, featureless landscape, how far from the source must an observer be located to not exceed an absorbed dose of 10 mSv?

04-10 During a production run at the International Linear Collider, located in Frostbite Falls, Minnesota, a new particle, the R^+, having a mass of 388 GeV is produced. The R^+ decays into two additional particles, the r^+ and s^o. The r^+ is electrically charged and is rapidly attenuated by the accelerator's concrete shielding. The s^o is neutral and predominantly interacts with ^{40}Ar nuclei in an unusual manner. Its ^{40}Ar interaction products are shown in the table at two specific locations. Location A (B) is 5 m (25 m) from the accelerator's exterior shield wall.

	$s^o + {}^{40}$Ar reaction products		
Emitted particle	Average kinetic energy released per s^o capture (MeV)	Percentage of energy deposited at the specified location[a]	
		Location A	Location B
Neutrons	750	5	95
Protons	400	99	1
Heavier nuclear fragments	350	100	0
Gamma photons	75	10	90

[a]The deposition occurs in a 0.5-cm radius sphere composed of water.

(a) What is the average absorbed dose rate at location A if 10^7 s^o particles are captured in the 1.0-cm diameter sphere in 1 h?
(b) What is the ratio of average absorbed dose rates at locations A and B?
(c) What does the tabulated data at locations A and B reveal regarding the interaction properties of s^o particles?
(d) Assuming that the s^o particle interactions occur along the beam direction, how would you protect workers from these particles and their interactions? For specificity, calculate a stay time based on the 20-mSv effective dose recommendation of ICRP-60 for location A. Assume that the produced particles have the following properties: the neutrons have energies <10 keV, the protons have an energy of 7 MeV, the heavy fragments have an energy of 20 MeV/n, and the gamma-rays have an energy of 0.25 MeV. Finally, assume that the s^o particles are produced over a period of 1 month rather than the hour period in (a).

04-11 It is the year 2094 and the Rocket J. Moose Orbital Accelerator Laboratory (RJMOAL) has started initial beam operations. The RJMOAL focuses cosmic ray nuclei and collides them at energies up to 100 PeV. In the initial run, ^{232}Th nuclei are captured in dual storage rings and then the nuclei in each ring collide. When catalyzed with negative muons, three ^{232}Th nuclei fuse and form superheavy nucleus $^{696}_{270}$X. The initial accelerator run produces 1×10^{22} $^{696}_{270}$X nuclei that are stored in a third storage ring. These $^{696}_{270}$X nuclei are then collided with 10^{40} thermal neutrons and 10^{20}

fissions occur within a small retention chamber having a volume of 1000 cm³. The energy released per fission and the percent of this energy deposited in the chamber are

Species produced following fission	Total energy per fission event (MeV)	Average yield of species	Percent of energy deposited in the reaction chamber
γ	70	8	0.05
β	100	4	0.62
Neutrons	120	5	0.12
Protons	50	3	0.82
Heavier fragments	100	2	0.97

(a) If the reaction chamber has a mass of 5 kg, what is the average absorbed dose that it received?
(b) Which nuclear species contribute to the absorbed dose in the control room that is shielded by 4 cm of iron and 12 cm of a stable super heavy element (^{450}Bv). The control room is 500 m from the reaction chamber.
(c) As a result of a shielding review, polyethylene and concrete are added to the control room to reduce the neutron source term. What is the effective dose from gamma-rays and neutrons for control room personnel? The control room shielding has an attenuation factor of 0.72 for photons and 0.036 for neutrons. Assume that the dose conversion factor for the fission photons is 1 Sv/h = 5.5×10^7 γ/cm² s and 1 Sv/h = 8×10^5 n/cm² s for the fission neutrons.
(d) If the RJMOAL administrative limit for effective dose is 0.1 mSv per accelerator run, how much ^{450}Bv must be added to the control room. For the conditions of this problem, ^{450}Bv has a mass attenuation coefficient (μ) of 0.8 cm^{-1} for photons. The point source buildup factor has the form $B = 1.6 + 3.0\,\mu x$, where x is the shield thickness. For neutrons the attenuation coefficient is 2.6 cm^{-1}.

References

10CFR20 (2007) *Standards for Protection Against Radiation*, National Archives and Records Administration, US Government Printing Office, Washington, DC.

10CFR835 (2007) *Occupational Radiation Protection*, National Archives and Records Administration, US Government Printing Office, Washington, DC.

Adler, K. (1972) Coulomb Interactions with Heavy Ions, CONF-720669. Proceedings of the Heavy Ion Summer School-ORNL, Oak Ridge National Laboratory, Oak Ridge, TN.

Albright, J.R. and Semat, H. (1972) *Introduction to Atomic and Nuclear Physics*, 5th edn, Chapman and Hall, New York.

Autin, B., Blondel, A. and Ellis, J. (eds) (1999) *Prospective Study of Muon Storage Rings at CERN, CERN 99-02*, European Organization for Nuclear Research, Geneva, Switzerland.

Bauer, P., Darve, C. and Terechkine, I. (2002) *Synchrotron Radiation Issues in Future Hadron Colliders, Fermilab-Conf-01/434*, Fermi National Accelerator Laboratory, Batavia, IL.

Bevelacqua, J.J. (1995) *Contemporary Health Physics: Problems and Solutions*, John Wiley & Sons, Inc., New York, NY.

Bevelacqua, J.J. (1999) *Basic Health Physics: Problems and Solutions*, John Wiley & Sons, Inc., New York, NY.

Bevelacqua, J.J. (2004) Muon colliders and neutrino dose equivalents: ALARA challenges for the 21st century. *Radiation Protection Management*, **21** (4), 8.

Boag, J.W. (1987) Ionization dosimetry, in *The Dosimetry of Ionizing Radiation*, Vol. II (eds K. R., Kase, B.E., Bjarngard and F.H. Attix), Academic Press, New York.

Brient, J.C. and Yu, J. (2005) *FNAL-TM-2291, International Linear Collider Calorimeter/Muon Detector Test Beam Program (A Planning Document for Use of Meson Test Beam Facility at Fermilab)*, Fermi National Accelerator Laboratory, Batavia, IL.

Bryant, P.J. and Johnsen, K. (1993) *The Principles of Circular Accelerators and Storage Rings*, Cambridge University Press, Cambridge, UK.

C-AD Radiation Safety Committee (RHIC) (2001) Minutes of Radiation Safety Committee Meeting of July 3, 2001, *Light Ions from the Tandem to C-A Complex*, Brookhaven National Laboratory, Upton, NY.

Carrington, R.A., Jr, Huson, F.R. and Month, M. (eds) (1982) AIP Conference Proceedings No. 92. The State of Particle Accelerators and High Energy Physics, American Institute of Physics, New York.

CERN 84-02 (1984) *Radiation Problems in the Design of the Large Electron–Positron Collider (LEP)*, European Organization for Nuclear Research, Geneva, Switzerland.

Chao, A.W. and Tigner, M. (eds) (2002) *Handbook of Accelerator Physics and Engineering*, World Scientific Publishing Co., Singapore (2nd printing).

Chao, A.W. and Tigner, M. (eds) (2006) *Handbook of Accelerator Physics and Engineering*, World Scientific Publishing Co., Singapore (3rd printing).

Clapier, F. and Zaidins, C.S. (1983) Neutron dose equivalent rates due to heavy ion beams. *Nuclear Instruments & Methods*, **217**, 489.

Clements, E. (2007) Evolution of a collider. *Symmetry*, **04**, (01), 14.

Collar, J. (1996) Biological effects of stellar collapse neutrinos. *Physical Review Letters*, **76**, 999.

Cossairt, J.D. (2004) *Radiation Physics for Personnel and Environmental Protection*, Fermilab Report TM-1834, Revision 7, Fermi National Accelerator Laboratory, Batavia, IL.

Cossairt, J.D. and Marshall, E.T. (1997) Comment on "biological effects of stellar collapse neutrinos." *Physical Review Letters*, **78**, 1394.

Cossairt, J.D., Grossman, N.L. and Marshall, E.T. (1996) *Neutrino Radiation Hazards: A Paper Tiger, Fermilab-Conf-96/324* Fermi National Laboratory, Batavia, IL.

Cossairt, J.D., Grossman, N.L. and Marshall, E.T. (1997) Assessment of dose equivalent due to neutrinos. *Health Physics*, **73**, 894.

Cossairt, J.D., Grossman, N.L. and Marshall, E.T. (1997) *Assessment of Dose Equivalent due to Neutrinos, Fermilab-Conf-97/101* Fermi National Laboratory, Batavia, IL.

Cossairt, J.D. and Mokhov, N.V. (2002) *Fermilab-Pub-02/043-E-REV. Assessment of the Prompt Radiation Hazards of Trapped Antiprotons*, Fermi National Accelerator Laboratory, Batavia, IL.

Feder, A.T. (2005) Accelerator labs regroup as photon science surges. *Physics Today*, **58** (5), 26.

Feder, T. (2005) Europe to set particle physics strategy. *Physics Today*, **58** (9), 34.

Fermilab-TM-2149 (2001) *Design Study for a Staged Very Large Hadron Collider*, Fermi National Laboratory, Batavia, IL.

Gallmeier, F.X., Ferguson, P.D., Popova, I.I. and Iverson, E.B. (2004) The Spallation Neutron Source (SNS) Project: A Fertile Ground for Radiation Protection and Shielding Challenges, ICRS-10, May 9–13, 2004, Funchal, Madeira, Portugal.

Goldstein, H., Poole, C.P. and Safko, J.L. (2002) *Classical Mechanics*, 3rd edn, Prentice-Hall, Upper Saddle River, NJ.

Griffiths, D. (1987) *Introduction to Elementary Particles*, John Wiley & Sons, Inc., New York, NY.

Halzen, F. and Martin, A.D. (1984) *Quarks and Leptons: An Introductory Course in Modern Particle Physics*, John Wiley & Sons, Inc., New York, NY.

ICRU Report 28 (1978) Basic Aspects of High Energy Particle Interactions and Radiation Dosimetry, International Commission on Radiation Units and Measurements, Bethesda, MD.

ICRU Report 34 (1982) The Dosimetry of Pulsed Radiation, International Commission on Radiation Units and Measurements, Bethesda, MD.

Johnson, K. (1987) Future Possibilities for e^+e^- Colliders, in *6th Topical Workshop on Proton–Antiproton Collider Physics* (eds K., Eggert, H., Faissner and E. Radermacher), Aachen, Germany, June 30–July 4, 1986, World Scientific, Singapore.

Johnson, C., Rolandi, G. and Silari, M. (1998) *Radiological Hazard Due to Neutrinos from a Muon Collider*, Internal Report CERN/TIS-RP/IR/98-34, European Organization for Nuclear Research, Geneva, Switzerland.

King, B.J. (1997) Neutrino Physics at Muon Colliders. Fourth International Conference on the Physics Potential and Development of Muon Colliders, San Francisco, CA, December 10–12.

King, B.J. (1999) Parameter Sets for 10 TeV and 100 TeV Muon Colliders, and their Study at the HEMC'99 Workshop. Proceedings of the HEMC'99 Workshop – Studies on Colliders and Collider Physics at the Highest Energies: Muon Colliders at 10 TeV to 100 TeV, Montauk, NY.

King, B.J. (1999) Neutrino Radiation Challenges and Proposed Solutions for Many-TeV Muon Colliders. Proceedings of the HEMC'99 Workshop – Studies on Colliders and Collider Physics at the Highest Energies: Muon Colliders at 10 TeV to 100 TeV, Montauk, NY.

Knoll, G.F. (2000) *Radiation Detection and Measurement*, 3rd edn, John Wiley & Sons, Inc., New York.

Lefevre, P. and Pettersson, T. (eds) (1995) *CERN/AC/95-05(LHC), The Large Hadron Collider Conceptual Design, 1995*, European Organization for Nuclear Research, Geneva, Switzerland.

Levinger, J.S. (1951) The high-energy nuclear photo-effect. *Physical Reviews*, **84**, 43.

Mason, T.E. (2006) Pulsed neutron scattering for the 21st century. *Physics Today*, **59** (5), 44.

Mokhov, N.V. and Cossairt, J.D. (1999) Radiation Studies at Fermilab. Proceedings of the Fourth Workshop on Simulating Accelerator Radiation Environments (SARE4), Knoxville, TN, September 14–16, 1998 (ed. T. Gabriel).

Mokhov, N.V., Drozhdin, A.I., Rakhno, I.L., Gyr, M. and Weisse, E. (2001) Protecting LHC Components against Radiation Resulting from an Unsynchronised Beam Abort. LHC Project Report 478, Presented at the 2001 Particle Accelerator Conference (PAC 2001), June 18–22, 2001, Chicago.

Mokhov, N.V., Striganov, S. and van Ginneken, A. (2001) Muons and Neutrinos at High Energy Accelerators. Proceedings of Monte Carlo 2000 – Advanced Monte Carlo for Radiation Physics, Particle Transport Simulation and Applications, Lisbon, Portugal, October 23–26, 2000 (eds A. Kling,

F. Barão, M. Nakagawa, L. Távora and P. Vez), Springer-Verlag, Berlin, Germany.

NCRP Report No. 144 (2003) *Radiation Protection for Particle Accelerator Facilities*, National Council on Radiation Protection and Measurements, Bethesda, MD.

Nuhn, H.D. (2004) From storage rings to free electron lasers for hard X-rays. *Journal of Physics: Condensed Matter*, **16**, S3413.

Particle Data Group (2004) Review of particle physics. *Physics Letters*, **B592**, 1.

Particle Data Group (2006) Review of particle physics. *Journal of Physics G: Nuclear and Particle Physics*, **33**, 1.

RHIC ASE (2006) Relativistic Heavy Ion Collider – Accelerator Safety Envelope, Brookhaven National Laboratory, Upton, NY, January 16.

Quigg, C. (1997) Neutrino Interaction Cross Sections. Fermilab-Conf-97/158-T, Fermi National Laboratory, Batavia, IL.

Schopper, H., Fassò, A., Goebel, K., Höfert, M., Ranft, J. and Stevenson, G. (eds) (1990) *Landolt–Börnstein Numerical Data and Functional Relationships in Science and Technology, Group I: Nuclear and Particle Physics Volume II: Shielding against High Energy Radiation*, Springer-Verlag, Berlin.

Silari, M. and Vincke, H. (2002) *Neutrino Radiation Hazard at the Planned CERN Neutrino Factory*, Technical Note TIS-RP/TN/2002-01, European Organization for Nuclear Research, Geneva, Switzerland.

SLAC-I-720-0A05Z-002-R001 (2006) *Radiation Safety Systems Technical Basis Document*, Stanford Linear Accelerator Center, Stanford, CA.

SLAC-R-0521 (1998) *LCLS Design Study GroupLinac Coherent Light Source (LCLS) Design Study Report*, Stanford Linear Accelerator Center, Stanford, CA.

USDOE Order 420.2A (2001) *Safety of Accelerator Facilities*, United States Department of Energy, Washington, DC.

van Ginneken, A., Yurista, P. and Yamaguchi, C. (1987) *Shielding calculations for multi-TeV hadron colliders, Fermilab Report FN-447*, Fermi National Laboratory, Batavia, IL.

Wiedemann, H. (2003) *Synchrotron Radiation*, Springer-Verlag, Berlin.

Zimmerman, F. (1999) Final Focus Challenges for Muon Colliders at Highest Energies. Proceedings of HEMC'99 Workshop – Studies on Colliders and Collider Physics at the Highest Energies: Muon Colliders at 10 TeV to 100 TeV, Montauk, NY.

5
Light Sources

5.1
Overview

The term light source is used to describe the collection of photon-generating devices. This chapter examines photon light sources and focuses on synchrotron radiation sources, free-electron lasers (FELs), and selected photon-generating approaches including Compton backscatter, laser ion acceleration (LIA), wake-field acceleration, laser accelerators, X-ray induced isomeric transitions, and gamma-ray lasers (GRA-SERs). The application of these devices is expected to enter a number of new areas in the twenty-first century.

Light source applications continue to expand as the brightness and range of wavelengths increase. These applications occur in a wide array of fields including biology, chemistry, condensed matter physics, geology, material science, medicine, and solid-state physics. Light sources probe the structure of matter over a broad frequency range. Specific applications of light sources include the research and development of pharmaceuticals, computer chips, motor oils, new materials, and manufacturing techniques. Given the range of practical applications, light source use and application will expand and require additional health physics attention as the twenty-first century progresses.

Both synchrotron light sources and free-electron lasers are based on the concept that an accelerated charge radiates photons. The output from these photon sources is quite intense and presents an external radiation hazard that must be carefully managed.

This chapter examines the nature of radiation from synchrotron light sources, free-electron lasers, and other photon-generating approaches and addresses their health physics implications. Prior to reviewing these sources of radiation, the concepts of bremsstrahlung and synchrotron radiation, and the physics underlying these radiation processes are addressed.

5.2
Physical Basis

Prior to reviewing the various photon radiation sources, a brief summary of the primary means of generating this radiation is presented. These production methods utilize the concepts of bremsstrahlung and synchrotron radiation.

5.2.1
Bremsstrahlung

Bremsstrahlung or breaking radiation occurs when an ion's velocity changes. It is a general term applied to the radiation from an accelerated charged particle. Radiation is produced from either a positive or negative change in velocity. This section describes the bremsstrahlung process by focusing on the radiation from accelerated electrons.

A change in velocity or acceleration occurs when an electromagnetic (EM) field alters an electron's trajectory. During the change in trajectory, electron energy is lost, and its velocity decreases. Energy is conserved during this process through the emission of photon radiation as the electron decelerates. The spectrum of the bremsstrahlung photons is a continuous function of energy.

The total instantaneous power (P) radiated by the accelerated electron is

$$P = \frac{2e^2 a^2}{3c^3}, \tag{5.1}$$

where e is the charge of the electron, a is the instantaneous acceleration of the electron, and c is the speed of light. Equation 5.1 provides no information regarding the distribution of photon energy relative to the direction of the electron's velocity. If θ is defined as the angle between the electron's initial velocity and the final electron direction resulting from the trajectory change, the variation of the radiated power as a function of the spherical solid angle Ω is

$$\frac{dP}{d\Omega} = \frac{e^2 a^2 \sin^2 \theta}{4\pi c^3 (1 - \beta \cos^5 \theta)}, \tag{5.2}$$

where

$$\beta = \frac{v}{c} \tag{5.3}$$

and v is the initial electron velocity.

At low energies, $\beta \ll 1$ and the radiated power distribution has a $\sin^2 \theta$ dependence that peaks near 90° relative to the direction of motion. However, at high energies ($v \to c$), the angular distribution of radiated power is tipped in the beam direction and increases in magnitude. The angle θ_{max} for which the radiated power intensity is a maximum is

$$\theta_{max} = \cos^{-1}[\{(1 + 15\beta^2)^{1/2} - 1\}/3\beta]. \tag{5.4}$$

For $\beta = 0.5$, corresponding to electrons of about 80 keV energy, $\theta_{max} = 38.2°$. Table 5.1 summarizes the variation of θ_{max} with β.

Table 5.1 Angle of peak bremsstrahlung intensity as a function of energy.

β	θ_{max} (°)
0.001	89.86
0.01	88.57
0.1	76.04
0.2	63.80
0.3	53.69
0.5	38.16
0.7	25.86
0.9	13.42
0.99	4.07
0.999	1.28

5.2.2
Synchrotron Radiation

A general discussion of synchrotron radiation was provided in Chapter 4. When discussing light sources, synchrotron radiation is the term normally applied to the radiation emitted from a relativistic electron $\gamma \gg 1$,

$$\gamma = \frac{1}{\sqrt{1-\beta^2}} \approx 1957 E[\text{GeV}], \qquad (5.5)$$

where γ is the Lorentz factor. If the electrons are moving at a speed close to that of light, two effects alter the nature of the radiation. First, a particle moving with a Lorentz factor γ toward an observer emits radiation into a cone of opening angle θ:

$$\theta \approx \frac{1}{\gamma}. \qquad (5.6)$$

Second, if a source moves at a velocity near c and emits photon pulses, the photon emitted at the end of the pulse almost overtakes the photon from the start of the pulse. This shortens the pulse from a single accelerated particle. The net result of these two effects is the production of very high-frequency synchrotron radiation with a continuous spectrum emitted into the narrow cone.

5.3
Overview of Photon Light Sources – Insertion Devices

Synchrotron light sources are found throughout the world, including facilities in Armenia, Australia, Brazil, Canada, China (PRC), China (ROC, Taiwan), Denmark, France, Germany, India, Italy, Japan, Jordan, Korea, Russia, Singapore, Spain, Sweden, Switzerland, Thailand, United Kingdom, Ukraine, and the United States. Table 5.2 lists synchrotron radiation sources that have electron energies of 3 GeV or more.

Table 5.2 Synchrotron radiation sources with energies of 3 GeV or greater.

Country	Location	Ring institute	Electron energy (GeV)
Armenia	Yerevan	Center for the Advancement of Natural Discoveries using Light Emission	3.2
Australia	Melbourne	Australian Synchrotron	3
China (PRC)	Shanghai	Institute for Nuclear Research	3.5
Germany	Bonn	Electron Accelerator – University of Bonn	1.5–3.5
Germany	Hamburg	DORIS III – Synchrotronstrahlungslabor – Deutsches Elektronen-Synchrotron	4.5
Germany	Hamburg	PETRA II – Synchrotronstrahlungslabor – Deutsches Elektronen-Synchrotron	7–14
Japan	Nishi Harima	SPring-8	8
Japan	Tsukuba	Accumulator Ring – High Energy Accelerator Research Organization	6.5
Russia	Novosibirsk	VEPP-4M Budker Institute for Nuclear Physics	5–7
Spain	Barcelona	Consortium for the Exploitation of the Synchrotron Light Laboratory	3
Sweden	Lund	MAX IV University of Lund	1.5–3
United Kingdom	Oxfordshire	Diamond Light Source Rutherford Accelerator Laboratory	3
United States	Argonne, IL	Advanced Photon Source – Argonne National Laboratory	7
United States	Ithaca, NY	Cornell High Energy Synchrotron Source – Cornell University	5.5
United States	Stanford, CA	Stanford Positron Electron Accelerating Ring – Stanford Linear Accelerator Center	3

Photon light sources have evolved considerably since the discovery of X-rays. These various light sources are based on the applications of the EM field acceleration of electrons and include bending magnets, wigglers, undulators, free-electron lasers, and planned X-ray free-electron lasers (XFELs) and gamma-ray free-electron lasers (GRFELs). Bending magnets, wigglers, and undulators are often associated with storage rings. Free-electron lasers and the planned X-ray free-electron lasers and gamma-ray free-electron lasers are based on linear accelerators rather than on storage rings. Each of these synchrotron radiation sources is discussed.

5.4
X-Ray Tubes

The initial photon light source was the X-ray source built by Röntgen in 1895. X-ray sources in the form of X-ray tubes were available soon after this discovery, and these sources function by decelerating electrons in a metal cathode.

In X-ray tubes, the electron kinetic energy produces three dominant effects. About 50% of the electron energy produces characteristic radiation, about 50% generates a continuous bremsstrahlung spectrum, and <1% heats the anode. Anode heating limits the brightness of an X-ray tube. To date, the maximum X-ray tube brightness has been achieved with a rotating anode, with a peak value of about 10^8 photons s^{-1} mm^{-2} mr^{-2} (0.1% bandwidth)$^{-1}$, where mr is the milliradian unit. This intensity is too low for many important experiments and applications, and this limitation led to the development of more potent photon light sources.

As an alternative to radiation generation from the physical impact of electrons on a metal anode, photon radiation is produced when electrons are accelerated. Acceleration also occurs when the electron trajectory is altered by a magnetic field. As noted in Equation 5.2, low-energy electrons produce a radiation angular distribution that does not have a dominant direction. However, as the velocity increases, the radiation becomes highly focused on the beam direction and the radiated power rapidly increases.

In 1945, the synchrotron radiation was first directly observed. Cyclic electron synchrotrons were the first sources used for practical applications and were developed in the 1950s. In the 1960s, storage ring operations began. Before reviewing details of the various synchrotron radiation sources, a historical overview of these sources is provided.

5.5
Overview of Synchrotron Radiation Sources and Their Evolution

Synchrotron radiation sources are characterized in terms of four generations. The generations represent the historical development of these light sources and their advancing complexity and brightness.

The first generation of synchrotron radiation sources includes storage rings designed for high-energy physics research. Many of the first-generation devices used the residual synchrotron radiation produced during high-energy experiments. Once these accelerators declined in usefulness for high-energy physics research, they became partly and eventually fully dedicated as synchrotron radiation sources. As fully dedicated synchrotron sources, these accelerators were modified to bring their performance to the second-generation level.

Second-generation devices include storage rings that were designed to be fully dedicated synchrotron radiation sources. The initial second-generation machines were designed in the late 1970s. They were primarily designed to exploit bending magnets, with a few straight sections for possible future implementation with wigglers and undulators.

The third generation includes storage rings that were constructed after 1990. These devices were optimized for the use of insertion devices and were specifically designed for undulators. The third-generation machines incorporate numerous straight sections for insertion devices and have a lower electron beam emittance than that of the first- and second-generation devices to maximize the brightness from the undulator sources.

The fourth generation is the next sequence of synchrotron sources that will be predominantly based on linear accelerators. They will be extremely bright and produce short pulse radiation. Advances in the creation, compression, transport, and monitoring of bright electron beams make it possible to base the next generation of synchrotron sources on linear accelerators rather than on storage rings. These sources will produce coherent radiation, that is, orders of magnitude greater in peak power and peak brightness than that of the existing third-generation sources. The main fourth-generation candidates are free-electron lasers and energy recovery LINACS (ERLs). These fourth-generation synchrotrons are examined in more detail in subsequent discussions.

5.6
X-Ray Radiation from Storage Rings

In contemporary light sources, a storage ring synchrotron radiation source has a configuration in which electrons are generated using a device such as an electron gun. Following their generation, the electrons are spatially bunched and preaccelerated in a linear accelerator and a booster synchrotron. The electron bunches are then transferred to a storage ring. As the electron's trajectory is altered by the magnetic fields of the storage ring magnets, synchrotron radiation is produced. A portion of the synchrotron radiation is transferred to beamlines for delivery to experimental areas. As the electrons produce synchrotron radiation, their power decreases. Radio frequency (RF) cavities are used to boost the electron's power using a variety of sources including high-power klystron amplifier tubes.

The radiation from storage rings is produced in bunches that are typically spaced by about 2 ns and have a full width at half-maximum (FWHM) of about 50 ps. Storage ring radiation spans a broad spectral range, is polarized, has a limited physical extent, is partially coherent, and is highly stable.

The intensity of storage ring radiation sources is characterized as flux or brightness. Brightness is the number of photons in a particular phase space volume and is a measure of the radiation concentration. The phase space volume is defined in a variety of units including combinations of length and angular coordinates. Phase space volume is not the conventional three-dimensional volume and does not have units of length cubed. Specific relationships defining brightness are provided in subsequent discussion.

5.6.1
Bending Magnets

The bending magnets in a storage ring severely alter the electron's trajectory and spread the radiation cone over a large angular range. Only a small portion of this angular range is useful for experimental purposes. Most of the radiation from a bending magnet is not utilized for experimental or applied purposes, but it must be evaluated in the shielding design.

5.6.2
Insertion Devices

Undulators and wigglers are insertion devices that produce a spatially periodic field variation. These field variations cause accelerated electrons (or positrons) to emit radiation of unique characteristics. The term insertion device is related to the fact that wigglers and undulators are used as devices that supply electrons or can operate independently of electron and positron storage rings. When compared to bending magnets, insertion devices have higher photon energies, higher flux, higher brightness, and different polarization characteristics. Insertion devices also form the basis for coherent output radiation and are a key element of the free-electron laser.

Insertion devices utilize pairs of magnets with an electron beam passing between the North (N) and South (S) poles of the pair or the dipole. Adjacent dipoles alternate the order of the poles such that the electron beam passes between the dipoles with the upper (lower) poles arranged in the order NSNS ... (SNSN ...). This periodic magnetic structure is utilized to produce quasi-monochromatic radiation over a broad energy range. For example, a 1-MeV electron beam produces microwave radiation, and a 1-GeV beam leads to X-rays. Undulators were utilized in the initial application of the periodic magnetic structure concept to produce quasi-monochromatic radiation.

5.6.3
Wigglers

Radiation utilization was significantly improved with the introduction of wigglers in the 1970s. A wiggler includes a linear array of very short, strong dipole magnets with alternating field directions. This array generates a magnetic force that causes the electron's trajectory to wiggle without producing a significant deflection of the beam. Wigglers are not part of the beam confinement/storage system. They are added to the storage ring to generate synchrotron radiation. In a wiggler, the radiation cone of each individual dipole is directed to the forward direction, but the radiation from the dipoles is superimposed incoherently. The radiation output is larger than a bending magnet output, and there is the utilization of most of the radiation of a wiggler.

5.6.4
Undulators

In the 1990s, undulators were developed to improve further the output intensity from wigglers, and their magnets form a periodic structure. An undulator reduces the wiggler's amplitude such that the radiation output from each dipole interferes coherently. Undulators produce extremely intense line spectrum superimposed on a continuous bremsstrahlung spectrum.

5.7
Brightness Trends

The brightness of X-ray sources has dramatically increased since Röntgen's discovery. Between 1900 and 1950, sources primarily in the form of X-ray tubes did not increase measurably in their brightness. X-ray tubes only realized a small increase in brightness with the advent of Cu-K_α rotating anode generators. In the 1970s, when electron storage rings came into existence, X-ray source brightness increased dramatically with a doubling time of about 10 months. This trend of increase has been sustained for almost four decades with no obvious limiting factor emerging. The rapid increase in brightness began with bending magnet radiation, continued with wigglers and undulators, and will be further sustained with FELs, which are discussed in more detail later in this chapter.

A summary of X-ray source output is provided in Table 5.3. Table 5.3 summarizes the brightness of a variety of X-ray devices including X-ray tubes, bending magnets, wigglers, undulators, FELs, X-ray FELs, and gamma-ray FELs. The results of Table 5.3 permit an overview assessment of the health physics considerations of the emerging FEL light sources.

5.8
Physics of Photon Light Sources

Prior to reviewing FELs, it is necessary to outline the basic physics that supports the operation of wigglers and undulators. This physics background facilitates a full understanding of the basis for the radiation characteristics of photon light sources.

5.8.1
Brightness of a Synchrotron Radiation Source

The usefulness of a synchrotron radiation source depends on the number of photons per second that can be directed to the desired location. Brightness is the relevant parameter that describes the emission property of a radiation source. The term

Table 5.3 X-ray source brightness.

X-ray production mode	Brightness photons $s^{-1} mm^{-2} mr^{-2}$ (0.1% bandwidth)$^{-1}$
X-ray tubes	10^7–10^8
Bending magnets	10^{10}–10^{14}
Wigglers	10^{13}–10^{16}
Undulators	10^{15}–10^{21}
FELs	10^{20}–10^{24}
X-ray FELs[a]	10^{31}–10^{35}
γ-ray FELs[a]	$>10^{35}$

Derived from Nuhn (2003, 2004).
[a] Author's extrapolation.

5.8 Physics of Photon Light Sources

radiance or illuminance may also be used. Brightness is defined as the radiated flux per unit area of the source and per unit solid angle of emission:

$$\text{Brightness} = \frac{d^4 F}{dx\, dz\, d\theta\, d\phi} = \left(\frac{\gamma}{\text{s mm}^2 \text{ mr}^2 \text{ 0.1\% bandwidth}}\right), \quad (5.7)$$

where F is the radiated photon flux in a narrow 0.1% bandwidth. The definition assumes that the flux does not vary over this bandwidth.

Integrating F over all wavelengths yields the following expression for the total power (P) output in kW of the synchrotron radiation spectrum:

$$P[\text{kW}] = \frac{[(88.5\,\text{kW m})/(\text{GeV}^4\,\text{A})]\,E^4[\text{GeV}]\,I_o[\text{A}]}{R[\text{m}]}, \quad (5.8)$$

where E is the electron energy in GeV, I_o the average beam current in A, and R the radius of the electron trajectory in m. As an example of the application of Equation 5.8, consider the Large Electron Positron (LEP) facility at Centre (Organisation) European pour la Recherche Nucleaire (CERN) (see Chapter 4). At LEP, 50 GeV electron beam, 6 mA beam current, and 3096 m bending radius lead to a total power output of 1072 kW.

The synchrotron radiation spectrum is often described with respect to a characteristic or critical wavelength (λ_c) and associated energy (ε_c):

$$\lambda_c[\text{Å}] = \frac{[(5.59\,\text{Å GeV}^3)/\text{m}]\,R[\text{m}]}{E^3[\text{GeV}]} = \frac{18.6\,\text{Å T GeV}^2}{B[\text{T}]\,E^2[\text{GeV}]}. \quad (5.9)$$

$$\varepsilon_c[\text{keV}] = \frac{12.39\,\text{Å keV}}{\lambda_c[\text{Å}]}. \quad (5.10)$$

Using the previously noted LEP parameters and the magnetic induction (B) of 0.054 T leads to a critical wavelength of 0.14 Å and a critical energy of 88.5 keV.

Once the critical wavelength is determined, Equation 5.7 is integrated over dx, dz, and $d\phi$ to obtain the spectral flux density ($dF/d\theta$):

$$\frac{dF}{d\theta} = 2.46 \times 10^{13} I_o[\text{A}]\,E[\text{GeV}] \left(\frac{\lambda}{\lambda_c}\right)^2 G\left(\frac{\lambda}{\lambda_c}\right) \frac{\gamma}{\text{s mr 0.1\% bandwidth Å GeV}}, \quad (5.11)$$

where $G(\lambda/\lambda_c)$ is a function that governs the shape of the spectrum.

The source brightness, as defined above, is a function whose value depends on the source density distribution and the observation angle. It is often more convenient to use an average brightness (AB). For dipole sources, AB is defined as

$$\text{AB} = \frac{(dF/d\theta)}{(2.36\sigma_x)(2.36\sigma_z)(2.36\sigma'_\gamma)}, \quad (5.12)$$

where $dF/d\theta$ is defined in Equation 5.11, $2.36\sigma_x$ is the FWHM of the horizontal electron beam size, $2.36\sigma_z$ is the FWHM of the vertical electron beam size, and $2.36\sigma'_\gamma$

is the FWHM of the photon emission angle in the vertical plane. The FWHM of the photon emission angle in the vertical plane is a combination of the electron beam vertical divergence and the photon emission angle (Equation 5.6).

$$\sigma'_\gamma = \sqrt{(\sigma'_z)^2 + 0.41 \frac{\lambda}{\lambda_c} \frac{1}{\gamma^2}}. \tag{5.13}$$

The brightness of an undulator is calculated in a similar manner, but the procedure must account for the physical differences between synchrotron and insertion devices. The flux in the central radiation cone of an undulator F_n at a given wavelength is averaged over the emission angle of that cone to give the average on-axis brightness (AOAB). In view of the small source size and the divergence in an undulator, diffraction effects must be considered. An undulator's AOAB is

$$\text{AOAB} = \frac{F_n}{(2.36)^4 \sigma_{\gamma x} \sigma'_{\gamma x} \sigma_{\gamma z} \sigma'_{\gamma z}}. \tag{5.14}$$

The average on-axis brightness has units of photons-s^{-1}-mm^{-2}mr^{-2}-(0.1% bandwidth)$^{-1}$. In Equation 5.14, $\sigma_{\gamma x(\gamma z)}$ are the photon source sizes in both the $x(z)$-directions, and $\sigma'_{\gamma x (\gamma z)}$ are the photon source divergence in both the $x(z)$-directions, including diffraction effects.

Associated with the on-axis brightness is the undulator flux. The flux in the central radiation cone of the undulator is

$$F_n = \frac{1.43 \times 10^{14} \gamma}{\text{s A 0.1\% bandwidth}} \frac{L[\text{m}]}{\lambda_o[\text{m}]} I_o[\text{A}] Q_n(K), \tag{5.15}$$

where L is the undulator length, λ_o is the undulator period, and $Q_n(K)$ has the form

$$Q_n(K) = \left(1 + \frac{K^2}{2}\right) \frac{f_n}{n}. \tag{5.16}$$

$$K = \frac{93.4}{\text{m T}} \lambda_o[\text{m}] B_o[\text{T}]. \tag{5.17}$$

In Equation 5.16, K is the deflection parameter, B_o is the peak magnetic induction, and f_n is a spatial harmonic factor related to K.

The flux provides the photon output per unit time per 0.1% bandwidth but does not characterize the details of the output spectrum. This output has a characteristic wavelength as well as its associated harmonics. For the typical condition that $\gamma \gg 1$, the wavelength (λ_n) of the radiation produced in the nth harmonic in an undulator is given by the relationship

$$\lambda_n = \frac{\lambda_o}{n}\left(\frac{1+K^2}{2\gamma^2}\right). \tag{5.18}$$

5.9
Motion of Accelerated Electrons

The motion of an electron in a bending magnet, insertion device, or FEL depends on a number of parameters including the physical configuration of the device, the magnetic field strength and profile, and the electron energy. Although the formalization is applicable to light sources using accelerated electrons, this discussion focuses on insertion devices. For specificity, assume that the electron moves in the z-direction, and the magnetic field has a sinusoidal variation in the y-direction:

$$B_y = B_o \sin(kz), \qquad (5.19)$$

where the wave number k is

$$k = \frac{2\pi}{\lambda_o}, \qquad (5.20)$$

and λ_o is the device period producing the electron acceleration (e.g., bending magnet, wiggler, undulator, or free-electron laser). This period is the distance between two North or two South poles on the same side of the beam. Considering the insertion device configuration noted previously, Newton's second law, and the magnetic force ($\vec{F}_m$) equation leads to the result

$$\vec{F}_m = \gamma m_e \vec{a} = e\vec{v} \times \vec{B}, \qquad (5.21)$$

where $\vec{v}$ is the electron's velocity, $\vec{a}$ is the electron's acceleration, $\vec{B}$ is the magnetic induction, m_e is the electron's rest mass, and e is the electron's charge.

For simplicity, the transverse motion in the y-direction is omitted from the discussion, but it still retains the essential elements needed for the health physics presentation. Equation 5.21 is written in terms of the x- and z-components of the electron acceleration:

$$\gamma m_e \vec{a} = \begin{vmatrix} \hat{i} & \hat{j} & \hat{k} \\ \dot{x} & 0 & \dot{z} \\ 0 & B_y & 0 \end{vmatrix}, \qquad (5.22)$$

or

$$\ddot{x} = -\frac{e}{\gamma m_e} \dot{z} B_y, \qquad (5.23)$$

$$\ddot{z} = \frac{e}{\gamma m_e} \dot{x} B_y, \qquad (5.24)$$

where the double dot over the coordinates x and z represents the acceleration or the second derivative of distance with respect to time in that coordinate, and the single dot represents the velocity or the first derivative of distance with respect to time.

Using Equation 5.19, Equation 5.23 is integrated to yield the velocity ($\dot{x}$):

$$\dot{x} = \frac{eB_o}{\gamma m_e} \frac{\cos(kz)}{k}. \qquad (5.25)$$

Using Equation 5.20, Equation 5.25 is written as

$$\beta_x = \frac{\dot{x}}{c} = \frac{K}{\gamma}\cos(kz), \quad (5.26)$$

where K is the dimensionless undulator or deflection parameter introduced in Equation 5.17.

$$K = \frac{eB_o\lambda_o}{2\pi m_e c} = \frac{93.4}{\text{m T}}\lambda_o[\text{m}]B_o[\text{T}]. \quad (5.27)$$

The gap height (distance between the individual dipole N–S faces through which the beam passes) does not enter the undulator description, and the undulator field strength is represented using the dimensionless parameter K. Traditionally, insertion devices with low K-values are called undulators and devices with high K-values are called wigglers.

The total electron velocity (β) is comprised of both x- and z-components:

$$\beta^2 = \beta_x^2 + \beta_z^2. \quad (5.28)$$

Using Equations 5.5 and 5.20, Equation 5.28 is solved for the z-component of the velocity:

$$\beta_z^2 = \beta^2 - \beta_x^2 = \beta^2 - \left(\frac{K\cos(kz)}{\gamma}\right)^2 = \beta^2 - (1-\beta^2)K^2\cos^2(kz), \quad (5.29)$$

which can be simplified using trigonometric identities:

$$\beta_z \cong \beta\left(1 - \frac{K^2}{4\gamma^2} - \frac{K^2}{4\gamma^2}\cos 2kz\right). \quad (5.30)$$

The average velocity in the z-direction is

$$\bar{\beta}_z = \bar{\beta} \cong \beta\left(1 - \frac{K^2}{4\gamma^2}\right) \cong 1 - \frac{1}{2\gamma^2} - \frac{K^2}{4\gamma^2}. \quad (5.31)$$

The essential information regarding the characteristics of the electron's motion is obtained by considering the limiting (relativistic) case in which

$$\frac{\gamma}{K} \gg 1. \quad (5.32)$$

With this restriction, a reasonable approximation to z and kz results:

$$z = \bar{\beta}ct, \quad (5.33)$$

$$kz = \Omega t, \quad (5.34)$$

where the oscillation frequency (Ω) is

$$\Omega = \frac{2\pi\bar{\beta}c}{\lambda_o}. \quad (5.35)$$

Equations 5.32–5.35 lead to the values of the electron velocity in the x- and z-directions:

$$\dot{x} = \frac{Kc}{\gamma}\cos\Omega t. \tag{5.36}$$

$$\dot{z} = \bar{\beta}c - \frac{K^2 c}{4\gamma^2}\cos 2\Omega t. \tag{5.37}$$

Equations 5.36 and 5.37 can be integrated to provide the x- and z-coordinates of the electron as it traverses the device:

$$x = \frac{Kc}{\gamma\Omega}\sin\Omega t. \tag{5.38}$$

$$z = \bar{\beta}ct - \frac{K^2 c}{8\gamma^2\Omega}\sin 2\Omega t. \tag{5.39}$$

The displacement of the electron or positron from its initial trajectory is relatively small. For example, Problem 05-01 considers an insertion device with a 50-mm period and $K = 2$ in a 2-GeV electron storage ring. For these parameters, the electron has a maximum deflection angle of 0.5 mr and maximum oscillation amplitude of 4 μm. The displacement in the z-direction is even smaller with an amplitude of 2.6 Å. The emitted wavelength of this device is 49 Å.

5.10
Insertion Device Radiation Properties

The radiation characteristics of electrons accelerated by the magnetic fields in an insertion device can be described in terms of electron wave fronts. The interference of these electron wave fronts as they traverse the insertion device determines the device's radiation characteristics. In the time for the electron to travel through one period length, from location A to B having the same phase, the wave front from A has advanced by a distance of $\lambda_o/\bar{\beta}$. The wave front originating at A is ahead of the radiation emitted at location B by a distance d:

$$d = \frac{\lambda_o}{\bar{\beta}} - \lambda_o \cos\theta, \tag{5.40}$$

where θ is the angle of emission with respect to the electron beam direction. When the distance d is equal to an integral number of radiation wavelengths (n), constructive interference occurs:

$$\frac{\lambda_o}{\bar{\beta}} - \lambda\cos\theta = n\lambda. \tag{5.41}$$

Using the expression for the average electron velocity

$$\frac{1}{\bar{\beta}} = 1 + \frac{1}{2\gamma^2} + \frac{K^2}{4\gamma^2}, \tag{5.42}$$

the interference condition is

$$\lambda = \frac{\lambda_o}{2\gamma^2} \frac{1}{n} \left(1 + \frac{K^2}{2} + \gamma^2 \theta^2\right), \tag{5.43}$$

where n is the harmonic number ($n = 1, 2, 3, \ldots$). Expressions for the output photon wavelength (λ) and photon energy (ε) of the insertion device are obtained using these results:

$$\lambda[\text{Å}] = \left(\frac{1305 \text{ GeV}^2 \text{ A}}{\text{m}}\right) \frac{\lambda_o[\text{m}]}{E^2[\text{GeV}]} \frac{1}{n} \left(1 + \frac{K^2}{2} + \gamma^2 \theta^2\right), \tag{5.44}$$

$$\varepsilon[\text{eV}] = \left(\frac{9.498 \text{ eV m}}{\text{GeV}^2}\right) n \frac{E^2[\text{GeV}]}{\lambda_o[\text{m}] \left(1 + \frac{K^2}{2} + \gamma^2 \theta^2\right)}, \tag{5.45}$$

where λ_o is the period of the insertion device in m and E is the electron energy in GeV.

Equations 5.43–5.45 suggest several characteristics of the radiation emitted from an insertion device. First, the fundamental wavelength of the radiation is much shorter than the period length as the Lorentz factor γ is large (see Equations 5.43 and 5.5). Second, the output harmonic wavelengths ($n = 1, 2, 3, \ldots$) are altered by changing the electron beam energy and/or the magnetic field strength of the accelerating device. In addition, the output wavelength varies with the observation angle and encompasses a spectrum of values. However, if the range of observation angles is restricted, the output wavelength is limited to the harmonic values.

The interference model also leads to information regarding the spread of angles and wavelengths. If the insertion device has N periods in a length L of the insertion device, the condition for constructive interference over its length is

$$\frac{L}{\beta} - L\cos\theta = nN\lambda. \tag{5.46}$$

Given this condition, there is a wavelength λ', where the interference is destructive. Destructive interference occurs when there is an additional complete wavelength separation of the wave fronts over the length of the insertion device:

$$\frac{L}{\beta} - L\cos\theta = nN\lambda' + \lambda'. \tag{5.47}$$

Equation 5.47 corresponds to an additional 2π phase advance over the length of the device.

Subtracting Equations 5.46 and 5.47 leads to a range of wavelengths at a given angle θ:

$$\frac{\Delta\lambda}{\lambda} = \frac{1}{nN}. \tag{5.48}$$

In a similar fashion, changing the angle θ at a fixed wavelength results in destructive interference when

$$\frac{L}{\beta} - L\cos\theta' = nN\lambda + \lambda. \tag{5.49}$$

5.10 Insertion Device Radiation Properties

If Equations 5.46 and 5.49 are subtracted and the trigonometric functions are expanded in a power series that only retains the first two terms, the angular width ($\Delta\theta$) is obtained:

$$\theta^2 - \theta'^2 = \Delta\theta^2 = \frac{2\lambda}{L}. \tag{5.50}$$

Using Equation 5.43, the angular relationship becomes

$$\Delta\theta = \sqrt{\frac{2\lambda}{L}} = \frac{1}{\gamma}\sqrt{\frac{\lambda_o}{nL}\left(1+\frac{K^2}{2}+\gamma^2\theta^2\right)}. \tag{5.51}$$

For the case in which the radiation is emitted along the beam direction, the angular width becomes

$$\Delta\theta = \frac{1}{\gamma}\sqrt{\frac{\lambda_o}{nL}\left(1+\frac{K^2}{2}\right)}. \tag{5.52}$$

Equation 5.52 suggests that the radiation cone is smaller in opening angle compared to that of conventional synchrotron radiation sources. This indicates the increased brightness that results from insertion devices compared to the other sources of synchrotron radiation (e.g., bending magnets).

5.10.1 Power and Power Density

With the basic physics properties of insertion devices established, the output power and power density can be addressed. These parameters are important from a health physics perspective because they directly affect the radiation characteristics of these devices.

The power density (power output per unit solid angle) is

$$\frac{dP}{d\Omega}[\text{W/mr}^2] = \frac{10.84\,\text{W}}{\text{mr}^2\,\text{GeV}^4\,\text{T A}} E^4[\text{GeV}] B_o[\text{T}] N I_b[\text{A}] G(K) f_K(\theta_x,\theta_y), \tag{5.53}$$

where I_b is the average beam current and

$$G(K) = \frac{K\left(K^6 + \frac{24}{7}K^4 + 4K^2 + \frac{16}{7}\right)}{(1+K^2)^{7/2}}. \tag{5.54}$$

The function $f_K(\theta_x,\theta_y)$ has a peak value of unity on-axis. The angular distribution approaches the output of a bending magnet as $K \to \infty$. For large K, $G(K) \to 1$, and the on-axis power density has a relatively simple form

$$\frac{dP}{d\Omega}[\text{W/mr}^2] = \frac{10.84\,\text{W}}{\text{mr}^2\,\text{GeV}^4\,\text{T A}} E^4[\text{GeV}] B_o[\text{T}] I_b[\text{A}] N. \tag{5.55}$$

The instantaneous rate of total power (P) emitted by a single electron is

$$P = \frac{2}{3} \frac{e^2 c}{4\pi\varepsilon_0 \rho^2} \beta^2 \gamma^4, \tag{5.56}$$

where ρ is the radius of curvature of the electron's trajectory. The total power output is obtained by integrating over the insertion device length. For an insertion device of length L with sinusoidal field amplitude B_o, the total power output is

$$P_{\text{total}}[\text{W}] = \frac{633 \text{ W}}{\text{GeV}^2 \text{ m A T}^2} E^2[\text{GeV}] B_o^2[\text{T}] L[\text{m}] I_b[\text{A}]. \tag{5.57}$$

Table 5.4 provides the total power and peak power density output for typical undulator and wiggler parameters in storage rings with energies of 0.8, 2.0, and 6.0 GeV. The tabulated values utilize an insertion device length of 3 m and an electron beam current of 200 mA.

Table 5.4 indicates that the power and the power density increase rapidly as the electron beam energy increases. This is especially true for the power density as it scales as E^4. Multipole wigglers emit significantly more total power than undulators if both devices have the same length. However, the peak power density values are similar because of an approximate inverse scaling of the field strength with period length. Care must be exercised in designing beamline components to accommodate the high power and power density values.

Designs must also prevent damage to the electron beam vacuum chamber during a beam misdirection event. Therefore, automatic interlock systems are required to detect any beam misdirection and to dump the beam within a sufficiently short time. In addition to design issues, these high total power and power density values challenge health physicists to ensure that effective doses are maintained in an as low as reasonably achievable (ALARA) manner. These values require a careful health physics monitoring and control during off-normal events.

With a physical description and basis for insertion devices, established free-electron lasers can now be addressed in a consistent manner. FELs are first reviewed in a general manner, and then their scientific basis and health physics characteristics are provided.

Table 5.4 Total power and peak power density for typical undulator and multipole wigglers.

	$\lambda_o = 0.05$ m, $B_o = 0.5$ T Undulator		$\lambda_o = 0.15$ m, $B_o = 1.6$ T Multipole wiggler	
E (GeV)	P_{total} (kW)	$d^2P/d\Omega$ (kW/mr²)	P_{total} (kW)	$d^2P/d\Omega$ (kW/mr²)
0.8	0.06	0.03	0.62	0.03
2.0	0.38	1.0	3.9	1.1
6.0	3.4	84.3	35.0	89.9

[a] Derived from Walker in CERN 98-04 (1998).

5.11
FEL Overview

The next generation of synchrotron radiation sources is primarily based on linear accelerators rather than on storage rings. These sources, including the free-electron laser, produce coherent radiation orders of magnitude greater in peak power and brightness than that of the bending magnets, wigglers, and undulators.

FELs are capable of producing coherent, monochromatic output over a wide region of the EM spectrum. These lasers have operated from the microwave to the ultraviolet region. Average and peak power levels of several kilowatts and about a gigawatt, respectively, have been achieved.

Electron linear accelerators (LINACs) that incorporate undulators are the basis for FELs. As the electron beam enters the undulator, it encounters a magnetic field that alters its trajectory, thereby causing it to wiggle. The change in the trajectory and the associated acceleration produces photon radiation. The intensity of this radiation increases linearly along the undulator length. After exiting the undulator, the electron beam is removed from the photon beam and directed toward a beam dump. The photon beam is guided by optical components toward an experimental area. For a short undulator, this spontaneous radiation is effectively decoupled from the electron beam. However, the electron beam and the photon begin to couple as the undulator length increases.

If the undulator is long and the electron beam quality is sufficiently high, the photon radiation and the electron beam couple or interact as they travel along the undulator axis. This interaction causes the electron beam intensity to be increased or bunched at the FEL's resonant photon wavelength. This process is called self-amplified spontaneous emission (SASE), and it causes the photon pulse intensity to increase exponentially until saturation occurs.

During SASE, the electron beam bunches generate an electromagnetic field that interacts with the electron beam. This field couples with the electron bunches and forms individual bunches separated by energy (wavelength). Radiation of a given wavelength is coherently amplified along the trajectory of the electron beam. The radiation is emitted with a small angular divergence on the order of 100 μr. This divergence is considerably smaller than the mr range encountered in conventional chemical lasers.

The SASE process generates a line spectrum superimposed on the spontaneous emission photon spectrum. SASE increases the intensity over spontaneous emission approximately by a factor of 10^5.

FELs are becoming quite common, but their characteristics and properties vary considerably. Table 5.5 provides a current listing of operating FELs with wavelengths shorter than 1 μm, including their geographic location, operating wavelength, and type.

As illustrated in Table 5.5, a variety of accelerators are used to drive FELs. In addition to the diverse modes of operation, FELs function over a large electron energy and parameter space. Rather than addressing all FEL types, subsequent discussion is limited to the amplifier or "single-pass" configuration. A single-pass

Table 5.5 Free-electron lasers with wavelengths $\leq 1\,\mu m$.[a]

Location	Name	Wavelengths	Type
France – Centre Universitaire – Orsay	Super-ACO	350 nm	Storage ring
Germany – Deutsches Elektronen-Synchrotron	FLASH-FEL	13–45 nm	Superconducting LINAC
Germany – Deutsches Elektronen-Synchrotron	TESLA-FEL	80 nm	Superconducting LINAC
Germany – University of Dortmund	Felicita 1	470 nm	Storage ring
Italy – Elettra (Trieste)	VUV-FEL	180–660 nm	Storage ring
Japan – Harima Institute	SPring-8	49 nm	LINAC
Japan – Electrotechnical Laboratory	NIJI-IV	228 nm	Storage ring
Japan – Institute for Molecular Research	UVSOR	239 nm	Storage ring
Japan – Institute of Free Electron Lasers (IFEL)	IFEL-3	230 nm–1.2 μm,	LINAC
	IFEL-2	1–6 μm	
USA – Duke University	OK-4	217 nm	Storage ring

[a] The information in his table was obtained from the University of California Santa Barbara summary located at: http://sbfel3.ucsb.edu/www/fel_table.html. Accessed on September 17, 2006.

FEL is represented by a few basic components: an electron accelerator, electron beam optics to focus and direct the beam to an undulator, electron beam diagnostics, undulator, output photon beam, photon diagnostics, and the electron beam dump.

From a global standpoint, a FEL is a device that produces coherent (or partially coherent) radiation from an electron beam by extracting energy from the beam through a self-amplified spontaneous emission process. As noted in the previous discussion, an undulator is a device producing a periodic magnetic field that alters the electron trajectory in a predetermined manner. Electrons, colinear with an optical (radiation) field, enter an undulator, have their trajectory altered, and radiate. This action transfers energy to the optical field and alters its phase. The optical field advances relative to the particle field by one wavelength per undulator period. With this FEL overview established, a more physical description is established by introducing the concepts of resonant energy and phase bunching.

5.12
Physical Model of a FEL

Previous discussion provided a description of synchrotron radiation and the various methods of its production. These synchrotron devices produce radiation by altering the electron's trajectory using a magnetic field. A FEL is more complex in that it utilizes an undulator as an input device, and the resulting coherent radiation output depends on coupling the synchrotron radiation from the undulator to the optical FEL radiation that is amplified by the resonant cavity of the device.

An insight into the physics of a FEL is gained by considering the interaction characteristics of a single electron. A single-electron model of a FEL considers the interaction of an electron with the optical electric field and the undulator's magnetic field. The model is defined in terms of five stages. In Stage 1, the electron is below the undulator axis and feels no force from the radiation field as the electric field is zero. The only force on the electron is derived from the undulator's magnetic field.

In Stage 2, the electron is on the undulator's axis. It has advanced one-fourth of an undulator period relative to Stage 1, while the radiation field has advanced one-fourth of a period relative to the electron. The electric field is at its maximum value and the electron experiences a retarding force, radiates, and loses energy to the radiation wave.

In Stage 3, the electron is above the undulator axis. It has advanced one-half of an undulator period relative to Stage 1, while the radiation field has advanced one-half of a period relative to the electron. The electric field is once again zero; so the electron feels no force and loses no energy.

In Stage 4, the electric field reaches a maximum value. The electron is again on the undulator's axis. It experiences a retarding force opposite to the force direction of Stage 2 and again loses energy to the radiation field.

In Stage 5, the configuration is the same as in Stage 1. However, the radiation field has "slipped" one wavelength and gained energy equal to the energy lost by the electron.

Slippage defines the relative relationship of the phase of the EM field with respect to the radiation wave propagating with a phase velocity c. The concept of slippage is essential to understand a physical FEL. Through slippage the radiation wave extracts energy from the EM field and intensifies as it progresses from Stage 1 through Stage 5.

The single-electron FEL model indicates that the phase of the electron determines whether it gains or loses energy. Therefore, for a distribution of electrons, some gain energy whereas others lose energy. Although the single-electron model provides information about FEL dynamics, it is important to realize that a physical electron beam is also distributed over energy and position.

The single-electron model does not incorporate the concept of resonant energy. At the resonant energy, electrons exchange no energy with the radiation field because a constant phase between the radiation field and the electron oscillations exist. Electrons with energy greater (less) than the resonant energy lose (gain) energy to (from) the radiation field. This FEL process is analogous to stimulated emission in a chemical laser that utilizes atomic transitions to generate monochromatic, coherent photons.

The force responsible for the energy change is called the ponderomotive force, which tends to bunch the electron beam. High gain ($\gg 1$) is produced through the instability of bunching, and phase bunching leads to coherent emission. Low gain can be achieved by injecting the beam at energy above the resonant energy.

The concept of bunching or having the electrons achieve a common phase (coherence) is an important factor in the efficient operation of a FEL. This effect causes the radiation from an ensemble of beam electrons to display collective effects. These collective phenomena increase the total number of radiated photons and influence their frequency and angular spectrum. Coherence occurs because the emitted radiation of separated electrons is in phase. Distributions that lead to coherence are referred to as bunched. The coherent enhancement of a radiative process is determined from the number of electrons located within a longitudinal half-wavelength and a transverse half-wavelength divided by the Lorentz factor. With the physical model defined and the concepts of resonant energy and phase bunching described, the basis for deriving the FEL equations is established. These equations are presented in the next section of this chapter.

5.12.1
FEL Physics

In distilling the discussion to this point, a free-electron laser is a device that converts the kinetic energy of an electron into electromagnetic energy. This conversion occurs in a manner that creates coherent, monochromatic radiation emitted in a specified direction. The two basic elements in a FEL are a component that generates energetic electrons and a device that converts the electron's kinetic energy into EM radiation.

Most of the diversity in FEL design is associated with the device for generating energetic electrons. A number of approaches for electron beam generation are utilized, including radio frequency linear accelerators, superconducting accelerators, microtrons, van de Graaff accelerators, induction accelerators, pulse-line accelerators, and storage rings.

The conversion of electron kinetic energy into EM radiation requires an energy transfer between the electron beam and the standing wave over an extended distance. To achieve the requisite energy transfer, synchronization between the electron velocity and the phase velocity of the output wave must occur. Efficient energy transfer must overcome two challenges. First, in a vacuum, an EM wave has a phase velocity equal to the speed of light (c), but the electrons have a velocity (v) that is less than c. Second, as the beam electrons are uniformly distributed along the direction of motion, bunching must occur for the electrons to be in a phase of the EM field in which the photon wave extracts energy from the electron beam. This condition is necessary to facilitate the sustained interchange of energy from the electron beam to the EM field. The manner of extraction determines the wavelength and the energy of the resultant EM field.

To transfer energy most efficiently, an oscillating or periodically cycling EM field is used to produce an undulating beam trajectory. This trajectory is readily achieved using insertion device magnets with flux concentrating poles. The desired EM wave is produced in an optical resonator. This condition is similar to the buildup of energy in the optical cavity of a conventional laser. The insertion device magnet pole faces are used to concentrate the energy flux in the plane perpendicular to the direction of the electron beam.

The rate of change of an electron beam's energy in an electric field ($\vec{E}$) is

$$mc^2 \frac{d\gamma}{dt} = e\vec{v} \cdot \vec{E}, \tag{5.58}$$

where m is the electron rest mass, e is the electron charge, and $\vec{v}$ is the electron velocity. To extract energy from the electron beam, the right-hand side of Equation 5.58 must be positive when averaged over an extended time interval:

$$\vec{v} \cdot \vec{E} > 0. \tag{5.59}$$

To further define the FEL physics, consider an electron beam propagating in the z-direction with the electric field extracting energy residing in the (x, z) plane. For specificity, the EM field consists of electric and magnetic fields with components

$$E_x = E_o \cos \frac{2\pi z}{\lambda_o}, \tag{5.60}$$

$$B_y = B_o \sin \frac{2\pi z}{\lambda_o}, \tag{5.61}$$

where λ_o is the wiggler period or the distance between the flux concentrating poles. If the electron beam is assumed to wiggle in the (x, z) plane, Equation 5.59 must be met for energy to be extracted from the electron beam. Over a wiggler period, the transverse velocity changes direction (sign). Therefore, the electron slips half an optical wavelength in traversing the wiggler period to preserve the Equation 5.59 condition:

$$\begin{aligned}\vec{v} \cdot \vec{E} > 0 \Rightarrow v E_o \cos \frac{2\pi z}{\lambda_o} &= (-v) E_o \cos\left(\frac{2\pi z}{\lambda_o} - \left(\frac{2\pi}{\lambda_o}\right)\left(\frac{\lambda_o}{2}\right)\right) \\ &= (-v) E_o \left(-\cos\left(\frac{2\pi z}{\lambda_o}\right)\right) = v E_o \cos \frac{2\pi z}{\lambda_o}.\end{aligned} \tag{5.62}$$

It can be shown that the condition of Equation 5.62 is equivalent to a synchronization condition:

$$\frac{\lambda}{\lambda_o} = \frac{c}{\bar{v}_z} - 1, \tag{5.63}$$

where λ is the output wavelength of the FEL and $\bar{v}_z$ is the z-component of the electron velocity averaged over the distance $\lambda_o/2$. Equation 5.63 simplifies for relativistic electron velocities ($v \gg 1$):

$$\frac{\lambda}{\lambda_o} = \frac{1+K^2}{2\gamma^2}, \tag{5.64}$$

where K is the wiggler parameter defined by Equation 5.27.

Equation 5.64 is an important result because it specifies how the output FEL wavelength can be altered. Changes in the FEL wavelength occur by either changing the electron beam energy or altering K by varying the strength of the insertion device magnetic field or the device period (λ_o).

For a relativistic beam, it is a somewhat slow process to tune the FEL by changing γ because this requires adjusting magnets in the beam transport system. The strength of the magnetic field is altered by changing the wiggler period (i.e., the pole spacing in the wiggler magnet), but this is an expensive wiggler modification.

From the discussion to date, the FEL has a number of unique features. It operates over a wide range of frequencies as no physical resonance is involved. The output wavelength is selected by specifying the electron beam energy, wiggler period, and wiggler parameter (K). In addition, the FEL is capable of delivering high peak power and average output power because it transfers energy from a high-power, relativistic electron beam.

With a basic understanding of the major components of the FEL and its underlying physical principles, specific characteristics of these devices are addressed. Subsequent discussion outlines the optical gain, cavity design, optical klystron (OK), and accessible output of a FEL.

5.13
FEL Characteristics

There are a variety of approaches for calculating the FEL gain, but in all procedures the necessary input relationships are based on the force equation

$$\vec{F} = \frac{d\vec{p}}{dt} = e(\vec{E} + \vec{v} \times \vec{B}) \tag{5.65}$$

and the Maxwell equations. The optical gain is obtained from the simultaneous solution of these equations.

5.14
Optical Gain

The optical gain (G) is defined as the optical power output of the FEL divided by the input power. For most infrared and visible output FELs, $G \ll 1$ and the gain has the form

$$G = \frac{j}{4} \frac{d}{d\theta} \left(\frac{\sin\theta}{\theta} \right)^2, \tag{5.66}$$

where the form for the function j depends on the wiggler configuration:

$$j = \left[\frac{\pi Z e}{mc^2} \right] \frac{IL^3 K^2 (1+K^2)}{A \lambda \gamma^5} \text{ (for a helical wiggler)}, \tag{5.67}$$

$$j = \left[\frac{\pi Z e}{mc^2} \right] \frac{IL^3 K^2 (1+K^2)}{A \lambda \gamma^5} [J_0(\xi) - J_1(\xi)]^2 \text{ (for a planar wiggler)}. \tag{5.68}$$

In Equations 5.67 and 5.68, Z is the impedance of free space, L is the interaction length, I is the beam current, A is the beam area plus optical mode area, J_0 and J_1 are Bessel functions of the first kind, and

$$\xi = \frac{K^2}{2(1+K^2)}. \tag{5.69}$$

The parameter θ is one-half the phase slip from synchronism over a distance L. For an interaction length L, the phase slip is

$$2\theta = L\left[k\left(\frac{1}{\bar{\beta}_z}-1\right)-k_o\right], \tag{5.70}$$

where

$$\frac{1}{\bar{\beta}_z} = 1 + \frac{1+K^2}{2\gamma^2}, \tag{5.71}$$

$$k = \frac{2\pi}{\lambda} \quad \text{and} \quad k_o = \frac{2\pi}{\lambda_o}. \tag{5.72}$$

If the gain per unit length is large, the output power (P) is a function of the input power (P_0) and is written in terms of the j-value defined in Equations 5.66–5.68:

$$P = \frac{P_0}{9}\exp\left[\sqrt{3}\left(\frac{j}{2}\right)^{1/3}\right]. \tag{5.73}$$

5.14.1
Cavity Design

For a radio frequency linear accelerator beam, the electron bunches are only a few millimeters in length. The cavity length L_c is selected such that the round trip time of the wave in the cavity is an integer (n) multiplied by the time between pulses to ensure coherence within the cavity. If λ_m is the distance between pulses, the coherence requirement is written as

$$2L_c = n\lambda_m. \tag{5.74}$$

The cavity increases the power of the electromagnetic wave. If P_0 is the initial power in the cavity, the buildup of power (P) in the cavity is written as

$$P = P_0(1+G_n)^N, \tag{5.75}$$

where G_n is the net cavity gain and N is the number of round trips the EM wave completes in the cavity. The net gain is the electronic gain in the cavity minus cavity losses. The number of round trips can be written in terms of the buildup time (t_b) or the time needed to complete N round trips in the cavity:

$$N = t_b\left(\frac{c}{2L_c}\right). \tag{5.76}$$

Using Equations 5.75 and 5.76, the buildup time is

$$t_b = \frac{2L_c}{c}\frac{\log(P/P_0)}{\log(1+G_n)}. \tag{5.77}$$

These gain and cavity relationships are important because they provide restrictions on cavity parameters for a FEL to operate efficiently. Equation 5.66 indicates that a FEL exhibits electronic gain for a wide range of phase slip angles:

$$0 > \theta > \pi. \tag{5.78}$$

Equation 5.78 imposes a limit on the energy dispersion in the beam. The relationship between the changes in the phase slip ($\Delta\theta$) and Lorentz factor ($\Delta\gamma$) derived from Equations 5.70 and 5.71 leads to

$$\Delta\theta = -\frac{kL}{2}\frac{1+K^2}{\gamma^3}\Delta\gamma, \tag{5.79}$$

where $\Delta\gamma$ is the full width of the beam energy spread.

Equations 5.64, 5.78, and 5.79 lead to an additional Lorentz factor constraint:

$$\left| kL\frac{\lambda}{\lambda_o}\frac{\Delta\gamma}{\gamma} \right| \leq \pi, \tag{5.80}$$

or

$$\left|\frac{\Delta\gamma}{\gamma}\right| \leq \frac{\lambda_o}{2L} = \frac{1}{2N_w}, \tag{5.81}$$

where N_w is the number of wiggler periods. There is also a limit on the angular divergence of the beam. If an electron is traveling at a half-divergence angle ψ relative to the beam direction (z-axis), then $\bar{\beta}_z$ is changed by an amount

$$\Delta\bar{\beta}_z = -\frac{\psi^2}{2}. \tag{5.82}$$

When Equation 5.82 is combined with Equation 5.70, a limit for ψ results:

$$\psi \leq \sqrt{\frac{\lambda}{2L}}. \tag{5.83}$$

Associated with the divergence angle is the beam emittance (ε):

$$\varepsilon = w\psi, \tag{5.84}$$

where w is the spot size of the output photon radiation. Maximum energy photon output from the beam is obtained for

$$w \leq \left(\frac{\lambda R}{\pi}\right)^{1/2}, \tag{5.85}$$

where R is the distance in which the area of the diffracted wave doubles (i.e., the Raleigh length for the optical mode). In contemporary FELs, R is selected as $L/2$, where L is the interaction length. Using Equations 6.83–6.85, an additional expression for the emittance is obtained:

$$\varepsilon \leq \frac{\lambda}{2\pi^{1/2}}. \tag{5.86}$$

5.14.2
Optical Klystron

An optical klystron converts an electron's energy change to a change in position. The use of an OK in a FEL enhances the electron bunching and increases the gain without increasing the interaction length. This application is important if the available interaction length is restricted or if beam divergence is a limiting factor.

If the optical klystron, including a dispersive magnet, is incorporated into a FEL, its configuration consists of three basic components. The electron beam enters the first section that performs input modulation and changes the electron energy. The second section is the dispersive magnet that bunches the electrons. The final section is the output section that extracts energy from the bunched electron beam. Although the length of the OK sections can vary, subsequent discussion assumes that the lengths (L) of the modulation and output sections are the same and that the dispersive magnet has a length S.

Bunching in a dispersive magnet is facilitated through the use of a specific magnetic field profile. In the initial distance of $S/4$ in the dispersive magnet, a magnetic field of strength B is present. The field direction is reversed over the next $S/2$ distance. Over the final $S/4$ distance, the field again has the original direction and the magnitude. The net effect of the dispersive magnet configuration is an enhanced gain.

If the modulation and output sections have identical length ($S/2$) and undulator parameter values and all the bunching occurs in the dispersive magnet, then the gain of the OK (G_{OK}) is given by

$$G_{OK} = \frac{16.5}{T^2 \, A \, m^2} \delta L^2 S^3 B^2, \tag{5.87}$$

where B is the dispersive magnet's magnetic induction and δ is given by

$$\delta = \frac{IK^2}{A\lambda\gamma^5}(J_0 - J_1)^2. \tag{5.88}$$

The terms in Equation 5.88 were defined with Equation 5.68, and δ has units of A/m^3. The maximum gain of a FEL (G_{FEL}) without using an OK can be derived from Equation 5.66:

$$G_{FEL} = \frac{3.12 \times 10^{-4}}{A} \delta(1+K^2) L_{FEL}^3, \tag{5.89}$$

where L_{FEL} is the length of the FEL.

One disadvantage of the use of an OK is that the allowed energy spread is less than that for a FEL. Using Equation 5.82, the energy acceptance of a FEL and an OK are

$$\left(\frac{\Delta\gamma}{\gamma}\right)_{FEL} = \frac{1}{2N_w}, \tag{5.90}$$

$$\left(\frac{\Delta\gamma}{\gamma}\right)_{OK} = 0.7 \times 10^{-4} \, m^2 \, T^2 \, \frac{\lambda\gamma^2}{S^3 B^2}. \tag{5.91}$$

The gain and the energy acceptance are combined to yield a simple relationship between these quantities in a FEL and an OK:

$$\frac{\left(G\frac{\Delta\gamma}{\lambda}\right)_{OK}}{\left(G\frac{\Delta\gamma}{\lambda}\right)_{FEL}} = \left(\frac{2L}{L_{FEL}}\right)^2. \tag{5.92}$$

5.15
Accessible FEL Output

The results of Table 5.5 suggest that contemporary FEL output is diverse and will become increasingly more diverse as the twenty-first century emerges. It is also possible that FELs will compete aggressively with chemical lasers as the twenty-first century advances. This is particularly likely in the far infrared region (10–1000 µm), where chemical laser options are limited. FELs also offer the advantage of potentially higher output powers and shorter pulse widths. The extent to which FELs displace chemical lasers as the primary source of coherent, monochromatic radiation depends on relative costs and technology advancements.

FELs also offer the potential for the production of shorter wavelengths. X-ray FELs are on the horizon and gamma-ray FELs are also a possibility. X-ray and gamma-ray FELs are addressed in the subsequent discussion.

5.16
X-Ray Free-Electron Lasers

X-ray lasers were initially conceived for defense applications (e.g., antimissile systems), but they also have nonmilitary applications. Much of the X-ray laser technology originally developed for strategic defense can be used in research applications in biotechnology, materials science, and materials analysis. Specific applications include destruction of toxic materials and spallation of radioactive materials such as fission products or actinides.

Conceptual XFELs are driven by a radio frequency electron gun. Electron acceleration is accomplished in a number of LINAC sections that typically cover a distance of 1–2 km. Two chicane-type bunch compressors are located along the LINAC. The first compressor is positioned at a low-energy location and the second is at a medium-energy location. After exiting the LINAC, the electron beam enters a 100–150-m-long undulator. The intensity of the radiation output increases linearly along the undulator. Upon exiting the undulator, the electron beam is deflected by bending magnets toward a beam dump. The FEL radiation is directed by optical components to an experimental area.

If the undulator length increases to about a kilometer and the electron quality is sufficiently high, self-amplified spontaneous emission occurs. In SASE, the bremsstrahlung radiation and the electron beam interact as they traverse the undulator, which causes the electron beam to be bunched at the resonant X-ray wavelength of the system. SASE causes the X-ray pulse intensity to increase in an exponential manner until saturation is reached. Typically, SASE yields an intensity gain of 10^5 over spontaneous emission.

5.17
Threshold X-Ray Free Electron

In 2005, the Deutsches Elektronen-Synchrotron (DESY) in Hamburg, Germany, opened the first vacuum ultraviolet free-electron laser (VUV-FEL) for user operation. Given the DESY FEL output radiation in the range of 6–100 nm, it is a transition facility and serves as a threshold X-ray FEL. The electron energy through the various system components is illustrated in Table 5.6.

The DESY FEL permits fundamental research in a variety of areas including the structural analysis of biological molecules, the study of light-matter interactions at new energy and intensity regimes, and the new applications in microscopy, holography, and semiconductor technology. The output of the VUV-FEL ranges from 6 to 100 nm, with pulse durations of 20–200 fs and a peak power exceeding current synchrotron output by about 7 orders of magnitude. Using linear accelerators as the electron source, DESY and the Stanford Linear Accelerator Center plan to decrease the output wavelength to about 0.1 nm.

At DESY, a 260-m LINAC accelerates electrons to a maximum energy of 1 GeV. The electrons then enter a 30-m magnetic undulator. The electron bunches are formed in bunch trains containing up to 7200 bunches with a 9-MHz pulse repetition frequency. A bunch train repetition rate of 10 Hz can be achieved. The first laser output was obtained in 2005 at 32 nm with a pulse length of 20 fs.

The VUV-FEL experience will be incorporated into an X-ray free-electron laser. This future DESY device is scheduled for completion in 2012. This FEL will produce 0.085 nm radiation using a 20-GeV linear accelerator of 2.1 km length with up to a 250-m undulator. XFELs are discussed in more detail in the following section.

Table 5.6 Major DESY FEL components and associated energies.

Component	Function	Electron energy (MeV)
Radio frequency electron gun	Initial energy	4
Initial accelerator modules/ bunch compressor	Increase energy and modify output	125
Final accelerator modules/ bunch compressor	Increase energy and modify output	380[a]
Collimator	Confine beam profile	380
Undulator	Increase energy	440
Photon beam	Useful output radiation	6–100 nm
Electron beam dump	Terminate electron beam	0

[a] The maximum energy is 1000 MeV.

Table 5.7 X-ray free-electron laser projected parameters.

Parameter	LCLS (United States)	DESY XFEL (Germany)	SCSS (Japan)
Pulse duration (fs)	<230	100	80
Wavelength (Å)	1–64	1–15	1–50
Repetition rate (Hz)	120	10	60
Electron bundles per pulse	1	≤3000	1
Electron beam energy (GeV)	4–14	≤20	≤8
Photons per pulse ($\times 10^{12}$)	1.2@1.5 Å	1.2@1 Å	0.76@1 Å
Linac length (km)	1	2	0.35
Estimated cost (M$)[a]	379	1000	330
Estimated start date	2009	2012	2010

Derived from Feder (2005).
[a] Estimates include varying amounts of instrumentation and different methods of accounting.

5.18
Near-Term X-Ray FELs

Three X-ray free-electron laser facilities are currently planned for the 2010 time frame. These include the LINAC Coherent Laser Source (LCLS) at the Stanford Linear Accelerator Center, the XFEL at DESY in Hamburg, and Japan's SPring-8 Compact self-amplified spontaneous emission (SCSS).

The XFELs produce coherent, intense, ultrafast pulses of hard X-rays with temporal and spatial pulses that are on the time and wavelength scale of atomic and molecular processes. These planned devices will also have about 10 billion times the peak brightness of synchrotron radiation sources. The expected characteristics of these XFELs are summarized in Table 5.7.

At all the three facilities of Table 5.7, the FEL output wavelengths are in the angstrom range, with electron beam energies in the GeV range. Each pulse provides on the order of 10^{12} photons and pulse widths will initially be in the 100–200-fs range. It is expected that the pulse widths will shorten as the facilities mature.

From a health physics perspective, the XFEL's peak brightness will be a formidable radiation challenge. The 10^{10} brightness enhancement over synchrotron radiation sources demands care in shield design and presents the likelihood of extreme radiation levels in a variety of normal and off-normal events. These brightness levels demand careful health physics task planning, task monitoring, and management of off-normal events. Additional FEL health physics issues are presented in the subsequent discussion.

5.19
Gamma-Ray Free-Electron Lasers (GRFEL)

Once XFELs are established, the next logical step in the FEL development is the generation of gamma rays. The generation of gamma rays in a FEL occurs with the increase in output energy and brightness. A gamma-ray FEL not only provides a powerful research tool but also presents significant technological challenges to an engineer. In addition, a gamma-ray FEL offers significant health physics challenges.

A number of parameters are available to optimize the GRFEL output. These parameters include device length, pulse duration, repetition rate, number of electron bunches per pulse, and number of photons per pulse. These parameters set can be adjusted to optimize gamma-ray output.

The principle radiation hazards in a GRFEL facility are direct and secondary radiation, tritium contamination in beam lines and other beam locations, and activated accelerator components. In a GRFEL, the direct radiation arises from the GeV energy electron beam, and secondary particles are produced from interactions with the beam. These particles include neutrons, pions, and muons. Scattering causes the neutrons to govern the shielding considerations at large angles. Pions are produced in the direction at which the electrons strike various components. Muons result from pion decays. Photons are also generated by electron beam interactions including bremsstrahlung and nuclear excitation.

Personnel exposure to the direct electron beam or GRFEL output should be prevented. This is accomplished using audible and visible alarms indicating when beam operations are ongoing, interlocks to disable the accelerator if personnel entry to undulator or beam line areas occurs during beam operations, audible and visual alerting methods and procedures prior to accelerator startup, and clearly marked exit routes from the accelerator and research areas.

As a facility will likely have multiple target areas, some of which may be occupied for experimental setup or maintenance, methods to terminate the photon beam from within target areas in the event the beam is inadvertently introduced into that area should be provided. In addition, radiation monitors should clearly indicate the dose rates during beam operations. Radiation monitor interlocks are also beneficial to detect unanticipated beam entry into occupied or potentially occupied areas.

Components and adjacent structures are activated during GRFEL operation following interactions with the primary beam or the secondary radiation. The type of radioactive material produced depends on the GRFEL output characteristics, the beam energy, and the materials used in facility construction. In addition to the activation of structural materials, air, water, and soil are activated.

5.20
Other Photon-Generating Approaches

In addition to FELs, other photon sources either have been proposed or are likely to emerge in the twenty-first century. These sources are based on Compton backscattering, laser ion acceleration, wake-field acceleration, laser accelerators, X-ray induced isomeric transitions, and gamma-ray lasers. These photon sources are addressed in subsequent discussion.

5.20.1
Compton Backscattering

High-energy photons can be produced using Compton backscattering. Compton backscattering is particularly effective if it is accomplished with a FEL.

The head-on collision of relativistic electrons and photons in a FEL creates a beam of outgoing γ-rays having an energy E_g that depends on the angle θ between the direction of the incident electrons and the output γ-rays. This method of γ-ray production has a number of positive aspects that include the following: (1) the alignment of the electron and the photon beams required for FEL operation ensures the alignment for the outgoing γ-ray beam, (2) electron bunches and optical pulses are naturally synchronized, and (3) continuous tuning of the FEL provides for a smooth variation of the outgoing γ-ray energy.

In the limit that the electrons have high energy ($\gamma \gg 1$), the outgoing gamma-ray energy is

$$E_g \cong \frac{4\gamma^2 E_{ip}}{1+(\gamma\theta)^2+(4\gamma E_{ip}/m_e c^2)}, \tag{5.93}$$

where E_{ip} is the energy of the incident photon beam, E_e is the energy of the incident electron beam, and

$$\gamma = \frac{E_e}{m_e c^2}. \tag{5.94}$$

The maximum gamma-ray energy occurs at $\theta=0$, and the dependence of the output photon energy on the angle θ suggests that a collimator can be utilized to produce nearly monoenergetic gamma rays. In addition to producing nearly monoenergetic photon beams by the use of a collimator, a FEL having high intracavity power is expected to produce a photon flux enhancement of greater than 10^3 compared to that produced in conventional lasers. This flux output presents a significant external radiation source that must be carefully managed.

Compton backscatter is not a theoretical abstraction as it has successfully produced gamma-ray photons. A nearly monochromatic beam of 12.2 MeV γ-rays has been produced via Compton backscattering inside a free-electron laser optical cavity. The 12.2-MeV γ-rays are obtained by backscattering 379.4 nm free-electron laser photons from 500 MeV electrons circulating in a storage ring.

Gamma-ray beams produced from Compton backscattering have basic research and commercial applications. The high flux, energy spread, and polarization of the output gamma-ray beam are superior to that of photon sources currently used in nuclear physics research. Commercial applications of the Compton backscattering of the photon beam include precision gamma-ray transmission radiography, cancer therapy, and positron beam production.

5.20.2
Laser Accelerators

Laser acceleration of charged particles is based on the concept that an intense photon pulse generates a wake of plasma oscillations that accelerate electrons to high energy. This approach offers the potential for useful acceleration over much shorter distances than conventional accelerators. The key aspect of this technology is the development of extremely high peak powers in an ultrashort pulse. Recent experimental effort suggests that laser ion accelerator technology can be successfully applied to electrons, protons, and heavy ions.

A significant advantage of the laser approach is its capability to generate considerably larger ion currents than conventional accelerators. In a conventional accelerator, the mutual repulsion of ions limits the beam current. A laser accelerator achieves kiloampere pulses because it generates a neutral beam that contains electrons, protons, and heavy ions. The beam is essentially a quasi-stable plasma.

5.20.2.1 Basic Theory

Charged particles are usually accelerated using alternating electromagnetic fields applied to a series of segments. Increased energy is attained by the particle with increasing accelerator length. Cutting edge accelerators have a length (circumference) of several kilometers or more because current technology only increases ion energy by tens of MeV/m. Therefore, very high energies require the construction of large, expensive accelerator facilities.

Laser acceleration generates considerably higher fields than that is used in conventional accelerators. An ultrashort laser pulse directed to plasma or solid media produces peak electric fields in the teravolt per meters range. These intense fields ionize medium and accelerate ions and electrons to high energies over short distances. Optimally accelerating protons and electrons require separate processes. The wake-field approach noted below works best for electrons.

A better approach for accelerating protons and heavy ions is to direct laser pulses to thin-film targets. The large energy deposited into the target causes its explosion, which results in spallation of the target material, freeing protons and heavy ions for acceleration by the associated electromagnetic fields.

5.20.3
Laser Wake-Field Acceleration (LWFA)

LWFA is analogous to surfing an ocean wave. When a laser pulse strikes plasma, a density wave of free electrons is created. The electrons pull protons and positive ions and create a density wave as they pass through the plasma. The density wave facilitates the acceleration of free-electrons producing 100 MeV/mm energy gradients. LWFA requires only about 1/5000th of the distance to achieve the energy obtained in a conventional accelerator.

LWFA applicability to practical situations depends on the generation of a monoenergetic output beam. Recent experiments fired 30–55 fs pulses with peak powers of 10–30 TW into 2 mm long gas jets. By creating plasma channels or adjusting the laser beam to direct the density waves through the jets, the energy spread was about 24% for electron energies up to a few GeV. Additional reductions in the energy spread are likely, and LWFA offers a significant potential for future development and application.

5.20.4
Laser Ion Acceleration (LIA)

Protons and positive ions are too massive to be effectively accelerated by LWFA. The preferred approach is to strike a thin, dense foil target with a laser pulse having a power density $>10^{18}$ W/cm^2 that effectively detonates the foil. The pulse's intense electric field is sufficient to eject the electrons from the exploding foil. The charge of the accelerating electrons pulls the protons and heavy ions and accelerates them over micrometer-scale distances. However, the energy spread of the accelerated particles tends to be unacceptably large for practical applications.

The energy spread has recently been reduced to generate protons and heavy ions using a somewhat different approach. In the experiment, 10 TW, 80 fs pulses from a Ti:sapphire laser produced a power density of 3×10^{19} W/cm^2 in a 5-µm-thick titanium foil. On the opposite side of the foil was an array of polymer dots 0.5 µm thick and 20 µm in diameter. When the laser pulse was directed toward the metal side of the foil behind a polymer dot, it expelled a cloud of hot electrons on the side of the dot. The pulse produced plasma in the foil that ejected electrons from the dot. These electrons produced a strong electric field that accelerated the protons liberated from the polymer. In the experiment, 10^8 protons were produced with an energy spectrum having a narrow 1.2 MeV peak.

Heavy ions can also be accelerated using the LIA method. In a similar experiment, 30 TW, 600 fs pulses of a 10-µm diameter laser beam struck a 20-µm palladium foil target with a thin graphite layer on the opposite side of the foil. A power density of 10^{19} W/cm^2 ejected relativistic electrons from the surface of the rear of the foil that accelerated the ionized carbon atoms. There was a 17% spread in the mean energy of about 36 MeV for C^{+5} ions. Further improvements to minimize the beam's energy width are expected to enhance the usefulness of the LIA approach.

5.20.5
Future Possibilities

In the near term, laser accelerators will not replace conventional particle accelerators. Laser acceleration is also unlikely to power the next generation of proton or heavy ion accelerators. However, laser acceleration has the capability to produce high-energy electrons.

Laser acceleration has two additional strengths. First, the generation of very intense fields to accelerate particles over short distances permits the use of laser acceleration on a laboratory scale. The second advantage is the ability of laser acceleration to produce high-power beams. These characteristics promote the use of laser acceleration in future applications.

Given these strengths, laser techniques have potential twenty-first century applications. These applications include serving as a driver for future generations of electron accelerators, treating tumors with heavy ions, and providing fast ignition in inertial confinement fusion.

5.21
X-Ray Induced Isomeric Transitions

Physicists continue to search for unique energy sources that produce powerful output. Nuclear isomers hold significant potential as an energy source if the isomer can be forced to spontaneously deexcite in a controlled manner. If an isomer were to deexcite instantaneously, it would produce a pulse of gamma-ray energy, rather than distribute energy over time that follows a conventional decay process. The simultaneous decay pulse would be a potent photon source.

A recent example of the application of isomeric decays involves the ^{178}Hf system. The ^{178}Hf results noted below have been challenged, and the majority of experimental evidence suggests against the viability of spontaneous isomeric deexcitation for nuclei investigated to date.

The ^{178}Hf results noted below are presented as an illustration of the spontaneous isomeric transition concept. This discussion is not intended to validate the ^{178}Hf claims presented to date. However, a number of exotic nuclear systems, isomers, and superheavy nuclei will likely be discovered as the twenty-first century advances. Subsequent discussion is directed at the possibility of discovering a spontaneous isomeric transition in one of these systems.

The X-ray induced release of energy stored in the long-lived (31 years) isomer ^{178}Hf^{m2} has been suggested. It has been reported that a 10-keV X-ray photon initiated a prompt 2.45-MeV gamma-ray cascade as ^{178}Hf^{m2} decayed to its ground state. A variety of photon sources, including low-energy X-ray tubes and synchrotron radiation, could drive the isomer's deexcitation. Potential drivers also include X-ray free-electron lasers.

A photon-initiated decay of a long-lived isomeric state presents the potential for a new, high-intensity photon source. Given these characteristics, ^{178}Hf^{m2} stores approximately 1.3 GJ/g.

The total angular momentum (J) of the ^{178}Hf^{m2} isomer is high ($J = 16$), and its projection on the nuclear symmetry axis (K) is also high ($K = 16$). This $K = 16$ state decays to a state in the $K = 8$ band. However, selection rules for low-multipole electromagnetic transitions severely inhibit transitions that change K. This is a major reason for the 31-year lifetime of ^{178}Hf^{m2}. The lowest energy state of the $K = 8$ band is another isomer, with a half-life of 4 s, that briefly inhibits the decay to the $K = 0$ ground-state band.

Some authors suggest that X-rays excite the $^{178}\text{Hf}^{m2}$ isomer to a mixing level from which electromagnetic transitions are no longer forbidden. Although specific (J, K) transition values for the mixing level are not yet determined, a number of transitions from excited states to lower members of the ground-state band are theoretically possible. Experimental results suggest that some newly observed photons arise from transitions through these induced cascades, but intensities in any one of the lines is insufficient to fully explain the scheme of induced decay. This may be an indication that the X-ray induced decay from the mixing level proceeds through a number of transitions and that none of them produce a dominant photon transition.

Although the physical mechanisms are not yet fully established, the net result is the production of a strong photon source. Given the intensity of the source, health physics issues need to be addressed. In particular, the X-ray induced isomeric transition has been suggested as a weapon as well as an energy source. In either application, health physics controls are warranted.

From a practical standpoint, high-intensity photon sources currently exist. Such sources arise in nuclear fission and accelerator applications, and the techniques for shielding reactors and accelerators also apply to X-ray induced isomeric transitions.

5.22
Gamma-Ray Laser/Fission-Based Photon Sources

In principle, a gamma-ray laser is similar to a conventional laser except the output is a more energetic photon rather than an optical photon. A conceptual gamma-ray laser utilizes a rod containing nuclei in a metastable or isomeric state. Photon transitions occur between the metastable state and a lower energy state involved in the transition. The energy difference between the upper and lower energy states is the output wavelength of the gamma-ray laser. The rod serves as the GRASERs optical cavity. Gamma rays emerge from the end of the rod with an intensity $I(l)$:

$$I(l) = I(0) \frac{\exp[(\beta-\delta)l]-1}{(\beta-\delta)l}, \tag{5.95}$$

where l is the length of the rod, $I(0)$ is the intensity of spontaneous radiation, β is the amplification factor, and δ is the rod's absorption coefficient. The absorption coefficient is energy dependent and includes contributions from photon interactions including nuclear resonant absorption, Compton scattering, and photoelectric effect. The amplification factor is

$$\beta = \left(\frac{\pi \hbar c}{E_o}\right)^2 \frac{\Gamma_o}{\Gamma} \frac{f}{1+\alpha} \eta \zeta, \tag{5.96}$$

where E_o is the gamma-ray energy resulting from the transition, α is an internal conversion coefficient, η is the density of metastable nuclei, f is the fraction of nuclei radiating into the Mössbauer line that is near E_o (Mössbauer factor), Γ_o is the energy-level width of the transition determined from the lifetimes of the upper (τ_1) and lower states (τ_2), ξ is the population difference between the upper and lower states involved in the transition, and Γ is the actual energy-level width resulting from all broadening mechanisms. The energy level width Γ_o is

$$\Gamma_o = \hbar \left(\frac{1}{\tau_1} + \frac{1}{\tau_2} \right). \tag{5.97}$$

The necessary condition for GRASER operation is $\beta > \delta$, which gives the critical density for the metastable nuclei:

$$\eta \geq \left(\frac{E_o}{2\hbar c} \right)^2 \frac{\Gamma}{\Gamma_o} \frac{1+\alpha}{f\xi} \delta. \tag{5.98}$$

An examination of Equation 5.96 suggests that a successful GRASER must achieve large metastable-state densities and optimum lifetimes to enhance the amplification factor.

A review of possible alternatives suggests that short-lived metastable states offer the best opportunity for successfully constructing a GRASER. However, technical issues arise when producing isotopes with short-lived metastable states. The optimum metastable state has a short lifetime to facilitate lasing. However, production, chemical separation, and manufacturing appear to be mutually exclusive for isomers with sufficiently short lifetimes. To minimize the technical issues associated with constructing a GRASER, it is best to use isomers with very short lifetimes and broad homogeneous widths, and to obtain a large density of isomeric nuclei by manufacturing them in the rod prior to laser activation in an intense neutron field. Currently, the best available source of intense neutron radiation is a nuclear detonation.

If the GRASER is to be a credible device, the density condition of Equation 5.98 must be satisfied concurrent with establishing acceptable energy-level widths. The width result is most easily achieved by keeping the laser rod cool. This is a challenge because a number of energy sources exist that increase the rod's temperature. These energy sources are associated with the neutron capture reaction that creates the desired isomer and include (1) recoil following neutron capture, (2) recoil from the gamma rays resulting from the decay of the initial state to the desired isomeric state, (3) heating from the absorption of the cascade gamma rays from the decay of the isomer, and (4) heating by gamma rays from the nuclear detonation. There are a number of options for reducing the impact of these energy sources and making a gamma-ray laser a reality. Although the gamma-ray laser appears complex, it is based on the conventional laser concept and has the three basic components of a conventional laser.

The three basic components of a laser are the amplifying medium, a source of energy to pump this medium to create the population inversion, and an optical cavity to support the amplification and stimulated emission. The GRASER amplification medium is the rod containing the nuclei that will become metastable states. The energy source is the high-intensity neutron radiation from the nuclear detonation that produces the population inversion through the creation of the metastable states. Once the desired metastable states are created, a population inversion results with more nuclei residing in the metastable state than in the ground state. With an appropriately designed optical cavity, gamma-ray amplification and stimulated emission occurs. For the GRASER, the optical cavity consists of the rod and its enclosure that promote the amplification and stimulated emission.

5.23
Photon Source Health Physics and Other Hazards

A variety of photon sources have been addressed in this chapter. This section reviews the health physics hazards of synchrotron light sources, free-electron lasers, and other photon-generating approaches.

Photon sources produce external hazards that include both ionizing and nonionizing radiation. Toxic gases are another potential hazard. These hazards arise from the various modes of producing synchrotron radiation (bending magnets, storage rings, insertion devices, and free-electron lasers), electron beams used in these devices, lasers, and other photon sources. The general hazards associated with electron accelerators were addressed in Chapter 4 and will not be repeated. Radiation hazards and toxic gases associated with the specific photon source applications addressed in this chapter are discussed.

5.23.1
Ionizing Radiation

Ionizing radiation is produced from electron energy losses including the interaction of the electron beam with the beam line vacuum chamber, its components, and residual gas molecules residing within it (gas bremsstrahlung). Synchrotron radiation beams are maintained within shielded enclosures or vacuum vessels.

Absorbed dose rates from the primary beam within these enclosures reach 10^9 Gy/h or greater. The dose rates from scattered beams are at least 3–4 orders of magnitude less. Dose rates from synchrotron radiation beams outside the shielded enclosures should be negligible.

Electron losses occur during injection into a storage ring. These losses produce high-energy gamma and neutron radiation outside primary shielding. For example, at the Daresbury Laboratory Synchrotron Radiation Source in the United Kingdom, the maximum dose rates outside the shielding are typically 25 μSv/h, although higher dose rates can be experienced for short times (<60 s).

Gas bremsstrahlung radiation is high-energy photon radiation resulting from the interaction of circulating electrons with residual gas molecules in the machine's beam line. Thick beam stops and shutters are used to absorb the primary beam. Usually, the absorbed dose rate from gas bremsstrahlung is negligible compared to the primary and secondary radiation sources.

Other ionizing radiation hazards include neutron, gamma, X-ray, muon, pion, and kaon radiation resulting from interactions with the electron beam. These radiation types were discussed in Chapter 4.

Activation sources and X-ray sources resulting from RF components are additional hazards. These radiation sources are addressed in subsequent discussion.

5.23.2
Nonionizing Radiation

Nonionizing radiation arises from various radio frequency sources including klystrons, RF cavities, RF amplifiers, waveguides, and FEL output. RF leakage measurements of the power and energy density ensure that radiation levels are maintained within the American National Standards Institute (ANSI) recommendations for laser light. The recommendations of ANSI Standard Z136.1 are summarized at the end of this chapter.

Table 5.8 Properties of common synchrotron accelerator activation products.

Radionuclide	Common source	Decay mode	Decay energy (MeV)	Half-life	Biological hazard
^{3}H	Water Soil	β^-	0.0186	12.3 yr	Internal
^{7}Be	Water Oil Plastic	γ	0.477	53.3 d	External[a]
^{11}C	Air Water	β^+	0.960	20.3 min	External
^{13}N	Air Water	β^+	1.20	9.97 min	External
^{15}O	Air Water	β^+	1.73	122 s	External
^{16}N	Air Water	β^- γ	10.4 6.13	7.14 s	External
^{22}Na	Water Concrete Soil	β^+	0.546	2.6 yr	External[a]
^{54}Mn	Steel	γ	0.835	312 d	External[a]
^{55}Fe	Steel	EC	Mn X-rays	2.73 yr	External[a]
^{57}Ni	Steel	β^+	0.843	35.6 h	External[a]
^{64}Cu	Copper	β^+ β^-	0.653 0.578	12.7 h	External[a]
^{65}Zn	Copper	β^+	0.329	244 d	External[a]

[a] These activation products are also an internal hazard if mobilized during maintenance operations.

X-rays are generated from RF components including the operation of RF cavities, klystrons, and their associated power supplies. The magnitude of these X-ray hazards needs to be monitored as accelerator energies and output powers increase.

Shielding and proper grounding reduce the radiation emitted from RF devices. Photon measurements ensure that the RF devices and their associated high-voltage components do not produce unidentified X-ray radiation. Periodic surveys and safety reviews of proposed modifications are essential for maintaining radiation at acceptable levels.

5.23.3
Activation of Accelerator Components

Activity is induced when the electron beam or secondary radiation strikes machine components, structural members, shielding, air, water, and surrounding soil. The effective dose rate owing to induced activity varies with the facility configuration and machine energy. Table 5.8 provides a summary of the activation products that are likely to be encountered at a synchrotron facility.

5.23.4
Shielding Design and Safety Analysis

Shielding is a key consideration in limiting occupational radiation doses. The shielding design must consider beam confinement failures and beam loss events. Table 5.9 summarizes the risk analysis used by the Stanford Linear Accelerator Center for analyzing beam loss events. A design limit of 0.01 Sv/year for normal beam loss is mandated by the USDOE that regulates SLAC.

Table 5.9 illustrates a risk matrix method used to evaluate beam loss events in terms of the probability and consequence of an event. The method utilizes several levels (high, medium, low, extremely low, negligible, and incredible) to characterize the event probability, consequence, and risk. For the beam loss example, these levels are used to evaluate the effectiveness of the beam loss mitigation systems.

Following US Department of Energy guidance, the consequence, probability, and risk levels are interpreted as follows:

(1) A system is acceptable if the risk estimate is low or negligible.
(2) If either the probability level or the consequence level of a failure is low, the risk level will be low or negligible.
(3) If both the probability and consequence levels are low, or one is low whereas the other is medium, then the system is acceptable.
(4) Other combinations of probability and consequence levels are not acceptable, and modifications to the system to reduce the probability, consequence, and risk levels are required.

In the risk matrix approach, the consequence level should be determined by the dose equivalent and not by the dose equivalent rate. This requires the use of two

Table 5.9 Risk analysis for a beam loss scenario.

Beam loss scenario[a]	Probability level (annual frequency)	Dose equivalent limit	Consequence level	Risk level[b]
Normal	High $>10^{-1}$	0.01 Sv/yr	Extremely low	Low
Missteered[c]	Medium (10^{-2} to 10^{-1}) to extremely low (10^{-4} to 10^{-6})	4 mSv/h	Extremely low	Negligible
Failure of one interlock protection device[c]	Low (10^{-2} to 10^{-4})	0.05 Sv/h	Low	Negligible
Accident	Extremely low (10^{-4} to 10^{-6})	0.25 Sv/h (or 0.03 Sv)	Medium	Negligible
Failure of all interlock protection devices[d]	Incredible ($<10^{-6}$)	—	Not considered in shielding design	—

Derived from SLAC-TN-93-3 (1993).
[a] Beam shutoff ionization chambers can be used to terminate the beam if any beam loss scenario is detected.
[b] A system with a negligible or low risk level is acceptable.
[c] A flexible limit should be used based on the true probability level.
[d] The failure of all interlock protection devices is not normally considered as it is judged to be incredible.

additional factors. The first is the occupancy factor, which can also be included in the estimation of the event probability. The second factor is the event duration.

In addition to beam loss events, other radiological events merit attention and can be evaluated using a risk matrix approach. Radiological events that require health physics attention include the loss of control of radioactive material including sources; unanticipated radiation exposure events; personal contamination events; and fire, chemical, and other energetic events that mobilize radioactive material.

5.24
Evaluation of Radiation Dose

The determination of doses to accelerator personnel is derived from a variety of data sources. These include the installed radiation monitoring system of the facility, personal dosimetry such as personal ionization chambers and electronic dosimeters, and thermoluminescent dosimeters. Internal doses are assessed using bioassay primarily whole-body counting and urine sampling.

Procedures must be available to determine both doses occurring during normal operational and doses occurring during off-normal and emergency events. The dose assessment may also involve the use of computer codes summarized in Appendix J.

5.25
General Safety Requirements

In addition to the specific hazards previously identified, there are a number of general safety issues associated with photon sources. Procedures and equipment-operating guidelines exist for the safe operation of the primary and secondary beam lines and ancillary equipment. Special work permits and health physics coverage are required for work that affects integrity of radiation shielding, personal safety system, beam ports, beam stops, beam lines, beam shutters, beam interlocks, warning systems, target areas, radiation detectors, viewing ports, and large radioactive materials sources.

Radiological response plans should be developed for industrial events involving radioactive material. Industrial events that merit attention include fire, chemical releases, steam releases, and personal injury events. Toxic materials are discussed in subsequent discussion.

5.26
Radioactive and Toxic Gases

Accelerator-induced activation products are generated during facility operations. These products include both radioactive gases and toxic gases that result from the activation of air. For a constant rate of production (P), the change in activity (or mass) per unit time ($\overset{\circ}{A}$) of these gaseous products is given by the relationship

$$\overset{\circ}{A} = P e^{-kt}, \tag{5.99}$$

where k is the total removal rate and t is the time in which the production term is active. The concentration (C) of the gas is obtained by dividing the quantity of gas produced by the confining volume such as a target or irradiation cell volume (V):

$$C = \frac{A}{V}. \tag{5.100}$$

The concentration of the gas as a function of time is obtained by integrating Equation 5.99:

$$C(T, \tau) = \frac{1}{V} \frac{P}{k} (1 - e^{-kT}) e^{-k\tau}, \tag{5.101}$$

where T is the activation time and τ is the decay time. The activation time is the time in which the beam is active and irradiating the air, and the decay time is the time following termination of the beam. During production operations, the time for beam operation is usually much longer than the half-life of the radioactive gas or mean lifetime of the toxic gas. Therefore, equilibrium conditions normally exist and Equation 5.101 can be simplified. The concentration of radioactive gas

$C(\tau)$ and toxic gas $Z(\tau)$ can be explicitly written in terms of the time following beam shutdown:

$$C(\tau) = C(0)e^{-\left(\frac{F}{V}+\lambda\right)\tau}, \tag{5.102}$$

$$Z(\tau) = Z(0)e^{-\left(\frac{F}{V}+\frac{1}{\overline{T}}\right)\tau}, \tag{5.103}$$

where $C(0)$ is the equilibrium concentration of radioactive gas, $Z(0)$ is the equilibrium concentration of toxic gas, F is the ventilation rate, $\overline{T}$ is the mean lifetime of the toxic gas, and λ is the radioactive decay constant of the radioactive gas.

Toxic gases include ozone, nitrous oxides, or sulfur oxides that arise from electron ionization of the air or from materials in the irradiation cell. Radioactive gases arising from the activation of air are provided in Table 5.8.

5.27
Laser Safety Calculations

Free-electron laser output will span the infrared, visible, ultraviolet, X-ray, and gamma-ray regions of the electromagnetic spectrum. The ionizing X-ray and gamma-ray regions are addressed using the conventional tools summarized in the previous discussion and Appendix J. The nonionizing regions require a different approach.

A common approach for performing a laser safety calculation is to use the methodology of the American National Standards Institute publication ANSI Z136.1. To perform a laser safety calculation, a number of parameters must be specified. These include the limiting aperture, exposure time, and maximum permissible exposure (MPE). The selection of laser safety parameters depends on the format of the laser output (pulsed or continuous). The basis for the selection of these parameters is described in subsequent discussion.

5.27.1
Limiting Aperture

ANSI Z136.1 defines the limiting aperture based on the wavelength and the pulse duration. If the beam area is larger than the limiting aperture, the standard recommends the use of the actual area in the laser safety calculation. However, if the beam area is smaller than the limiting aperture, use the limiting aperture in the calculation. For radiation within the range of 400–1400 nm and for pulse durations between 1.0×10^{-13} and 3.0×10^{4} s, the limiting aperture diameter for the eye is 7 mm and 3.5 mm for the skin.

5.27.2
Exposure Time/Maximum Permissible Exposure

For continuous wave (CW) lasers, the MPE is determined from the exposure time and the wavelength of the laser radiation. The exposure time is determined by the laser wavelength. For visible CW lasers (400–700 nm), the exposure duration is the maximum time of anticipated exposure. If purposeful staring into the beam is not intended or anticipated, the aversion (blinking) response time of 0.25 s is used as the exposure duration.

For nonvisible CW lasers (less than 400 nm or greater than 700 nm), the exposure duration is the maximum time of anticipated exposure. For the hazard evaluation of retinal exposures in the near infrared (700–1400 nm), an exposure duration of 10 s provides an adequate hazard criterion for either incidental viewing or purposeful staring conditions. Eye movements provide a natural exposure time limitation. For special applications, such as health care or experimental conditions, longer exposure durations may be appropriate.

For pulsed lasers, the MPE is derived from three pulsed laser calculation rules. The selected MPE is the minimum value derived from these rules. Rule 1 addresses the single-pulse MPE and specifies that the exposure time is the pulse width and that the MPE is determined from the laser wavelength and the exposure duration.

The second rule protects against the average power MPE for thermal and photochemical hazards:

$$\text{MPE-2} = \frac{\text{CW MPE for the same wavelength}}{\text{PRF}}, \tag{5.104}$$

$$\text{MPE-2} = \frac{\text{W/cm}^2}{\text{pulses/s}} = \frac{\text{J}}{\text{cm}^2 \text{ pulse}}, \tag{5.105}$$

where PRF is the pulse repetition frequency.

The third rule addresses a multiple-pulse MPE for thermal hazards. Rule 3 protects against subthreshold pulse cumulative thermal injury:

$$\text{MPE-3} = (\text{individual pulse MPE})n^{-1/4} \text{ with } n^{-1/4} \leq 1, \tag{5.106}$$

where n is the number of pulses during the exposure duration (T):

$$n = \text{PRF} \times T. \tag{5.107}$$

The exposure duration is determined as noted above for CW lasers.

The application of ANSI Z136.1 to laser safety calculations is illustrated in Problem 05-08. Application of photon output from various light sources are also included in the Chapter 5 problems.

Problems

05-01 An insertion device with a 50-mm period and an undulator parameter of 2 is located in a 2-GeV electron storage ring. In the insertion device, the electron travels in the z-direction. For this configuration, calculate the following insertion device characteristics: (a) total electron energy, (b) electron velocity, (c) electron's Lorentz factor, (d) the average electron velocity along the direction of motion, (e) the electron's maximum deflection angle, (f) the maximum oscillation amplitude of the electron, (g) the electron's maximum displacement in the z-direction, and (h) the emitted wavelength of the fundamental ($n=1$) of this device along the electron beam direction.

05-02 The University of Western Rhode Island has commissioned a new free-electron laser that produces a 0.5-MeV photon beam. Your boss reviews the target-shielding design documentation and questions not using gamma-ray buildup factors. Is he correct? Why?

05-03 The Black Hole University Physics Department will upgrade a laboratory currently holding a rotating anode X-ray source to house a new X-ray free-electron laser operating at the same energy. The brightness values for the old and new X-ray sources are summarized in the following table. Assume that each device is required to meet the same kerma rate standard. How many additional half-value layers of shielding are needed for the new X-ray source?.

	X-ray Source Brightness
Production mode	Brightness ($\gamma \cdot s^{-1} \cdot mm^{-2} \cdot mr^{-2} \cdot (0.1\% \text{ bandwidth})^{-1}$)
Rotating anode X-ray tube	10^8
X-ray FEL	10^{35}

05-04 A dipole magnet upgrade has been installed at the James T. Kqirk Synchrotron Facility (JTKSF). The JTKSF upgrade has an electron beam energy of 6 GeV and current of 200 mA. (a) What is the spectral flux density at the critical wavelength? Assume that the spectral shape function $G(1) = 0.65$. (b) Given the electron beam standard deviations ($\sigma_x = 0.07$ mm, $\sigma_z = 0.032$ mm, and $\sigma'_z = 0.055$ mr), what is the average dipole brightness?

05-05 The Elmer T. Fuddwell Memorial Synchrotron (ETFMS) utilizes a 50-GeV electron beam. During operations, the accelerator has an average current of 6 mA, a bending radius of 3096 m, and a magnetic induction of 0.054 T. (a) What is the total power in the synchrotron spectrum? (b) What is the critical wavelength? (c) What is the critical energy? (d) A new accelerator is planned that will upgrade the ETFMS to 500 GeV with a 75-mA beam and a bending radius of 10 km. If the dose rate scales with the total power, what increase in dose rate is expected following this upgrade?

05-06 The University of South Eastern Alaska (USEA) has installed an undulator that utilizes a 6-GeV electron beam. USEA's undulator has a period of 46 mm, an effective magnetic field of 0.233 T, and an average beam current of 200 mA. (a) What is the wavelength of the fundamental? (b) If the undulator has a length of 1.66 m, what is the fundamental's flux in the central radiation cone? Assume that the spectral harmonic factor f_1 is 0.37. (c) What is the average on-axis brightness? Assume the photon source sizes $\sigma_{\gamma x(\gamma z)}$ in the $x(z)$ direction are 0.06 mm (0.013 mm) and the photon source divergence ($\sigma_{\gamma x'(\gamma z')}$) in the $x(z)$ direction including diffraction effects are 0.12 mr (0.012 mr). (d) Assume that the flux in the central cone expands to fully illuminate 1.0 sr at a distance of 500 m and that the bandwidth multiplier is $2.0 \times 0.1\%$bandwidth. What is the absorbed dose rate at 500 m if the flux-to-dose conversion factor is 2.2×10^{-10} (Gy/h)/(γ/m^2 s)? (e) The USEA is reviewing options for a subsequent upgraded machine. The three options under consideration are the following:

USEA synchrotron upgrade options

Option	Number of steradians illuminated at the experimental location (r)	Experimental location r (m)	Average beam current I_o (mA)	Undulator length L (m)
1	0.5	600	500	2.5
2	2	400	1000	3.0
3	π	500	300	5.0

Assume that the undulator parameter, electron beam energy, bandwidth multiplier, undulator period, and dose-conversion factors are the same for all three options. If all three the options meet the scientific objectives of the USEA, which one is the preferred option from a radiological perspective?

05-07 A 10-GeV electron beam is utilized at the Black Hole FEL. The FEL has a wiggler parameter value of 1.0 and a wiggler period of 2.5 cm. (a) What is the output wavelength of the first harmonic for the FEL? (b) If an output wavelength of 0.1 Å is desired for the first harmonic, what is the required electron beam energy?

05-08 During startup testing of the Australian National FEL, beams are created for calibration purposes to simulate a HeNe laser (633 nm) and a ruby laser (694.3 nm). The beams are collimated using a test aperture having an associated divergence. Base your answers to the following questions on ANSI Z136.1-2000:

Maximum permissible exposure for direct ocular exposure intrabeam viewing from a laser beam[a]

Wavelength λ (μm)	Exposure time t (s)	MPE
0.400–0.700	10^{-9} to 1.8×10^{-5}	5×10^{-7} J/cm^2
0.400–0.700	1.8×10^{-5} to 10	$1.8\, t^{3/4} \times 10^{-3}$ J/cm^2
0.400–0.550	10 to 1×10^4	1×10^{-2} J/cm^2
0.550–0.700	10 to T_1	$1.8\, t^{3/4} \times 10^{-3}$ J/cm^2
0.550–0.700	T_1 to 1×10^4	$10\, C_B \times 10^{-3}$ J/cm^2
0.550–0.700	10^4 to 4×10^4	$C_B \times 10^{-6}$ W/cm^2

[a] $C_B = 1$ for $\lambda = 0.400$–$0.550\,\mu m$; $C_B = 10^{15(\lambda - 0.550)}$ for $\lambda = 0.550$–$0.700\,\mu m$; $T_1 = 10 \times 10^{20(\lambda - 0.550)}$ s for $\lambda = 0.550$–$0.700\,\mu m$ [from ANSI Z136.1-2000].

(a) The 694.3-nm beam has a pulse energy of 20 J, a pulse repetition frequency of 2 pulses/min, a pulse duration of 10 μs, a beam divergence of 15 mr, and an aperture diameter of 2 mm. What is the required optical density for protective goggles to reduce the FELs radiant exposure to the maximum permissible exposure

at 1 m from the outlet aperture? Assume that purposeful staring into the beam does not occur.
(b) The 633-nm beam has an output of 50 mW, an aperture diameter of 3 mm, and a beam divergence of 0.3 mr. Calculate the emergent irradiance.
(c) For the conditions of the previous problem, calculate the hazardous intrabeam viewing distance. Ignore atmospheric attenuation. For this problem, assume that the MPE is determined by continuous viewing.

05-09 A 6-GeV linear accelerator is used to supply electrons to the University of North Brazil Free Electron Laser Laboratory. (a) What activation products of air are expected? (b) What activation products of water will likely be produced? (c) What activation products of soil are anticipated? (d) Are any activation products expected to occur in the iron structures? For each of these questions list the likely production modes of the activation products.

05-10 For the previous FEL system, what radiation types of health physics significance are produced by interactions with the 6-GeV electron beam?

05-11 During FEL startup testing, a 25-MeV electron beam is established with an average beam current of 400 µA. The electron beam strikes a temporary concrete beam stop. From a radiological perspective assume that the dominant elements of interest in the concrete are sodium, potassium, and iron. The concrete has a density of $2.37 \, \text{g/cm}^3$. The characteristics of the radiologically significant elements are summarized in the following table:

	Target elements in concrete	
Element	Atomic mass	Mass density (g/cm^3)
Na	22.99	0.012
K	39.10	0.008
Fe	55.85	0.018

The FEL test chamber has a thin tungsten-niobium-hydride exit window that results in a thermal neutron yield of 0.001 neutrons per electron. The thermal neutron activation cross section for the target elements and the characteristics of the activation products from these interactions are illustrated in the following table:

		Target element thermal neutron reactions		
Target isotope	Abundance (%)	Activation product	Half-life	Cross section (cm^2/g)
^{23}Na	100	^{24}Na	15.0 h	1.39×10^{-2}
^{41}K	6.77	^{42}K	12.4 h	1.22×10^{-3}
^{58}Fe	0.31	^{59}Fe	45.6 d	3.01×10^{-5}

(a) What is the thermal neutron fluence rate (flux) at a distance of 3.0 m from the electron exit window? (b) For a thermal neutron fluence rate of 2×10^7 n/cm² s at the concrete surface, calculate the ^{24}Na activity in 1 cm³ of concrete at saturation. (c) Calculate the ratio of saturation activities for the ^{42}K and ^{24}Na.

05-12 The International Brotherhood of Nuclear Workers has filed a grievance regarding task completion times for the electron beam dump area of a FEL at the University of Eastern Siberia. The beam dump resides in a cubicle having a volume of 500 m³ and an exhaust velocity of 4 m³/s. As part of the arbitration process, you have been asked to answer the following questions:

(a) Calculate the time following the shutdown of the electron beam for the radioactive gas concentration to be reduced to 2 Bq/cm³. The equilibrium radioactive gas concentrations in the beam dump area are provided in the following table:

Radioactive gas concentration in the electron beam dump area

Radionuclide	Half-life (min)	Equilibrium concentration (Bq/cm³)
^{16}N	0.12	6.5×10^5
^{15}O	2	3.9×10^4
^{13}N	10	7.3×10^4

(b) The equilibrium toxic gas concentration in the beam dump cubicle is 5.5 ppm. If the mean lifetime of the toxic gas is 30 min, calculate the time for the concentration to be reduced to 0.1 ppm.

05-13 You are the senior health physicist at the Orsay National Laboratory's 12 GeV Free Electron Laser Laboratory. The laboratory administrator is concerned that residual ozone (O_3) and oxides of nitrogen (NO_x) will delay access to electron beam support equipment after the electron beam is terminated. The facility has a peak power of 50 MW, a duty factor of 0.12, a support area volume of 95 m³, a support area ventilation rate of 5 m³/s, and an electron beam path length in air of 10 m.

Ozone production (P) is characterized in terms of the following empirical relationship:

$$P(\text{molecules}/\text{cm}^3 \text{ s}) = (600 \text{ eV}/\text{cm}^4 \text{ A s}) \times GId,$$

where G has the value of 10.3 molecules/100 eV for ozone, I is the average beam current (A), and d is the path length traveled by the electron beam in air (cm).

(a) What is the ozone production rate in molecules/cm³ s?
(b) For a production rate of 150 molecules/cm³ s, calculate the steady-state NO_x concentration in the support area. Assume that the mean lifetime of NO_x is 1800 s.

05-14 You are responsible for health physics at the Top Quark Free Electron Laser Laboratory (TQFELL). The facility is designed to operate over a broad range of output wavelengths and gain values. In its initial configuration, the beam current is 10 A, the interaction length is 1 m, the undulator parameter is 0.7, the electron energy corresponds to $\gamma = 100$, the output wavelength is 1 μm, and the beam plus optical mode area is 2×10^{-6} m^2. (a) If the impedance of free space (Z) is 377 Ω, what is the j-value for the TQFELL? Assume the FEL uses a helical wiggler. (b) What is the maximum gain, assuming that the FEL operates in a low-gain mode? (c) A second FEL configuration is under evaluation. This configuration has a beam current of 100 A, an interaction length of 5 m, and an undulator parameter of 1.2. What is the ratio of the new and original FEL absorbed doses?

05-15 A free-electron laser has an output to input power ratio of 10^{10}, a net gain of 0.2, and a cavity length of 2 m. (a) What is the cavity buildup time? (b) A buildup time of 4×10^{-7} s results from a proposed FEL modification. If the cavity length and the gain are unchanged, what is the power ratio for the modified FEL? (c) What is the half-divergence angle of an output beam having a wavelength of 1 μm? Assume that the interaction and cavity lengths are the same. (d) For the parameters of Question (c), what is the beam emittance?

05-16 You have been tasked with assessing the effect of incorporating an optical klystron into the Boston University FEL design. The proposed modification has a 0.2-m-long dispersive magnet, equal length modulation and output sections with each having a length of 0.4 m, a dispersive magnet field strength of 0.3 T, a wiggler parameter of 1.0, and a free-electron laser length of 1 m. (a) What is the ratio of FEL gain values with and without the optical klystron? (b) What is the ratio of energy acceptance values ($\Delta\gamma/\gamma$) of the OK modification relative to that of the FEL?

References

ANSI Z136.1-2000 (2000) American National Standard for Safe Use of Lasers, American National Standards Institute, New York.

Bauer, P., Darve, C. and Terechkine, I. (2002) FERMILAB-Conf-01/434, *Synchrotron Radiation Issues in Future Hadron Colliders*, Fermi National Accelerator Laboratory, Batavia, IL.

Bevelacqua, J.J. (1995) *Contemporary Health Physics: Problems and Solutions*, John Wiley & Sons, Inc., New York.

Bevelacqua, J.J. (1999) *Basic Health Physics: Problems and Solutions*, John Wiley & Sons, Inc., New York.

Bonifacio, R., McNeil, B.W.J. and Pierini, P. (1989) Slippage and superradiance in the high-gain free electron laser. *Physical Reviews*, **A40**, 4467.

Brinkmann, U. (2006) VUV free electron laser in operation at DESY. *Laser Focus World*, **42** (1), 37.

Bryant, P.J. and Johnsen, K. (1993) *The Principles of Circular Accelerators and Storage Rings*, Cambridge University Press, Cambridge, UK.

Chao, A.W. and Tigner, M. (eds)(2006) *Handbook of Accelerator Physics and Engineering*, World Scientific Publishing, Co., Singapore (3rd printing).

Ciocci, F. (2000) *Insertion Devices for Synchrotron Radiation and Free Electron*

Lasers, World Scientific Publishing, Co., Singapore.

Clarke, J.A. (2004) *The Science and Technology of Undulators and Wigglers*, Oxford University Press, New York.

Collins, C.B., Rusu, A.C., Zoita, N.C., Iosif, M.C., Camase, D.T., Davanloo, F., Ur, C.A., Popescu, I.I., Pouvesle, J.M., Dussart, R., Kirischuk, V.I., Strilchuk, N.V. and Agee, F.J. (2001) Gamma-ray transitions induced in nuclear spin isomers by X-rays. *Hyperfine Interactions*, **135**, 51.

Collins, C.B., Zoita, N.C., Davanloo, F., Emura, S., Yoda, Y., Uruga, T., Patterson, B., Schmitt, B., Pouvesle, J.M., Popescu, I.I., Kirischuk, V.I. and Strilchuk, N.V. (2004) Accelerated decay of the 31-year isomer of Hf-178 induced by low-energy photons and electrons. *Laser Physics*, **14** (2), 154.

Cossairt, J.D. (2004) Radiation Physics for Personnel and Envitonmental Protection, Fermilab Report TM-1834, Revision 7, Fermi National Accelerator Laboratory, Batavia, IL.

Cossairt, J.D. and Mokhov, N.V. (2002) Fermilab-Pub-02/043-E-Rev, Assessment of the Prompt Radiation Hazards of Trapped Antiprotons, Fermi National Accelerator Laboratory, Batavia, IL.

Dattoli, G., Renieri, A. and Torre, A. (1993) *Lectures on the Free Electron Laser – Theory and Related Topics*, World Scientific Publishing, Co., New York.

Fassò, A., Goebel, K., Höfert, M., Ranft, J. and Stevenson, G. (1990) Landolt-Börnstein Numerical Data and Functional Relationships in Science and Technology New Series; Group I: Nuclear and Particle Physics Volume II: Shielding Against High Energy Radiation (O. Madelung, Editor-in-Chief, Springer-Verlag, Berlin, Heidelberg, 1990).

Fassò, A., Goebel, K., Höfert, M., Rau, G., Schönbacher, H., Stevenson, G.R., Sullivan, A.H., Swanson, W.P. and Tuyn, J.W.N. (1984) CERN 84-02, *Radiation Problems in the Design of the Large Electron–Positron Collider (LEP)*, CERN, European Organization for Nuclear Research, Geneva, Switzerland.

Feder, T. (2005) Accelerator labs regroup as photon science surges. *Physics Today*, **58** (5), 26.

Hazi, A. (2006) UCRL-TR-218474, *Testing the Physics of Nuclear Isomers*, Lawrence Livermore National Laboratory, Livermore, CA.

Hecht, J. (2006) Short pulses speed particles. *Laser Focus World*, **42** (4), 79.

LCLS Design Study Group(1998) SLAC-R-0521, *Linac Coherent Light Source (LCLS) Design Study Report*, Stanford Linear Accelerator Center, Stanford, CA.

Litvinenko, V.N., Burnham, B., Emamian, M., Hower, N., Madey, J.M.J., Morcombe, P., O'Shea, P.G., Park, S.H., Sachtschale, R., Straub, K.D., Swift, G., Wang, P., Wu, Y., Canon, R.S., Howell, C.R., Roberson, N.R., Schreiber, E.C., Spraker, M., Tornow, W., Weller, H.R., Pinayev, I.V., Gavrilov, N.G., Fedotov, M.G., Kulipanov, G.N., Kurkin, G.Y., Mikhailov, S.F., Popik, V.M., Skrinsky, A.N., Vinokurov, N.A., Norum, B.E., Lumpkin, A. and Yang, B. (1997) Gamma-ray production in a storage ring free-electron laser. *Physical Review Letters*, **78**, 4569.

Liu, J.C. (1993) SLAC-TN-93-3, *The Personal Protection System for a Synchrotron Radiation Accelerator Facility: Radiation Safety Perspective*, Stanford Linear Accelerator Center, Stanford, CA.

National Academy of Sciences Report(1994) Free Electron Lasers and Other Advanced Sources of Light: Scientific Research Opportunities, The National Academy of Sciences, Washington, DC.

Nuhn, H.-D. (2003) SLAC-PUB-10300, *From Storage Rings to Free Electron Lasers for Hard X-rays*, Stanford Linear Accelerator Center, Stanford, CA.

Nuhn, H.-D. (2004) From storage rings to free electron lasers for hard X-rays. *Journal of Physics Condensed Matter*, **16**, S3413.

O'Shea, P.G. and Freund, H.P. (2001) Free electron lasers: status and applications. *Science*, **292**, 1853.

Pantell, R.H. (1989) Free Electron Lasers, in Proceedings of the Beijing FEL Seminar, Beijing University, August 11–23, 1988, (eds C. Jiaer, X. Jialin, D. Xiangwon and Z. Kui), World Scientific Publishing, Co., Singapore.

Particle Data Group(2006) Review of particle properties. *Journal of Physics G: Nuclear and Particle Physics*, **33**, 1.

Saldin, E.L., Schneidmiller, E.V. and Yorkov, M.V. (2000) *The Physics of Free Electron Lasers*, Springer, New York

Schwarzschild, B. (May 2004) Conflicting results on a long-lived nuclear isomer of hafnium have wider implications. *Physics Today*, **21** (5).

Sessler, A.M. and Vaughan, D. (1987) Free-electron lasers. *American Scientist*, **75** (1), 34.

Shiozawa, T. (2004) *Classical Relativistic Electrodynamics: Theory of Light Emission and Application to Free Electron Lasers*, Springer, New York.

Solem, J.C. (1979) *LA-7898-MS, On the Feasibility of an Impulsively Driven Gamma-Ray Laser*, Los Alamos Scientific Laboratory, Los Alamos, NM.

Suller, V.P. (1998) Introduction to Current and Brightness Limits, in CERN 98-04, *Synchrotron Radiation and Free Electron Lasers*, European Organization for Nuclear Research, Geneva, Switzerland,77.

Tavish, G. (1995) Experimental Requirements for a Self Amplified Spontaneous Emission Test System: Design, Construction, Simulation, and Analysis of the UCLA High Gain Free Electron Laser, PhD thesis, University of California Los Angeles, Los Angeles, CA.

USDOE Guide (2001) DOE O 420.2A, *Accelerator Safety Implementation Guide for Safety of Accelerator Facilities*, U.S. Department of Energy, Washington, DC.

USDOE Order(2004) DOE O 420.2B, *Safety of Accelerator Facilities* U.S. Department of Energy, Washington, DC.

Walker, R.P. (1998) Insertion Devices: Undulators and Wigglers, in CERN 98-04, *Synchrotron Radiation and Free Electron Lasers*, European Organization for Nuclear Research, Geneva, Switzerland, p. 129.

Weinberger, S. (2006) *Imaginary Weapons – A Journey through the Pentagon's Scientific Underworld*, Nation Books, New York.

Wiedemann, H. (2003) *Synchrotron Radiation*, Springer-Verlag, Berlin.

IV
Space

Part Four examines planetary and deep space missions. A manned Mars mission is planned for the third decade of the twenty-first century. Missions to the outer planets have been conceptually discussed, but are largely undefined. Travel to neighboring star systems are mostly topics for science fiction, but are conceptually feasible considering recent theoretical efforts. These efforts involve unique space–time geometries that challenge established perceptions of space and time.

6
Manned Planetary Missions

6.1
Overview

Following a period of successful low-Earth orbit and limited lunar missions, manned spaceflight is poised to enter a period of expansion and exploration. The next phase of manned space missions includes a return to the moon, with subsequent missions to Mars, outer planetary missions, and the possibility of interstellar missions occurring during the late twenty-first century. This chapter provides an overview of planetary missions and addresses their likely health physics challenges. The next chapter reviews deep space missions beyond our Solar System including missions to nearby star systems.

6.2
Introduction

Numerous low-Earth orbit (LEO) and lunar missions have been successfully accomplished, and the general characteristics of the LEO and lunar radiation environment are relatively well established. However, we know considerably less about planetary and deep space radiation environments.

This chapter reviews available radiation data for LEO and lunar missions. In particular, Mercury, Gemini, Apollo, Skylab, Space Transport Shuttle (STS), International Space Station (ISS), and NASA–Mir data are reviewed to obtain average effective dose rates applicable to a portion of more complex missions. In evaluating planetary missions, effective dose values are determined for planetary transit from Earth, planet surface, and return Earth transit. For each mission phase, the various sources of radiation, the radiation types comprising those sources, and the magnitude of the associated radiation doses are discussed. This chapter also summarizes the health physics considerations for planetary missions with a focus on the initial manned Mars mission.

Missions beyond Mars can encounter unique physical and radiation environments, but these missions are not well defined. Mission durations, destinations, and

Health Physics in the 21st Century. Joseph John Bevelacqua
Copyright © 2008 WILEY-VCH Verlag GmbH & Co. KGaA, Weinheim
ISBN: 978-3-527-40822-1

trajectories have a significant impact on the radiation profile. These aspects and other relevant health physics considerations involved in manned planetary missions are also discussed in this chapter.

6.3
Terminology

In describing planetary missions, it is necessary to introduce terminology to facilitate the presentation of the space radiation environment. Relevant terms include the following:

- *AU (astronomical unit)* – The average distance from the Sun to the Earth (1.5×10^8 km).

- *Blood-forming organs (BFO)* – A term defined in NCRP 132 to denote the dose equivalent at a depth of 5 cm.

- *Gray equivalent (Gy Eq)* – A dose weighted by the relative biological effectiveness (RBE). Following NCRP 132, dose limits for deterministic effects are expressed as the organ dose in gray multiplied by the relative biological effectiveness for the specific organ and radiation.

- *Hohmann orbit* – An elliptical trajectory, named after German rocket engineer Walter Hohmann. With this trajectory, a spacecraft moves from one orbit to another with a minimum expenditure of energy. It involves two firings of the spacecraft's engine: one to transition from the original orbit and another to enter the destination orbit. Its chief disadvantage is that it requires relatively long flight times.

- *Inclination* – The acute angle that the trajectory of an orbit makes with a planet's equator.

- *Sol* – The name of Earth's sun.

- *Solar minimum* – The portion of the 11-year Solar cycle during which the Solar wind is least intense resulting in higher levels of galactic cosmic radiation about the Earth.

- *Solar maximum* – The portion of the 11-year Solar cycle during which the Solar wind is most intense resulting in lower levels of galactic cosmic radiation about the Earth.

- *Solar wind* – The plasma flowing into space from the Solar corona. The ionized gas and its associated electromagnetic fields alter the intensity of the interplanetary radiation.

- *Unrestricted linear energy transfer (L_∞)* – The quotient of dE and dl where dE is the mean energy lost by the particle because of collisions with electrons traveling a distance dl (i.e., $L_\infty = dE/dl$). The incident particles are not restricted to originate from a fixed direction. The unit for linear energy transfer (LET) is J/m. Conventional units for LET are keV/μm.

The literature cited in this chapter provides radiation dose information in terms of a variety of quantities including the absorbed dose, dose equivalent, gray equivalent, effective dose, whole-body dose, skin dose equivalent, eye dose equivalent, and blood-forming organ dose equivalent. The uncertainties in quality factors, relative biological effectiveness values, and radiation-weighting factors, and quantification of the space radiation fields do not require a rigorous distinction between these quantities (e.g., dose equivalent and effective dose). Accordingly, various dosimetry units appear in this chapter for consistency with the literature. Appropriate comments regarding dosimetry information are provided in subsequent discussion.

6.4 Basic Physics Overview

Space radiation is grouped into three broad components or source terms. These components involve the source of the radiation, namely, (1) a planet's equivalent of the Earth's van Allen belts (VABs) in which charged particles are trapped by the planet's magnetic field, (2) galactic cosmic rays (GCRs), and (3) solar particle events (SPEs). Table 6.1 summarizes the impact of the fundamental physics interactions on the three space radiation components. The radiation characteristics of planetary VABs, GCRs, and SPEs are addressed in subsequent sections of this chapter.

Table 6.1 also lists each of the four fundamental interactions. For each interaction, the space radiation component impacted by the fundamental interaction and specific reaction types affected by each of the fundamental interactions are listed. In addition, comments regarding the physical impact of the fundamental interactions on the space radiation component are provided. Emphasis is placed on characteristics that affect important health physics parameters (e.g., reaction rates, radiation buildup in shields, particle stability, particle trajectories, energy levels, and yields).

The description of the gravitational interaction in Table 6.1 is the most speculative one. Extra spatial dimensions are an area of active research, but they have yet to be experimentally verified. Subsequent discussion in Chapter 7 illustrates how the emergence of extra spatial dimensions affects the calculation of dose rates and the magnitude of nuclear energy levels.

The four fundamental interactions govern nucleosynthesis, the process through which elements heavier than hydrogen are produced in stars. Stars are initially composed of hydrogen, but sufficient stellar mass compresses the hydrogen gas and fusion reactions occur. Fusion interactions that produce selected elements up to ^{16}O include

$$n+p \rightarrow {}^2H, \tag{6.1}$$

$$^1n + {}^2H \rightarrow {}^3H, \tag{6.2}$$

Table 6.1 Fundamental interactions and their impact on space radiation components.

Interaction	Affected component	Reactions	Comment
Strong	VAB GCR SPE	• Nucleosynthesis • Fission • Fusion • Other nuclear reactions	• Impacts reaction rates • Governs secondary radiation buildup in shields • Impacts particle stability
Electromagnetic	VAB GCR SPE	• Nucleosynthesis • Coulomb Interactions	• Influences reactions of charged particles • Governs the trajectory of ions
Weak	GCR SPE	• Nucleosynthesis • Nuclear reactions (particularly beta decay)	• Impacts particle stability • Impacts reaction rates • Impacts secondary radiation
Gravitational	Uncertain	Potentially all interactions are affected depending on the number and scale of any extra spatial dimensions. Local spatial anomalies and interaction fields contribute to the emergence of extra spatial dimensions beyond the expected three-dimensional space	The impact depends on local spatial conditions. Depending on the scale of any extra dimensions, energy levels and associated yields and the functional form of basic dosimetry equations could be affected

$$^{2}H + ^{3}H \rightarrow ^{4}He + n, \tag{6.3}$$

$$^{2}H + ^{2}H \rightarrow ^{3}H + p \rightarrow ^{4}He + \gamma, \tag{6.4}$$

$$^{2}H + ^{2}H \rightarrow ^{3}He + n \rightarrow ^{4}He + \gamma, \tag{6.5}$$

$$^{4}He + ^{4}He \rightarrow ^{8}Be, \tag{6.6}$$

$$^{8}Be + ^{4}He \rightarrow ^{12}C, \tag{6.7}$$

$$^{12}C + ^{4}He \rightarrow ^{16}O. \tag{6.8}$$

In the next step of the fusion process, oxygen combines with an alpha particle to form neon. This step is somewhat more difficult because quantum effects, including a number of selection rules, inhibit the formation of neon. Difficulties in producing neon create a situation in which stellar nucleosynthesis produces relatively large amounts of carbon and oxygen, but only a small fraction of these elements is converted into neon and heavier elements. Fusion only produces elements up to iron. Heavier elements are created with other processes including those resulting from a supernova explosion.

The nucleosynthesis mechanisms summarized in Equations 6.1 through 6.8 are interrelated, and each production reaction dependents on previous reactions. As the production mechanisms favor light nuclei and the universe started with predominantly hydrogen, the abundance of nuclei in the universe tends to decrease as the mass of the nucleus increases. This effect is important because it governs the composition of GCRs and SPEs.

6.5
Radiation Protection Limitations

The space radiation environment and dose limit recommendations such as those published in NCRP 132 are considerations for limiting the duration of space missions. The NCRP 132 recommendations only apply to activities in low Earth orbit. No recommendations currently exist for planetary or deep space missions, but NCRP 153 outlines information needed to make radiation protection recommendations for space missions beyond low Earth orbit. Given these considerations, the expected mission length for various destinations is provided in Table 6.2.

The NCRP 132 LEO recommendations are established for short-term exposure, limiting health effects, and career doses. Included in the NCRP 132 recommendations are career whole-body exposure limits for lifetime excess risk of total cancer of 3% (Table 6.3), 10-year career limits based on 3% excess lifetime risk of cancer mortality (Table 6.4), and dose limits for all ages and both genders (Table 6.5).

The NCRP 132 risk estimates are subject to large uncertainties. Part of this uncertainty is inherent in the nature of SPEs. These uncertainties include limits of scientific knowledge, risk model limitations, and lack of data to adequately characterize the risk. In addition, these uncertainties lead to shielding requirements that place significant limitations on space vehicle design and mission duration. Given these uncertainties, risk estimates suggest that for each week in space outside the Earth's magnetosphere there is a 1 in 500 chance that unshielded astronauts will receive a lethal dose from Solar flare radiation. Missions on the order of 2 years would correspond to approximately a 20% chance for exceeding a lethal dose. Considering these doses and their associated probability of occurrence, the potential for adverse health effects owing to radiation exposure threatens man's ability to develop long-duration space missions unless appropriate shielding or other protective measures are provided.

6.6
Overview of the Space Radiation Environment

The characteristics of the three dominant space radiation source terms (i.e., trapped radiation (VAB), GCR, and SPE) are summarized in Table 6.6. The VAB data of

Table 6.2 Length of different types of space exploration missions.[a]

Destination	Mission duration (d)
LEO	180
Earth's moon	100
Mars and asteroids	500–1000
Other planets	>2000

[a]Derived from Schimmerling (2003).

Table 6.3 Career whole-body exposure limits for a lifetime excess risk of total cancer of 3% as a function of age at exposure.[a]

Age (yr)	Female (Sv)	Male (Sv)
25	1.0	1.5
35	1.75	2.5
45	2.5	3.25
55	3.0	4.0

[a]Derived from NCRP 132 (2000).

Table 6.4 Ten-year career limits based on 3% excess lifetime risk of cancer mortality.[a]

Age at exposure (yr)	Effective dose (Sv)	
	Female	Male
25	0.4	0.7
35	0.6	1.0
45	0.9	1.5
55	1.7	3.0

[a]Derived from NCRP 132 (2000).

Table 6.5 Recommended dose limits for all ages and both genders.[a]

Time frame	Blood-forming organs (Gy Eq)	Eye (Gy Eq)	Skin (Gy Eq)
Career	[b]	4.0	6.0
1 y	0.50	2.0	3.0
30 d	0.25	1.0	1.5

[a]Derived from NCRP 132 (2000).
[b]The career stochastic limits in Table 6.4 are adequate for protection against deterministic effects.

Table 6.6 are applicable to Earth. A more detailed description of the LEO environment is provided in subsequent discussion.

An examination of Table 6.6 suggests that the space radiation environment is complex, and the individual source terms are dominant at different spatial locations.

Table 6.6 Characteristics of space radiation.[a]

Characteristic	Trapped radiation [VAB (Earth)]	GCR	SPE
Proton energy range (MeV)	Up to several hundred	Up to several thousand	Up to several hundred
Energetic, highly charged nuclei (MeV/nucleon)	No significant contribution	Up to several thousand	No significant contribution
LET range (keV/μm)	0.25–10	0.25–1000	0.25–10

[a]Derived from Schimmerling (2003).

The van Allen belts are important for low Earth orbit. At higher latitudes GCRs are also important. Both SPEs and GCRs are important for missions outside the Earth's magnetosphere. These source terms also vary with Solar and extra-solar conditions.

The sun is constantly releasing protons and electrons as a result of fusion reactions. These particles are trapped in the Earth's magnetic field where they circulate between the north and south magnetic poles. This physical process produces zones of radiation or belts of trapped radiation referred to as the VABs.

Protons, with smaller admixtures of helium ions and heavy ions, are the dominant components of GCRs and SPEs. There are a number of differences in the particle types, energy distribution, and emission frequency of GCRs and SPEs.

The first difference between GCRs and SPEs is the energy distribution of the emitted particles. GCRs are extremely high-energy events that originate outside the Solar System. SPEs are lower energy events governed by Solar dynamics. SPEs may be produced by coronal mass ejections or Solar flare events.

The second difference lies in the relative periodicity of SPEs and GCRs. SPEs are sporadic and governed by stellar dynamics and Solar plasma instabilities. It is difficult to forecast the onset, duration, and magnitude of the Solar mass ejection event. This uncertainty is likely to remain until advances in Solar physics and observational capabilities improve. In addition, GCRs are usually more slowly varying events because the initial violence of the *big bang* is damped by the long time period since that event. However, violent, short-lived events (e.g., supernova explosions) occur and are addressed in Chapter 7.

A final difference involves the presence of high-Z particles. Energetic, highly charged nuclei (HZE particles) are principally found as part of GCRs. The range of GCR nuclei extend from protons to isotopes of iron and heavier nuclei. Additional discussion regarding trapped radiation or VABs, GCR, and SPE radiation sources are described in subsequent commentary.

6.6.1
General Characterization

Although the GCR background is reasonably well characterized by existing models, the amount of shielding that can be incorporated into a space vehicle is only of limited effectiveness against HZE particles. The biological effectiveness of HZE particles is not well understood, but they are known to be one of the most significant radiological hazards in space.

Physics issues also exist for trapped radiation. The trapped radiation environment is dynamic. The short-term variation in the VABs is not well understood and is difficult to accurately forecast. However, average values of trapped radiation are reasonably well understood.

On the positive side, shielding is effective in attenuating low-energy SPE protons and trapped radiation. Judicious scheduling and the selection of low-dose trajectories effectively manage the hazards of LEO activities.

6.6.2
Trapped or van Allen Belt Radiation

Charged particles are influenced by the electromagnetic fields of planets, moons, and other bodies. These particles originating either within or outside a Solar System can be trapped by these electromagnetic fields. This trapped radiation forms radiation belts that have a significant influence on objects orbiting these planetary bodies. Rapid transit through trapped radiation belts is an effective means to limit crew doses.

For the Earth, there are two VABs (i.e., an inner and an outer belt). In terms of the Earth's radius ($R_e = 6.4 \times 10^3$ km), the inner belt extends to about 2.8 R_e. The outer belt occupies the region between 2.8 R_e and 12 R_e.

The inner belt is composed primarily of protons having energies up to several hundreds of MeV. Protons with energies of 400 MeV peak in intensity at about 1.3 R_e and protons with energies of 4 MeV peak at about 2 R_e. The time-averaged spatial distribution of protons, with energies greater than 100 MeV, exhibit a sharp rise at about 1.16 R_e and reach a maximum at about 1.5 R_e with an integral fluence of 10^4 protons/cm^2. Beyond 1.5 R_e, the proton fluence decreases slowly, then rises to a maximum at about 2.2 R_e and again drops to about 100 protons/cm^2 at about 2.8 R_e.

Electrons with energies greater than several MeV dominate the outer belt. Electrons are also found in the inner belt, but their intensity is only about 10% of the outer belt intensity and their energy is lower. The electron fluence, for energies greater than 40 keV, peaks at about 3.5 R_e with a value of 10^9 electrons/cm^2. In the outer belt, large variations in the electron fluence (2–4 orders of magnitude) occur over periods of hours to days.

The VABs have a distorted toroidal shape and lie in the plane of the geomagnetic equator. The energy and spatial distribution of the particles in the belts, particularly the lighter electrons, vary with time.

The fluences for both protons and electrons are each a strongly varying function of altitude and location above the Earth. In the vicinity of 35° South latitude and 325° East longitude, the fluence of trapped particles is largest. This region is known as the South Atlantic Anomaly (SAA). At 370 km altitude, the proton fluence in the SAA for energies between 40 and 100 MeV is as much as 1000 times larger than other proton fluences in the belt at the same altitude. Similar spatial behavior is observed for electrons.

6.6.3
Galactic Cosmic Ray Radiation

Crews on missions in LEO are not exposed to the full intensities of the GCR and SPEs. The Earth and its atmosphere provide shielding to attenuate GCR and SPE radiation. Earth's geomagnetic field deflects lower energy protons and heavier ions into deep space. Therefore, particle fluence rates from GCR and SPE sources are much lower in LEO than will be encountered in missions beyond LEO.

Galactic cosmic rays constitute a major radiation source outside the magnetosphere and consist of protons (88%), alpha particles (10%), electrons and gamma rays (1%), and heavy ions or HZE particles (1%). In view of its abundance and high linear energy transfer, iron is one of the most important HZE particles. Table 6.7 summarizes the distribution of radiation types by intensity and abundance for galactic cosmic radiation. The maximum GCR total particle fluence is about 4 particles/cm^2 s.

The GCR energy spectra decrease rapidly with increasing energy with energies extending to 10^{20} eV. As GCR sources lie well outside the Solar System, their spatial distribution is essentially isotropic.

The GCR proton energy spectrum exhibits a broad maximum at about 1 GeV, and the spectrum of alpha particles and HZE particles peaks at about 300 MeV/n. At Solar maxima, the GCR intensity is a minimum and slowly increases until its maximum value is reached during Solar minimum conditions.

6.6.4
Solar Flare Radiation or Solar Particle Events

Solar flare radiation or Solar particle events are ejections of matter from the Sun. Their composition reflects the mass constituents characteristic of Solar plasmas. Therefore, they are composed predominantly of protons with admixtures of alpha particles and heavier nuclei. The intensity and composition of Solar flare radiation varies with the specific event. Carbon, nitrogen, and oxygen dominate the $Z > 2$ particles and constitute about 1% of the Solar flare fluence rate.

Typical flare events last from 1 to 4 days, although somewhat longer durations have been observed. Annually, 8–11 significant Solar flares occur. Solar physics models are

Table 6.7 Distribution of galactic cosmic radiation.[a]

Radiation type	Fluence rate (particles/cm^2 s)	Abundance (%)
Protons	3.6	88
Alpha particles	0.4	9.8
Electrons and gamma rays ($E > 4$ GeV)	0.04	1
C, N, O, and F nuclei	0.03	0.75
Li and B nuclei	0.008	0.2
$10 \leq Z \leq 30$ nuclei	0.006	0.15
$Z \geq 31$ nuclei	0.0005	0.01

[a] Derived from Santora and Ingersoll (1991).

not yet able to predict the timing, duration, and intensity of a Solar flare event. This uncertainty and the magnitude of these SPEs present a significant radiation hazard to astronauts in low Earth orbit, during moon missions, and on missions to other planets.

When the Sun is very active, such as the periods near sunspot maxima that occur about every 11 years, SPEs can deliver doses of 0.3–3.0 Gy over a period of about 3 days. These absorbed doses are significant and merit attention.

Given the general characteristics of the trapped radiation (VAB), GCR, and SPE radiation sources, the radiological consequence of LEO, planetary, and deep space missions can be addressed. Prior to considering these missions, the historical radiation doses received by astronauts in LEO and during lunar missions are outlined.

6.7
Calculation of Absorbed and Effective Doses

The analytical methodology and numerical approaches used to calculate absorbed and effective doses from photons and neutrons are summarized in Appendixes C and J, respectively. Dose calculations for heavy charged particles are more complex and require the utilization of computer codes (see Appendix J). An analytical discussion of the calculation of doses from heavy charged particles is provided in Appendix K. Readers not familiar with these techniques should consult Appendixes C, J and K.

6.8
Historical Space Missions

Manned space missions have occurred in low Earth orbit in a variety of space vehicles. Lunar missions have also been accomplished. This section addresses the radiation environment in low Earth orbit and in the spatial region between the Earth and its moon. Historical crew radiation doses for both LEO and lunar missions are also addressed.

6.8.1
Low-Earth Orbit Radiation Environment

Manned LEO missions are influenced by all three dominant components of the space radiation environment. The relative importance of each of the components depends on the specific LEO parameters including the spacecraft trajectory (e.g., altitude, orientation, and orbital characteristics), mission timing relative to periodic Solar activity, mission duration, and spacecraft shielding characteristics.

LEO environments are normally dominated by energetic charged particles including electrons, protons, and heavy ions such as alpha particles and ions with $Z \leq 92$. The environment is also significantly influenced by large emissions of Solar particles and the temporal and spatial fluctuations of the particles trapped by the Earth's magnetic field.

The LEO radiation environment is influenced by spatial and temporal factors including the 11-year Solar cycle, the Solar wind, and the Earth's magnetic field that traps some particles and deflects others. The Earth's magnetic field varies in strength and configuration over timescales from days to years. Solar events also alter the distribution of trapped particles in the Earth's VABs.

Nuclear interactions of neutrons, protons, and heavy ions with the spacecraft, space suit, Earth's atmosphere, and the human body produce secondary particles that contribute to the astronaut's effective dose. In contrast, most of the electrons do not penetrate the wall of a spacecraft, but could penetrate suits worn during extravehicular activity (EVA) resulting in eye and skin equivalent doses.

Table 6.8 summarizes the LEO radiation environment by particle type, source of the particle, particle energy, and possible impact during EVA or inside the spacecraft. The unrestricted linear energy transfer (L_∞) in water is also provided.

6.8.2
The Space Radiation Environment Outside Earth's Magnetic Field

Prior to outlining dosimetric data from historical missions, the projected dose equivalent rates outside the Earth at 1 AU and in LEO are reviewed. Table 6.9 provides dose equivalent values that could be experienced at 1 AU outside the Earth's magnetic field. The values in Table 6.9 quantify the GCR and SPE radiation fields. Table 6.10 summarizes dose equivalent values appropriate for entry into the Earth's VABs.

Doses of the type presented in Tables 6.9 and 6.10 exhibit variation in the literature. As an illustration of the derivation of these values from basic data, the Table 6.7 and proton fluence-to-dose conversion factors (k) can be used to compute the dose equivalent rate from GCR protons (P) using the relationship,

$$\dot{H}_P = k\phi, \tag{6.9}$$

where ϕ is the proton fluence rate. Using a value of 3×10^3 pSv cm^2/proton applicable to GCR protons leads to a predicted proton dose equivalent rate

$$\dot{H}_P = 3000 \frac{\text{pSv cm}^2}{\text{proton}} \frac{1\,\text{Sv}}{10^{12}\,\text{pSv}} \frac{1000\,\text{mSv}}{\text{Sv}} \frac{3.6\,\text{proton}}{\text{cm}^2\,\text{s}} \frac{3600\,\text{s}}{\text{h}} = 0.04\,\text{mSv/h}. \tag{6.10}$$

Considering that no spectral averaging was performed, this value is in reasonable agreement with the unshielded positive ion value of 0.02 mSv/h provided in Table 6.9.

Table 6.8 Characterization of the LEO radiation environment.[a]

Particle type	Source	Energy (MeV)	L_∞ (keV/μm)	Ability to penetrate EVA suit	Ability to penetrate Spacecraft	Comments
Electrons	Trapped particles	0.5–6	~0.2	Yes	Yes	Electrons dominate the dose equivalent for aluminum shields with areal densities <0.15 g/cm^2
Electrons	Decay products from interactions with trapped GCR ions Atmospheric scattering	1 to >1000	0.2 to >3	Yes	Yes	The dose equivalent contribution is about 10 times greater than the trapped electron dose equivalent
Protons	Trapped particles	<10	>5	No	No	At low energies, protons pose a limited dose equivalent concern
Protons	Trapped particles SPEs	10–400	0.3–5	Yes	Yes	As proton energies increase, the dose equivalent increases
Light ions	SPE	10–400	0.3–5	Yes	Yes	The dose equivalent depends on the nature of the SPE and the specific ions and their energies
Ions ($Z>1$) and charged secondary fragments	GCR	>50 MeV/nucleon	1–1000	Yes	Yes	Pion production occurs, but the pion contribution to the dose equivalent is not well characterized
Charged target fragments	Nuclear interactions from all sources	<10 MeV/nucleon	2–1200	Yes	Yes	Large dose equivalents are possible
Neutrons	Nuclear interactions	0.1–500	[b]	Yes	Yes	Large dose equivalents are possible

[a]Source: NCRP 142 (2002).
[b]Neutrons interact with atomic nuclei to produce highly ionizing charged secondary particles.

Table 6.9 Sol system radiation fields exterior to the Earth at 1 AU.[a]

		Dose equivalent or dose equivalent rate		
Source	Radiation type	Unshielded	Space suit	Spacecraft
GCR	Positive ions	0.02 mSv/h	0.02 mSv/h	0.02 mSv/h
Solar wind	Positive ions	10^{-4} mSv/h	0	0
Medium solar flare	Positive ions	1000 mSv	500 mSv	3 mSv
Maximum solar flare	Positive ions	10^6 mSv	5×10^5 mSv	3500 mSv

[a]Derived from IPS Radio and Space Services (2005)

6.8.3
Radiation Data from Historical Missions

In evaluating LEO dose rates, mission data from Gemini, Skylab, Space Transport Shuttle, the Mir space station, and the International Space Station are considered. Apollo lunar mission's data are also presented. As all data do not contain sufficient information to determine the quality factor or radiation-weighting factor, the focus is on the average absorbed dose. An examination of NCRP 132 suggests an average quality factor or radiation-weighting factor of about 2 would be appropriate for LEO. This appears to be at least qualitatively correct based on the International Commission on Radiological Protection accepted radiation-weighting factors.

6.8.4
Gemini

Given the variability in the data, it is reasonable to approximate the Gemini mission trajectories as circular orbits with a constant 300 km altitude above the Earth. The subsequent results are relatively insensitive to the altitude value.

For a stable circular orbit, the centripetal force (F_C)

$$F_C = \frac{mv^2}{r}, \tag{6.11}$$

Table 6.10 Low-Earth orbit van Allen belt radiation fields.[a]

		Dose equivalent rate (mSv/h)		
Source	Radiation type	Unshielded	Space suit	Spacecraft
van Allen's belt	Positive ions/protons	600	300	3
van Allen's belt	Electrons	10^6	100	10

[a]Derived from IPS Radio and Space Services (2005)

and gravitational force (F_G)

$$F_G = G\frac{mM_e}{r^2}, \qquad (6.12)$$

are equal

$$\frac{mv^2}{r} = G\frac{mM_e}{r^2}, \qquad (6.13)$$

where m is the mass of the Gemini vehicle, v is the velocity of the Gemini vehicle in orbit, r is the orbital radius of the Gemini vehicle relative to the Earth's center, M_e is the mass of the Earth (5.96×10^{24} kg), and G is the gravitational constant (6.67×10^{-11} nt m^2/kg^2).

As dosimetry data provide an average absorbed dose by Gemini mission in terms of the number of orbits, the absorbed dose is determined from the orbital period (T). The period is just the distance traveled divided by the orbital velocity

$$T = \frac{2\pi r}{v}. \qquad (6.14)$$

The period is determined by solving Equation 6.13 for v

$$v = \left(\frac{GM_e}{r}\right)^{1/2}. \qquad (6.15)$$

Inserting Equation 6.15 into Equation 6.14, leads to the desired orbital period

$$T = \frac{2\pi r^{3/2}}{(GM_e)^{1/2}}. \qquad (6.16)$$

All terms appearing in Equation 6.16 have been assigned numerical values except for r, which is the sum of the Gemini spacecraft altitude (h) and the Earth's radius (R_e):

$$r = h + R_e, \qquad (6.17)$$

where R_e is the Earth's mean radius (6.4×10^3 km), and a typical value for h is about 300 km. Using these values in Equation 6.17, leads to the nominal period of a Gemini orbit

$$T = \frac{(2\pi)(6.7 \times 10^6 \text{ m})^{3/2}(1 \text{ min}/60 \text{ s})(1 \text{ h}/60 \text{ m})}{[(6.67 \times 10^{-11} \text{ nt}-\text{m}^2/\text{kg}^2)(5.96 \times 10^{24} \text{ kg})]^{1/2}} = 1.5 \text{ h}. \qquad (6.18)$$

The 1.5-h orbital value is the expected orbital period and similar results would be obtained for other reasonable orbital altitudes.

The orbital period is used to define the average Gemini absorbed dose rate

$$\dot{D} = \frac{D}{nT}, \qquad (6.19)$$

where D is the Gemini mission absorbed dose and n is the number of orbits in the Gemini mission. The Gemini absorbed dose rates are summarized in Table 6.11 for missions III–XII. Gemini crew absorbed doses were measured with instrumentation

6.8 Historical Space Missions

Table 6.11 Gemini mission parameters and average absorbed dose.[a]

Gemini mission	n	D (mGy)	$\dot{D}$ (mGy/h)
III	3	0.25	0.056
IV	62	0.46	0.0049
V	120	1.76	0.0098
VI-A	16	0.24	0.010
VII	206	1.64	0.0053
VIII	7	0.10	0.0095
IX-A	45	0.19	0.0028
X	8	0.28	0.023
XI	2	0.28	0.093
XII	59	0.20	0.0023
			Average = 0.022

[a]Derived from NCRP 132 (2000).

packages containing lithium fluoride and nuclear emulsions. Four packages were used for each crewmember.

The results of Table 6.11 suggest the average absorbed dose during Gemini missions is about 0.022 mGy/h. The average is the sum of the mission absorbed dose rates divided by the total number of missions listed in Table 6.11.

6.8.5
Skylab

Table 6.12 summarizes Skylab crew absorbed dose rates. The Skylab was placed into a 435-km circular orbit at an inclination of 50° in July 1972. The average mission absorbed dose rate of 0.029 mGy/h was measured by crew thermoluminescent dosimeters.

6.8.6
Space Transport Shuttle

Space Transport Shuttle missions provide an additional source of LEO data. NCRP 132 tabulates data for 67 successful STS missions with durations between 2 and 16

Table 6.12 Skylab crewmember absorbed doses.[a]

Skylab mission	Duration (d)	Mean absorbed dose (mGy)	Average absorbed dose rate (mGy/h)
2	28	17.0 ± 1.0	0.025
3	59	38.7 ± 3.0	0.027
4	90	73.9 ± 6.1	0.034
			Average = 0.029

[a]Derived from NCRP 132 (2000).

days. Mission altitudes lie between 215 and 617 km with inclinations between 28.5° and 62°. Crew thermoluminescent dosimeters were used to determine the average absorbed dose rates.

For altitudes greater than 450 km, higher absorbed dose rates were measured in flights with inclinations <38°. Below 450 km, GCRs contribute a significant fraction of the absorbed dose. Above 450 km, trapped protons become the dominant source of radiation dose. Measured absorbed dose rates vary between 0.0015 mGy/h at 215 km (STS-38) and 0.068 mGy/h at 617 km (STS-31). Averaging these values yields an approximate STS absorbed dose rate of 0.035 mGy/h.

6.8.7
Mir Space Station

The Russian Mir space station has a slightly elliptical orbit with a mean altitude of about 400 km and an inclination of 51.6°. The dosimetry information for Mir is summarized in Table 6.13. The average Mir absorbed dose rate is 0.015 mGy/h.

6.8.8
International Space Station

The International Space Station operates between 360 and 450 km and at an inclination of 51.6°. The average absorbed dose rate measured during the Mir-18 and -19 missions to the ISS with a tissue equivalent proportional counter is 0.013 mGy/h. The contributions from trapped protons and GCR to this absorbed dose rate are about equal.

6.8.9
Apollo Lunar Missions

The Apollo lunar flights are not LEO missions, but the measurements made while traversing the VABs are of interest. Lunar flights demonstrated that the trapped radiation belts are of negligible importance in total crew dose because of the short time spent in these belts relative to the total time of the lunar mission. This assumes the SAA or other regions of elevated dose equivalent are avoided.

Table 6.13 Mir space station dosimetric information.[a]

Time period	Absorbed dose rate (mGy/h)
December 1988	0.013
March and April 1989	0.019
September 1994	0.012
	Average = 0.015

[a]Derived from NCRP 132 (2000).

Table 6.14 Summary of average absorbed dose rates for LEO.

Vehicle	Average absorbed dose rate (mGy/h)
Gemini	0.022
Skylab	0.029
STS	0.035
Mir	0.015
ISS	0.013
	Average = 0.023

Once the Apollo spacecraft leaves the protection of the Earth's magnetic field, it becomes vulnerable to SPEs. Accordingly, a more sophisticated dosimetry system than utilized in Mercury and Gemini missions was required, because more significant crew doses were possible, particularly from a large SPE.

Apollo missions had an average duration of about 15 days. The average absorbed dose from the Apollo 10–17 missions was 3.9 mGy and the maximum dose was 7.3 mGy (Apollo 12). These values lead to average and maximum absorbed dose rates of 0.011 and 0.020 mGy/h, respectively. These values are of the same order of magnitude as the LEO values summarized in Table 6.14.

6.8.10
Validation of LEO and Lunar Mission Absorbed Dose Rates

The results of Table 6.14 summarize the average absorbed dose rates that were derived for the Gemini, Skylab, STS, Mir, and ISS missions. The results are very similar as would be expected from the nature of the averaging process described in the previous sections. The average of all LEO data is 0.023 mGy/h.

In Table 6.15, the average and individual dose rates from NASA missions through June 2002 are summarized. These results are derived from records of passive

Table 6.15 Tabulation of NASA crew doses.[a,b]

Program	$\dot{D}(mGy/h)^a$	$\dot{D}(mGy/h)$ This Work	$\dot{H}(mSv/h)^a$	$\bar{Q}$
Gemini	0.020	0.022	0.036	1.8
Apollo	0.018	0.011–0.020[b]	0.050	2.8
Skylab (50° × 430 km)	0.030	0.029	0.058	1.9
STS (28.5° × >400 km)	0.05	0.035	0.088	1.8
STS (28.5° × <400 km)	0.0042	0.035	0.0075	1.8
STS (39–40°)	0.0042	0.035	0.0088	2.1
STS (>50° × >400 km)	0.018	0.035	0.046	2.6
STS (>50° × <400 km)	0.0083	0.035	0.019	2.3
NASA–Mir (51.6° × 390 km)	0.015	0.015	0.035	2.3
				Average = 2.2

[a] Derived from Cucinotta et al. (2003).
[b] NCRP Report No. 132 (2000).

dosimeters worn on NASA missions and are compared to the results of Table 6.14. Radiation transport codes and flight spectrometers were used to estimate absorbed doses and average quality factors.

Table 6.15 provides the average absorbed dose rate ($\dot{D}$) and dose equivalent rate ($\dot{H}$). In addition, an average quality factor ($\bar{Q}$) is provided:

$$\bar{Q} = \frac{\dot{H}}{\dot{D}}. \tag{6.20}$$

The results of Table 6.15 suggest that the dose rates increase at high altitudes because of the longer residence times in the VABs with the highest doses occurring on the Hubble Space Telescope launching and repair missions at altitudes near 600 km. In addition, the use of an average quality factor of 2 in dose equivalent calculations is justified as the tabular average of Table 6.15 is 2.2.

6.9
LEO and Lunar Colonization

The historic LEO and lunar missions provide a scientific basis for the increased utilization of these environments in the twenty-first century. Lunar exploration and colonization are receiving increased attention following the possible detection of water at the lunar poles. There are scientific as well as commercial motivations for establishing a lunar colony. Commercial applications include mining raw materials, manufacturing structures for use in Earth orbit and deep space, training for deep space missions, low-gravity manufacturing, and energy generation. Scientific motivation includes biological research, Solar System research, and the deep space observation. A lunar colony could also be a base for space exploration.

Earth orbit colonization also offers a number of scientific and commercial benefits. These include advancing orbital space station science, astronomical research, remote sensing, Earth science, space station commercialization, low-Earth orbit tourism, space ports, power generation and power transfer to Earth, Earth orbit manufacturing, and Earth orbit tethers.

A tether or space elevator is a lifting structure that includes (1) a tower/cable system rising from a point on the Earth's equator to an orbital height and (2) a counterweight in geostationary orbit where the tower terminates. Payloads destined for orbit would ride the tower/cable elevator to the counterweight location where they could be offloaded into low Earth orbit.

The space elevator provides the potential for an inexpensive mode for transporting orbital payloads. However, numerous engineering issues exist before the tether becomes a viable alternative to rockets. Accordingly, rockets are considered as the most credible alternative for taking payloads into space.

Both lunar and LEO activities subject humans to increased ambient radiation. Without the protective shielding of the Earth's atmosphere, colonists receive an increased radiation dose from GCR and SPE events. Measures must be implemented to maintain these doses ALARA.

A variety of ALARA techniques are available, but their viability depends on the particular application. For example, the use of shielding could be implemented in a lunar environment easier than in LEO. Accessible lunar shielding materials include crust and rocks that are readily available on the Moon's surface. Shielding is not readily accessible in LEO and would need to be staged for use. However, the LEO colonists have the flexibility to change their orbital trajectory to minimize the impact of trapped radiation. Both lunar and LEO environments could effectively utilize electromagnetic deflection techniques to alter the trajectory of charged particles and avoid their dose consequences. Radioprotective chemical usage is available for both lunar and LEO environments.

6.10
GCR and SPE Contributions to Manned Planetary Missions

Planetary and deep space missions transit the LEO and lunar environments. Once outside the influence of the Earth's magnetosphere, these missions encounter unshielded GCR and SPE radiation fields.

6.10.1
GCR Doses

Secondary neutrons and charged particles are the major sources of radiation exposure in an interplanetary spacecraft. The annual bone marrow GCR dose is normally limited to about 15 cGy/year at Solar minimum behind 2 cm of Al shielding. The effective dose at Solar minimum is in the 45–50 cSv/year range. At Solar maximum, the effective dose is about 15–18 cSv/year. At present, there are currently no dose limit recommendations for planetary and deep space missions. The NCRP 132 guidance only applies to LEO missions.

In converting absorbed dose to gray equivalent, specific RBE values are required. Applicable RBE values for GCR particles are provided in Table 6.16.

The effectiveness of water shields in attenuating GCR during the transit from Earth to Mars and on the Martian surface is summarized in Tables 6.17 and 6.18, respectively. These tables provide the annual GCR dose equivalent for the skin, eye, and BFO. Tables 6.17 and 6.18 note that the unattenuated GCR BFO dose equivalent

Table 6.16 RBE values for converting absorbed dose to Gy-Eq.[a]

Radiation type	Energy (MeV)	RBE
Neutrons	1–5	6.0
	5–50	3.5
Heavy ions ($A \geq 4$)	—	2.5
Protons	>2	1.5

[a]Derived from NCRP Report No. 132 (2000) and Townsend (2004).

Table 6.17 Annual GCR dose equivalent (Sv/yr) during transit from Earth to Mars as a function of shield thickness.[a]

Tissue/organ	Water shield thickness (g/cm^2)		
	0	5	10
Skin	0.94	0.73	0.60
Eye	0.96	0.74	0.61
Blood-forming organs	0.70	0.58	0.51

[a]Derived from Saganti et al. (2005).

rate is 0.70 Sv/year during transit to Mars and 0.19 Sv/year on the Martian surface. These values merit attention from both risk and ALARA perspectives.

The GCR values of Table 6.17 are validated by comparison with NASA Conference Publication 3360 that estimates a GCR transit BFO dose equivalent of 0.31 Sv for a 500-day Mars mission. In NASA 3360, the shielding for nominal crew operations is specified at 2 g/cm^2. The crew is also assumed to have shelter protection of 20 g/cm^2 for 8 h/day. These dose equivalent values are consistent considering their uncertainties and associated assumptions.

The Martian surface dose equivalent values of Table 6.18 also compare favorably to the NASA values. The dose equivalent values differ by approximately a factor of 2, which is well within the uncertainties in the respective values and their associated assumptions. This comparison provides reasonable confidence in the GCR values used in this chapter.

6.10.2
SPE Doses

Given the current level of knowledge of Solar physics, it is not possible to forecast key SPE parameters with any degree of accuracy. These key parameters include predicting the timing, magnitude, duration, and fluence rate of the SPE. The unpredictability of SPEs adds to their inherent radiation hazard.

The intensity, energy spectra, and angular distributions of SPE protons and alpha particles vary considerably with individual Solar flares and are a function of time within any given event. A typical flare has a duration of about 1–4 days although

Table 6.18 Annual GCR dose equivalent (Sv/yr) on the Mars surface as a function of shield thickness.[a]

Tissue/organ	Water shield thickness (g/cm^2)		
	0	5	10
Skin	0.19	0.19	0.18
Eye	0.20	0.19	0.18
Blood-forming organs	0.19	0.19	0.18

[a]Derived from Saganti et al. (2005).

6.10 GCR and SPE Contributions to Manned Planetary Missions

longer duration flares have been observed. Normally, a flare's intensity increases rapidly over the first few hours and then decreases.

The strength of the Solar flare event is characterized in terms of the time integral of the SPE energy spectrum. These time-integrated spectra are written in terms of exponential functions of magnetic rigidity (P_j) defined by

$$P_j = \frac{p_j c}{z_j e}, \qquad (6.21)$$

where p_j is the momentum of the jth particle in the Solar flare or SPE, c is the speed of light, z_j is the atomic number of the jth particle, e is the charge of a proton or electron, and $j=1$ for protons and $j=2$ for alpha particles. The particles in the SPE are primarily protons ($z=1$) and alpha particles ($z=2$).

The omnidirectional fluence of SPE particles above kinetic energy E_o [$\Phi_j(E > E_o)$] is defined as

$$\Phi_j(E > E_o) = \Phi_{oj} exp(-P_j(E)/P_o), \qquad (6.22)$$

where Φ_{oj} and P_o are parameters that characterize a particular SPE, and P_o varies between 50 and 200 MV. The proton fluence for energies greater than 30 MeV is typically in the range of 10^6–10^{10} protons/cm^2. Table 6.19 provides a summary of SPEs from Solar cycles 19–22 that are likely to exceed the NCRP 132 recommendations.

In space, SPE doses can be quite large. An August 1972 SPE was the largest dose event of the space era and it occurred between two Apollo missions. However, ice core data from Antarctica indicate that the largest SPE in the past 500 years was probably the Carrington Flare of 1859. Its total fluence was approximately 20 times larger than the August 1972 SPE.

Selecting the worst-case SPE depends on the perspective of the individual analyst. As there is no consensus in the selection of the worst-case event, three options are presented.

Option 1 is the August 1972 SPE event. From a different perspective, Option 2 uses the 1859 Carrington Flare as the worst-case SPE. Option 3 defines the worst-case SPE as an event that is 10 times the flare of September 29, 1989. This selection is a reasonable basis for the worst-case event that would occur on about a 50-year frequency. Given differences in spectra, modeling assumptions, and dose conversion coefficients, the bounding values using both the September 29, 1989 (Option 3) and Carrington Flare (Option 2) SPEs are further evaluated. A limited discussion of Option 1 is also provided.

The effective dose from the August 1972 SPE (Option 1) is summarized in Table 6.20 as a function of the aluminum shield thickness. Bone marrow doses of about 1 Gy can be delivered in a 24-h period. These doses produce a physiological response characterized by the classic acute radiation syndrome. Skin doses of 15–20 Gy could result in erythema and possibly desquamation. Dose equivalent values comparable to those from the August 1972 SPE are of significant concern for planetary missions.

Absorbed doses from Carrington-type SPEs (Option 2) as a function of Al shield thickness are summarized in Table 6.21. For the Carrington Flare, bone marrow

Table 6.19 Proton fluence levels of significant solar events of cycles 19–22 likely to exceed the NCRP 132 recommendations.[a]

Date	Fluence (protons/cm^2)	
	$E > 10$ MeV	$E > 30$ MeV
February 23, 1956	2×10^9	1×10^9
July 10–11, 1959	5×10^9	1×10^9
July 14–15, 1959	8×10^9	1×10^9
July 16–17, 1959	3×10^9	9×10^8
November 12–13, 1960	8×10^9	2×10^9
November 15, 1960	3×10^9	7×10^8
July 18, 1961	1×10^9	3×10^8
November 18, 1968	1×10^9	2×10^8
April 11–13, 1969	2×10^9	2×10^8
January 24–25, 1971	2×10^9	4×10^8
August 4–9, 1972	2×10^{10}	8×10^9
February 13–14, 1978	2×10^9	1×10^8
April 30, 1978	2×10^9	3×10^8
September 23–24, 1978	3×10^9	4×10^8
May 16, 1981	1×10^9	1×10^8
October 9–12, 1981	2×10^9	4×10^8
February 1–2, 1982	1×10^9	2×10^8
April 25–26, 1984	1×10^9	4×10^8
August 12, 1989[b]	8×10^9	2×10^8
September 29, 1989[b]	4×10^9	1×10^9
October 19, 1989[b]	2×10^{10}	4×10^9
November 26, 1989[b]	2×10^9	1×10^8

[a] Wilson et al. (2005).
[b] The listed 1989 SPEs had an extended duration.

Table 6.20 Effective dose for the August 1972 SPE[a] (Option 1).

Shielding (g/cm^2 Al)	Effective dose (Sv)	Average dose equivalent to BFO (Sv)
1	3.38	1.11
2	2.00	0.91
5	0.89	0.56
10	0.40	0.31

[a] Derived from Townsend (2004).

Table 6.21 Carrington Flare (Option 2) absorbed dose estimates.[a]

Shielding (g/cm^2 Al)	Skin (Gy)	Eye (Gy)	BFO (Gy)
1	35.4	23.4	2.81
2	6.65	6.02	1.71
5	2.82	2.73	1.09

[a] Derived from Townsend (2004).

Table 6.22 Skin dose equivalent (Sv) from the September 29, 1989 (Option 3) worst-case SPE.[a–c]

Location	Free space	Lunar surface	Martian surface
Space suit	295	148	0.45
Helmet/pressure vessel	64.4	32.2	0.44
Equipment room	6.48	3.24	0.38
Shelter	2.62	1.31	0.33

[a]Derived from Wilson et al. (1999).
[b]Includes contributions from ions with $Z \leq 28$.
[c]NCRP 132, 30-day recommended limit of 1.50 Gy Eq.

doses of 1–3 Gy are possible inside a spacecraft. A shielded room with about 18 cm Al is needed to reduce the Carrington Flare absorbed doses to the applicable NCRP 132 recommended deterministic limits (30 days BFO limit of 0.25 Gy Eq).

Tables 6.22–6.24 provide summaries of the Option 3 dose equivalents for the skin, eye, and BFO, respectively. These dose equivalents are calculated for free space, the lunar surface, and the Martian surface for shielding afforded by a spacesuit (0.28 g/cm^2 Al), helmet/pressure vessel (1 g/cm^2 Al), equipment room (5 g/cm^2 Al), and a shelter (10 g/cm^2 Al).

Prior to reviewing the details of Tables 6.22–6.24, a review of the free space, lunar surface, and Martian surface dose equivalent assumptions is provided. This is important because the assumptions affect the resultant dose equivalents.

The free space dose equivalent values present the unmodified, Solar particle flux. Low-energy protons and alpha particles are the most important components of the dose equivalent especially for lightly shielded tissues. The HZE particles are a less important contributor to the total dose equivalent.

The lunar surface offers more biological protection than free space. Operations on the lunar surface result in a factor of two reduction in the free space dose equivalent values. This reduction is derived from a number of factors with the major effect arising from the shadow shielding produced by the moon itself. However, the dominant sources of the dose equivalent are very similar to the free space values.

Table 6.23 Eye dose equivalent (Sv) from the September 29, 1989 (Option 3) worst-case SPE.[a–c]

Location	Free space	Lunar surface	Martian surface
Space suit	81.3	40.7	0.44
Helmet/pressure vessel	35.5	17.8	0.42
Equipment room	5.54	2.77	0.37
Shelter	2.43	1.22	0.32

[a]Derived from Wilson et al. (1999).
[b]Includes contributions from ions with $Z \leq 28$.
[c]NCRP 132, 30-day recommended limit of 1 Gy Eq.

Table 6.24 Dose equivalent (Sv) to blood-forming organs from the September 29, 1989 (Option 3) worst-case SPE.[a-c]

Location	Free space	Lunar surface	Martian surface
Space suit	4.21	2.11	0.32
Helmet/pressure vessel	3.52	1.76	0.31
Equipment room	1.93	0.97	0.28
Shelter	1.26	0.63	0.25

[a]Derived from Wilson et al. (1999).
[b]Includes contributions from ions with $Z \leq 28$.
[c]NCRP 132, 30-day recommended limit of 0.25 Gy Eq.

Astronauts on the Martian surface receive radiation that is attenuated by the atmosphere. On the Martian surface, the most important contributions to the dose equivalent are energetic protons having energies on the order of 100 MeV and low-energy alpha particles with a peak energy of 1 MeV/nucleon.

If 1 Sv is assumed to equal 1 Gy Eq, Table 6.22 indicates that a single worst-case Option 3 SPE exceeds the ICRP 132 recommended LEO 30-day skin dose limit of 1.5 Gy Eq for all evaluated configurations in free space. On the lunar surface, only the shelter prevents the NCRP 132 recommended skin dose limits from being exceeded. The recommended LEO limit is not jeopardized on the Martian surface because the atmosphere provides sufficient attenuation.

The career recommended LEO skin dose limit of 6.0 Gy Eq is exceeded in free space for all configurations except the shelter. On the lunar surface, the equipment room and the shelter provide sufficient shielding to protect the career skin dose limit. The values of Table 6.22 are sufficiently large so that deterministic skin effects would be observed.

Eye dose equivalent values from a worst-case SPE are provided in Table 6.23. The recommended 30-day LEO eye dose limit of 1 Gy Eq of NCRP 132 is exceeded in free space and on the lunar surface. The career eye dose recommendation of 4 Gy Eq is exceeded in free space for all configurations except the shelter and on the lunar surface for spacesuit and helmet/pressure vessel configurations. The values of Table 6.23 are sufficiently large to produce deterministic effects.

The recommended 30-day LEO BFO dose limit of 0.25 Gy Eq is reached or exceeded for all configurations of Table 6.24. The LEO recommended 10-year career limit varies with both age and sex. For example, a male astronaut exposed at age 25 has a 10-year career limit of 0.7 Sv for 3% excess lifetime risk of cancer mortality. This value is exceeded for all configurations in free space and for all configurations on the lunar surface, except the shelter. In addition, the $LD_{50,30}$ dose of 3–4 Gy is exceeded for the spacesuit and helmet/pressure vessel configurations in free space. The values of Table 6.24 also suggest deterministic effects occur for a number of configurations.

The Options 2 and 3 SPE events are used to determine the doses for a Mars mission. This mission and associated doses are outlined in subsequent discussion.

6.10.3
Planetary Mission to Mars

Current planning suggests that the initial planetary excursion is a Mars mission. Three distinct radiation environments are encountered during this mission. These include the radiation trapped by the planet's magnetic field (i.e., Mars and Earth), GCRs, and SPEs. The contribution from each of these sources depends on the spacecraft shielding as well as secondary radiation produced in the spacecraft shell and internal structures. The mission profile, trajectory, and duration dictate the specific contribution of each radiation source to personnel and equipment.

Before considering the specific contributions of trapped radiation, GCRs, and SPEs, it is necessary to quantify the mission profile including the spacecraft's trajectory and the duration of its various phases. The trajectory and duration of the various mission phases are considered in the following section.

6.10.4
Mars Orbital Dynamics

A key consideration in a manned mission to Mars is the trajectory used for the Earth–Mars transit. The important aspects of the trajectory are illustrated by making a number of simplifying assumptions in the formulation of a trajectory model. These assumptions provide sufficient accuracy (about 20%) to illustrate important health physics considerations without significantly altering the orbital mechanics and include the following:

- The Earth and Mars orbit the Sun in coplanar circular orbits.
- The spacecraft follows a Hohmann transfer orbit during its transit from Earth to Mars and in the return from Mars to Earth.
- The gravitational interaction between the spacecraft and nearby masses, excluding the Earth, Mars, and their moons is ignored.
- Atmospheric drag on the spacecraft when in the vicinity of the Earth and Mars is neglected.

The orbital parameters for the Earth and Mars are provided in Table 6.25. The orbital period (T) and orbital radius (r) are related by Kepler's third law:

$$T^2 = kr^3, \tag{6.23}$$

Table 6.25 Orbital periods and effective radii for the Earth and Mars relative to Sol.

Planet	Period (d)	Radius (AU)
Earth	365.3	1.000
Mars	687.2	1.524

where k is a constant having a value of 1 year^2/AU^3 for the Sol system. In Kepler's formulation, r is the length of the semimajor axis, which is approximated by the orbital radius using the assumptions noted previously.

The Hohmann trajectory assumptions lead to the average orbital radius of the spacecraft (r_{sc}) relative to the Sun:

$$r_{sc} = \frac{1}{2}(r_{earth} + r_{mars}) = \frac{1}{2}(1.000 \text{ AU} + 1.524 \text{ AU}) = 1.262 \text{ AU}. \tag{6.24}$$

The period of the spacecraft (T) is determined using Kepler's third law:

$$T = \left(\frac{1 \text{ year}^2}{\text{AU}^3}(1.262 \text{ AU})^3\right)^{1/2} = 1.418 \text{ year} \frac{365.3 \text{ days}}{\text{year}} = 517.9 \text{ day}. \tag{6.25}$$

For simplicity, assume the Earth and spacecraft are initially at the Cartesian coordinates $(x, y) = (1.0 \text{ AU}, 0)$. This is equivalent to the polar coordinate assignment $(r, \theta) = (1.0 \text{ AU}, 0°)$. The spacecraft requires a time $T/2$ to travel from its initial position on the Earth where it is closest to the Sun to its intersection with the Mars orbit where it is furthest from the sun. During this time, Mars undergoes an angular displacement (θ). Assuming uniform motion, the time derivative of the angular displacement ($\dot{\theta}$) is a constant angular velocity (ω). The angular displacement is integrated to find its time variation,

$$\theta = \int \omega \, dt = \omega t + \theta_o, \tag{6.26}$$

where θ_o is the initial ($t = 0$) angular position, and

$$\omega = \frac{360°}{T}. \tag{6.27}$$

As Mars completes an orbit (360°) during its period, the angular displacement of Mars during the time $T/2$ ($T = 517.9$ days) is $\theta = (360°/687.2 \text{ days})(517.9 \text{ days}/2) = 135.6°$. Using this angular displacement, Mars must be ahead of the Earth by $180 - 135.6° = 44.4°$. Therefore, at the time of spacecraft launch, the Earth and spacecraft are at $\theta_o = 0°$ and Mars is at $\theta_o = 44.4°$, respectively.

For the return trip, Earth and Mars must have the proper orientation to minimize the spacecraft's energy requirements. Given the assumed Hohmann orbit, the spacecraft requires the same time for the return trip to Earth as it did for the Earth–Mars trip (i.e., 258.9 days). During this time, the Earth moves through an angle of $258.9 \text{ days}(360°/365.3 \text{ days}) = 255.1°$. This result requires that the Earth lag Mars by an angle of $255.1 - 180° = 75.1°$ for the return trip from Mars to Earth. The time to reach this planetary orientation is determined from the information noted above.

Using Equation 6.26 and previous results, the time-dependent displacements for the Earth and Mars are

$$\theta_{mars}(t) = 44.4° + \frac{360°}{687.2 \text{ days}} t = 44.4° + \frac{0.5239°}{\text{day}} t, \tag{6.28}$$

$$\theta_{earth}(t) = \frac{360°}{365.3 \text{ days}} t = \frac{0.9855°}{\text{day}} t. \tag{6.29}$$

where t is the transit time from Mars to Earth ($T/2$) plus the time on Mars (T_{Mars}):

$$t = \frac{T}{2} + T_{mars}. \tag{6.30}$$

Using these results, the proper orientation for the return trip from Mars to Earth occurs when

$$\theta_{earth} + 75.1° = \theta_{mars} + 360° n, \tag{6.31}$$

where n is an integer.

Equations 6.28–6.30 are solved for the total transit time from Mars landing to the return trip to Earth. The first solution occurs for $n = 1$ and allows the minimum time on the Martian surface

$$\frac{0.9855°}{\text{day}} t + 75.1° = 44.4° + \frac{0.5239°}{\text{day}} t + 360°(1), \tag{6.32}$$

which leads to the total transit time $t = 713$ days.

The time on Mars is determined from Equations 6.30 and the total transit time

$$T_{mars} = t - \frac{T}{2} = 713 \text{ days} - \frac{518 \text{ days}}{2} = 454 \text{ days}. \tag{6.33}$$

Equations 6.25 and 6.33 permit the calculation of the total Mars mission time ($T_{mission}$) as the sum of the time from Earth to Mars ($T/2$), the time on the Mars surface (T_{Mars}), and the time from Mars to Earth ($T/2$):

$$T_{mission} = \frac{T}{2} + T_{mars} + \frac{T}{2} = \frac{518 \text{ days}}{2} + 454 \text{ days} + \frac{518 \text{ days}}{2} = 972 \text{ days}. \tag{6.34}$$

The times for Earth orbit and Mars orbit are not included in Equation 6.34. These times are within the errors described previously and would not significantly affect the Hohmann orbit results.

The Equation 6.34 value is approximately the same as NASA's projected Mars duration of 919 days. The NASA values are summarized in Table 6.26. The differences are because of the assumptions used in formulating the Mars trajectory model.

Table 6.26 Comparison of Mars mission duration.

Mission phase	NASA (d)[a]	This work (d)
Earth to Mars transit	224	259
Mars surface	458	454
Mars to Earth transit	237	259
Total mission	919	972

[a] Turcotte (2005).

The Hohmann orbit model has the advantage that it can be used to calculate the mission times for transit between any two locations in a Solar System. The input to the calculation only requires the period and radius of the origin and destination planets. Therefore, it provides a general formulation to obtain a reasonably accurate estimate of the mission transit time, the planetary residence time, and total mission duration without the need to perform a detailed orbital mechanics calculation.

Given the duration of the various mission phases, the dose equivalent for these phases can now be determined. These doses are the subject of the next sections of this chapter.

6.10.5
Overview of Mars Mission Doses

The results of Tables 6.14–6.24 are used to estimate bounding Mars mission doses. Prior to calculating these doses, the results of a previous Mars mission assessment are outlined. Although the mission times presented in the Oak Ridge study are shorter than current mission plans, the results are worth presenting.

6.10.6
Oak Ridge National Laboratory (ORNL) Mars Mission

ORNL provided an estimate of doses based on a 480-day Mars mission that includes a 220-day flight to Mars, 30-day Mars orbit/surface exploration, and 230-day return trip and reentry. The ORNL calculations include contributions from trapped radiation, GCRs, and SPEs.

6.10.7
Trapped Radiation Contribution

After a spacecraft is launched, it assumes an Earth orbit with an equatorial trajectory and 1.5 R_e altitude. In this orbit, the spacecraft encounters about 1×10^{12} electrons/cm^2 ($E_e > 40$ keV) and 6×10^6 protons/cm^2 ($E_p > 100$ MeV) prior to the transit to Mars. A generic spacecraft design in a spherical shape with an inner radius of 5.5 m is assumed. If the spacecraft is uniformly shielded with 5 g/cm^2 (1.85 cm) of aluminum, the crew's proton and electron dose equivalents following the launch are 14 and 2.7 mGy, respectively. The VAB dose on the return leg is higher owing to the time in orbit required to determine an appropriate Earth reentry trajectory. The total VAB (Earth orbit) dose equivalent is 36 mSv.

6.10.8
GCR Contribution

The GCR dose rate depends on the time the mission is initiated relative to the Solar maximum or minimum. If the mission is initiated at the peak of the Solar maximum cycle, the mission dose equivalent is about 260 mSv corresponding to about

200 mSv/year. For a flight beginning at the Solar minimum, the dose equivalent is about 660 mSv based on a dose equivalent rate of about 500 mSv/year. For the assumed 1.85 cm Al shield, the dose equivalent is primarily due to protons with a relatively small contribution arising from secondary particles produced in the shield by primary protons. Contributions from HZE particles are not included in the GCR dose equivalent.

The combined VAB and GCR dose equivalent is nearly 700 mSv. This value exceeds the current US regulatory limit of 50 mSv/year for occupational exposures. However, the largest contribution to the Mars mission has not been included in this value. The largest contribution is from SPEs and is discussed in the following section.

6.10.9
SPE Contribution

Outside the Earth's magnetosphere the weekly probability that an astronaut will receive a lethal dose from Solar flare radiation is 0.002. Assuming, 60 SPEs during the mission, there is approximately a 10% probability that astronauts protected by a 1.85-cm Al shield will receive a dose of more than 6 Sv from protons and alpha particles during the postulated 500 days Mars mission.

A total dose from the VABs (40 mSv), GCR (660 mSv), and SPEs (>6000 mSv) in excess of 6700 mSv is of concern. The 6700-mSv value varies with the assumed SPE properties. However, the magnitude of the total dose equivalent warrants additional shielding beyond the 1.85 cm Al assumed in the ORNL Mars mission. The results support the Hohmann orbit assumptions of ignoring the transit time through the VABs. The VAB contributes only 0.6% (i.e., 40/6700 mSv) of the total Mars mission dose.

6.10.10
Mars Mission Doses

The SPE results are combined with the GCR results to obtain the Mars mission doses. As the Mars trapped radiation dose is small, relative to the GCR and SPE components, the total mission dose (H_{Mars}) is calculated in terms of the dose equivalent to reach and return from Mars ($H_{Mars}^{transit}$) and the dose on the Mars surface ($H_{Mars}^{surface}$),

$$H_{Mars} = H_{Mars}^{transit} + H_{Mars}^{surface}, \qquad (6.35)$$

where

$$H_{Mars}^{transit} = H_{SPE}^{transit} + \dot{H}_{GCR}^{transit}\, t_{transit}, \qquad (6.36)$$

$$H_{Mars}^{surface} = H_{SPE}^{surface} + \dot{H}_{GCR}^{surface}\, t_{surface}. \qquad (6.37)$$

In view of the uncertainties in the SPE magnitude, a single worst-case event is selected to represent the collective SPE mission dose. This is bounding based on the Solar flare characteristics noted previously.

In Equations 6.36 and 6.37, $H_{SPE}^{transit}$ is derived from Tables 6.21–6.24, $\dot{H}_{GCR}^{transit}$ is obtained from Table 6.17, $t_{transit} = 518$ days (Table 6.26), $t_{surface} = 454$ days (Table 6.26), $H_{SPE}^{surface}$ is derived from Tables 6.22–6.24, and $\dot{H}_{GCR}^{surface}$ is obtained from Table 6.18. An estimate of the Mars surface SPE dose equivalent values for the skin, eye, and BFO for the Carrington Flare (Option 2) is obtained by multiplying the Mars surface SPE values from the September 29, 1989 SPE (Option 3) by the ratio of the free space transit SPE values from the Carrington Flare and September 29, 1989 events.

The absorbed dose results could be converted to Gy Eq using the values in Table 6.16. Given the nature of the data presented in Table 6.8 and the values in Tables 6.15 and 6.16, an RBE value of 2 is selected for use to determine the Gy Eq. For simplicity, we make the further assumption that 1 Gy Eq = 1 Sv in the Table 6.27 summary. All Table 6.27 entries are shielded by 5 g/cm². The results of calculations using the large magnitude Solar flare events of Options 2 and 3 are summarized in Table 6.27.

The results summarized in Table 6.27 suggest that the calculated Mars mission deep dose equivalent values, in excess of 3 Sv, are possible. These values are derived from a wide range of mission parameters and conditions that suggest the result is more significant than a single, isolated calculation. The magnitude of the dose equivalent values suggests the need for additional shielding or other measures to mitigate the dose equivalent. These mitigation measures are discussed in a subsequent section of this chapter.

6.11
Other Planetary Missions

The formulation of this chapter is used to calculate the duration and dose equivalents for other planetary missions. These mission characteristics depend on a variety of

Table 6.27 Comparison of Mars mission dose equivalent values.[a]

		Dose equivalent (Sv)		
			This work	
Organ/tissue	Evaluation depth (mg/cm²)	ORNL (1991)	Carrington Flare[b]	September 29, 1989 event[c]
Skin	7	—	7.2	8.1
Eye	300	—	7.1	7.2
Whole body[d]	1000	6.7	—	—
BFO	5000	—	3.6	3.3

[a] All calculated values assume 5 g/cm³ shielding.
[b] Derived from the Townsend (2004) Carrington Flare (Option 2) values and the Saganti et al. (2005) GCR values.
[c] Derived from the Saganti et al. (2005) GCR and Wilson et al. (1999) September 29, 1989 SPE (Option 3) values.
[d] Deep dose equivalent.

parameters including the planet's orbital radius, the orbital period of the planet, its orbital eccentricity, and its surface gravity. The values of these factors for the Sol System are provided in Table 6.28.

The first three of these parameters (i.e., mean distance from the sun, eccentricity, and period) govern the timing of the various mission phases. Surface gravity influences the time spent on the planetary surface. Table 6.28 also provides a qualitative reflection of the accuracy of our assumed circular orbit model.

A circle is a conic section that is a curve traced by a point P moving in a plane so that the distance PF of the point from a fixed point or focus (F) is in a constant ratio to the distance PM of the point P from a fixed line or directrix in the plane of the curve. The eccentricity (e) is defined as the ratio

$$e = \frac{PF}{PM}. \tag{6.38}$$

If $e < 1$ the curve is an ellipse; $e = 1$ is a parabola; and $e > 1$ is a hyperbola. A circle is a special case of an ellipse which has $e = 0$.

Within the trajectory model, the Hohmann orbit approximation most accurately represents planetary orbits when $e \to 0$. The model will be less accurate as e increases. Therefore, it will be least accurate within the Sol system for Mercury and Pluto.

As missions move to more distant regions of the Sol system, longer mission times result and higher dose equivalent values are possible. These results suggest that a practical method for reducing these doses is needed. Possible dose reduction approaches are addressed in a subsequent section.

As previously noted for Mars and Earth, planetary doses are governed by the three dominant radiological source terms (i.e., trapped radiation, GCR, and SPE). Trapped radiation depends on the strength and configuration of the planet's magnetic field.

Table 6.28 Planetary data for Sol system.[a]

Planet	Mean distance from Sun (AU)	Eccentricity	Period (yr)[b]	Equatorial surface gravity (cm/s^2)
Mercury	0.387	0.206	0.241	370
Venus	0.723	0.007	0.615	890
Earth	1.000	0.017	1.000	978
Mars	1.524	0.093	1.88	371
Jupiter	5.203	0.049	11.86	2288
Saturn	9.523	0.053	29.46	905
Uranus	19.164	0.046	84.01	830
Neptune	29.987	0.012	164.1	1115
Pluto[c]	39.37	0.249	247	~20

[a]Derived from Anderson (1981).
[b]Relative to Earth.
[c]For simplicity, Table 6.28 retains the original nine planet Solar System convention. The 2006 International Astronomical Union assigned Pluto as a minor planet. Ceres, Charon, and Xena were also assigned minor planet status in the 2006 Sol system reclassification.

A characterization of the trapped radiation environment is complex and requires direct planetary observation.

GCR dose equivalent rate ($\dot{H}_{GCR}$) values in the vicinity of a planet are relatively constant throughout any given Solar System. This is shown by considering the GCR dose equivalent for a planet a distance d_{planet} from the source of the GCR radiation:

$$\dot{H}_{GCR} = \frac{S_{GCR}}{d_{planet}^n}, \tag{6.39}$$

where S_{GCR} is the strength of the GCR radiation in Sv AU^n/year and n is a positive real number that may differ from the value $n = 2$ because of the presence of other forces acting on the GCR particles (e.g., Solar wind, Solar conditions, and electromagnetic fields). As noted in Table 6.28, the scale (diameter) of the Sol system is on the order of 80 AU. The distance d_{planet} may be written in terms of a vector relationship,

$$\vec{d}_{planet} = \vec{d}_{Sun} - \vec{r}_{planet}, \tag{6.40}$$

where d_{sun} is the distance from the GCR source to the sun and r_{planet} is the distance of the planet from its sun.

To evaluate Equation 6.40 in more detail, consider the source of generation of GCR. GCRs are produced in deep space normally outside the boundaries of the Milky Way Galaxy that has a radius of 4.1×10^4 light years (LY). Therefore, the source of the GCR radiation lies greater than 10^4 LY beyond the Sun:

$$\begin{aligned} LY &= \left(3 \times 10^8 \frac{m}{s}\right)(1 \text{ year})\left(\frac{365.3 \text{ days}}{\text{year}}\right)\left(\frac{24 \text{ h}}{\text{day}}\right)\left(\frac{3600 \text{ s}}{\text{h}}\right)\left(\frac{1 \text{ AU}}{1.5 \times 10^{11} \text{ m}}\right) \\ &= 6.3 \times 10^4 \text{ AU}. \end{aligned} \tag{6.41}$$

Given the values of Table 6.28 and the distance from the GCR source,

$$d_{Sun} \gg r_{planet}. \tag{6.42}$$

Therefore,

$$d_{planet} \approx d_{Sun}. \tag{6.43}$$

Using Equations 6.39 and 6.43 imply that within a Solar System unattenuated GCR radiation is constant:

$$\dot{H}_{GCR} = \frac{S_{GCR}}{d_{Sun}^n} = K_{GCR} \exp(-\mu z), \tag{6.44}$$

where K_{GCR} is the constant, unattenuated GCR dose equivalent rate in a given Solar System. The second term in Equation 6.44 accounts for any attenuation provided by the spacecraft or planet. In free space, the second term represents the shielding provided by the spacecraft. On a planet's surface, the attenuation term consists of any shielding afforded by the habitat and the planetary atmosphere. Therefore, the attenuation factor consists of multiple terms

$$\mu z = \sum_{i=1}^{N} \mu_i z_i(t), \qquad (6.45)$$

where N is the total number of elements shielding the individual, μ_i is the energy-dependent macroscopic removal coefficient for the ith element, and z_i is the equivalent thickness of the ith element. The thickness of any element can vary with time as the space traveler changes location.

The total GCR dose equivalent is the integral of Equation 6.44:

$$H_{GCR} = \int_0^T \dot{H}_{GCR}\, dt = K_{GCR} \int_0^T \exp(-\mu z)\, dt, \qquad (6.46)$$

where T is the mission duration determined from the Hohmann orbit methodology and z and μ are functions of time.

In contrast with GCR, the SPE dose equivalent rate ($\dot{H}_{SPE}$) varies with the distance from a sun (R) because the Sun is the source of the SPE radiation:

$$\dot{H}_{SPE} = \frac{S_{SPE}}{R^m}, \qquad (6.47)$$

where S_{SPE} is the strength of the SPE radiation in Sv AU^m/year and m is a positive real number that may differ from the value $m=2$ because of the presence of other forces acting on the SPE particles (e.g., Solar wind, Solar conditions, and electromagnetic fields). As Sol has a diameter (D_{Sun}) of 1.4×10^9 m, a point source approximation is applicable to within about 1% for

$$R \geq 3 D_{Sun} \geq 3 \times \frac{1.4 \times 10^9 \text{ m}}{1.5 \times 10^{11} \text{ m/AU}} = 0.028 \text{ AU}. \qquad (6.48)$$

Therefore, a point source is applicable for most locations within the Solar System including all planets.

Table 6.29 summarizes the results of using the point source approximation to calculate the SPE dose equivalent at the location of the conventional Solar System's

Table 6.29 SPE dose equivalent values at conventional Solar System planet locations normalized to the Earth's SPE dose equivalent.

Planet	Distance to the sun (AU)	Relative SPE dose equivalent[a]
Mercury	0.387	6.68
Venus	0.723	1.91
Earth	1.000	1.00
Mars	1.524	0.43
Jupiter	5.203	0.037
Saturn	9.523	0.011
Uranus	19.164	0.0027
Neptune	29.987	0.0011
Pluto	39.37	0.00065

[a] Based on an inverse square assumption.

planets. Using Equation 6.47 and $m = 2$, the SPE dose equivalent rate from any given event at a distance R from the Sun is related to the SPE dose equivalent in the vicinity of Earth (r_{Earth}):

$$\dot{H}_{\text{SPE}}(R) = \dot{H}_{\text{SPE}}^{\text{Earth}} \left(\frac{r_{\text{Earth}}}{R}\right)^2, \tag{6.49}$$

where $\dot{H}_{\text{SPE}}^{\text{Earth}}$ is the dose equivalent rate measured in the vicinity of the Earth at 1 AU ($r_{\text{Earth}} = 1$ AU). As noted in Table 6.29, the SPE dose equivalent rates become less of a concern as the spacecraft moves away from the Sun.

The total SPE dose equivalent received during a mission of duration T is obtained by integrating Equation 6.49:

$$H_{\text{SPE}} = \int_0^T \dot{H}_{\text{SPE}} \, dt = \int_0^T \dot{H}_{\text{SPE}}^{\text{Earth}} \left(\frac{r_{\text{Earth}}}{R(t)}\right)^2 = r_{\text{Earth}}^2 \int_0^T \frac{\dot{H}_{\text{SPE}}^{\text{Earth}}(t) \, dt}{R^2(t)}, \tag{6.50}$$

where $R(t)$ is the spacecraft trajectory or its position as a function of time (t). $\dot{H}_{\text{SPE}}^{\text{Earth}}$ is also a function of time because multiple SPEs will occur during a planetary mission. In addition, the SPE component is attenuated in a manner analogous to Equation 6.46.

The inverse square dependence of the SPE fluence is a reasonable first approximation. However, variations in the radial dependence are expected because of inhomogeneities in the interplanetary magnetic field, anisotropies in the emitted Solar flux distribution, and the complexities of the SPE energy spectrum. The absolute radial dependence of the SPE fluence as a function of distance from the Sun is still an open issue. On the basis of the current data, the radial dependence of an SPE depends on the distance region and quantity of interest:

- For fluence rate extrapolations from 1 AU to >1 AU, use a functional form of $r^{-3.3}$ and expect variations ranging from r^{-4} to r^{-3}.
- For fluence rate extrapolations from 1 AU to <1 AU, use a functional form of r^{-3} and expect variations ranging from r^{-3} to r^{-2}.
- For fluence extrapolations from 1 AU to other distances, use a functional form of $r^{-2.5}$ and expect variations ranging from r^{-3} to r^{-2}.

Given these results, a general equation for the unshielded dose equivalent rate ($\dot{H}$) within our Solar System can be written as in terms of the trapped, GCR, and SPE dose equivalent components defined previously:

$$\dot{H} = \dot{H}_{\text{trapped}} + \dot{H}_{\text{GCR}} + \dot{H}_{\text{SPE}}. \tag{6.51}$$

The total mission dose equivalent is the integral of Equation 6.51 from launch ($t = 0$) until mission completion ($t = T$):

$$H = \int_0^T \dot{H} \, dt = \int_0^T (\dot{H}_{\text{trapped}} + \dot{H}_{\text{GCR}} + \dot{H}_{\text{SPE}}) \, dt. \tag{6.52}$$

Again, the integral is highly dependent on the trajectory of the spacecraft and any shielding provided by the spacecraft and planetary atmosphere.

In Equation 6.52, $\dot{H}_{\text{trapped}}$ is the dose equivalent rate from the radiation trapped by a planet's magnetic field. The trapped radiation contribution strongly depends on the characteristics of the planet's magnetic field and the trajectory of the spacecraft used to traverse the region of space governed by the magnetic field. The total dose equivalent from trapped radiation is formally written as

$$H_{\text{trapped}} = \int_0^T \dot{H}_{\text{trapped}}(r, \theta, \phi) dt, \quad (6.53)$$

where the coordinates (r, θ, ϕ) define the position of the spacecraft relative to the planet in terms of standard spherical coordinates. On the basis of previous discussions, a planet's trapped radiation term is likely to be smaller than either the GCR or the SPE contributions to the dose equivalent.

6.11.1
Planetary Atmospheric Attenuation

The attenuation terms in Equation 6.42 include atmospheric attenuation. In the continental United States, the average annual GCR dose equivalent is about 0.26 mSv at sea level. The dose equivalent doubles for each 1500–2000 m increase in altitude in the lower atmosphere. The dose equivalent also varies by about 10% as a function of latitude.

Atmospheric shielding of GCR and SPE radiation is not unique to the Earth. It also occurs on any planet with an atmosphere including Mars.

The amount of protection provided by the low-density and high-density Martian atmosphere models as a function of altitude is provided in Table 6.30. Altitudes up to 12 km are included in Table 6.30 to accommodate the topographical features on the Martian surface. Both atmosphere models are considered to estimate the possible variation in the radiation intensities.

Total dose calculations for BFO are summarized in Table 6.31 for both the high- and the low-density Mars atmosphere models at altitudes of 0, 4, 8, and 12 km. Results are provided for GCRs at Solar minimum and maximum conditions and for

Table 6.30 Martian atmospheric protection in the vertical direction.[a]

Altitude (km)	Low-density model[b] (g CO_2/cm^2)	High-density model[c] (g CO_2/cm^2)
0 (surface)	16	22
4	11	16
8	7	11
12	5	8

[a] Derived from NASA Conference Publication 3360 (1997).
[b] Corresponds to a surface pressure of 5.9 mb.
[c] Corresponds to a surface pressure of 7.8 mb.

Table 6.31 Total BFO dose equivalent (Sv) on Mars as a function of altitude using both high- and low-density atmospheric models.[a–c]

Radiation source	BFO dose equivalent (Sv) at 0 km	BFO dose equivalent (Sv) at 4 km	BFO dose equivalent (Sv) at 8 km	BFO dose equivalent (Sv) at 12 km
GCR at solar minimum (annual)	0.105–0.119	0.120–0.138	0.137–0.158	0.156–0.180
GCR at solar maximum (annual)	0.057–0.061	0.062–0.068	0.067–0.074	0.073–0.081
February 1956 SPE	0.085–0.099	0.100–0.118	0.117–0.136	0.134–0.153
November 1960 SPE	0.050–0.073	0.075–0.108	0.106–0.148	0.144–0.191
August 1972 SPE	0.022–0.046	0.048–0.099	0.095–0.185	0.174–0.303
August 1989 SPE	0.001–0.003	0.003–0.006	0.006–0.013	0.012–0.026
September 1989 SPE	0.010–0.020	0.020–0.038	0.037–0.065	0.061–0.106
October 1989 SPE	0.012–0.027	0.028–0.059	0.057–0.114	0.106–0.205

[a]High-density model dose estimate – low-density model dose estimate.
[b]Derived from NASA Conference Publication 3360 (1997).
[c]Dose equivalent values are annual doses for GCRs. For SPEs, the dose equivalent is the dose incurred during the flare duration.

significant SPEs that occurred in 1956, 1960, 1972, and 1989 (see Table 6.19). The GCR dose equivalent during Solar maximum conditions is approximately half the dose incurred during Solar minimum conditions. GCR dose equivalent values are relatively constant with altitude, but SPEs have considerably more variation.

6.12
Mars and Outer Planet Mission Shielding

The 3–8 Sv values of Table 6.27 exceed any acceptable mission dose equivalent value and would result in a serious health detriment to personnel. These values mandate that space crewmembers be protected from SPE radiation by thicker shielding than the 5-g/cm^2 assumed in Table 6.27 or mitigated in another manner. The shielding could be part of the spacecraft surface structure, shadow shielding, or incorporated into a shielded shelter located inside the spacecraft.

The 480-day ORNL mission determined that the 6-Sv SPE dose equivalent with 5-g/cm^2 aluminum shielding could be reduced to about 2 Sv if the shield thickness was doubled to 10 g/cm^2. However, this would increase the shield weight from 19 to 38 metric tons (MT).

Shielding the spacecraft to 20-g/cm^2 aluminum or equivalent increased the spherical spacecraft's weight to about 78 MT if the shielding were uniformly distributed over the spacecraft. However, an internal shelter concept could be added at a fraction of the weight. Spacecraft weight is an important consideration because launch rockets have limited payload capability.

To resolve this problem, an internal shelter was placed within the spherical spacecraft shell. This option included the addition of a 2-m sphere shelter surrounded by 15-g/cm² of aluminum inside the 5.5-m spherical spacecraft that has a 5-g/cm² shielding as part of its basic design. The shelter adds 7.8 MT to the spacecraft weight of 19 MT. This configuration, providing a total of 20 g/cm² aluminum shielding within the shelter decreases the Solar flare dose by a factor of about 100 to 0.06 Sv.

The ORNL shielding analysis is based on a high intensity SPE data profile from 1956 to 1961. As noted in Table 6.19, this cycle is representative of the results that would be obtained by utilizing more recent Solar data. Using these data for the 480-day Mars mission, the dose equivalent with 1 g/cm² shielding varied by a factor of 10–140 depending on the launch date within the assumed SPE profile. The absorbed dose values as a function of Mars mission duration are summarized in Table 6.32. To achieve an absorbed dose of 0.15 Gy or less during a 1-year mission within the 1956–1961 SPE profile, requires the launch occur during a 6-week period when Solar conditions are similar to those that occurred between mid-July and the end of August 1959. It is likely that similar low-dose windows will occur for planetary and deep space launches in the future. However, predicting the specific timing of these future windows is beyond the current capability of Solar physics. The ability of a twenty-first century physicist to accurately forecast Solar conditions would be a powerful ALARA tool, which would significantly reduce LEO, lunar, and planetary mission doses.

Space travelers will encounter 8–13 SPEs during the 480-day Mars mission. If a single Solar flare with a magnetic rigidity of 125 MV were to occur during the mission, astronauts shielded by 5-g/cm² of aluminum would receive a total dose of 2.85 Sv comprised of about 2.00 Sv from protons and 0.85 Sv from alpha particles.

Table 6.32 Minimum and maximum Mars mission SPE absorbed doses for various mission durations.[a]

Mission duration (months)	Absorbed dose (Gy)[b]	
	Minimum	Maximum
0.25	0	14.9
0.5	0	14.9
1	0	14.9
1.5	0	14.9
3	0	19.6
6	0	19.6
9	0.02	19.6
12	0.15	21.1
18	1.76	24.2
24	5.26	27.8
36	9.74	32.3
48	24.4	34.9

[a]Based on the solar cycle 19 (1956–1961) SPE profile, Santora and Ingersoll (1991).
[b]Surface absorbed dose inside 1 g/cm² uniform aluminum shielding.

A 20-g/cm² shield would reduce the dose equivalent to about 0.14 Sv. Shielding effectiveness depends on the SPE energy spectrum and fluence of the various particle types that define the event.

Although the intensity distribution of Solar flares is unpredictable, it is likely that the total SPE dose from a Mars or deep space mission will be significant. This dose equivalent arises from one or more large SPEs such as the 125-MV event noted above with the remaining flares contributing little additional dose equivalent. However, given the potential magnitude of these flares, additional shielding or another means of reducing the SPE dose should be considered. In the following section, an alternative to bulk shielding is considered.

6.13
Electromagnetic Deflection

Reduction of the SPE charged particle fluence is usually addressed with conventional shielding materials such as aluminum. However, providing sufficient shielding presents a problem because of launch weight constraints on planetary or deep space missions. An alternative to conventional shielding materials for charged particle radiation is the use of electromagnetic (EM) fields to alter the charged particle trajectory.

The EM force ($\vec{F}$) imposed on a charged particle or ion of mass (m) and charge (q) is

$$\vec{F} = q(\vec{E} + \vec{v} \times \vec{B}), \tag{6.54}$$

where $\vec{E}$ is the electric field, $\vec{B}$ is the magnetic induction, and $\vec{v}$ is the ion velocity. The magnetic induction is related to the magnetic field $\vec{H}$ through the relationship:

$$\vec{B} = \mu \vec{H}, \tag{6.55}$$

where μ is the permeability of free space ($4\pi \times 10^{-7}$ N/A²).

In subsequent discussions, the motion of the charged particle is presented in the reference frame of the spacecraft. A number of assumptions are made to simplify the underlying physics, but the essential health physics elements are preserved. These assumptions include the following:

- The deflector's electric and magnetic fields are static (constant) in the reference frame of the spacecraft. These conditions can be achieved by design.
- Bremsstrahlung radiation from an ion, accelerated by the deflector, is small. This is valid for reasonable deflector fields and the SPE ions and their expected energy spectra.
- The electric and magnetic fields of the individual SPE ions are small and do not affect their trajectory once it is subjected to the deflector's EM field. Given the spectra and flux density of SPE particles, this is a credible assumption.
- The ion's bremsstrahlung does not significantly modify its trajectory.

- In the absence of deflector fields, the SPE particles form a parallel beam that would strike the circular face of the cylindrical spacecraft.

For specificity, assume the ion's velocity is in the positive x-direction and the spacecraft is also traveling in the positive x-direction. Equation 6.54 suggests the ion's trajectory is altered such that it will miss the spacecraft through the use of either an electric field in the y- or z-direction or a magnetic field in the y- or z-direction.

Before defining the specific characteristics of the EM deflector, it is necessary to utilize selected elements of special relativity that are provided in Appendix F. Readers not familiar with this topic should consult Appendix F before proceeding further. Relativistic considerations are needed to properly describe the kinematics of HZE particles that accompany a SPE event.

With this background, consider a cylindrical spacecraft of radius Y, traveling in the positive x-direction and towing an electromagnetic field generator to a distance X from the spacecraft. Using Cartesian coordinates, the SPE particles are assumed to enter the deflector located at $(x, y, z) = (0, 0, 0)$ along the spacecraft axis. The circular spacecraft face closest to the detector resides at $(X, 0, 0)$ and the face's periphery is at $(X, Y, 0)$. A charged particle that enters the detector and travels a distance X in the x-direction will not impact the spacecraft if the deflection is greater than a distance Y in the y-direction.

6.13.1
EM Field Deflector Physics

For an ion of charge q and kinetic energy (T), its total energy (W) is the sum of the ion's rest energy (E_o) and its kinetic energy:

$$W = E_o + T, \tag{6.56}$$

where

$$E_o = m_o c^2, \tag{6.57}$$

$$T = \frac{1}{2}\gamma m_o v^2. \tag{6.58}$$

In Equations 6.57 and 6.58, m_o is the ion's rest mass, c is the speed of light, v is the ion's velocity, and γ is the Lorentz factor defined by the relationship

$$\gamma = \left(1 - \frac{v^2}{c^2}\right)^{-1/2}. \tag{6.59}$$

The ion's total energy is related to the particle's momentum (p) through the relativistic relationship:

$$W^2 = p^2 c^2 + m_o^2 c^4. \tag{6.60}$$

Equation 6.60 can be solved for the momentum

$$p = \left(\frac{W^2 - m_o^2 c^4}{c^2}\right)^{1/2}, \tag{6.61}$$

which is also be written as

$$p = \gamma m_o v. \tag{6.62}$$

Equations 6.59 and 6.62 are used to express the momentum as

$$p = \frac{m_o v}{\left(1 - \frac{v^2}{c^2}\right)^{1/2}}, \tag{6.63}$$

Equation 6.63 can be solved for v

$$v = \frac{p}{\left(m_o^2 + \frac{p^2}{c^2}\right)^{1/2}}. \tag{6.64}$$

With the determination of v and γ, the necessary parameters to define the EM deflector field strengths are determined. Using these values, the distance X traveled by the particle from the EM field generator to the periphery of the spacecraft is written in terms of the ion's velocity and travel time (t):

$$X = vt, \tag{6.65}$$

which is solved for the time for the ion to travel a distance X following its initial deflection:

$$t = \frac{X}{v}. \tag{6.66}$$

Assuming the particle is initially traveling in the *x*-direction on the cylinder axis before deflection, its total deflection in the *y*-direction (Y) is written in terms of its acceleration in the *y*-direction (a) caused by the EM field and the travel time:

$$Y = \frac{1}{2} a t^2 \tag{6.67}$$

Equation 6.67 is solved for the acceleration

$$a = \frac{2Y}{t^2}. \tag{6.68}$$

Equation 6.67 assumes that there is no initial velocity in the *y*-direction and that the deflection initially occurs at the $y = 0$ location. Smaller field strengths would be required if the particle was of the axis ($y > 0$) of the spacecraft.

With the definition of acceleration, the EM fields can be determined. Subsequent sections review two specific cases:

- *Case I:* Static Magnetic Field Deflector
- *Case II:* Static Electric Field Deflector.

6.13.2
Case I – Deflection Using a Static Magnetic Field

In the first case, the deflector generates a magnetic induction of constant strength B in the $-z$-direction. Using these conditions and Equation 6.54 yields a magnetic force in the y-direction:

$$\vec{F}_m = q\vec{v} \times \vec{B} = q \begin{vmatrix} \hat{i} & \hat{j} & \hat{k} \\ v & 0 & 0 \\ 0 & 0 & -B \end{vmatrix} = qvB\hat{j}. \tag{6.69}$$

Equation 6.69 can be equated to the force because of the accelerated ion in the y-direction. Using Newton's second law,

$$\vec{F} = m_o \gamma a \hat{j}. \tag{6.70}$$

The equality of Equations 6.69 and 6.70 yields

$$qvB = m_o \gamma a. \tag{6.71}$$

Equation 6.71 is solved for the magnetic induction that is necessary to deflect an ion, traveling along the spacecraft centerline, a distance Y such that it will miss the spacecraft:

$$B = \frac{m_o \gamma a}{qv}. \tag{6.72}$$

6.13.3
Case II – Deflection Using a Static Electric Field

The electric field required for the ion to miss the spacecraft is governed by the electric force component of Equation 6.54:

$$\vec{F}_e = q\vec{E}. \tag{6.73}$$

For an electric force in the positive y-direction, an expression analogous to Equation 6.71 can be written as

$$qE = m_o \gamma a. \tag{6.74}$$

Equation 6.74 is solved for the electric field necessary to deflect an ion on the spacecraft axis a distance Y such that it will miss the spacecraft:

$$E = \frac{m_o \gamma a}{q}. \tag{6.75}$$

The required electric and magnetic fields that cause an ion to miss the spacecraft are provided in Tables 6.33–6.35. The following assumptions are utilized in the Tables 6.33–6.35 results:

- The EM field generator is located behind the spacecraft on its axis at evaluation distances (X) of 100, 1000, and 10 000 m. This distance can be achieved in a variety of ways including towing the EM field generator behind the spacecraft.
- The spacecraft has a cylindrical radius (Y) of 15 m.

Table 6.33 Required electric and magnetic fields for incident protons to miss the spacecraft.

	E (V/m)			H (A/m)		
E (MeV)	X = 100 m	X = 1000 m	X = 10 000 m	X = 100 m	X = 1000 m	X = 10 000 m
1	6000	60.0	0.60	350	3.5	0.035
50	2.9×10^5	2900	29.0	2500	25.0	0.25
500	2.5×10^6	2.5×10^4	250	8700	87.0	0.87

The required fields for protons with energies of 1, 50, and 500 MeV are provided in Table 6.33. Tables 6.34 and 6.35 summarize the required fields for alpha particles and $^{12}C^{+6}$ ions, respectively. The alpha particles and $^{12}C^{+6}$ are assumed to possess the same total energy as the protons.

The results of Tables 6.33–6.35 suggest that the required deflection fields to miss the spacecraft decrease as the distance between the spacecraft and the deflector (X) increases. Following Equations 6.66 and 6.68, the field strength decreases as the inverse square of the distance between the spacecraft and the deflector. For example, 50 MeV protons require magnetic fields of 2500, 25, and 0.25 A/m for deflector-to-spacecraft distances of 100, 1000, and 10 000 m, respectively. Therefore, placing the deflector further from the spacecraft minimizes the field requirements. In addition, higher energies for a given ion and deflector distance require additional field strength to achieve the required deflection.

Table 6.34 Required electric and magnetic fields for incident alpha particles to miss the spacecraft.

	E (V/m)			H (A/m)		
E (MeV)	X = 100 m	X = 1000 m	X = 10 000 m	X = 100 m	X = 1000 m	X = 10 000 m
1	3000	30	0.30	340	3.4	0.034
50	1.5×10^5	1500	15	2400	24	0.24
500	1.4×10^6	1.4×10^4	140	8000	80	0.80

Table 6.35 Required electric and magnetic fields for incident $^{12}C^{+6}$ ions to miss the spacecraft.

	E (V/m)			H (A/m)		
E (MeV)	X = 100 m	X = 1000 m	X = 10 000 m	X = 100 m	X = 1000 m	X = 10 000 m
1	1000	10	0.10	200	2.0	0.02
50	5.0×10^4	500	5.0	1400	14	0.14
500	4.9×10^5	4900	49	4500	45	0.45

Tables 6.33–6.35 also suggest that an EM deflector can utilize either electric fields or magnetic fields or a combination of both. For example, for 500 MeV protons an electric field of about 2.5 MV/m is required for an $X = 100$ m deflector. These same conditions require a magnetic field strength of 8700 A/m.

Using Equation 6.55, a magnetic field strength of 8700 A/m corresponds to a magnetic induction of

$$B = \mu H = \left(4\pi \times 10^{-7} \frac{N}{A^2}\right)\left(8700 \frac{A}{m}\right) = 0.01\ \text{T}, \qquad (6.76)$$

or about 100 gauss (1 Tesla (T) = 10^4 gauss). These magnetic field/induction values are achieved using conventional or superconducting technology.

No attempt has been made to minimize the deflector field strength by determining an optimum admixture of the electric and magnetic fields. Such an optimization is required to minimize deflector power requirements. However, deflector optimization requires a detailed SPE spectral information, a better definition of the spacecraft geometry, and specification of mission parameters. Defining and integrating this information is beyond the scope of this textbook.

Table 6.36 considers the radial dependence of the required EM fields to deflect an ion such that it misses the spacecraft. For specificity, we consider a 500-MeV proton and present the fields required to deflect an ion as a function of its distance from the axis (r) of the spacecraft. For specificity, $r = 0$ corresponds to the spacecraft centerline and $r = 15$ m corresponds to an ion that just misses the spacecraft periphery. In Table 6.36, the deflector is assumed to reside at $X = 1000$ m.

Table 6.36 Radial profile of the EM fields as a function of the spacecraft off axis distance.[a]

Off axis distance (m)	E (V/m)	H (A/m)	B (T)
0	2.5×10^4	87	1.1×10^{-4}
1	2.3×10^4	81	1.0×10^{-4}
2	2.1×10^4	75	9.5×10^{-5}
3	2.0×10^4	69	8.7×10^{-5}
4	1.8×10^4	64	8.0×10^{-5}
5	1.7×10^4	58	7.3×10^{-5}
6	1.5×10^4	52	6.5×10^{-5}
7	1.3×10^4	46	5.8×10^{-5}
8	1.2×10^4	41	5.1×10^{-5}
9	9.9×10^3	35	4.4×10^{-5}
10	8.3×10^3	29	3.6×10^{-5}
11	6.6×10^3	23	2.9×10^{-5}
12	5.0×10^3	17	2.2×10^{-5}
13	3.3×10^3	12	1.5×10^{-5}
14	1.7×10^3	5.8	7.3×10^{-6}
15	0	0	0

[a] Tabulated values assume a 500-MeV proton and an EM deflector located 1000 m behind the spacecraft.

The reader should note that these profiles are linear functions of the off-axis distance. This linear relationship of the field strength follows from Equations 6.68, 6.72, and 6.75.

As noted previously, an electromagnetic deflector significantly reduces the SPE contribution to the dose equivalent. It should also be feasible to design a deflector that also alters the trajectory of the GCR source ions. Assuming deflectors can be constructed to deflect both SPE and GCR ions, the total mission dose equivalent is defined as

$$H = (1-\varepsilon_{SPE})\int_0^{t_{transit}} \dot{H}_{SPE}^{transit}\, dt + (1-\varepsilon_{GCR})\int_0^{t_{transit}} \dot{H}_{GCR}^{transit}\, dt \\ + \int_0^{t_{surface}} (\dot{H}_{SPE}^{surface} + \dot{H}_{GCR}^{surface})\, dt, \quad (6.77)$$

where ε_{SPE} is the SPE EM deflector efficiency and ε_{GCR} is the GCR EM deflector efficiency. In Equation 6.77, the EM deflector is only utilized during transit and not on the Martian surface. It should also be possible to build a deflector on the Martian surface to further reduce the mission dose.

The impact of an EM deflector is illustrated in Table 6.37 in which a constant deflector efficiency is assumed to be applicable to all particle types and locations. Given deflector design and SPE and GCR spectrum differences, the SPE and GCR deflectors are not expected to have the same efficiency. However given the uncertainties in the calculation, Table 6.37 was determined assuming the SPE and GCR deflectors have the same efficiency (ε). Although an oversimplification, it is sufficient for the scoping calculations of this chapter.

EM deflectors have the potential to significantly reduce the total mission dose equivalent. The results of Table 6.37 suggest that an EM field deflector is a promising alternative to conventional shielding, but it represents a more significant engineering challenge.

Table 6.37 Impact of an electromagnetic deflector on the total Mars mission dose equivalent.[a]

EM deflector efficiency	Dose equivalent (Sv)		
	Skin	Eye	BFO
0.00	8.1	7.2	3.3
0.10	7.4	6.5	3.0
0.20	6.6	5.9	2.7
0.30	5.9	5.2	2.4
0.40	5.1	4.6	2.2
0.50	4.4	3.9	1.9
0.60	3.6	3.2	1.6
0.70	2.9	2.6	1.3
0.80	2.1	1.9	1.1
0.90	1.4	1.3	0.8
1.00	0.6	0.6	0.5

[a]Based on Wilson et al. (1999) and Table 6.27 values for the September 29, 1989 SPE.

6.13.4
Engineering Considerations for EM Field Generation

An electromagnetic deflector requires considerable engineering analysis before it could become a practical alternative to conventional shielding. Although shielding functions passively, an EM deflector must be activated. Moreover, the deflector must function when required during an SPE. Since the timing, duration, and intensity of an SPE cannot be predicted, the operability and availability of the EM deflector must be very high. In addition, there are uncertainties in the inherent risk of space radiation exposures that mandate that the EM deflector be reliable.

To ensure its availability and operability, a number of design and engineering requirements must be met to ensure that the EM deflector provides

- defense in depth to generate fields of the required intensity, orientation, and duration during an SPE;
- primary and backup power to energize the EM fields;
- redundancy in EM field generating devices;
- the ability to maintain key components via spacecraft maintenance or through extravehicular activity;
- radiation hardening for the EM deflector field generators to survive SPE and GCR radiation damage.

The availability and operability requirements are more difficult to satisfy in space than on the surface of a planet. Until EM deflector design is sufficiently engineered and hardened, it would be advisable to incorporate shielded shelters on space missions to mitigate the effects of a large SPE event.

6.14
Space Radiation Biology

The biological effects of ionizing radiation from the various sources encountered during a space mission are not completely understood. These effects are broadly categorized as either deterministic or stochastic. Deterministic effects include threshold effects such as acute somatic effects resulting from Solar particle events. Stochastic effects include cancer and hereditary effects. The major risks of space flight are cancer induced by HZE particles, immune system detriment, and neurological and behavioral effects that would jeopardize an extended duration mission. The cancer risk resulting from these effects is compounded because synergistic effects may exist with other hazards encountered in space. As an example, the possible synergistic effect of low gravity and high dose rates resulting from an SPE is not fully understood.

Historically, mitigating the early effects of a crew's radiation exposure was addressed by shielding the spacecraft and the focus was on estimating the risk of late radiation effects such as cancers. However, the true risk includes both early and late radiation effects.

Mission radiation risks are difficult to assess, because SPEs can be large and unpredictable. In addition, SPE events offer the potential for rapid and progressive exposure to the primary charged particles representing a wide array of atomic numbers, atomic masses, energies and dose rates, and any resulting secondary radiation. In spite of this risk, countermeasures can be used to prevent or reduce the detriment from exposure to protons or HZE particles. These countermeasures are discussed in the next chapter.

6.15
Final Thoughts

The three dominant sources of the space radiation dose equivalent are trapped radiation such as the Earth's van Allen belts, galactic cosmic radiation, and Solar particle events. Dose equivalents received in low-Earth orbit are larger than terrestrial dose equivalents but generally tolerable. For planetary space missions, the GCR dose equivalent increases with mission duration and approaches values on the order of 1 Sv for a Mars mission. Even larger Mars mission dose equivalent values (>3 Sv) can be received from large SPEs. These SPE dose equivalent values are hazardous to crews of vehicles on planetary missions. Doses that are large but survivable with medical intervention on Earth may not be survivable in space.

Reducing the effect of an SPE can be accomplished with conventional bulk shielding, but launch weight restrictions limit the shielding thickness. A potentially attractive alternative to bulk shielding is the electromagnetic deflection of ions, but engineering issues must be addressed to ensure the reliability and effectiveness of this approach.

Once a spacecraft moves beyond Mars toward Jupiter, the SPE dose equivalent is <4% of the value received on Earth, and the GCR component becomes a more significant contributor to the source term. The distance limitations of travel to the outer planets will likely require the introduction of nuclear powered spacecraft. This propulsion system's radiation output also contributes to the crew's dose equivalent during an outer planet mission. Nuclear powered spacecraft are addressed in the following chapter.

Problems

06-01 An ion of charge q and mass m is trapped in the Earth's magnetosphere. At the location of interest, the magnetic induction is oriented in the positive z-direction and has magnitude B_0. The ion's velocity is $\vec{v} = a\hat{z} + b\hat{w}$ where $\hat{z}$ is a unit vector in the z-direction, w is a unit vector in the (x, y) plane, a is the magnitude of the velocity in the z-direction, and b is the magnitude of the velocity in the (x, y) plane. (a) Describe the ion's trajectory. (b) What is the radius of the ion's motion in the (x, y) plane?

06-02 You are an assistant professor of health physics and one of your students concludes that trapped radiation in the Earth's van Allen radiation belts originates from cosmic ray interactions in the upper atmosphere. At the altitude of interest, the cosmic ray flux is 2 protons/cm² s and the atmospheric density is 5×10^5 atoms/cm³. If the mean cross section describing the interaction of protons with atoms in the upper atmosphere is 0.2 b, is the student correct?

06-03 Cosmic ray protons collide with oxygen and nitrogen nuclei in the Earth's atmosphere and produce neutrons. For example, a 5-GeV cosmic ray proton produces about seven neutrons when it collides with oxygen or nitrogen nuclei in an atmospheric collision. Neutrons are also produced in the Sun and reach the Earth during a Solar flare event. (a) Assuming the energy of the Solar neutrons is 1 MeV, what fraction of the neutrons released with the flare reach the Earth's atmosphere. (b) What products result from the decay of a neutron? (c) What is the radiological significance of cosmic ray generated neutrons that are produced in the upper atmosphere of the Earth?

06-04 Cosmic radiation contributes to the natural radiation environment. (a) What is the contribution of cosmic rays to the background radiation level at sea level? (b) Given the value in part (a), what is the expected cosmic ray dose contribution at 10 000 m?

06-05 In 1962, the "Starfish" nuclear test with a yield of 1.4 MT occurred above Johnson Island in the Pacific Ocean. In 1958, the "Argus I" test occurred in the South Atlantic with a 1-kT yield at an altitude of 500 km. (a) Artificial radiation belts are produced by high-altitude nuclear weapon detonations. What radiation type dominates these artificial belts? (b) Compare the expected radiation belts produced by the "Argus I" and "Starfish" detonations.

06-06 Cosmic radiation interacts with atmospheric atoms to produce a variety of radiation types. (a) Describe the primary constituents of cosmic radiation in the upper atmosphere. (b) Describe the constituents of cosmic radiation at sea level. (c) Describe the interaction of secondary particles as they penetrate the atmosphere and reach the surface of the Earth.

06-07 A Solar particle event has occurred with a massive release of protons. The projected Solar proton flux at the spacecraft is 8.0×10^5 p/cm² s and the effective duration of the event is 1 h. The proton fluence factor (dose conversion factor) is 3×10^3 pSv cm²/proton.

Your spacecraft normally deflects Solar protons using its electromagnetic deflector. However, it is out-of-service and cannot be repaired before the protons reach your craft. The event is sufficiently large to pose a lethal radiation threat to the crew.

As an option of last resort, you suggest utilizing the aft antiproton rail-gun to reduce the radiological impact of the incoming Solar protons. The proton–antiproton interaction reduces the proton effective dose received by the spacecraft crew. Assume that each proton–antiproton interaction produces an effective dose that is equal to the unattenuated proton dose reduced by a factor of 100.

The output of the antiproton rail-gun decreases during its operation. In particular, the rail-gun's antiproton flux during the first 15 min of the event is 8×10^5 p/cm² s, 6×10^5 p/cm² s during the second 15 min, and 4×10^5 p/cm² s during the final 30 min of the event.

(a) What is the unattenuated proton effective dose? (b) What is the effective proton dose using the antiproton beam? (c) What other options are available to reduce the crew's effective dose?

06-08 You are the radiation safety officer for Global Space Dynamics Corporation (GSDC), a manufacturing organization that utilizes microgravity as part of its industrial operations. Workers are assigned to GSDC's LEO operations for a period of 1 year of uninterrupted service. The average effective dose for the LEO orbit in occupied areas is 0.06 mSv/h. GSDC follows the recommendations of NCRP 132. (a) What is the annual effective dose for GSDC's workers? (b) Assuming the effective doses from part (a) apply to the whole body and blood-forming organs, what limitations are imposed by NCRP 132 for a 25-year-old male worker and how do these limits affect the number of LEO tours the worker is allowed?

06-09 During a Mars mission, several SPEs occur and irradiate the spacecraft crew. The SPEs can be treated as pure proton events and the characteristics of those flare events are noted in a subsequent table. (a) Assuming no EM deflector is available, what absorbed dose is received by the crew? (b) Assume the electromagnetic deflector functioned at 90% efficiency during January 1–14, 2022, was out of service on January 15, and functioned at 60% during the period after January 15, 2022. Given the deflector's availability, what is the crew's absorbed dose during the flare events? Characteristics of relevant flare events are provided in the following table:

Date	Total fluence inside Mars mission spacecraft (protons/cm^2)	Dose conversion factor (pGy cm^2/proton)
January 12, 2022	8×10^7	2500
January 13, 2022	2×10^8	3000
January 15, 2022	4×10^8	3200
January 16, 2022	1×10^9	3100
January 20, 2022	5×10^8	4500
January 21, 2022	8×10^7	5800

06-10 During transit from Neptune to Pluto, the USS Pittsburgh encounters a dark energy cloud, which interacts with the aluminum shell of the Pittsburgh and produces energy that is absorbed per unit volume per unit time in tissue according to the relationship:

$$\xi = a(1 - bx + c\,e^{-dx})\,e^{-\lambda t},$$

where x is the penetration depth into tissue, t is the time since the cloud struck the spacecraft shell, and a, b, c, d, and λ are constants with a value of 0.2 mJ/cm^3 h, 0.025/cm, 0.4, 0.1/cm, and 1/h, respectively. Given these values calculate (a) the average absorbed dose rate and the total absorbed dose delivered to the outer 5 cm of tissue during the first hour following impact with the cloud, and (b) the average dose rate and the total absorbed dose delivered to the outer 5 cm of tissue during the first 8 h following impact with the cloud. (c) On the basis of these results, would you recommend that the Pittsburgh accelerate to leave the dark energy field in 1 h? Without accelerating, the Pittsburgh would remain in the field for 8 h.

06-011 During a manned mission a supernova event, originating at the far edge of the Milky Way Galaxy, is detected while the HMS Goodship is in orbit around Jupiter. The ship's spectrometers provide particle fluence and energy information and estimate dose conversion factors based on particle type and energy. The Goodship has a shell that is 3 cm of equivalent aluminum and is equipped with an emergency shelter having 8 cm of aluminum shielding. The spectrometer output for the event duration (2 h) and same time interval prior to the event are summarized in the following table. Relevant attenuation coefficients are also provided.

	Spectrometer output by particle type			
	Average values prior to event		Average event values	
Particle type	Integrated fluence (particles/cm^2)	Dose conversion factor (pGy cm^2/particle)	Integrated fluence (particles/cm^2)	Dose conversion factor (pGy cm^2/particle)
Protons	3×10^5	3000	5×10^9	4500
Heavy ions	4×10^4	7000	8×10^9	8800

	Applicable attenuation coefficient for aluminum (1/cm)	
Particle type	Prior to event	During event
Protons	0.20	0.15
Heavy ions	0.35	0.30

(a) If the distance from the Goodship to Earth is 4.2 AU and the average velocity of the particles is 0.1 c, how much warning will the Earth have if an alert is immediately broadcast upon event detection?
(b) What is the absorbed dose received by the crew during the 2-h period prior to the event? Assume the crew is not in the shelter during this period.
(c) What is the absorbed dose received by the crew during the extra-solar event? Assume the crew does not relocate to the shelter during the event.
(d) On the basis of the result of question (c), should the crew relocate to the shelter during the event?
(e) What is the absorbed dose on the Earth's surface at sea level resulting from this event? Assume that the atmosphere's effective thickness is 25 km. For the spectrum striking the Earth, the atmosphere decreases the absorbed dose by a factor of 2 for every 2000 m for protons and 1500 m for heavy ions.

(f) If the event occurred over a 3-week period, what is the effective dose on the Earth's surface? Assume the radiation-weighting factors for protons and heavy ions are 5 and 20, respectively.

(g) For the effective dose calculated in the previous question, what recommendation would you make to the UN's Secretary General regarding protection of the Earth's population during the 3-week event?

06-012 You are the health physics officer on a ship involved in a 2-year survey of Uranus. The ship has a spherical shape and a 3-cm-thick structural shell with an inner radius of 10.0 m. It is equipped with a 15-cm-thick, shielded shelter with an outer radius of 2.0 m. All shielding thicknesses are in terms of water equivalent.

The spacecraft carries 2000 kg of water for consumption and for fuel cell operations. You have the option of storing the water in a spherical shell outside the shelter or inside the spacecraft shell in a spherical annulus, but the decision must be made before launch. For the Uranus mission, the applicable attenuation coefficient for water is $0.1\ \mathrm{cm}^{-1}$.

Solar scientists predict the free space radiation profile for your mission includes a GCR component and a large Solar flare. During the mission, the average GCR effective dose rate is 0.7 Sv/year and the expected SPE effective dose is 12 Sv.

Assume that the crew will spend 100% of their time in the shelter during the SPE, but no time is spent in the shelter during normal operations. Given these circumstances, where do you store the water to minimize the crew's dose?

06-013 Your grant proposal to build an electron–positron collider around the Earth has been approved by the International Science Foundation. The ring will have a radius of 6600 km and a constant magnetic induction of $5\ \mu\mathrm{T}$. (a) Assuming the magnetic induction is perpendicular to the direction of motion, what is the center of mass energy of the machine? (b) What is the total power output of the synchrotron radiation spectrum from the electron beam produced in the accelerator? Assume the electrons are configured as a single bunch. The bunch contains 10^{18} electrons.

References

10CFR20 (2007) Standards for Protection Against Radiation, National Archives and Records Administration, US Government Printing Office, Washington, DC.

10CFR835 (2007) Occupational Radiation Protection, National Archives and Records Administration, US Government Printing Office, Washington, DC.

Anderson, H.L. (ed.) (1981) *Physics Vade Mecum*, American Institute of Physics, New York.

Aravind, P.K. (2007) The physics of the space elevator. *American Journal of Physics*, **75**, 125.

Badhwar, G.D. and O'Neill, P.M. (1992) An improved model of galactic cosmic radiation for space exploration missions. *Nuclear Tracks and Radiation Measurements: International Journal of Radiation Applications and, Instrumentation, Part D*, **20**, 403.

Bailey, D.K. (1962) Time variations of the energy spectrum of Solar cosmic rays in relation to the radiation hazard in space. *Journal of Geophysical Research*, **67**, 391.

Bate, R., Mueller, D. and White, J. (1971) *Fundamentals of Astrodynamics*, Dover, New York.

Bevelacqua, J.J. (1995) *Contemporary Health Physics: Problems and Solutions*, John Wiley & Sons, Inc., New York.

Bevelacqua, J.J. (1999) *Basic Health Physics: Problems and Solutions*, John Wiley & Sons, Inc., New York.

Bevelacqua, J.J. (2004) Muon colliders and neutrino dose equivalents: ALARA challenges for the 21st century. *Radiation Protection Management*, **21** (4), 8.

Bevelacqua, J.J. (2004) Point source approximations in health physics. *Radiation Protection Management*, **21** (5), 9.

Bevelacqua, J.J. (2005) An overview of the health physics considerations at a 21st century fusion power facility. *Radiation Protection Management*, **22** (2), 10.

Bozkurt, A. and Xu, X.G. (2004) Fluence-to-dose conversion coefficients for monoenergetic proton beams based on the VIP-man anatomical model. *Radiation Protection Dosimetry*, **112**, 219.

Burrell, M.O., Wright, J.J. and Watts, J.W. (1968) NASA-TN-D-4404, An Analysis of Space Radiation and Dose Rates. National Aeronautics and Space Administration.

Calzetti, D., Livio, M. and Madau, P. (eds)(1995) *Extragalactic Background Radiation*, Cambridge University Press, New York.

Caron, R.P. *Radiation Shielding for Manned Missions to Mars.* (2004) Worcester Polytechnic Institute, Worcester, MA, http://users.wpi.edu/~aiaa/reports/armor.pdf (accessed on August 22, 2005).

Cucinotta, F.A., Schimmerling, W., Wilson, J.W., Peterson, L.E., Badhwar, G.D., Saganti, P.B. and Dicello, J.F. (2001) Space radiation cancer risks and uncertainties for mars missions. *Radiation Research*, **156**, 682.

Cucinotta, F.A., Wu, H., Shavers, M.R. and George, K. (2003) Radiation dosimetry and biophysical models of space radiation effects. *Gravitational and Space Biology Bulletin*, **16**, 11.

Feynman, J. and Gabriel, S. (eds) (1988) *JPL 88-28, Interplanetary Particle Environment*, Jet Propulsion Laboratory, California Institute of Technology, Pasadena, CA.

Fitchell, C.E. and Guss, D.E. (1961) Heavy nuclei in Solar cosmic rays. *Physical Review Letters* **6**, 495.

IPS Radio and Space Services (2005) *A Guide to Space Radiation*, The Australian Space Weather Agency, Sydney, Australia, http://www.ips.gov.au/Category/Educational/Space%20Weather/Space%20Weather%20Effects/guide-to-space-radiation.pdf (accessed on August 22, 2005).

Jackson, J.D. (1999) *Classical Electrodynamics*, 3rd edn, John Wiley & Sons, Inc., New York.

Kennedy, A.R. and Todd, P. (2003) Biological countermeasures. *Gravitational and Space Biology Bulletin*, **16** (2), 37.

Letlaw, J.R., Silberberg, R. and Tsao, C.H. (1987) Radiation hazards on space missions. *Nature*, **330**, 709.

Marsden, J.E. and Ross, S.D. (2006) New methods in celestial mechanics and mission design. *Bulletin of the American Mathematical Society*, **43** (1), 43.

Miroshnichenko, L.I. (2003) *Radiation Hazard in Space*, Springer, New York.

NASA Conference Publication 3360 (1997) Shielding Strategies for Human Space Exploration. (eds J.W. Wilson, J. Miller, A. Konradi and F.A. Cucinotta). Proceedings of a Workshop sponsored by the National Aeronautics and Space Administration and held at Lyndon B. Johnson Space Center, Houston, Texas, December 6–8, 1995.

NCRP Report No. 132 (2000) *Radiation Protection Guidance for Activities in Low-Earth Orbit*, National Council on Radiation Protection and Measurements, Bethesda, MD.

NCRP Report No. 142 (2002) *Operational Radiation Safety Program for Astronauts in Low-Earth Orbit: A Basic Framework*, National Council on Radiation Protection and Measurements, Bethesda, MD.

NCRP Report No. 153 (2006) *Information Needed to Make Radiation Protection Recommendations for Space Missions Beyond Low-Earth Orbit*,

National Council on Radiation Protection and Measurements, Bethesda, MD.

Ness, W.N. (ed.) (1965) *Introduction to Space Science*, Gordon and Breach Publishers, New York.

Ness, W.N. (1968) *The Radiation Belt and Magnetosphere*, Blaisdell Publishing Company, Waltham, MA.

Ney, E.P. and Stein, W.A. (1962) Solar protons, alpha particles, and heavy nuclei in November 1960. *Journal of Geophysical Research*, **67**, 2087, http://nssdc.gsfc.nasa.gov/planetary/Mars/Marsprof.html (accessed on August 22, 2005).

National Research Council (2000) *Radiation and the International Space Station: Recommendations to Reduce Risk*, National Academy Press, Washington, DC.

Saganti, P.B., Cucinotta, F.A., Wilson, J.W., Simonsen, L.C. and Zeitlin, G. (2005) Radiation Climate Map for Analyzing Risks to Astronauts on the Mars Surface from Galactic Cosmic Rays, http://marie.jsc.nasa.gov/Documents/Mars-Flux-Paper.pdf (accessed on September 20, 2005).

Santora, R.T. and Ingersoll, D.T. (1991) *ORNL/TM-11808, Radiation Shielding Requirements for Manned Deep Space Missions*, Oak Ridge National Laboratory, Oak Ridge, TN.

Schimmerling, W. (2003) Overview of NASA's space radiation research program. *Gravitation and Space Biology Bulletin*, **16**, 5.

Townsend, L.W. (2004) Implications of the Space Radiation Environment for Human Exploration in Deep Space. Joint Organization of the Tenth International Conference on Radiation Shielding (ICRS-10) and Thirteenth Topical Meeting on Radiation Protection and Shielding (RPS-2004), May 9–14, Funchal, Madeira Island (Portugal), http://www.itn.mces.pt/ICRS-RPS/oralpdf/Friday14/ClosingPlenarySession/Townsend03.pdf (accessed on August 22, 2005).

Townsend, L.W., Shinn, J.L. and Wilson, J.W. (1991) Interplanetary crew exposure estimates for the August 1972 and October 1989 Solar particle events. *Radiation Research*, **126**, 108.

Tribble, A.C. (2003) *The Space Environment: Implications for Spacecraft Design*, Princeton University Press, Princeton, NJ.

Turcotte, S.B. (2005) Orbital timing for a mission to Mars. *The Physics Teacher*, **43**, 293.

Vette, V.I. (1967) Models of the Trapped Radiation Environment, Vol. I, Inner Zone Electrons and Protons, NASA-SP-3024, National Aeronautics and Space Administration.

Wilson, J.W., Simonsen, L.C., Shinn, J.L., Kim, M.-H.Y., Cucinotta, F.A., Badavi, F.F. and Atwell, A. (1999) Paper Number 1999-01-2173, *Astronaut Exposures to Ionizing Radiation in a Lightly-Shielded Spacesuit*, http://techreports.larc.nasa.gov/ltrs/PDF/1999/mtg/99-29ices-jww.pdf (accessed on September 22, 2005).

7
Deep Space Missions

7.1
Introduction

In the previous chapter, health physics aspects of space travel within our Solar System were reviewed. This chapter reviews the health physics aspects of deep space missions that leave the Sol System and travel to nearby stars such as the Alpha Centauri system or to a Solar System in another galaxy.

A number of factors complicate the health physics evaluation of deep space missions. These include the length of time to accomplish the mission, uncertain stellar radiation characteristics, and the uncertain trapped radiation environment of planets orbiting that sun. Uncertainties also exist in the radiation characteristics of the propulsion system such as the utilization of fission or fusion processes as the propulsion mechanism. This chapter addresses these and other issues associated with missions that leave the Sol System.

It is likely that the twenty-first century will not witness travel to other galaxies. The sheer size of the Milky Way Galaxy, the time required to travel to another galaxy, and the velocity limitations of projected propulsion systems suggest that a focus on nearby stars is likely limiting. However, this conclusion could change with the discovery of new physics or technology advances.

Using existing technology, a voyage to a star either within or outside the Milky Way Galaxy offers a number of significant challenges, which depend on the radiation characteristics of the star to be visited, its location within the galaxy of interest, and sources of radiation that affect the region of interest. Accordingly, the review of deep space radiation begins with an overview of the radiation characteristics of stars and galaxies.

7.2
Stellar Radiation

An interstellar mission encounters stars having radiation characteristics that may be quite different from those of our Sun. These characteristics are largely determined by the star's mass and age as well as the location of the star within a galaxy and the radiation environment in the vicinity of that star. As the direct radiation characteristics of a star depend on its mass, luminosity, temperature, and radius, these factors are reviewed in terms of the various star types.

7.2.1
Origin of Stars

Following the *big bang*, large clouds of hydrogen gas dominated the Universe. Overtime, these clouds expanded and cooled. Stars formed when the clouds of mostly hydrogen gas contracted as a result of the gravitational interaction. As the hydrogen gas was compressed, its temperature and pressures increased. Once the gas temperature reached about 15×10^6 K, hydrogen nuclei fused to form helium with an associated release of energy. This energy release counteracted the gravitational contraction. Eventually, equilibrium was established and gravitational contraction was balanced by fusion energy release in the stellar core. The fusion energy flowed radially from the core through the star. When the fusion energy reached the surface, it radiated into space and the star was observed to shine in the heavens by a distant observer.

Specific details of the stellar evolution sequence and the associated radiation release depend on the initial mass of the star. To quantify the stellar evolution process, stars are classified as having either low mass or high mass. As a matter of specificity, stellar regions are described in terms of an inner core or zone, a middle region, and an outer region.

7.2.2
Low Mass Stars

A low mass star has a mass less than one to five times the mass of our Sun. These stars begin their lives by fusing hydrogen into helium in their cores. The fusion process continues for billions of years until the hydrogen in the core is exhausted and hydrogen fusion ceases. Without the core's fusion energy to oppose the gravitational attraction, the star contracts and its temperature and pressure increase. The temperature and pressure increases facilitate the fusion of hydrogen in the star's middle region. Hydrogen fuses into helium in a shell around the star's core, and the shell's fusion energy heats the star's outer region. The heat energy input expands the star's outer region well beyond its original size. This expansion cools the outer stellar region. The expansion/cooling in the outer region of the star causes it to appear red in color and this characteristic leads to the standard classification designation of red giant.

With its hydrogen exhausted, the star's core continues to contract until temperatures reach about 1×10^8 K. This temperature is sufficient for helium to fuse into carbon and oxygen:

$$^{4}He + {}^{4}He \rightarrow {}^{8}Be, \tag{7.1}$$

$$^{8}Be + {}^{4}He \rightarrow {}^{12}C, \tag{7.2}$$

$$^{12}C + {}^{4}He \rightarrow {}^{16}O. \tag{7.3}$$

Energy released in the fusion reactions (Equations 7.1–7.3) inhibits further core collapse. Once the core's helium is exhausted, the star's core resumes its collapse. The heat generated by the core's collapse produces sufficient energy to expel the star's outer layers.

The ejection of the star's outer layers creates a planetary nebula – a shell of gas ejected from the star at the end of its lifetime. The gas continues to expand as it moves away from the star. The planetary nebula may contain up to 10% of the star's mass and disperses the various elements created in the star into space.

The final collapse that ejected the planetary nebula generates additional energy, but this energy is insufficient to foster additional fusion reactions. Without fusion, the gravitational interaction causes the star's remaining mass to contract. The star becomes a white dwarf with only the repulsive interaction between electrons opposing the gravitational interaction.

A white dwarf is a very small, hot star, with high density. An isolated white dwarf is stable. With its fuel exhausted, the hot, white dwarf radiates its residual energy into space for billions of years. Its endpoint is a cold, massive object that is referred to as a black dwarf.

If the white dwarf is close to another star (e.g., part of a binary star system), matter from the other star can be pulled into the dwarf. The high temperature and large gravitational force exerted by the white dwarf cause the captured matter to fuse in a rapid excursion or nova. A nova explosion temporarily increases the dwarf's brightness by a factor as large as 10^4. As a consequence of the nova, new elements are created and dispersed into space with helium, carbon, oxygen, nitrogen, and neon.

In rare cases, the white dwarf detonates in a massive explosion or supernova (Type Ia). A Type Ia supernova occurs if the white dwarf is part of a binary star system with matter accumulating in the dwarf. Once sufficient matter accumulates within the white dwarf, the gravitational interaction causes the star to collapse. This collapse heats the helium and carbon/oxygen and these rapidly fuse to produce elements with atomic numbers in the 26–28 range (e.g., iron, cobalt, and nickel). These fusion reactions occur very rapidly and the energy release from the fusion reactions cause the white dwarf to violently detonate. This violent event disperses all stellar contents into space and obliterates the white dwarf in the supernova event. Supernova events have significant radiological consequences, and their radiation characteristics are addressed in subsequent discussions.

7.2.3
High Mass Stars

Stars larger than about five times the mass of the Sun begin their life cycle by fusing hydrogen into helium. High mass stars burn hotter and faster than low mass stars and all the hydrogen in the core is fused into helium in less than a billion years. After exhausting its hydrogen, the star becomes a red supergiant that is similar to a red giant, but larger. The larger mass of the red supergiants produces higher gravitational forces that create a higher core temperature and pressure that facilitates the fusion of elements up to iron. The pressure and temperature decrease as the distance from the center of the star increases. This pressure model suggests that heavier elements predominate in the star's interior with the lighter elements residing in the outer regions. A somewhat less physical description is the onion-skin model, with iron concentrated in the core, and successive spherical shells of increasing radii being composed of lighter elements.

Fusion continues in red supergiants until iron is produced. Unlike lighter elements, iron fusion releases no energy because ^{56}Fe is the most stable nuclear system on the basis of its binding energy per nucleon. Elements beyond iron can be produced by neutron capture.

In a large star, neutron capture occurs over thousands of years. A large star can produce elements from cobalt to thallium ($Z = 69$) via neutron capture reactions. A portion of the these elements is dispersed into space via convection, stellar winds, and solar particle events (SPEs).

As the red supergiant's core fills with iron, fusion energy production deceases till there is no longer sufficient energy to counteract the gravitation-induced collapse. This collapse increases the temperature and pressure in the core with core temperatures exceeding 1×10^{11} K. At this temperature, a large number of protons and electrons interact to produce neutrons. Being uncharged, the neutrons pack tightly as they are compressed in the core. Upon reaching the core, the neutrons collide and very energetic shock waves emanate radially outward. These waves heat and accelerate the surrounding layers of the star. Sufficient energy is deposited in these layers to cause the majority of the star's mass to be violently ejected into space. This event is termed a Type II supernova.

In a Type II event, sufficient energy is released such that the supernova can briefly exceed the brightness of its host galaxy. Supernovas are an important mechanism for dispersing elements into space. In addition, sufficient energy is present to facilitate rapid neutron capture in the outer stellar layers. Under these circumstances, elements as heavy as uranium are produced before they are ejected into space. More details on supernovas and their radiation characteristics are provided in subsequent discussions.

Some, and possibly all, Type II supernovas become neutron stars, pulsars, or black holes. Neutron stars are produced by supernova explosions that expel the outer region of the star and leave a core about 1.4 Solar masses. In the remnant core, the remaining atoms are severely compressed. Nuclear transformations convert protons and electrons into neutrons such that only neutrons remain. The remnant core is termed a neutron star.

A pulsar is a photon source emitting short intense bursts of radio waves, X-rays, or visible electromagnetic radiation at regular intervals. Pulsars are generally believed to be rotating neutron stars with their rotation axis and magnetic axis not aligned.

If the original star was very massive (about 15 or more Solar masses), the neutron star becomes a black hole – the end state of a collapsed high mass star that often resides within the matter-rich center of galaxies. The gravitational field of a black hole is so intense that spacetime is altered. Once matter enters the black hole's event horizon, the boundary surrounding a black hole, it cannot escape its gravitational attraction.

7.2.4
Star Types

The stars, both low-mass and high-mass, can also be broadly classified as main sequence (MS) stars. MS stars account for about 90% of all known stars. The lower mass limit of MS stars is about 0.08 Solar masses. Stars that are less massive do not produce sufficient energy to fuse hydrogen into helium. The upper mass limit for MS stars is on the order of 100–200 Solar masses. Stars heavier than this mass range become violently unstable.

The introduction of the MS star concept is advantageous because it permits a comparison of various characteristics of stars that might be encountered during a deep space mission. In particular, the luminosity (L), radius (R), and temperature (T) of an MS star are compared to the properties of our Sun through the relationship

$$\frac{L}{L_{Sun}} = \left(\frac{R}{R_{Sun}}\right)^2 \left(\frac{T}{T_{Sun}}\right)^4. \tag{7.4}$$

Table 7.1 illustrates the application of Equation 7.4 to selected MS stars in the Milky Way Galaxy. A summary of the luminosity, mass, and radius of MS stars is provided in Table 7.1.

A star is classified in terms of its spectral type. Stars with the same mass have the same spectral class or type. As the temperature of a main sequence star increases, its spectral type changes through the following classification progression: M, K, G, F, A, B, and O. This stellar classification is based upon surface temperature and luminosity and ranges from the bluer, hotter, and early type stars (Type O) to redder, cooler, and older stars (Type M).

The MS stars can also be ordered in mass with blue stars having more mass than red stars. Sol is an MS star and all MS stars burn hydrogen in their cores. As the MS star burns hydrogen, its radiation characteristics change. These characteristics can be summarized in a Hertzsprung–Russell (HR) diagram that plots L/L_{Sun} as a function of the surface temperature of the star. As the MS star ages, its position on an HR diagram changes as it evolves and changes its luminosity and surface temperature.

The lifetime of an MS star is determined from its mass (M) and luminosity. A star's mass is proportional to the available fuel supply or elements that fuse and supply energy. As the rate of fuel consumption is proportional to the luminosity, the rate of

Table 7.1 Selected properties of main sequence stars.

Number of stars in the Milky Way Galaxy for each O type star	$\frac{L}{L_{Sun}}$	$\frac{M}{M_{Sun}}$	$\frac{R}{R_{Sun}}$	Example
1	2.6×10^5	20	10	Rigel
10^5	60	3	2.5	Vega
10^6	1	1	1	Sol and Capella
5×10^6	0.06	0.4	0.6	Barnard's Star

[a] Derived from Bennett et al. (2005).

consumption times the lifetime (τ) is the total fuel or mass consumed. This suggests that a star's mass is proportional to the product of its lifetime and the luminosity:

$$M \propto L\tau. \tag{7.5}$$

For MS stars, the luminosity is related to mass through a power law

$$L \propto M^4. \tag{7.6}$$

Using Equations 7.5 and 7.6, the lifetime is written in terms of the MS mass

$$\tau \propto \frac{M}{L} = \frac{M}{M^4} = \frac{1}{M^3}. \tag{7.7}$$

The lifetime of an MS star can be written more explicitly as

$$\tau_{MS} = 10^{10} \text{ year} \times \left(\frac{M_{Sun}}{M}\right)^3. \tag{7.8}$$

Equation 7.8 means that heavier MS stars burn more quickly and have a shorter lifetime. Lifetimes for selected MS spectral types are illustrated in Table 7.2. Lifetimes are determined relative to groupings of stars (e.g., classified as associations, open clusters, and globular clusters) that occur within a galaxy.

Table 7.2 MS spectral type characteristics.

Spectral types	Mass (M_{Sun})	MS lifetime (yr)	Comment
O	10	$<1.0 \times 10^7$	When grouped in star associations, these MS stars are younger than 1.0×10^7 yr. All stars more massive than G stars will eventually evolve to become red giants. During steady state conditions, O stars have the highest radiation output of MS stars.
A	3	$1 \times 10^7 - 4 \times 10^8$	When grouped in open clusters, these MS stars have lifetimes in the range of 10^7 to 4.0×10^8 yr. All stars more massive than G stars will eventually evolve to become red giants. During steady state conditions, A stars have a larger radiation output than G stars, but less than O MS stars.
G	1	$\sim 1.0 \times 10^{10}$	When grouped in globular clusters these MS stars have lifetimes on the order of 1.0×10^{10} yr. During steady state conditions, G stars have a larger radiation output than K stars, but less than A MS stars.
K	0.8	$\sim 2.0 \times 10^{10}$	Lowest mass and longest lifetime MS stars. During steady state conditions, K stars have a lower radiation output than G MS stars.

[a] Derived from Bennett et al. (2005).

7.2.5
MS Star Health Physics Considerations

The radiation hazard from MS stars is governed by their luminosity or the amount of energy radiated per unit time. For steady state output conditions, ambient radiation from the MS star is proportional to the star's luminosity (i.e., the dose is proportional to the star's luminosity). For MS stars, the highest dose occurs for O stars with decreasing radiation output from type B, A, F, G, K, and M, with type M having the lowest radiation output.

MS stars are also expected to periodically produce Solar particle events or ejections of stellar matter into space. Although the periodicity and intensity of the SPEs are uncertain, the mass ejections are the highest intensity source of radiation from an MS star.

An SPE output is likely to vary by orders of magnitude. Periods of Solar instability are a function of the star's mass, luminosity, and temperature profile. Considering the inadequacies in understanding the dynamics of Sol, stellar science needs to significantly advance to characterize the instability of MS stars and to predict the magnitude, periodicity, and intensity of their SPEs. These uncertainties complicate the radiation evaluation of future missions to MS star systems.

Other radiation producing processes occur in MS stars. These processes include supernova events and gamma-ray bursts (GRBs) that occur either in the Milky Way or in another galaxy. Each of these sources, their radiation and physical hazards are addressed in subsequent discussion.

7.2.6
Supernovas

Given the previous discussion, a more complete description of a supernova and their radiation characteristics can now be provided. Supernovas reach maximum visual luminosities of 10^7 to 10^{10} times that of Sol during its steady state condition, and then the nova's luminosity decreases. In addition to producing a spectacular visual effect, the supernova represents a significant radiation hazard. The degree of the radiation hazard depends on the characteristics of the event, and its proximity to life forms.

Supernovas can be broadly classified as either Type I or Type II events, distinguishable by their spectrum. A Type II supernova has hydrogen (H_α) in its spectrum, but a Type I does not.

Type I supernovas result either from the thermonuclear explosion of white dwarfs (I A) or from the core collapse in a massive star that lost its hydrogen envelope (I B). Type II supernovas come from the explosion of massive stars (e.g., blue or red supergiants) when their iron cores collapse. The rate of energy emission is enormous.

Type I events are found in all types of galaxies and their occurrence appears to have a random pattern. Type II supernovas only occur in galaxies where star formation is continuing and tends to occur at specific locations such as the arms of a spiral galaxy. A discussion of galaxy types will be forthcoming.

A Type I supernova occurs in a star with a mass of 1.2–1.5 Solar masses. At an advanced evolutionary stage, this star has produced quantities of ^{12}C, ^{16}O, and ^{20}Ne. When temperatures increase to about 2×10^9 K, these elements fuse (e.g., ^{12}C+^{12}C and ^{16}O+^{16}O) on a rapid timescale of 1–100 s, and on the order of 10^{33} to 10^{36} W are liberated in a supernova explosion. This explosion is sufficiently violent to obliterate the stellar mass. Prior to the explosion, the core's temperature and pressure are about 3.5×10^9 K and 10^8 g/cm^3, respectively. Following the supernova excursion, heavy nuclei are synthesized that are expelled with stellar matter into the space.

Conditions leading to a Type II supernova occur when a massive star reaches a stage where iron is synthesized in its core. As part of this synthesis, neutrinos are emitted and this emission accelerates an increase in the temperature and density of the central core. When temperatures reach about 6×10^9 K, iron dissociates through the reaction

$$^{56}\text{Fe} \rightarrow 13\,^4\text{He} + 4\text{n}, \tag{7.9}$$

which has a large negative Q-value (-124.4 MeV). This negative Q-value prevents subsequent temperature increases. Density increases occur because energy is not available to dissociate additional ^{56}Fe. Without additional fusion energy, the core is unable to support the stellar configuration and gravitational collapse occurs. The energy released during the collapse further drives the iron dissociation reaction, but the temperature rise is insufficient to overcome the collapse.

Gravitational collapse continues with an associated timescale τ

$$\tau \sim \frac{1}{\sqrt{G\rho}}, \tag{7.10}$$

where G is the gravitational constant (6.67×10^{-11} m^3/s^2 kg) and ρ is the stellar density. When $\rho \sim 10^{10}$ kg/m^3, the time of collapse is on the order of 1 s. In another second, ^{12}C, ^{16}O, and ^{20}Ne nuclei are heated to an ignition temperature of 2×10^9 K, and outer stellar regions are ejected at velocities on the order of 5000 km/s. Between the collapse (implosion) and explosion, iron and its neighboring elements are formed.

In a supernova event, the peak energy release continues from several weeks (Type IA) to several months (Type II). Both Type I and Type II events exhibit an exponential decay in their light curves with a decay rate of approximately 77 days. These events result in the release of enormous amounts of energy that can be lethal even at extreme distances. In our Galaxy, about one Type Ia Supernova event is expected to occur every 200 years and two Type II events occur every 100 years.

There have been a number of major supernova events in the past 1000 years in the Milky Way Galaxy. One of the brightest supernovas observed in over 400 years (Supernova 1987A) was detected on February 23, 1987 approximately 160 000 years after its occurrence. Its progenitor star was a previously catalogued blue supergiant located in the Large Magellanic Cloud. Recent evidence suggests that the largest known supernova occurred in 2006. Supernova 2006 gy, in the peculiar galaxy NGC (New General Catalogue) 1260, was about 2 orders of magnitude larger than any previous supernova event. Although infrequent, supernova events are catastrophic.

The matter expelled in a supernova can be viewed as a gigantic SPE that poses an extraordinary health physics hazard. This hazard is sufficiently large to potentially lead to the extinction of life on a planetary scale.

As an example, recent work suggests that a nearby supernova explosion could have caused one or more of the mass extinctions identified by paleontologists. The radiation hazard depends on the energy output of the supernova event and the proximity of the event to inhabited planets and space travelers.

A supernova explosion within 10 pc from Earth is expected every few hundred million years and can destroy the ozone layer for hundreds of years. The loss of the ozone layer increases the Solar ultraviolet radiation reaching the Earth's surface, and elevated levels of UV radiation have a detrimental effect on plants, animals, and the environment. In addition to its adverse effects on land ecology, the postulated supernova would result in the mass destruction of plankton and reef communities, and would also have disastrous consequences for marine life.

7.2.7
White Dwarfs, Pulsars, and Black Holes

In addition to the catastrophic supernova events, spacecraft may encounter other endpoints of MS stars including white dwarfs, pulsars, and black holes. White dwarfs have been identified as sources of soft X-rays.

Some supernovas become rotating neutron stars (pulsars) or black holes. Pulsars emit a wide range of radiation including the radio, optical, X-ray, and γ-ray frequency range.

Black holes are so massive that radiation does not escape once matter penetrates the event horizon. However, the matter in the accretion disks emits a variety of radiation types including X-rays.

The radiation from these sources is not as hazardous as supernova radiation but can be lethal to spacecraft personnel. The radiation output from these sources depends on specific circumstances, but it will likely be diverse and have a wide range of values. Mission planning needs to credit known radiation emitting celestial objects and have the flexibility to quickly react to the radiation emitted from a variety of celestial bodies.

7.2.8
Dark Matter/Dark Energy

Dark matter is not readily detected and its nature is not fully understood. It is inferred from astronomical phenomena that are best explained by gravitational interactions attributed to hidden or nonluminous matter. For example, dark matter is suggested by measuring the rotational characteristics of the Andromeda Galaxy.

By mapping the Doppler shift in the 21-cm hydrogen line, the velocity of rotation of stars in Andromeda is approximately determined as a function of the distance (r) from the galaxy's center. This rotation curve does not have the expected $r^{-0.5}$ behavior for large r if most of the mass were concentrated near the center of Andromeda.

Instead, the rotation curve is constant for the observed locations of visible matter indicating that Andromeda's mass is still increasing with r. These measurements suggest that Andromeda has a halo of dark matter that is as much as 10 times the mass of the galaxy's visible matter.

Dark energy is viewed as the driving mechanism for overcoming gravitational attraction that would minimize the expansion of the universe. However, the nature of dark energy is unknown.

It has been proposed that dark energy is a constant vacuum energy density inferred from the Einstein's cosmological constant. Other theories propose that dark energy is a time-dependent, dynamic energy that changed as the universe established an equilibrium configuration. In spite of the uncertainty of its nature, dark energy is needed to provide an explanation for cosmic expansion.

There are additional data to suggest that the visible matter and detectable radiation comprise only a small fraction (on the order of a few percent) of the mass of the universe. The postulated abundance of dark matter and dark energy has the potential to alter the anticipated radiation characteristics of the deep space. A defensible model of dark matter and dark energy is a central problem for cosmology and has challenged physicists, including the author, for years.

7.2.9
Gamma-Ray Bursts

Most gamma-ray bursts originate in distant galaxies. A large percentage of these events probably arise from explosions of stars having a mass greater than 15 Solar masses. A GRB creates two oppositely directed beams of gamma rays that propagate into space. In view of their large radiated power level, these beams present a significant hazard to living organisms and the environment.

Gamma-ray bursts are the most powerful events in the universe. Although roughly 1000 GRBs/day are presumed to occur in the observable universe, Earth-based telescopes detect only about 1 GRB/day. These spectacular bursts of gamma-ray energy are not predictable and their nature and origin are not definitely known. GRBs may be produced from the collision of two neutron stars or black holes. These bursts could also arise from a hypernova, which is the massive explosion hypothesized to occur when a supermassive star collapses into a black hole.

In a hypernova-induced formation of a black hole, jets of material are released during the core collapse and these jets limit the hole's angular momentum. The jets impact outer material of the collapsing star, create high temperatures, and produce gamma-rays. As the jets get further from the collapsing star's interior, they encounter increasingly less dense material in the star's outer regions. The emitted radiation becomes less energetic as the jets move outward from the star's interior.

Gamma-ray bursts are observed to have radiation emissions that are energy and time dependent. The initial energy burst consists of gamma rays and other radiation types including X-rays, UV light, visible light, and radio waves. All gamma-ray bursts are, however, not because of supernovas. There is considerable variability in their emission spectra and some gamma-ray bursts appear to sporadically release energy.

7.2 Stellar Radiation

Table 7.3 Power released by various terrestrial and extraterrestrial events.[a,b]

Source	Power (W)
Light bulb	10–100
Campfire	10^3
Nuclear power plant	10^9
Hydrogen bomb	10^{13}
Sun	10^{26}
Supernova	10^{33}–10^{36}
Gamma-ray burst	10^{46}

[a] Derived from Whitlock and Granger (2000).
[b] The values in this table are representative, but some variation exists in the literature.

Table 7.3 provides a comparison of GRBs with other power output conditions that occur on Earth as well as in the cosmos. Although GRBs are at least 10 orders of magnitude more powerful than a supernova, their durations are short, ranging from 30 ms to 1000 s. From a health physics perspective, these events merit attention. GRBs have the potential to extinguish life on the Earth and on spacecraft during deep space voyages. Although there are considerable consequences from a GRB, there is an extremely low probability of a planet or spacecraft being struck by a gamma-ray burst's beam.

Gamma-ray bursts in the Milky Way Galaxy are rare, but it is estimated that at least one struck the Earth in the past billion years. Research suggests that a gamma-ray burst may have caused the Ordovician extinction 450 million years ago, killing 60% of all marine invertebrates. Although most life forms existed in the sea, there is evidence for the existence of primitive land plants during this time period.

A gamma-ray burst striking the Earth would have devastating effects. These bursts would impact the Earth's ozone layer, sea and land creatures, and the climate. It is possible that mass extinction on the Earth could be triggered by a gamma-ray burst. For example, a gamma-ray burst from a relatively nearby star explosion within 6000 LY of only 10 s duration could deplete up to half of the atmosphere's protective ozone layer.

Ozone depletion arises from the fact that gamma-rays have sufficient energy to break molecular bonds of atmospheric gases. As an example consider the impact of gamma-ray initiated dissociation of nitrogen and oxygen on atmospheric ozone:

$$N_2 + \gamma \rightarrow N + N, \tag{7.11}$$

$$O_2 + \gamma \rightarrow O + O, \tag{7.12}$$

$$2N + O_2 \rightarrow 2NO, \tag{7.13}$$

$$3NO + O_3 \rightarrow 3NO_2, \tag{7.14}$$

$$NO_2 + N \rightarrow 2NO, \tag{7.15}$$

$$NO_2 + O \rightarrow NO + O_2. \tag{7.16}$$

The net result of Equations 7.11–7.16 is the destruction of ozone. Equations 7.11 and 7.12 illustrate the gamma-ray dissociation of atmospheric nitrogen and oxygen gases. Equation 7.13 forms the NO species that destroys ozone through reaction 7.14. The NO is replenished through reactions 7.15 and 7.16, which leads to further ozone depletion. Up to half the atmosphere's ozone layer could be destroyed within weeks of a gamma-ray burst striking the Earth. Even 5 years after the event, at least 10% of the ozone would still be depleted.

The gamma-ray burst also leads to elevated levels of ultraviolet radiation that would significantly impact life. For example, creatures living several feet below water are protected, but surface-dwelling plankton and other life forms near the surface would not survive. Plankton and surface sea creatures are key elements of the marine food chain.

The impact of UV radiation would persist beyond the initial impact of the GRB. During the recovery period, ultraviolet radiation from the Sun would kill much of the life on land and near the surface of oceans and lakes.

7.3
Galaxies

After exploring our Solar System, man will likely travel to nearby stars within our Galaxy, the Milky Way. A galaxy is an immense cloud of 10^8–10^{11} stars, each moving in an orbit governed by the collective gravitational forces of other stars in its galaxy. Galaxies, excluding the few that collide, are independent and isolated star systems.

Galaxies, excluding the Milky Way, are at extreme distances from the Earth. For example, the Andromeda galaxy is about 2.4×10^6 LY from Earth. Therefore, travel to another galaxy presents a significant challenge.

This section outlines the characteristics of various galaxies and the potential health physics issues associated with intergalactic travel. The discussion is somewhat limited, because it is difficult to predict the scientific and technological advances that will occur in the twenty-first century. These advances will alter the ability to travel vast distances and could dramatically reduce travel times.

7.3.1
Distance Scales

If the spectra of starlight from galaxies are measured, a redshift is observed. This frequency shift is interpreted as a Doppler effect in flat spacetime. The frequency shift means that galaxies are moving away from Earth with a recession velocity V. Assuming that $V \ll c$, the Doppler relationship is

$$\frac{V}{c} = \frac{\Delta\lambda}{\lambda} = z, \tag{7.17}$$

where λ is the wavelength, $\Delta\lambda$ is the shift in wavelength, c is the speed of light, and z is the astronomical designation for redshift. Galaxy recession is an artifact of the *big bang* and subsequent expansion of the universe.

Table 7.4 Astronomical distance scales.

Location	Distance from Earth
Sun	5 μpc
Proxima Centauri (nearest star)	1 pc
Distance to center of Milky Way	10 kpc
Sagittarius Dwarf Elliptical Galaxy (nearest Galaxy)	25 kpc
Large Magellanic Cloud	55 kpc
Andromeda	725 kpc
Virgo cluster of several thousand galaxies	20 Mpc
Distance to the edge of the visible universe	14 Gpc

Derived from Hartle (2003).

The recession velocity of a galaxy and its distance (d) from Earth are related through the Hubble's law

$$V = H_0 d, \tag{7.18}$$

where H_0 is the Hubble constant with a value of about 72 ± 7 km/s Mpc. The megaparsec (Mpc) is a distance unit that is convenient for intergalactic distances. One parsec (pc) is 3.086×10^{18} cm or 3.262 light-years. A comparison of astronomical distances is provided in Table 7.4.

The recession of galaxies does not suggest that Earth is the center of the universe. Hubble's law implies that no center can be determined from observation of the expansion. A life form in another galaxy would also observe that every galaxy is receding in accordance with Hubble's law.

Before considering the types of radiation that may be encountered in deep space travel, a consideration of galaxy classification and characterization is presented. These characteristics are essential to the subsequent discussion in this chapter.

7.3.2
Characteristics of Galaxies

Galaxies are maintained as integral structures through the attractive nature of the gravitational force. Gravity also ensures that the orbits of stars and their planets and residual matter revolve around the center of a galaxy.

A galaxy has two dominant visible components, a central bulge and an exterior disk. The bulge is the round or elliptical central region of a galaxy that is often uniform in brightness. The disk is the flat, circular region of a galaxy extending radially outward from the central bulge.

Galaxies have a variety of shapes and configurations. In the 1920s, Edwin Hubble studied the morphology of galaxies and characterized their basic shapes as spiral, barred spiral, elliptical, irregular, and peculiar. Some galaxies such as the Andromeda Galaxy (M31) appear as disks with geometric extensions or arms that contained stars, gas, and dust and also appear uniform in brightness. In some galaxies, the arms were tightly wound and Hubble called these spirals. The Milky Way is an example of a spiral galaxy.

Spiral galaxies have most of their stars in the disk within which stars cluster into open or galactic clusters containing 10–2000 stars. The disk also contains clouds of gas and dust called nebulae. The stars, clusters, and nebulae in the disk rotate around the center of the galaxy. In the Milky Way, the rotational period of our Sun is about 200 million years. Spiral galaxies also have a bulge composed of stars, dust, and gas.

In the Milky Way Galaxy, the bulge contributes about 20% of the total light output. The Milky Way's bulge consists primarily of older stars. Surrounding the bulge, stars are grouped into globular clusters that are collections of up to hundreds of thousands of stars bound in a tight spherical pattern.

Some spirals have a bright bar of gas through the center and these are referred to as barred spirals. Galaxies NGC 1365 and NGC 1530 are examples of a barred spiral with arms that emerge from the ends of an elongated central region or bar, rather than from the core of the galaxy. A description of galaxy classification schemes, such as NGC, is provided in subsequent discussion.

Hubble also noted galaxies that were slightly elliptical or circular in shape and referred to them as elliptical galaxies. In elliptical galaxies, only the bulge is present, and it contains old stars with a small amount of dust and gas. An elliptical galaxy has a smooth and featureless appearance. Galaxies NGC 205, M49, and M87 are examples of elliptical galaxies.

Irregular galaxies are neither spiral nor elliptical, but are irregular in shape. The irregular galaxies show neither arms nor a smooth uniform appearance. In general, their stars and gas and dust clouds appear to be scattered in a random manner. In irregular galaxies, the stars, nebulas, and clusters may or may not be visible. Irregular galaxies have a disk, but no spiral arms. These galaxies contain a mixture of young and old stars combined with large quantities of gas and dust. The Magellanic Clouds and NGC 6822 are examples of irregular galaxies.

Finally, some galaxies fit none of these descriptions and are called peculiar galaxies. Centaurus A is an example of a peculiar galaxy.

In modern terminology, there are two types of galaxies: the spirals and the ellipticals. Barred spirals are a subclass of spirals. Irregulars may be either spirals or barred spirals. Peculiars are galaxies in the process of colliding with the collision distorting their individual shapes.

Hubble's classification scheme is still used today. Spirals are denoted by "S" and barred spirals by "SB." The letters "a," "b," "c," and "d" denote how tightly the spiral arms are wound in a spiral or barred spiral galaxy. The "a" designation denotes a galaxy with a large bulge with tightly wound arms. "b" systems have a medium-sized bulge with arms not so tightly wound. The "c" systems have a small bulge and loosely wound arms. Finally, the "d" designation has a very small bulge and arms that are loosely wound.

Elliptical galaxies are designated by "En," with n being an integer from 0 to 7 indicating its shape. An E0 galaxy is circular and E7 is more elongated. The n value is defined in terms of the semimajor axis (x) and semiminor axis (y) of the ellipse as $10(x-y)/x$.

"Irr" denotes irregular galaxies, such as the Magellanic Cloud. "P" denotes peculiar Galaxies, such as Centaurus A.

Galaxies have a variety of naming conventions. Charles Messier, an eighteenth-century astronomer, compiled one of the earliest of catalogues of stellar objects, which are denoted by the letter M. Another common cataloguing system is the New General Catalogue that dates from the nineteenth century. The NGC numbers objects from west to east across the sky. Objects in the same area of the sky have similar NGC numbers. Other cataloguing systems include the European Southern Observatory (ESO), Infrared Astronomical Satellite (IR), Markarian (Mrk), and Uppsala General Catalogue (UGC). The galaxy's letter designation is followed by a number that indicates either its order in a list or the location of the galaxy in the sky. Some galaxies are given descriptive names such as Andromeda if their location or appearance is distinctive.

7.4
Deep Space Radiation Characteristics

Previous discussion suggests that the time and distance scales of the universe are extremely large. Given the age of the universe of roughly 13×10^9 years and assuming constancy of the speed of light, Earth could observe a region of the universe on the order of 10^{23} km in diameter. Observations suggest that on the largest scales, the universe is nearly homogeneous and isotropic.

The temperature (T) profile of cosmic background radiation is important because it provides an indication of the radiation distribution in the universe. This is essentially a restatement of the Stefan–Boltzmann law that quantifies the fact that the power (radiation) output of a pure blackbody emitter is proportional to T^4.

The *big bang* proceeded from a singularity characterized by infinite mass density, infinite temperature, and infinite spacetime curvature. Shortly after the *big bang*, matter and radiation had not yet condensed into galaxies, but were in equilibrium in a hot, constant temperature fluid.

As the universe expanded, the matter cooled and radiation decreased in energy. After roughly 10^5 years, temperature further decreased and free electrons combined with nuclei to form a neutral matter. The dominant atomic species were hydrogen and helium. Subsequent expansion cooled the radiation to 2.73 K, which corresponds to the microwave region of the electromagnetic spectrum.

Measurements made by the Cosmic Background Explorer (COBE) satellite confirm the near isotropy in the cosmic microwave background radiation. The COBE data provides an indication of the universe at a previous epoch before galaxies formed. Fluctuations from exact isotropy are important, because they denote density differences that evolved as a result of the gravitational interaction. Averaging over the vast distances between galaxies confirms that the universe is essentially isotropic.

From a health physics perspective, this isotropic distribution is consistent with the uniform microwave background that permeates the universe. Any given location in the universe is affected by four basic sources of radiation: the GCR, SPE, and trapped radiation components introduced in Chapter 6 and the microwave background radiation.

This microwave background is not of health physics concern. Various locations in the universe are also subjected to radiation from supernova and hypernova events, and gamma-ray bursts. These sources collectively contribute to the galactic cosmic radiation source term. A third source of radiation is limited to the vicinity of a given sun and results from Solar particle events. The last source of radiation is the particles trapped by a planet's geomagnetic field.

The sun in the star system being investigated produces a specific and unique SPE source term. Therefore, a future space voyage must consider GCR, SPE from the star system being explored, trapped radiation from the planet being investigated, and any radiation sources inherent to the spacecraft. Other sources of radiation may also be encountered during a deep space mission. These sources include (1) radiation generated from dark matter/dark energy, (2) radiation generated from the emergence of extra dimensions beyond the three spatial plus time dimensions of conventional spacetime, (3) radiation generated by gravitational anomalies, and (4) radiation produced by new forces and particles that are not yet known. These radiation sources are examined in more detail in subsequent discussion and in the problems at the end of this chapter.

The previous discussion is somewhat generic because it applies to all galaxy types. The discussion of specific galaxy configurations (e.g., spiral or elliptical) and star, gas, and dust characteristics provides additional specificity to the radiation environment.

The classification of galaxies involved various configurations of dust, gas, and stars. From a radiation protection perspective, stars present a more significant radiation source than dust and gas.

Younger stars will tend to be main sequence stars. Of these stars, the radiation output depends on the specific spectral class. As noted in Table 7.2, the highest radiation output occurs for Type O stars and with Type M being the lowest.

Older stars depart from the main sequence and reach end states including white dwarfs, red giants, supernovas, and hypernovas. Therefore, galaxies with end state stars tend to represent a greater risk from infrequent but high dose events such as supernovas and gamma-ray bursts. Galaxies with younger stars represent a greater risk from likely events in terms of transient SPE radiation.

The bulge in a galaxy tends to contain older stars, which are more likely to produce high dose end state events. Therefore, elliptical galaxies are more likely to experience supernova and hypernova events. The disk region tends to be dominated by younger stars with their associated SPE hazard.

The specific galaxy configuration, for example, spiral or elliptical, could also affect the radiation environment particularly upon approaching a galaxy. However, the geometry effects resulting from the spatial distribution of stars within the galaxy are less significant than the age and spectral class of MS stars comprising the galaxy or the frequency of high dose end state events.

After establishing the characteristics and structures that comprise stars and galaxies, specific challenges applicable to the health physics of deep space voyages can be addressed. The first challenge is the vast distance associated with a stellar mission.

7.5
Overview of Deep Space Missions

For the purpose of this chapter, deep space is defined to be the region lying outside the major planets of the Sol System. Deep space missions include transit to the nearest star, Proxima Centauri, in the Alpha Centauri star group that resides at a distance of 4.3 LY from Earth. The distances involved in an Alpha Centauri mission lead to prohibitive transit times. In addition, an Alpha Centauri mission is not practical using conventional chemical propulsion. Nuclear propulsion provides the possibility of reaching velocities on the order of $0.1\,c$. However, even at $0.1\,c$, the time for a one-way trip to Alpha Centauri is 43 years.

In deep space, the trapped radiation dose is minimal. SPEs contribute to the total dose equivalent during the mission phase within the Solar System. However, as noted in Table 6.29, the SPE contribution rapidly decreases as the spacecraft moves away from the Sun. The dominant contribution to the deep space mission's effective dose outside the Solar System arises from GCR and radiation from the spacecraft's propulsion system.

Galactic sources such as supernovas and gamma-ray bursts present extremely unlikely events that have the potential to deliver enormous absorbed doses. If the radiation from these galactic sources reached the spacecraft, life-threatening consequences would result. The supernova output includes photons, protons, and heavy ions. Gamma-ray bursts involve primarily a photon radiation that includes gamma-rays, X-rays, ultraviolet, and visible radiation. In addition, radiation contributions could arise from alterations of the fundamental interactions. As noted in Table 6.1, alterations of the fundamental interactions governing the health physics considerations could occur as a result of spatial anomalies. These anomalies may have been observed and are considered in a subsequent section.

7.6
Trajectories

In deep space, the Hohmann orbit assumptions are no longer applicable. As the space vehicle leaves the Sol System, trajectories no longer rely on the planetary orbital characteristics around the Sun. Instead of the Hohmann orbit assumptions, trajectories are spacetime geodesics or the minimum distance between the launch point and destination. Alterations in a geodesic path are required if the path intersects areas having large radiation sources.

Accordingly, geodesic trajectories are utilized to characterize deep space missions. However, determination of the geodesic depends on the spacetime geometry. To understand the complexity of geodesic determination, spacetime metrics and associated coordinate systems are introduced.

Before introducing these concepts, the reader should understand that the purpose of the geodesic discussion is necessary to determine the relevant dosimetric quantity such as the absorbed dose (D) that is defined as the product of the absorbed dose rate

($\dot{D}$) and the time of exposure to the source of radiation. The geodesic defines the path length (s) of the spacecraft in a given region of space. Knowing the velocity of the spacecraft (v), the travel time (time of exposure) (t) is determined:

$$t = \frac{s}{v}. \tag{7.19}$$

7.6.1
Spacetime and Geodesics

Conventional four-dimensional (4D) spacetime refers to a set of coordinates that include the three spatial dimensions and time. In describing spacetime and its associated geometric characteristics, emphasis is placed on aspects particularly relevant to health physics considerations that are associated with the trajectories utilized in deep space exploration. Readers will find additional information regarding differential geometry, curvature systematics, and general relativity in Appendix L.

Spacetime consists of a set of spatial coordinates (x^1, x^2, x^3) and time coordinate (x^0), and is defined in terms of a metric tensor ($g_{\mu\nu}$). The metric tensor defines a line element giving a squared distance ds^2 connecting adjacent spacetime points:

$$ds^2 = g_{\alpha\beta} dx^\alpha dx^\beta. \tag{7.20}$$

In formulating the discussion associated with the metric and its associated quantities, relationships use a set of units in which major constants (i.e., the speed of light and gravitational constant) are equal to unity. These units are consistent with conventional general relativity literature.

In this text, the standard Einstein summation convention is used. Using this convention, any repeated index is summed. Therefore, Equation 7.20 requires sums over $\alpha = 0, 1, 2, 3$ and $\beta = 0, 1, 2, 3$. The index "0" specifies time and "1, 2, and 3" define the space components.

The metric is a 4×4 symmetric, position-dependent tensor. As a symmetric position-dependent tensor, $g_{\alpha\beta}$ has 10 independent components. The specific form of $g_{\alpha\beta}$ depends on the coordinate system used to describe the spacetime geometry. As there are four arbitrary functions in transforming four coordinates, there are only 6 (10 − 4) independent functions associated with a metric.

To illustrate the physical meaning of the metric tensor or metric, consider flat spacetime described in terms of spherical polar coordinates (r, θ, φ, t). For these conditions the metric tensor ($g_{\mu\nu}$) is

$$g_{\mu\nu} = \begin{bmatrix} 1 & 0 & 0 & 0 \\ 0 & r^2 & 0 & 0 \\ 0 & 0 & r^2 \sin^2\theta & 0 \\ 0 & 0 & 0 & -1 \end{bmatrix}. \tag{7.21}$$

This result is expected based solely on the nature of the spherical polar coordinate system and the flat spacetime limitations.

Flat spacetime characterizes the motion of a particle in terms of its three spatial and one temporal coordinates in the absence of a significant mass (e.g., a star). The motion of a particle in the absence of modifying fields is given by its geodesic equations. Mass and energy deform flat spacetime and lead to complex spacetime geometries that generate structures such as black holes or wormholes. These structures are further defined in subsequent discussion and in Appendix L.

The geodesic equations that define the motion of a particle in arbitrary spacetime are given by

$$\frac{d^2 x^\alpha}{d\tau^2} = -\Gamma^\alpha_{\beta\gamma} \frac{dx^\beta}{d\tau} \frac{dx^\gamma}{d\tau}, \quad (7.22)$$

where τ is the proper time and $\Gamma^\alpha_{\beta\gamma}$ is a set of terms known as Christoffel symbols. The Christoffel symbols are geometric objects constructed from the metric and its first derivatives,

$$\Gamma^\alpha_{\beta\gamma} = \frac{1}{2} g^{\alpha\sigma} (\partial_\beta g_{\sigma\gamma} + \partial_\gamma g_{\sigma\beta} - \partial_\sigma g_{\beta\gamma}), \quad (7.23)$$

where ∂_α stands for the partial derivative $\partial/\partial x^\alpha$ and $g^{\alpha\beta} = 1/g_{\alpha\beta}$.

With the specification of the metric, the geodesic trajectory is determined. Although the mathematical operations are straightforward, calculations of the geodesic are often facilitated using symbolic algebra programs such as Mathematica® or Maple®. Once the geodesic is determined, mission phases and associated mission times are determined, and candidate missions may be evaluated from a radiological perspective.

7.7
Candidate Missions

Deep space missions must consider the various radiological source terms that could be encountered along the planned trajectory. Although a geodesic trajectory offers the shortest mission distance and duration, it may not be the optimum path from a radiological perspective. Both the strength of the radiological source term and the duration of exposure to that source term must be considered. The various source terms encountered during a deep space mission include radiation contributions from the propulsion system, galactic cosmic radiation, Solar particle events, trapped radiation, radiation emanating from spatial anomalies, dark matter/dark energy, and radiation from unlikely sources such as supernovas and gamma-ray bursts.

As an example of a deep space mission consider a voyage from the Sol System to another Solar System within the Milky Way Galaxy. The nearest three Solar Systems (distance to Sol) are Alpha Centauri (4.3 LY), Barnard's Star (6.0 LY), and Wolf 359 (7.5 LY).

Alpha Centauri is a three-star system. The star nearest to Earth is Proxima Centauri, which is similar to Sol. As the distance from Earth increases, single star systems similar to Sol are encountered. As of the date of this text, the nearest Solar System candidates with the potential for an Earth-like planet are Epsilon Eridani (10.8 LY) and Tau Ceti (11.8 LY).

Deep space missions are most often discussed in terms of fusion or fission reactor propulsion systems. Current planning suggests that sending a scientific payload 10 000 AU in the 2050 timeframe to the Oort Cloud would be a logical precursor to subsequent missions to Alpha Centauri and beyond.

As noted in Table 7.4, the distances involved in deep space travel are enormous. A conventional chemical rocket travels at velocities on the order of 10^4 to 10^5 km/h. The time (t) required for a round trip to Alpha Centauri (about 100 million times the distance between the Earth and the moon) while traveling at a velocity (v) of 10^5 km/h is

$$t = \frac{s}{v} = \frac{(2)(4.3 \text{ LY})(9.46 \times 10^{15} \text{ m/LY})}{(10^5 \text{ km/h})(1000 \text{ m/km})(24 \text{ h/day})(365 \text{ day/year})} = 93\,000 \text{ year}.$$

(7.24)

The distances between stars mandate a more rapid means of transit. A candidate propulsion system for star travel must have the potential to reach velocities approaching the speed of light.

7.8
Propulsion Requirements for Deep Space Missions

Chemical propulsion systems are well-established technology, but are characterized by low nozzle exhaust velocities. Although interstellar space can be reached using chemical propulsion, aided by the gravitational assist from the Sol System planets, no mission to interstellar space can be performed in a reasonable time. Therefore, chemical systems are not a good candidate for an interstellar propulsion system.

From a theoretical perspective, deep space missions only require an escape velocity of approximately 16.5 km/s to exit the Solar System from the surface of the Earth. However, a far higher velocity is required to avoid the very long mission times required for interstellar travel. A number of spacecraft including the *Voyager* and *Pioneer* probes are traveling into interstellar space, with speeds up to 16.6 km/s. This performance was achieved using planetary gravity assist to supplement the initial chemical propulsion. However, the use of gravity assist has negative health physics aspects when an inner planet such as Venus is used. These negative aspects include an increase in the mission duration and the necessity of traveling through the higher radiation regions of the inner Solar System. In addition, SPEs become more hazardous when the spacecraft's distance to the Sun decreases.

A number of alternatives to chemical propulsion should become available in the twenty-first century. Solar electric propulsion is often mentioned as a well-established alternative. Although Solar electric propulsion can be used near a sun, it is not a viable candidate for an interstellar mission, as its effectiveness decreases with increasing distance from a star. Solar sails, nuclear electric propulsion and nuclear thermal propulsion are viable alternatives to chemical propulsion, but they require additional development. Other concepts, including those based on laser or microwave beamed energy systems, also require additional technological developments. Fusion thermal rockets and antimatter devices require even greater effort to develop and implement. It is also likely that other propulsion concepts based on new physics discoveries will be available in the future. Their use in interstellar missions depends upon their date of discovery, effort to implement, and public support for an interstellar mission.

7.9 Candidate Propulsion Systems Based on Existing Science and Technology

Although a wide variety of propulsion systems are possible, the most promising twenty-first century candidates are addressed. These systems include propulsion technologies derived from antimatter annihilation, fission reactions, fusion processes, an interstellar ramjet, and unique nuclear reactions. Alternative propulsion concepts based on new physics are addressed later in this chapter.

7.9.1 Antimatter Propulsion

Antimatter propulsion is based on the observation that a particle and its antiparticle release their entire mass–energy equivalent when they collide. The antimatter could be in the form of antiprotons or antihydrogen atoms. When a proton and an antiproton collide at low center-of-mass energy most of the annihilation energy appears as charged pions. The created pions have velocities of about $0.94\,c$ and they travel about 21 m before decaying. Following production, charged pions could be channeled via magnetic fields to a nozzle to produce thrust. It is estimated that velocities of 20 000 km/s are achievable using antimatter propulsion. This velocity is $<0.1\,c$ and still leads to extended travel times to nearby star systems.

The radiation hazard from antimatter propulsion includes charged pions and their decay products (muons, electrons, and neutrinos) and photons. The charged particles could be deflected by electromagnetic fields to avoid occupied areas, but the photons and neutrons are uncharged and would require shielding. Neutron activation products also require evaluation. As the neutrino energies do not reach the TeV–PeV range, they would not be a radiation hazard.

Mitigation for photons and neutrons is provided by either bulk shielding or locating the propulsion at a remote location with respect to occupied areas. However, the radiation source term for an antimatter propulsion system requires careful evaluation to ensure it does not become a limiting factor during interstellar travel. In addition, considerable research and development would be required for the implementation of an antimatter propulsion system.

7.9.2 Fission Driven Electric Propulsion

Electric propulsion is one of the most advanced flight-tested technologies that are currently available. An energy source such as a fission reactor is used to produce electrical energy and this energy drives the expulsion of reaction mass to produce thrust. Using this technology in an interstellar mission requires a power source independent of external factors and fission provides a credible source to produce electrical energy. However, the large mass of the fission reactor and support components and the electrical converters are limiting aspects of electric propulsion. In addition, the velocities generated from a fission-electric system appear to be too slow for interstellar travel.

7.9.3
Fusion Propulsion

Fusion processes occur in a plasma state. If a portion of this plasma reaction products were directed to a nozzle, it would provide thrust for a space vehicle. Fusion propulsion is another candidate for interstellar propulsion, but achievable velocities are likely to be $<0.1\,c$, which is less than desirable for interstellar travel. In addition, considerable research and development are required to implement this technology.

Fusion reactions were considered in Chapter 3, and the general radiological characteristics of a fusion reactor also apply to deep space fusion propulsion systems. The DD, DT, D^3He, or other processes could provide fusion propulsion. Given the physics, research has focused on the DT fusion process. For long-duration missions, the decay of tritium (12.3 year half-life) must be considered. This suggests a tritium production method, tritium purification system, or tritium fuel supply must be available on board the spacecraft.

For a fusion-powered spacecraft, direct neutron and gamma radiation must be shielded. The neutrons also activate structural materials and the associated beta–gamma activity is an additional health physics concern. Therefore, occupied areas of the spacecraft must consider neutron, beta, and gamma radiation types. Either shielding or distance is utilized to reduce those levels.

Neutron shielding is most effective if it contains a significant hydrogen content. Photons are most effectively shielded using bulk quantities of material such as iron or lead. If the spacecraft design does not accommodate conventional shielding techniques, then distance must be utilized to reduce radiation levels in occupied areas.

7.9.4
Interstellar Ramjet

The interstellar ramjet collects free space hydrogen for fusion reactor fuel. Upon achieving a DT fusion reaction, energy is channeled to the reaction product ^4He nuclei to provide thrust for the vehicle. As there is a small hydrogen density throughout the universe, the ramjet has the advantage that it would not run out of fuel. Its fuel is inherent in the medium in which it travels. The health physics hazards of the ramjet are similar to those of fusion reactor noted previously.

7.9.5
Unique Nuclear Reactions

One of the disadvantages of the antimatter, fission, and fusion propulsion methods is the production of ionizing radiation that introduces an additional hazard to the spacecraft's crew. A nuclear reaction that produces particles capable of generating thrust without the introduction of penetrating radiation would simplify the spacecraft's radiological design requirements.

There are nuclear reactions that could be used to power a spacecraft without the production of penetrating nuclear radiation. As an example, consider proton capture by ^{11}B

$$^{11}B + p \rightarrow 3\alpha. \tag{7.25}$$

This reaction is unique because nonpenetrating alpha radiation is produced and hydrogen/protons can be collected in free space. Minimal shielding would be required, but the ^{11}B fuel component needs to be stored or produced on the spacecraft. The generated alpha particles are directed via a nozzle to provide thrust.

However, none of the proposed interstellar propulsion technologies is currently available, and these require significant research and development to implement. Only antimatter propulsion approaches the minimum velocity requirements ($\sim 0.1\ c$) for interstellar travel. The practicality of implementing any of the candidate interstellar propulsion methods depends on the level of technology that society can achieve. In the following section, both current and possible technology levels of advancement are reviewed.

7.10
Technology Growth Potential

The reader should not conclude that interstellar travel is precluded by propulsion system limitations. There are two significant reasons for optimism.

First, physics knowledge and subsequent technological advances develop in an accelerating manner as evidenced by the progress made in the twentieth century. In a span of less than 100 years, physics progressed from a limited knowledge of the fundamental interactions to the initial stage of the unification of the fundamental interactions through the Standard Model. Emerging theories extending the Standard Model offer significant promise to provide additional insight into the nature of the universe.

As physics knowledge advanced, propulsion advances progressed from wind and sails to internal combustion engines and steam powered craft, to jet aircraft, and rockets that facilitate the travel to and exploration of space. To date, the Moon has been visited and nuclear powered probes are exploring the Solar System and beyond.

Second, our civilization is relatively primitive and significant technological growth possibilities exist. The potential for future technological growth is illustrated by comparing the current level of technology to the level that existed 100 or even

1000 years ago. In 1964, the Russian astronomer, Nikolai Kardashev, proposed a Civilization Type scale that provides insight into the nature of Earth's technological status and its growth potential.

The Kardashev scale is a method of quantifying the technological advancement of a civilization. This scale has three categories (Type I, II, and III civilizations) that are defined in terms of the usable power a civilization has at its disposal. The selected power levels were based on the view of technology that existed in 1964. Although these power levels are somewhat arbitrary, the Kardashev scale does reflect the relative technological level of a civilization.

A Type I civilization has the capability to utilize all power available on a single planet. Kardashev's original definition assigned a maximum planetary power value of 4×10^{12} W. Today, it is estimated that Earth has an available power of about 2×10^{17} W.

In a Type I civilization, the basic laws of physics are essentially understood. Propulsion methods include chemical, nuclear (fission and fusion), Solar sail, and laser sail technologies.

Type II civilizations are sufficiently advanced to harness all power output from a single star. Kardashev's original single star output was selected to be 4×10^{26} W based on Sol's characteristics. A Type II civilization has the capability to construct a Dyson Sphere around a star to capture its power output or the capability to feed stellar matter into a black hole and utilize the output power. Its propulsion options include improvements to the Type I propulsion systems and antimatter technologies.

Civilizations categorized as Type III utilize the power available from a single galaxy. Kardashev's original definition assigned a value of 4×10^{37} W to a galaxy's power availability.

Type III civilizations colonize their home galaxy. Available propulsion methods include the Type I and II methods and their improvements and gravitational drives, warp drives, and wormhole technologies. Elements of these theoretical methods are discussed in subsequent sections of this chapter.

Currently, Earth's civilization is below Type I since it is able to utilize only a portion of the available power on the Earth. The current state of our civilization can be designated as Type 0. Although intermediate values were not discussed in Kardashev's original proposal, they can be defined by interpolating his power values. Earth's current technology level is roughly 0.7. The Kardashev rating of a civilization (K) can be parameterized as

$$K = \frac{\log_{10} P - 6}{10}, \qquad (7.26)$$

where P is the usable power in W.

7.10.1
Dyson Spheres

As an example of the technological capability of a Type II civilization, consider the concept of a Dyson Sphere. In 1959, Freeman Dyson developed the concept of constructing a spherical shell around a star. This concept allows billions of people to live inside the shell and to harness the star's energy output. A civilization acquiring a Type II level of technology would have the scientific and technical capability to construct and fully utilize a Dyson Sphere.

For example, a Dyson Sphere with a radius of 1 AU around Sol would create an Earth climate with a huge surface area for growth. All of the Sun's output is available for use.

The engineering challenges of constructing a Dyson Sphere are significant. Structural designers would need to select appropriate construction materials that not only provide structural integrity but also absorb external impacts of space debris, comets, and asteroids. The sphere's design also needs to account for external GCR radiation and internal SPE radiation from the surrounded star.

7.11
Sources of Radiation in Deep Space

The total effective dose (E) inside a solar system (SS) is the sum of the contributions from galactic cosmic ray (GCR), solar particle event (SPE), trapped radiation (TR), and propulsion system (PS) source terms:

$$E_{SS} = E_{GCR} + E_{SPE} + E_{TR} + E_{PS}, \tag{7.27}$$

where E_{PS} is the effective dose derived from the propulsion system that could include fission, fusion, matter-antimatter interaction, nuclear reaction propulsion, or propulsion derived from a modification of spacetime. In deep space (DS) outside the Solar System boundary, the effective dose simplifies because the trapped radiation and SPE components are small relative to the GCR and PS doses:

$$E_{DS} = E_{GCR} + E_{PS}. \tag{7.28}$$

As the vehicle moves away from the Sun or toward nearby stars (e.g., Alpha Centauri (AC)), the SPE dose decreases as $1/r^2$ where r is the distance from the star producing the SPE. As the vehicle approaches AC, the crew effective dose (E_{AC}) takes the form

$$E_{AC} = E_{GCR} + E_{SPE}^{AC} + E_{TR}^{AC\ planet} + E_{PS}. \tag{7.29}$$

The individual terms in this equation are addressed in subsequent discussions. Equation 7.29 omits contributions from supernovas and gamma-ray bursts. These improbable radiation sources should be acknowledged in all interstellar travel. However, given their extremely unlikely occurrence, they are not included in the effective dose relationships.

7.12
Mission Doses

During a mission to a nearby star system, the effective dose varies with the specific mission segment. For deep space travel, the mission segments include (a) transit

Table 7.5 Contributions of various radiation sources to the effective dose for a mission to a nearby star system.

Mission phase	Trapped radiation	GCR	SPE	Propulsion system
Transit from Earth to the Sol System boundary	Yes[a]	Yes	Yes[b]	Yes
Transit from the Sol System boundary to the new star system boundary	No	Yes	No	Yes
Transit within the new star system	Yes[c]	Yes	Yes[d]	Yes
Transit from the new star system boundary to the Sol System boundary	No	Yes	No	Yes
Transit from entering the Sol System and returning to Earth	Yes[a]	Yes	Yes[b]	Yes

[a] Trapped radiation from the Earth's VABs.
[b] SPE source is Sol.
[c] Trapped radiation from any planet visited in the new solar system.
[d] SPE radiation from the visited star.

from Earth to the Sol System boundary, (b) exiting the Sol System to entrance into the new star system, (c) mission duration within the new star system, (d) transit from the new Solar System to entering the Sol System boundary, and (e) transit from entering the Sol System to Earth.

The mission also includes a brief period of orbiting the Earth, but this effective dose is small relative to the other source terms. Therefore, it is not included in subsequent discussion. This choice is consistent with the source-term discussion in Chapter 6.

Each of these mission segments includes contributions from a variety of radiation sources noted in Table 7.5. These radiation sources include trapped radiation, GCR, SPE from Sol, SPE from the visited star, and radiation from the propulsion system. Each of these sources is addressed in subsequent discussions.

7.12.1
Trapped Radiation

Trapped radiation was addressed in Chapter 6 for the Earth's VABs. Planets that are visited in the nearby star system also have a trapped radiation component. The characteristics of the trapped radiation associated with a new star system planet depend on the characteristics of the new sun, its radiation profile, and the characteristics of the new planet's geomagnetic field. The visited star may have radiation characteristics different from those of the Sol. It is also likely that the trapped radiation component varies significantly with time and spatial location above the planet.

The magnitude of the visited planet's trapped radiation effective dose depends on the time the spacecraft orbits the planet, the shielding provided by the spacecraft, and the availability and efficiency of a three-dimensional (3D) electromagnetic deflector. If the visited star/planet is similar to Sol/Earth, trapped radiation will not be a dominant mission source term unless anomalies like the South Atlantic Anomaly are encountered. However, the variability in star types and planet compositions could lead to significant variability in the trapped radiation effective dose. If this dose is significant, it requires monitoring and appropriate action (e.g., alteration of the spacecraft's orbital trajectory or use of an electromagnetic deflector) to mitigate the trapped radiation effective dose.

7.12.2
Galactic Cosmic Radiation

Given the variability of stars and local radiation conditions in the universe, the GCR source term has both time and position dependence. For example, measurements of ^{10}Be in polar ice cores and other data suggest that the cosmic ray intensity was significantly higher 50–100 years ago. The estimated radiation levels during these periods were approximately two times greater than during recent Solar minima. The initial analysis of this GCR data is summarized in Table 7.6. Since this data spans a relatively short period of time, the possibility of greater variation exists for Sol and other stars. This uncertainty must be recognized in any planned interstellar mission.

Uncertainties also exist because the visited star system may have characteristics that are quite different from the Sol system. The GCR radiation values presented in Chapter 6 and this chapter were based on the Sol output measured at 1 AU. The influence of the visited star's Solar wind on the GCR radiation remains a major uncertainty in determining the interstellar mission doses.

There is also the remote possibility of contributions from galactic sources such as supernovas and gamma-ray bursts. These events are difficult to predict but have significant dose consequences.

Table 7.6 GCR radiation at 1 AU.[a]

Time frame	Unshielded dose (cGy/yr)	Unshielded dose equivalent (cSv/yr)	Shielded BFO dose equivalent (cSv/yr)[b]
Current Solar maximum	6	39	27
Current Solar minimum	16	88	50
Circa 1954	19	109	62
Circa 1890	30	147	83

[a] Derived from Mewaldt et al. (2005).
[b] Shielding consists of 3 g/cm^2 of aluminum and the self-shielding of blood-forming organs by the body.

The results of Table 7.6 suggest that large GCR effective doses result from extended deep space missions. Using the 1890 time frame data suggests shielded GCR doses to BFO of 4.2, 8.3, 21, and 42 Sv result from 5, 10, 25, and 50 years interstellar mission durations, respectively,

These results indicate large effective dose values for extended mission durations. Reductions in the transient GCR effective dose is achieved through increased spacecraft shell shielding or increased shielding of normally occupied areas within the spacecraft. The use of an electromagnetic deflector is also a possibility, but the design must be different from the SPE deflector outlined in Chapter 6 because the GCR radiation is essentially isotropic.

7.12.3
SPE Radiation

Solar particle events or Solar flares from Sol were addressed in the previous chapter. The SPE characteristics of the new star may be very different from the characteristics of Sol and are largely determined by the age of the star and its spectral type.

Excluding the very unlikely gamma-ray burst or supernova events, SPEs produce the dominant effective dose when the spacecraft is traveling within the boundary of a star system. The large variability of star types noted in Table 7.2 suggests the SPE source term must be carefully monitored. Given the potential for effective doses to reach the lethal range, instrumentation needs to be developed to accurately predict the SPE magnitude before it occurs. In addition, once an SPE occurs, spacecraft systems must detect the event and initiate protective actions in a timely manner. These actions include engaging the propulsion system to move the spacecraft to a lower dose region of the star system, activating the deflector to change the direction of incident charged particle radiation, and shielding the incoming SPE radiation by relocating the crew to shielded shelters.

7.12.4
Radiation from a Fusion Reactor Propulsion System

A number of nuclear propulsion systems have been proposed for interstellar travel. From a technological development perspective, the most credible near-term systems utilize a DT fusion reactor. Accordingly, the radiation characteristics for a DT fusion reactor propulsion system are reviewed.

The Los Alamos National Laboratory proposed a reference DT propulsion system. It incorporates 1275 pulses with a 20-min duration and an output of 3.26×10^{23} n/pulse. For an unshielded propulsion device with these characteristics, a total effective dose of approximately 10^{10} Sv results at 10 m from the unshielded reactor. This effective dose is substantially larger than any tolerable dose and must be reduced through shielding or incorporating a larger distance between the reactor and the crew.

7.12.4.1 Distance

The practicality of distance to reduce the effective dose is examined using a point source approximation. This is a reasonable approach given the relative size of the DT reactor and the standoff distance. Using a point source approximation, the effective dose (E) resulting from DT fusion during the mission as a function of distance (r) from the DT reactor is written as

$$E(r) = \left(\frac{10\,\text{m}}{r}\right)^2 E(10\,\text{m}), \tag{7.30}$$

where $E(10\,\text{m})$ is the 10^{10} Sv value noted previously.

The use of Equation 7.30 indicates that distance alone is not a credible option for reducing the effective dose. A standoff distance of 10^6 m (1000 km) between the crew and the reactor is required to reduce the effective dose to 1 Sv and a 10^4 km distance is needed to achieve 0.01 Sv. Given current levels of technology, a ship having a length of 10^3–10^4 km is not a credible method to reduce the crew's effective dose. Therefore, the exclusive use of distance to reduce the effective dose does not lead to a viable spacecraft design.

7.12.4.2 Shielding

Since distance is not a likely solution to reducing the DT dose, shielding must be considered to reduce the crew's dose. Aluminum is used to estimate the thickness of required shielding since it is often quoted as the basis for Sol System shielding calculations. For a DT fusion source, the neutron energy is 14.1 MeV. In aluminum, the microscopic neutron absorption cross section (σ) at 14.1 MeV is 1.71 b/atom, the density (ρ) is 2.7 g/cm^3, and the gram molecular weight (GMW) is 27. The macroscopic cross section (Σ) is obtained from these values and Avogadro's number ($\bar{A}$):

$$\Sigma = N\sigma = \frac{\bar{A}\rho\sigma}{\text{GMW}} = \frac{(6.02\times10^{23}\,\text{atoms})}{(27\,\text{g})}\left(\frac{2.7\,\text{g}}{\text{cm}^3}\right)\left(\frac{1.71\,\text{b}}{\text{atom}}\right)\left(\frac{10^{-24}\,\text{cm}^2}{\text{b}}\right) \tag{7.31}$$
$$= 0.103\,\text{cm}^{-1}$$

For a given distance from the source, the effect of a shield of thickness x is written in terms of the unshielded values using the relationship

$$E(x) = E(0)\exp(-\Sigma x) \tag{7.32}$$

The shielding needed to reduce the unshielded dose equivalent to a predetermined target value is obtained from Equation 7.32. As an arbitrary target value, select 0.1 Sv that is the threshold for observing acute radiation-induced effects (e.g., chromosome aberrations) and also the allowable ICRP 60 effective dose over a 5-year period. Combining the distance and shielding terms leads to

$$E(x, r) = E(0, r_o)\left(\frac{r_o}{r}\right)^2 \exp(-\Sigma x), \tag{7.33}$$

where $E(0, r_o)$ is the unshielded effective dose at 10 m (10^{10} Sv), r_o is the distance from the reactor for the estimated effective dose (10 m), $E(x, r)$ is the desired 0.1 Sv effective dose, r is the distance from the reactor, and x is the aluminum shield thickness. The shielding thickness is obtained from Equation 7.33

$$x = -\frac{1}{\Sigma} \ln\left(\frac{r^2 E(x, r)}{r_o^2 E(0, r_o)}\right). \tag{7.34}$$

In formulating Equation 7.34, the total effective dose is assigned to the dominant neutron dose component. This is conservative since the total dose includes a smaller gamma-ray component. Although Equation 7.34 only includes the neutron source term, it is sufficient to provide a shielding estimate as there is considerable uncertainty in the proposed DT propulsion design parameters used in the calculation.

Table 7.7 summarizes the solution of Equation 7.34 for the shield thickness required to meet the 0.1 Sv target effective dose at distances between 10 and 10^6 m. For example, meeting the 0.1 Sv value requires an aluminum shield thickness of about 2 m when the effective dose is evaluated at a distance of 100 m. The shielding values of Table 7.7 are not unreasonable, since spacecraft construction would likely occur in space. Orbital construction is warranted in view of the mass of the shielding required and the limited launch weight capacity of Earth-based propulsion systems.

Using the Table 7.7 results, spacecraft lengths of 100–1000 m are not unreasonable. When combined with the 2 m of aluminum shielding, a viable radiological design envelope (e.g., size and shielding requirements) is obtained. These results suggest that a DT powered spacecraft is feasible for limited space exploration.

In addition to the spacecraft's size and shielding, other propulsion system characteristics require consideration. These include the mission duration and effects of the radiation environment on the spacecraft's crew.

Table 7.7 Aluminum shielding thickness required to reduce the DT reactor effective dose to 0.1 Sv as a function of distance from the source.

Crew standoff distance from the DT reactor (m)	Aluminum shield thickness (cm)
10	246
100	201
10^3	156
10^4	112
10^5	67.1
10^6	22.4

7.13
Time to Reach Alpha Centauri

As noted previously, the nearest star system beyond Sol is Alpha Centauri – a distance (s) of 4.3 LY from Earth. If a spacecraft travels at a velocity (v) of 0.1 c, the time (t) to reach Alpha Centauri (from the perspective of the spacecraft crew) is

$$t = \frac{s}{v} = \frac{(4.3 \text{ LY})(9.26 \times 10^{15} \text{ m/LY})}{(0.1)(3 \times 10^8 \text{ m/s})} \left(\frac{1 \text{ h}}{3600 \text{ s}}\right)\left(\frac{1 \text{ day}}{24 \text{ h}}\right)\left(\frac{1 \text{ year}}{365 \text{ day}}\right) = 43 \text{ years}. \tag{7.35}$$

The total travel time (assuming a round trip) is 86 years. From a human lifetime perspective, this time is limiting. Moreover, there is another problem associated with traveling long distances at high velocities. This problem is most easily outlined using the Special Theory of Relativity.

Using special relativity (see Appendix F for a detailed discussion), the elapsed time on Earth (T_{Earth}) is not the same as the time measured on the spacecraft ($T_{\text{spacecraft}}$). These times are related through the time dilation relationship

$$t_{\text{Earth}} = \frac{t_{\text{spacecraft}}}{\sqrt{1 - \frac{v^2}{c^2}}}. \tag{7.36}$$

Table 7.8 provides the results of applying Equation 7.36 for various velocities for a round trip to Alpha Centauri.

The results of Table 7.8 illustrate the time dilation phenomenon of special relativity. As the spacecraft velocity increases, the time as measured by the spacecraft crew decreases as expected. However, the elapsed time on Earth increases dramati-

Table 7.8 Times for round-trip travel to Alpha Centauri.

Spacecraft velocity (v/c)	$t_{\text{spacecraft}}$ (yr)	t_{Earth} (yr)
0.1	86	86.4
0.2	43	43.9
0.3	28.7	30.1
0.4	21.5	23.5
0.5	17.2	19.9
0.6	14.3	17.9
0.7	12.3	17.2
0.8	10.8	18.0
0.9	9.6	22.0
0.95	9.1	29.1
0.99	8.7	61.7
0.999	8.6	192
0.9999	8.6	608

cally as the spacecraft velocity approaches c. For example, at $0.99\,c$, the crew observes that 9 years have passed, but on Earth almost 62 years will elapse. This time dilation phenomenon has been experimentally verified.

The behavior of muons created by cosmic rays in the upper atmosphere confirms the time dilation effect. Given their velocity and lifetime, muons could not reach the Earth without the validity of the time dilation effect. The fact that muons created in the upper atmosphere reach the Earth's surface is part of the evidence for the validity of special relativity.

Time dilation also leads to social trauma for the crewmembers of an interstellar spacecraft upon their return to Earth. For example, consider a mission to Alpha Centauri for a spacecraft that travels at $0.9999\,c$. The crew would age 8.6 years, but they would return to an Earth that aged over 600 years and advanced significantly. Assuming that the crew's society survived, its level of technology and culture would be radically different from the society that the crew had left. The crew would be in a position similar to a European explorer that visited the New World in 1500, but returned to a twenty-second century European society. However, the crew's cultural shock would be even greater as evidenced by the social, religious, and national identity changes that occurred between 1500 and present day. In subsequent discussions, theoretical approaches to minimize the time dilation effect are explored.

7.14
Countermeasures for Mitigating Radiation and Other Concerns During Deep Space Missions

In Chapter 6, the biological effects of space radiation were addressed. The use of shielding and the possibility of the development of an EM deflector were noted as options available for reducing the effective dose during a deep space mission. A number of approaches for mitigating the radiation doses or their effects are feasible and include carcinogenic inhibitors, operational restrictions, shadow shielding, genetic screening, radioprotective chemicals, genetic enhancement, biomedical intervention, nanotechnology, and hibernation.

Numerous agents can either enhance or suppress the carcinogenic process induced by ionizing radiation. Recent research indicates that carcinogenesis, induced by HZE particles, is suppressed by chemical agents (e.g., retinyl acetate and tamoxifen), but significant toxicities and adverse side effects are associated with these agents. Other agents, classified as dietary supplements, are being evaluated as cancer preventive agents and have the advantage of not producing negative side effects. Vitamin E is one of the agents proposed for use since it reduces the amount of oxidative damage observed in astronauts. In addition, there are numerous studies indicating that vitamin supplements reduce the cancer incidence in various populations.

Operational restrictions limit the exposure duration or increase the margin to recommended dose limits. Specific restrictions include the use of older crewmembers, limiting extravehicular activity, and using trajectories that minimize the mission duration. Operational restrictions are a credible means to mitigate the effects of space radiation.

Shadow shielding reduces the effective dose to personnel by attenuating the incident radiation before it reaches the crewmembers. Optimizing the arrangement of structural materials and the equipment layout adds significant shielding without increasing the spacecraft weight. Shadow shielding is a viable approach to mitigate space radiation.

Screening crewmembers having a predisposition to higher cancer risk is of potential use. However, procedures to screen for radiation susceptibility are not fully developed. Accordingly, genetic screening is not currently a practical option, but could be an important tool in the future.

Radioprotective chemicals are available to mitigate the effects of space radiation. These chemicals often have side effects, but are a viable mitigating agent for space radiation. However, radioprotective chemicals may not be effective for protection against HZE particles.

Genetic methods to enhance the body's ability to withstand radiation damage are conceptually feasible. Although they are beyond existing technology, genetic enhancement methods may be available in the future.

In principle, biomedical intervention could be used to mitigate prompt radiation effects arising from high intensity radiation. This approach may be feasible in the future to repair cellular radiation damage using gene therapy or related techniques. At present, biomedical intervention is not a viable alterative for mitigating the effects of space radiation.

Nanotechnology offers the potential to repair radiation-induced cellular damage. Although this technology has yet to facilitate the repair of radiation-induced cellular damage in a consistent or complete manner, it could be available in the future.

In addition to radiation-related concerns, the large distances and times associated with interstellar travel place severe demands on human physiology to survive extended mission durations. Given these circumstances, it is logical to ask if it is possible for the spacecraft crew to hibernate during most of the interstellar voyage.

Hibernation offers a number of positive health physics aspects. The crew could be placed in small, shielded structures that would limit the transit radiation dose. These structures could also be engineered to limit the effects of reduced gravity. In addition, hibernation reduces environmental requirements (food, water, breathing air, heating, cooling, climate control, etc.) to sustain the crew during transit. Hibernation also permits a larger crew size to be accommodated because of the reduced demands to sustain the crew during transit, but it does not eliminate the potential cultural shock caused by extended interstellar missions.

7.15
Theoretical Propulsion Options

The travel time to star systems in proximity to Sol are prolonged and can easily exceed 10 years. These times become prohibitive as missions to even more distant stars are planned. Extreme distances warrant consideration of nontraditional propulsion methods that utilize the inherent properties of spacetime to alter the trajectory instead of traveling along the spacetime geodesic. Candidate approaches include modifying spacetime, the utilization of wormholes to traverse extreme distances, folding spacetime to shorten the spacetime distance, and mapping spacetime.

7.15.1
Modifying (Warping) Spacetime

It is theoretically possible to formulate a spacetime geometry, the warp drive metric that permits a spacecraft to travel a distance d to a star and return home, such that the elapsed time for stationary observers on Earth is less than $2\,d/c$. The term "warp" applies to the alteration of spacetime as a means to change the geodesic. This metric avoids the time dilation effect and its negative consequences for the spacecraft's crew.

The warp drive metric utilizes an approach that is consistent with the theory of general relativity (see Appendix L) to modify spacetime in a manner that allows the spacecraft to travel with an arbitrarily large velocity relative to an observer outside a localized warp bubble. By generating a local expansion of spacetime behind the spacecraft and a contraction in front of it, a velocity exceeding the speed of light, as observed outside the warp bubble, is theoretically possible. However, the local velocity of the spacecraft inside the bubble is less than the speed of light. The practical difficulty is generating and maintaining the postulated warp bubble.

In the theoretical warp drive metric, the spacecraft resides at rest inside a single, spatial bubble (SSB) and never locally travels faster than the speed of light. A ship traverses spacetime in a manner analogous to the motion of galaxies receding from each other at extreme speeds owing to the expansion of the universe, while locally they are at rest. The SSB makes use of this type of expansion (and contraction) to achieve the ability to travel faster than light (FTL).

Although SSB "warp drive" sounds appealing, it does have a number of theoretical assumptions that have yet to be confirmed. One of these assumptions includes the use of exotic matter with a negative energy density. An extremely large mass of exotic matter is required to generate the SSB. Although exotic matter is forbidden classically, it is permitted within quantum field theory. The need for exotic matter does not eliminate the warp metric as a propulsion scheme, but it is a complication. In addition to these SSB issues, there are other concerns that would arise as a consequence of FTL travel.

Assuming that a space vehicle generates an SSB, it would need to overcome a number of additional obstacles. One obstacle is the concern that it could collide with objects encountered during FTL operations. Such collisions would be extremely hazardous to the ship and its crew. Another concern is associated with the physics associated with radiation in the path of the ship. For example, photons arriving at the front of the ship are blue shifted to higher energies by the physical principles creating

7.15 Theoretical Propulsion Options

the FTL condition (i.e., the warped region of spacetime). These higher energy photons and other radiation types encountered could be lethal to the ship's crew and could also damage the ship and its systems. Therefore, it would be desirable to find solutions that protect the ship from various hazards while maintaining the FTL characteristics.

One possible solution to the SSB difficulties is the use of a second-generation "warp drive" that incorporates an optimized, spatial bubble (OSB). An OSB drive system requires a more complex modification of spacetime, but this configuration eliminates a number of issues created by the SSB drive. The OSB metric with respect to an observer exterior to the warp bubble is written in terms of Cartesian coordinates as

$$ds^2 = -dt^2 + B^2[dx - v_s(t)f(r_s)dt]^2 + dy^2 + dz^2, \tag{7.37}$$

where $v_s(t)$ is the spacecraft velocity with respect to an external observer, $x_s(t)$ is the x-coordinate of the central geodesic, and $r_s(t, x, y, z)$, and $v_s(t)$ are given by

$$r_s(t, x, y, z) = \sqrt{(x - x_s(t))^2 + y^2 + z^2}, \tag{7.38}$$

$$v_s(t) = \frac{dx_s}{dt}. \tag{7.39}$$

The function $f(r_s)$ is related to the first warp region in the range from $R - \delta$ to $R + \delta$:

$$f(r_s) = \frac{\tanh[\delta(r_s + R)] - \tanh[\delta(r_s - R)]}{2\tanh[\delta(R)]}, \tag{7.40}$$

and R is the radius of the warp bubble. The parameter B is defined in terms of a second warp region with a radius D

$$B = \left[\frac{1 + \tanh[\delta(r_s - D)]^2}{2}\right]^{-P}, \tag{7.41}$$

where the exponent P is a parameter used to alter spacetime in the desired manner. The warp region parameter (δ) is related to the shape of the OSB bubble.

OSB optimization is an improvement over the SSB drive system in a number of ways. First, it offers a conceptual means to travel in a FTL manner. Second, a more reasonable exotic matter requirement, on the order of 10 kg, is needed to power the OSB FTL propulsion drive.

The OSB approach offers other possibilities. For example, one application of a static warp field is the deflection of matter. In principle, the trajectory of large objects such as asteroids or meteors could be altered to avoid collisions with spaceships or even planets. In a similar manner, the OSB approach can be used to mitigate the effect of the incident blue-shifted radiation. However, these potential applications depend on the viability of producing a metric having OSB properties and finding an appropriate quantity of exotic manner. In the early part of the twenty-first century, the OSB is clearly a theoretical construct that requires significant refinement and development before it becomes a credible propulsion approach.

Exotic matter is also needed to produce spatial effects that establish wormholes. A wormhole is another theoretical construct that is used to alter spacetime by tunneling through it.

7.15.2
Wormholes

To illustrate the wormhole concept, the Morris–Thorne (MT) wormhole geometry is reviewed. The coordinates used to define the MT wormhole geometry are the spherical polar coordinates $\{r, \theta, \varphi, t\}$, and its metric tensor is

$$g_{\mu\nu} = \begin{bmatrix} 1 & 0 & 0 & 0 \\ 0 & b^2+r^2 & 0 & 0 \\ 0 & 0 & (b^2+r^2)\sin^2\theta & 0 \\ 0 & 0 & 0 & -1 \end{bmatrix}, \quad (7.42)$$

where b is a constant having the dimensions of length. An examination of the wormhole geometry indicates that it reduces to flat spacetime (Equation 7.21) in the limit $b \to 0$. Except for the $b=0$ metric, the MT geometry is not flat but curved. For $b \neq 0$, an embedding of the (r, φ) slice of the wormhole geometry produces a surface with two asymptotically flat regions connected by a region of minimum radius b. This region resembles a tunnel or wormhole connecting the two asymptotically flat regions. The tunnel represents a short cut in spacetime and minimizes the distance required to travel between spacetime locations. The difficulty is generating a wormhole structure connecting the origin and destination. Wormholes currently have no experimental validation. Additional discussion regarding the wormhole metric is provided in Appendix L.

7.15.3
Folding Spacetime

The concept of folding spacetime is analogous to the observation that light bends when it is in proximity of a massive star. Rather than traveling a large distance between points A and B, spacetime is altered to temporarily move A and B closer. This folding of space is a theoretical construct and has yet to be demonstrated.

7.15.4
Mapping Spacetime

Spacetime modification is a generalized concept that maps points A and B to two analogous points A' and B' that are close to each other. This mapping could be accomplished through spacetime modification including folding or modifying spacetime, wormholes, or other more generalized concepts. Although numerous options exist, none has been observed or demonstrated.

7.16
Spatial Anomalies

Although space exploration has been limited, Pioneer 10 and 11 spacecraft data suggest spatial anomalies may be present within our Solar System. The nature of these anomalies has yet to be explained.

At distances between 20 and 70 AU from the Sun, radiometric tracking data from both Pioneer 10 and 11, indicate the presence of an anomalous, small, constant Doppler frequency drift. The drift is a blueshift, uniformly changing at the rate of $(5.00 \pm 0.01) \times 10^{-9}$ Hz/s. This signal has been interpreted as a constant acceleration of each spacecraft toward the Sun. Similar anomalies were observed during the Galileo and Ulysses missions. This acceleration is unexpected and not consistent with the current understanding of the physics associated with the gravitational interaction. Given the extremely limited exploration of space, the Pioneer Anomaly is not likely to be the only spatial anomaly observed as exploration reaches deeper regions of space.

A number of possible explanations have been proposed for the Pioneer Anomaly. One of the more interesting explanations for this anomaly is the emergence of extra spatial dimensions. Although the emergence of extra spacetime dimensions is not the only possible explanation for the Pioneer Anomaly, it does suggest that such concepts are more than idle theoretical speculation.

7.17
Special Considerations

Deep space presents a unique environment for observing the effects of the four fundamental interactions. In deep space, the gravitational interaction may have a significant impact on the radiation environment. This impact can be manifested in a variety of ways including the emergence of additional spatial dimensions beyond the three spatial dimensions of conventional spacetime.

Additional spatial dimensions would not normally emerge at low energies in conventional flat spacetime. However, in the vicinity of gravitational or high-energy anomalies, extra dimensions could emerge and affect the radiation environment. The emergence of these extra dimensions could appear as a localized spatial anomaly that would manifest itself in a variety of ways.

The magnitude of the impact of extra spatial dimensions depends on their physical size or scale, with the greatest impact occurring if the scale of the extra dimensions is similar to the scale of the conventional three spatial dimensions. The scale of a dimension is addressed in more detail in subsequent discussion.

As an introduction to the impact of extra dimensions, consider their impact on nuclear energy levels. For simplicity, the energy levels are derived from a nonrelativistic Schröedinger equation using a representative nuclear interaction. The phenomenological nuclear interaction is reasonably well approximated by Woods–Saxon (WS) potentials. WS potentials can be approximated by square well potentials. Although square well potentials are not as accurate as WS potentials, they produce the essential elements of nuclear energy levels from a health physics perspective and are sufficient for the purpose of this chapter.

Consider a square well potential with a dimension or range b. Using this potential, the solution of the Schröedinger's equation leads to energy levels (E) having the following spectrum

$$E = \frac{\hbar^2}{2m}\left(\frac{K\pi}{b}\right)^2, \tag{7.43}$$

where $\hbar$ is Planck's constant divided by 2π, K is a quantum number defining the energy level spectrum ($K = 1, 2, 3, \ldots$), and m is the mass of the nuclear cluster confined by the square well potential. The dimension b can also be considered to represent the scale or characteristic length of the nuclear system within the constraints of conventional spacetime.

As a second example, consider the same square well but add a new spatial dimension that is curled up into a circle of radius R. R can be considered a measure of the scale of the extra dimension. The energy levels of the square well with the extra dimension are

$$E = \frac{\hbar^2}{2m}\left(\left(\frac{K\pi}{b}\right)^2 + \left(\frac{L}{R}\right)^2\right), \tag{7.44}$$

where K has the same values as in Equation 7.43 and $L = 0, 1, 2, \ldots$.

The emergence of the extra dimension can change the nuclear level spectrum dramatically. If $R \ll b$, then the low-energy portion of the spectrum and the radiation emitted from the nucleus is unchanged. However, if R is on the same order of magnitude as b, then the spectrum is dramatically altered. Since the radiation types emitted by a nucleus are dependent on its energy level spectrum, the energies, yields, and types of radiation emitted from a nucleus are dramatically altered. This affects the health physics considerations as noted in the next section.

7.18
Point Source Relationship

As an additional example of the impact of additional spatial dimensions, the dose rate from a point isotropic radiation source for spatial dimension (d) having a value $d \geq 3$ is derived. This derivation assumes the scale of all dimensions is the same. To evaluate the point source radiation relationship for general dimension d, consider conventional three-dimensional space ($\mathcal{R}^3$) with coordinates x_1, x_2, and x_3, and define a three-ball as the region defined by

$$\mathcal{B}^3(R) : x_1^2 + x_2^2 + x_3^2 \leq R^2, \tag{7.45}$$

where R is the radius of the three-ball. This region is enclosed by the two-sphere

$$\mathcal{S}^2(R) : x_1^2 + x_2^2 + x_3^2 = R^2. \tag{7.46}$$

The superscripts in $\mathcal{B}$ and $\mathcal{S}$ denote the dimensionality of the space under consideration.

In arbitrary dimensions, balls and spheres are defined as subspaces of $\mathcal{R}^d$:

$$\mathcal{B}^d(R) : x_1^2 + x_2^2 + \cdots + x_d^2 \leq R^2. \tag{7.47}$$

Equation 7.47 defines the region enclosed by the sphere $\mathcal{S}^{d-1}(R)$:

$$\mathcal{S}^{d-1}(R) : x_1^2 + x_2^2 + \cdots + x_d^2 = R^2. \tag{7.48}$$

For simplicity, the term "volume" is used to characterize the spatial extent. For example, if a space is one-dimensional, volume means length. In two-dimensions, volume means area. All higher dimensional spaces ($d \geq 3$) have d-dimensional volume. Accordingly, the volumes of one- and two-dimensional spheres are the circumference of a circle of radius R and surface area of a sphere of radius R, respectively:

$$\text{vol}(\mathcal{S}^1(R)) = 2\pi R, \tag{7.49}$$

$$\text{vol}(\mathcal{S}^2(R)) = 4\pi R^2. \tag{7.50}$$

Since volume has units of length to the power of the spatial dimension, the volume of a sphere of radius R is related to the volume of a unit radius sphere by

$$\text{vol}(\mathcal{S}^{d-1}(R)) = R^{d-1} \text{vol}(\mathcal{S}^{d-1}). \tag{7.51}$$

The volumes of the unit one-sphere and two-spheres are recognized from Equations 7.49 and 7.50:

$$\text{vol}(\mathcal{S}^1) = 2\pi, \tag{7.52}$$

$$\text{vol}(\mathcal{S}^2) = 4\pi. \tag{7.53}$$

The reason for calculating the volumes in multiple dimensions is that isotropic point sources are defined by the fact that equal flux penetrates all incremental areas on the surface of a sphere surrounding the source. These volumes must be known to generalize the point source relationship to multiple dimensions.

To calculate the volume of the sphere $\mathcal{S}^{d-1}$, consider $\mathcal{R}^d$ with r being the radial coordinate:

$$r^2 = x_1^2 + x_2^2 + \cdots + x_d^2. \tag{7.54}$$

The desired volume is obtained by evaluating the following integral using two different approaches

$$I_d = \int \exp(-r^2) dx_1 dx_2 \ldots dx_d. \tag{7.55}$$

Using Equation 7.54, Equation 7.55 becomes the product of d one-dimensional integrals:

$$I_d = \prod_{i=1}^{d} \int_{-\infty}^{+\infty} dx_i \exp(-x_i^2) = \pi^{d/2} \tag{7.56}$$

The integral in Equation 7.55 is also evaluated by breaking $\mathcal{R}^d$ into thin shells. Since the space of constant r is the sphere $\mathcal{S}^{d-1}(r)$, the volume of a shell lying between r and $r+dr$ equals the volume of $\mathcal{S}^{d-1}(r)dr$. Using Equation 7.51

$$I_d = \int_0^\infty dr \exp(-r^2)\text{vol}(\mathcal{S}^{d-1}(r)) = \text{vol}(\mathcal{S}^{d-1}) \int_0^\infty dr \exp(-r^2) r^{d-1}. \tag{7.57}$$

Equation 7.57 is simplified through a change of variables ($t=r^2$):

$$I_d = \frac{1}{2}\text{vol}(\mathcal{S}^{d-1}) \int_0^\infty dt \exp(-t) t^{((d/2)-1)}. \tag{7.58}$$

The integral of Equation 7.57 is written in terms of the gamma function $\Gamma(x)$ for $x > 0$:

$$\Gamma(x) = \int_0^\infty dt \exp(-t) t^{x-1}. \tag{7.59}$$

Using Equation 7.59, I_d becomes

$$I_d = \frac{1}{2}\text{vol}(\mathcal{S}^{d-1})\Gamma\left(\frac{d}{2}\right). \tag{7.60}$$

Comparing Equations 7.56 and 7.60, we obtain

$$\text{vol}(\mathcal{S}^{d-1}) = \frac{2\pi^{d/2}}{\Gamma(d/2)}. \tag{7.61}$$

Table 7.9 provides values of the gamma function and $\text{vol}(\mathcal{S}^{d-1})$ for $d=2, 3, 4,$ and 5.

With knowledge of the volumes of multidimensional spaces, the specific functional dependence of the radiation emitted from an isotropic point source can be

Table 7.9 Values of the gamma function and $\text{vol}(\mathcal{S}^{d-1})$ for $d=2$–5.

d	$\Gamma(d/2)$	$\text{vol}(\mathcal{S}^{d-1})$
2	1	2π
3	$\frac{\sqrt{\pi}}{2}$	4π
4	1	$2\pi^2$
5	$\frac{3\sqrt{\pi}}{4}$	$\frac{8\pi^2}{3}$

Table 7.10 Generalized isotropic point source relationship for $d = 3$, 4, and 5.[a]

		Units		
d	$\dot{D}_d$	A	G_d	r
3	$\dfrac{GA}{r^2}$	MBq	$\dfrac{\text{Gy m}^2}{\text{h MBq}}$	m
4	$\dfrac{2}{\sqrt{\pi}} \dfrac{G_4 A}{r^3}$	MBq	$\dfrac{\text{Gy m}^3}{\text{h MBq}}$	m
5	$\dfrac{3}{2} \dfrac{G_5 A}{r^4}$	MBq	$\dfrac{\text{Gy m}^4}{\text{h MBq}}$	m

[a] All dimensions have the same scale.

written. For simplicity, only photon-emitting radionuclides are considered. The formulation of the absorbed dose rate ($\dot{D}$) from an isotropic, point, photon-emitting source in three dimensions is

$$\dot{D} = \frac{AG}{r^2}, \tag{7.62}$$

where A is the source activity (MBq), G is the gamma constant (Gy m^2)/(h MBq), and r is the distance from the point source (m).

Equation 7.62 can be rewritten as a general d dimensional ($d \geq 3$) expression for the point source ($\dot{D}_d$):

$$\dot{D}_d = \frac{2\Gamma\left(\dfrac{d}{2}\right)}{\sqrt{\pi}} \frac{G_d A}{r^{d-1}}. \tag{7.63}$$

where the leading term is a normalization factor that ensures Equation 7.62 is recovered for $d = 3$ and r is defined in Equation 7.54. Table 7.10 provides the generalized point source relationship of Equation 7.63 for $d = 3$, 4, and 5 based on the restriction that all dimensions have the same scale.

Table 7.10 suggests that the characteristics of a point source are quite different in multidimensional spaces. The only caveat is that the scale of the extra dimensions impacts the derivation of Equation 7.63 and the results of Table 7.10. In conventional three-dimensional space, the scale of x_1, x_2, and x_3 are equal. If the fourth spatial dimension is of the same scale, different physics results. However, if the scale of x_4 is much less than x_1, x_2, and x_3, then the $d = 4$ result becomes essentially equivalent to the $d = 3$ result and the relevant health physics properties would be unchanged. The significance of the Table 7.10 results ultimately depends on the characteristics of the spacetime geometry in which the radiation originates.

For completeness, we also note that the gamma constant depends on the photon energy (E) and its associated yield (Y):

$$G = k \sum_i E_i Y_i, \tag{7.64}$$

where the sum over i includes all gamma rays emitted by the radionuclide and k is a constant that depends on the units defining G. As noted previously, the extra dimensions also impact the energies and associated yields of a radionuclide. As with the point source relationship, the impact of the extra dimensions depends on their scale relative to conventional three-dimensional space. Problem 07-01 explores the impact of an extra dimension on the energy levels, and yields of a nucleus and its associated gamma constant and absorbed dose.

Problems

07-01 The Starship Berlin is on the first mission to Proxima Centauri. After leaving the Sol System, the mission scientist, Sheza Really-Smart, reports that she has encountered a spatial anomaly. Sheza observes that a fourth spatial dimension has been detected and that its apparent scale (R) is b/π where b is the scale of conventional 3D space. Photon radiation is detected from a ^{60}Ar gas cloud, but it is not consistent with the values in the ship's reference library. Given the 4D space and its scale:

(a) Calculate the gamma-ray spectrum of the ^{60}Ar nucleus in both 3D and 4D spaces. Assume the energy levels for 3D space are based on $\Delta K = \pm 1$ transitions and the 4D energy levels are based on $\Delta K = \pm 1$ with L constant and $\Delta L = \pm 1$ transitions with K constant where K and L are quantum numbers that define the spectra in 3D and 4D spaces:

$$E_K = \frac{\hbar^2}{2m}\left(\frac{K\pi}{b}\right)^2 \quad K = 1, 2, 3, \ldots,$$

$$E_{K,L} = \frac{\hbar^2}{2m}\left[\left(\frac{K\pi}{b}\right)^2 + \left(\frac{L}{R}\right)^2\right] \quad K = 1, 2, 3, \ldots; \quad L = 0, 1, 2,$$

where m is the nucleon mass and $b = 36.4$ fm. Assume all allowed nuclear transitions have a yield of 1.0. Only consider energy levels with a value of 2.5 MeV or less.

(b) Calculate the 3D and 4D dose factors having the units Gy m^2/h MBq and Gy m^3/h MBq, respectively.

(c) Assuming the distance of interest is 5 m in both three- and four-dimensional space, calculate the absorbed dose rate assuming the source is isotropic and its activity is 4×10^5 MBq.

07-02 During transit to Alpha Centauri, a probe detects two bursts of neutron radiation. These radiation sources originate in Solar Systems A and B and have the characteristics noted in the following table.

(a) If the spacecraft is stopped for repairs and remains stationary during the passage of the neutron bursts, what is the unshielded effective dose received at the

Solar system	Neutron fluence (n/cm²)	D_{SP} (LY)	D_{SSP} (LY)
A	2×10^5	2	10
B	3×10^7	2	50

D_{SP} = Distance between the spacecraft (S) and probe (P).
D_{SSP} = Distance between the solar system (SS) and probe.

spacecraft's position? Assume the appropriate dose conversion factor is 3.5×10^{-10} Sv cm²/n; the probe, spacecraft, and star system reside on a free space geodesic; and the suns in Solar System A and B have the same energy spectrum.

(b) During the critique of the event, the ship's captain is concerned that the electromagnetic deflector did not reduce the neutron fluence. How do you resolve the captain's concern?

(c) The engineering officer suggests developing an antineutron source to eliminate neutron doses. Is this an effective ALARA measure?

07-03 Shielding is required for a DT propulsion system that produces a total mission dose equivalent (unshielded) of 10^{10} Sv at 10 m from the fusion chamber. Assume the effective dose criterion is 0.1 Sv for the mission dose from the propulsion system. The neutron removal coefficient for 14.1 MeV neutrons in aluminum is 0.103 cm^{-1}. (a) Derive Equation 7.34. (b) Calculate the required shielding for the following standoff distances from the fusion chamber: 10, 100, 1000, 10^4, 10^5, and 10^6 m.

07-04 You have been selected as the lead ALARA engineer for Project Zeus, a mission to Proxima Centauri using a DT fusion reactor propulsion system. As part of your ALARA review, (a) list the major radiation types of health physics concern associated with this propulsion system, (b) describe the type of hazard imposed by these radiation types, (c) list the various sources of radiation external to the spacecraft and possible ALARA measures to mitigate these exposures, (d) Repeat questions (a) and (b) for an antimatter propulsion system using protons and antiprotons.

07-05 During deep space operations a 250-kg mass of stellar matter is encountered. The mass has a specific activity of 2.5×10^8 MBq/g and is characterized by a gamma constant of 0.05 Gy m²/h MBq. How close can your spacecraft approach the stellar matter, but not exceed 0.01 Gy/h? In formulating the solution ignore any shielding provided by the spacecraft. The density of the stellar mass is 100 kg/cm³.

07-06 Your spacecraft is in orbit around an Earth-sized planet in the Tau Ceti Star System. Spacecraft radiation instruments indicate that Tau Ceti is in the process of a massive Solar particle event that is 10 times larger than the largest event detected on Earth. What actions would you recommend to minimize the crew's absorbed dose from this event?

07-07 Given the conditions of the previous problem, what actions would you recommend for crewmembers exploring the planet's surface?

07-08 As the chief ALARA engineer for Project Wolf 359, you are evaluating a proposal for two candidate propulsion systems. The specifications for these systems are provided in the following table:

Power system parameter/ characteristic	Propulsion system-1 tau catalyzed lithium fusion	Propulsion system-2 polarized muon catalyzed DLi fusion
Mean reactor radius (m)	12	20
Gamma source strength (γ/s)	4×10^{16}	9×10^{15}
Neutron source strength (n/s)	5×10^{15}	7×10^{14}
Distance from propulsion system to crew quarters (m)	1000	525
Mean gamma-ray energy (MeV/γ) and yield	4.5@0.94	6.7@0.65
Mean neutron energy (MeV/n)	11.0	6.0
Neutron flux to dose conversion factor (Sv cm^2/n)	4.3×10^{-8}	4.2×10^{-8}
Neutron attenuation factor	0.00001	0.00001
Gamma-ray attenuation factor (cm^2/g)	0.0005	0.0004
Gamma-ray mass energy absorption coefficient (cm^2/g)	0.05	0.06

(a) What is the annual effective dose rate for propulsion system-1?
(b) What is the annual effective dose rate for propulsion system-2?
(c) Which of these two propulsion systems would you recommend on the basis of their impact on the crew's effective dose? Assume that an annual effective dose of 100 mSv/year is established as the Wolf 359 dose limit.

07-09 You are asked to shield the 14.1 MeV neutrons from a DT fusion reactor. Lead and polyethylene are available. How are these materials to be arranged to shield the 14.1 MeV neutrons?

07-10 A Type II Civilization constructed a Dyson Sphere around a star very similar to Sol. The star's output is characterized in terms of three proton energy groups as noted in the table. Assume that the Type II Civilization species has a response to radiation very similar to human response. (a) What is the annual effective dose for typical years not having a major SPE? (b) What are the biological effects of this exposure? (c) What absorbed dose is received during a maximum SPE that occurs over a 90-min period? (d) What are the biological effects of the maximum SPE dose? (e) If the sphere's inhabitants have 10 min warning, what actions could be taken to minimize the absorbed dose from a maximum SPE that lasts 18 h?

			Fluence (proton/cm²)	
Proton energy group	Chronic dose conversion factor (pSv cm²)/proton	Acute dose conversion factor (pGy cm²)/proton	Typical annual magnitude	Maximum SPE
1	200	50	1.0×10^6	3.2×10^{11}
2	500	100	2.0×10^5	5.0×10^{10}
3	1500	500	5.0×10^4	2.2×10^9

07-11 Planet Cleveland orbits a massive star (A) but is in the vicinity of two additional stars B and C. All three stars have a similar energy spectrum and undergo simultaneous Solar flares that are described in the following table:

Star	Distance of star to planet Cleveland (AU)	Total output rate of the star (particles/s)	Duration of event (s)
A	1.5	1×10^{30}	2000
B	10	3×10^{31}	5000
C	400	5×10^{34}	1000

The radial dependence of the fluence for stars A, B, and C is r^{-2}. Assume all stars radiate isotropically. (a) What is the total fluence at Planet Cleveland resulting from the events on A, B, and C? (b) Cleveland's scientists know that seeding a star's corona with ^{293}Hf and negative muons decreases its total output by a factor of 5 but generates a spacetime anomaly that shifts the star's position such that its distance to Cleveland decreases by factor of 1.25. If only one star can be seeded, which one would you target to minimize the radiological impact of the SPEs? Assume the shift in the star's position creates no effects other than a change in the radiation levels on Planet Cleveland.

07-12 Show that the geodesics of flat spacetime in a two-dimensional plane are straight lines. Use the two-dimensional metric in polar coordinates (r, θ) and the differential geodesic equations to prove that the straight line conjecture is valid:

$$ds^2 = dr^2 + r^2 d\theta^2,$$

$$\frac{d^2 r}{ds^2} = r\left(\frac{d\theta}{ds}\right)^2,$$

$$\frac{d}{ds}\left(r^2 \frac{d\theta}{ds}\right) = 0.$$

07-13 During a spectrum analysis mission to a newly discovered star, your spacecraft intercepts a portion of the output beam from a gamma-ray burst. The estimated distance from your location to the burst is 750 kpc. The mean energy of the gamma-ray burst is 2 MeV, the total gamma-ray output is 1.9×10^{64} γ, and the flux to dose conversion factor for 2 MeV photons is 3.27×10^{-8} Gy/h per γ/cm^2 s. (a) What is the unshielded absorbed dose at the spacecraft location? Assume the spacecraft intercepts 0.2% of the total gamma-ray burst's fluence. (b) If the spacecraft is shielded by 20 cm of water equivalent, what is the shielded absorbed dose at the spacecraft location? The applicable attenuation coefficient is 0.05 cm^2/g, and a buildup factor of 1.8 is applicable to this situation. (c) What are the expected health effects from the absorbed dose calculated in part (b)?

07-14 You are assigned to perform an ALARA evaluation for Operation Ka-Boom that is designed to convert gas giant planets into stars. Ka-Boom focuses on planets whose sun's light output is decreasing. The following table lists candidate gas giant planets and their mean distance from Earth-like planets that the gas giant would illuminate. Assume the gas giants become main sequence stars following their conversion.

Candidate gas giant planet	Mean distance from gas giant to Earth-like planet (AU)	Mass of gas giant (M_{Sun})	Temperature of gas giant star (T_{Sun})
Aries-6	8.0	10.0	4.0
Cetus-4	58.0	14.0	5.0
Bootes-3	4.6	3.0	2.0

(a) What are the luminosities of the three gas giants relative to Sol after being converted into stars? Assume the density of all three stars will be the same as Sol.
(b) Assume that the value calculated in (a) is the luminosity at 1 AU and that the luminosity decreases as r^{-2}. Which of the three stars produces a luminosity at its companion Earth-like planet that is closest to the Earth's value?
(c) What are the consequences of colonizing these three Earth-like planets?

References

Abramowitz, M. and Stegun, I.A. (1972) *Handbook of Mathematical Functions with Formulas, Graphs, and Mathematical Tables*, Dover Publications, Inc., New York.

Alcubierre, M. (1994) The warp drive: hyper-fast travel within general relativity. *Classical and Quantum Gravity*, **11**, L73.

Anderson, H.L. (ed.) (1981) *Physics Vade Mecum*, American Institute of Physics, New York.

Anderson, J.L. (1999) Roadmap to a star. *Acta Astronautica*, **44**, (2–4), 91.

Anderson, J.D., Laing, P.A., Lau, E.I., Liu, A.S., Nieto, M.M. and Turyshev, S.G. (1998) Indication from pioneer 10/11, Galileo, and Ulysses data of an apparent anomalous, weak, long-range acceleration. *Physical Review Letters*, **81**, 2858.

Arny, T.T. (2004) *Explorations – An Introduction to Astronomy*, 3rd edn (Updated), McGraw-Hill, New York.

Balcomb, J.D., Booth, L.A., Cotter, T.P., Hedstrom, J.C., Robinson, C.P., Springer, T.E. and Watson, C.W. (1970) LA-4541-MS. in *Nuclear Pulsed Space Propulsion Systems*, Los Alamos Scientific Laboratory of the University of California, Los Alamos, NM.

Belayev, W.B. (2002) Five-Dimensional Gravity and the Pioneer Effect, http://arxiv.org/abs/

gr-qc/0209095 (accessed on November 9, 2006).

Bennett, J.O., Donahue, M., Schneider, N. and Voit, M. (2005) *Cosmic Perspective Volume 2, The: Stars, Galaxies and Cosmology*, 3rd edn, Benjamin Cummings, New York.

Bevelacqua, J.J. (1981) Zero-range Distorted Wave Born Approximation calculations for the $^{68}Zn(d,^{6}Li)^{64}Ni$ Reaction at 27.2 MeV. *Canadian Journal of Physics*, **59**, 35.

Bevelacqua, J.J. (1989) Alpha particle cluster states in ^{40}Ca, ^{114}Pd, ^{188}Os, and ^{238}U. *Fizika*, **21**, 267.

Bevelacqua, J.J. (1995) *Contemporary Health Physics: Problems and Solutions*, John Wiley & Sons, Inc., New York.

Bevelacqua, J.J. (1998) Reinterpretation of nuclear scattering experiments in terms of nucleon–nucleon interaction properties. *Physics Essays*, **11**, 337.

Bevelacqua, J.J. (1999) *Basic Health Physics: Problems and Solutions*, John Wiley & Sons, Inc., New York.

Bevelacqua, J.J. and Prewett, S.V. (1979) Form factor effects in the $^{18}O(p,t)^{16}O$ reaction. *Canadian Journal of Physics*, **57**, 1484.

Bevelacqua, J.J., Stanley, D. and Robson, D. (1977) Sensitivity of the reaction $^{40}Ca(^{13}C,^{14}N)^{39}K$ to the $^{39}K+p$ form factor. *Physical Review*, **C15**, 447.

Croswell, K. (25 May,1996) The Milky Way. *New Scientist*, **150**, S1.

Dyson, F.J. (1960) Search for artificial stellar sources of infrared radiation. *Science*, **131**, 1667.

Dyson, F.J. (2001) *Disturbing the Universe*, Basic Books, New York.

Ellis, J. and Schramm, D.N. CERN-TH.6805/93, Could a Nearby Supernova Explosion Have Caused a Mass Extinction? European Organization for Nuclear Research, Geneva, Switzerland (1993)arXiv:hep-ph/9303206 v1 2 March 1993, accessed on October 23, 2006.

Forward, R.L. (1986) Feasibility of interstellar travel: a review. *Journal of the British Interplanetary Society*, **39**, 379.

Gaidos, G., Smith, R.A., Smith, G.A., Dunmore, B. and Chakrabarti, S. Antiproton-Catalyzed Microfission/Fusion Propulsion Systems for Exploration of the Outer Solar System and Beyond,http://www.engr.psu.edu/antimatter/Papers/ICAN.pdf (accessed on November 9, 2006).

Hart, C.B., Held, R., Hoiland, P.K., Jenks, S., Loup, F., Martins, D., Nyman, J., Pertierra, J. P., Santos, P.A., Shore, M.A., Sims, R., Stabno, M. and Teage, T.O.M. On the Problems of Hazardous Matter and Radiation at Faster than Light Speeds in the Warp Drive Space–Time, arXiv:gr-qc/0207109 v1 (27 July, 2002) (accessed on November 9, 2006).

Hartle, J.B. (2003) *Gravity: An Introduction to Einstein's General Relativity*, Addison-Wesley, New York.

Herbst, W. and Assovsa, G.C. (August 1979) Supernovas and star formation. *Scientific American*, **241**, 138.

Hess, W.N. (1965) *Introduction to Space Science*, Gordon and Breach Science Publishers, New York.

Hurd, J.R., James, D.R., Bevelacqua, J.J. and Medsker, L.R. (1979) (^{7}Li, ^{8}Be) reaction on ^{40}Ca. *Physical Review*, **C20**, 1208.

Jastrow, R. (1990) *Red Giants and White Dwarfs*, Norton, New York.

Kaler, J.B. (1989) *Stars and their Spectra: An Introduction to the Stellar Spectral Sequence*, Cambridge University Press, New York.

Kardashev, N.S. (1964) Transmission of information by extraterrestrial civilizations. *Soviet Astronomy*, **8**, 217.

Kauffmann, G. and van den Bosch, F. (June 2002) The life cycle of galaxies. *Scientific American*, **286**, 46.

Kennedy, A.R. and Todd, P. (2003) Biological countermeasures. *Gravitational and Space Biology Bulletin*, **16**, (2), 37.

Krasnikov, S.V. (1988) Hyperfast travel in general relativity. *Physical Review*, **D57**, 4760.

Lamb, D.Q. and Reichart, D.E. (2000) Gamma-Ray bursts as a probe of the very high redshift universe. *The Astrophysical Journal*, **536**, 1.

La Parola, V., Mangano, V., Fox, D., Zhang, B., Krimm, H.A., Cusumano, G., Mineo, T., Burrows, D.N., Barthelmy, S., Campana, S., Capalbi, M., Chincarini, G., Gehrels, N., Giommi, P., Marshall, F.E., Mészáros, P.,

Moretti, A., O'Brien, P.T., Palmer, D.M., Perri, M., Romano, P. and Tagliaferri, G. (2006) GRB 051210: swift detection of a short gamma ray burst. *Astronomy & Astrophysics*, **454**, 753.

Leighton, R.B. (1959) *Principles of Modern Physics*, McGraw-Hill Book Company, New York.

Lewis, R.A., Meyer, K., Smith, G.A. and Howe, S. D. AIMStar: Antimatter Initiated Microfusion for Precursor Interstellar Missions, http://www.engr.psu.edu/antimatter/Papers/AIMStar_99.pdf. accessed on November 9, 2006.

Lochner, J.C., Rohrbach, G. and Cochrane, K. (2003) EG-2003-7-023-GSFC, *What is Your Cosmic Connection to the Elements?* National Aeronautics and Space Administration, Greenbelt, MD.

Lochner, J.C., Williamson, L. and Fitzhugh, E. (2001). EG-2000-08-003-GSFC, *The Hidden Lives of Galaxies*, National Aeronautics and Space Administration, Greenbelt, MD.

Melott, A., Lieberman, B., Laird, C., Martin, L., Medvedev, M., Thomas, B., Cannizzo, J., Gehrels, N. and Jackman, C. (2004) Did a gamma-ray burst initiate the late ordovician mass extinction? *International Journal of Astrobiology*, **3**, 55.

Murdin P. and Murdin, L. (1985) *Supernovae*, Cambridge University Press, New York.

Mewaldt, R.A., Davis, A.J., Binns, W.R., de Nolfo, G.A., George, J.S., Israel, M.H., Leske, R.A., Stone, E.C., Wiedenbeck, M.E. and von Rosenvinge, T.T. (2005) The Cosmic Ray Radiation Dose in Interplanetary Space – Present Day and Worst-Case Evaluations, 29th International Cosmic Ray Conference, Pune, India, August 3–10.

Morris, M.S. and Thorne, K.S. (1988) Wormholes and supersymmetry. *American Journal of Physics*, **56**, 395.

Müller, T. (2004) Visual appearance of a Morris–Thorne–Wormhole. *American Journal of Physics*, **72**, 1045.

O'Neill, G.K. (2000) *High Frontier: Human Colonies in Space*, Collector's Guide Publishing, Inc., Burlington, Ontario, Canada.

Patrick, H.J., Briel, U.G. and Böhringer, H. (December 1998) The evolution of galaxy clusters. *Scientific American*, **279**, 52.

Petrovich, F., Philpott, R.J., Robson, D., Bevelacqua, J.J., Golin, M. and Stanley, D. (1976) Comments on primordial superheavy elements. *Physical Review Letters*, **37**, 558.

Radiation the International Space Station: Recommendations to Reduce Risk, National Research Council, National Academy Press, Washington, DC. (2000).

Schimmerling, W. (2003) Overview of NASA's space radiation research program. *Gravitation and Space Biology Bulletin*, **16**, 5.

Smith, N., Li, W., Foley, R.J., Wheeler, J.C., Pooley, D., Chornock, R., Filippenko, A.V., Silverman, J.M., Quimby, R., Bloom, J.S. and Hansen, C. SN 2006GY: Discovery of the Most Luminous Supernova Ever Recorded, Powered by the Death of an Extremely Massive Star Like Eta Carinae, arXiv:astro-ph/0612617v3, accessed on May 24, 2007.

Tayler, R. (1989) The birth of elements. *New Scientist*, **124**, 25.

Turyshev, S.G., Nieto, M.M. and Anderson, J.D. (2005) Study of the pioneer anomaly: a problem set. *American Journal of Physics*, **73**, 1033.

Visser, M. (1995) *Lorentzian Wormholes: From Einstein to Hawking*, American Institute of Physics, Woodbury, NY.

Whitlock, L.A. and Granger, K.C. (2000) EG-1999-08-002-GSFC, *Gamma-Ray Bursts*, National Aeronautics and Space Administration, Greenbelt, MD.

Williams, C.H., Borowski, S.K., Dudzinski, L.A. and Juhasz, A.J. (1998) NASA/TM-1998-208831, *A Spherical Torus Nuclear Fusion Reactor Space Propulsion Vehicle Concept for Fast Interplanetary Travel*, National Aeronautics and Space Administration, Lewis Research Center, Cleveland, OH.

Zwiebach, B. (2004) *A First Course in String Theory*, Cambridge University Press, Cambridge, UK.

V
Answers and Solutions

Part Five of this book further defines and develops the material presented in Parts Two, Three, and Four. The answers and solutions presented in Part Five illustrate many of the practical difficulties that will be encountered in twenty-first century health physics applications. The reader is strongly encouraged to carefully examine these solutions to gain the maximum benefit from this text.

Solutions

Solutions for Chapter 2

02-01

(a) The effective dose rate at 10 m above the spill on the axis of the disk is given by the thin-disk relationship:

$$\dot{E} = \pi C_a \Gamma \ln\left(\frac{R^2 + h^2}{h^2}\right),$$

where C_a = activity per unit area = 4×10^7 MBq/$[\pi(5\,\text{m})^2]$ = 5.10×10^5 MBq/m²; Γ = dose factor for the activated coolant = 5.7×10^{-7} Sv m²/MBq h; R = radius of disk source = 5 m; h = distance above the disk on axis = 10 m.

$$\dot{E} = \pi(5.10 \times 10^5\,\text{MBq/m}^2)(5.7 \times 10^{-7}\,\text{Sv m}^2/\text{MBq h})$$
$$\times \ln\left(\frac{(5\,\text{m})^2 + (10\,\text{m})^2}{(10\,\text{m})^2}\right) = 0.204\,\text{Sv/h}.$$

(b) The effective dose rate received by an off-site individual submerged in a semi-infinite cloud of the metal aerosol is given by the relationship:

$$\dot{E} = Q\left(\frac{\chi u}{Q}\right)\frac{1}{u}(\text{DRCF}).$$

Using the values provided in the problem statement, the effective dose rate is

$$\dot{E} = (10^5\,\text{MBq/s})(5.0 \times 10^{-4}\,\text{m}^{-2})\left(\frac{1}{2\,\text{m/s}}\right)$$
$$\times (3 \times 10^{-8}\,\text{Sv m}^3/\text{MBq s})(3600\,\text{s/h}),$$

$$\dot{E} = 2.7 \times 10^{-3}\,\text{Sv/h}.$$

Health Physics in the 21st Century. Joseph John Bevelacqua
Copyright © 2008 WILEY-VCH Verlag GmbH & Co. KGaA, Weinheim
ISBN: 978-3-527-40822-1

02-02

(a) A decontamination factor (DF) is applicable for iodine, because iodine reacts with water. This interaction reduces the quantity of iodine available for release. For spent fuel pools, a DF value of 100 is appropriate.

Iodine can also be released from the fuel to the primary coolant. During a steam generator tube rupture event, the primary coolant flows into the secondary system. A release to the environment can occur either through an open secondary system relief valve or through the condenser air ejector. A DF of 100 is appropriate for a relief valve release and a DF of 10 000 is applicable for releases through the condenser air ejector. The increased DF for the condenser air ejector is attributable to the longer residence time of the secondary fluid through the main steam piping, high-pressure turbine, low-pressure turbines, condenser, and air ejector. This residence time enables iodine to plate out on secondary system piping and components.

A DF is not applicable to xenon. As xenon is an inert gas, its concentration is unaltered by the spent fuel pool water or transit through piping. For this reason, the DF for a noble gas is unity.

(b) The stability class applicable for the meteorology conditions at the APWR depends on the temperature gradient:

$$\frac{\Delta T}{50\,\text{m}} = \frac{T(60\,\text{m}) - T(10\,\text{m})}{60\,\text{m} - 10\,\text{m}} = \frac{21\,°\text{C} - 21.5\,°\text{C}}{50\,\text{m}} = \frac{-0.5\,°\text{C}}{50\,\text{m}}.$$

Using the table of meteorological data, this temperature gradient corresponds to Class D stability.

(c) A break in the cladding represents a mechanical defect or fault that would only release gap activity – activity residing between the fuel pellets and between the pellets and clad. The ^{131}I activity released is 2.8×10^7 MBq. The total fuel pin activity is only released if the fuel melts. Since only mechanical damage has occurred, gap activity is the appropriate source term.

(d) Since the activity was released through the stack with a height (h) of 65 m over a 2-h period, the maximum downwind ^{131}I concentration (χ) at the plume centerline at the site boundary is calculated using the Pasquill–Gifford equation:

$$\chi = \frac{Q}{\pi \sigma_y \sigma_z u} \exp\left[-\frac{1}{2}\left(\frac{y^2}{\sigma_y^2} + \frac{h^2}{\sigma_z^2}\right)\right],$$

where Q is the iodine release rate, σ_y is the horizontal standard deviation at the site boundary (100 m), σ_z is the vertical standard deviation at the site boundary (40 m), u is the mean wind speed (2.5 m/s), y is the cross-wind distance, which is zero at the plume centerline, and h is the release (stack) height (65 m). All quantities are available to calculate the ground-level iodine concentration except the release rate and it is determined from the relationship:

$$Q = \frac{\frac{A_{gap}}{DF}}{t},$$

where A_{gap} is the ^{131}I gap activity (2.8×10^7 MBq), DF is the iodine decontamination factor through water (100), and t is the release duration (2 h). Using these values, the release rate is

$$Q = \frac{(2.8 \times 10^7 \text{ MBq}/100)}{(2 \text{ h})(3600 \text{ s/h})} = 38.9 \text{ MBq/s}.$$

With these values, the maximum ground-level concentration of ^{131}I at the site boundary is

$$\chi = \frac{(38.9 \text{ MBq/s})}{\pi(100 \text{ m})(40 \text{ m})(2.5 \text{ m/s})} \exp\left[-\frac{1}{2}\left(\frac{0}{(100 \text{ m})^2} + \frac{(65 \text{ m})^2}{(40 \text{ m})^2}\right)\right],$$

$$\chi = (1.24 \times 10^{-3} \text{ MBq/m}^3)(0.267) = 3.31 \times 10^{-4} \text{ MBq/m}^3,$$

02-03

(a) The unshielded effective dose rate ($\dot{H}_0$) at the worker's location is determined by treating the valve as a point isotropic source. These assumptions are reasonable given the size of the valve (5 cm), the distance of the worker from the valve (200 cm), and the nature of accumulation of activity on internal valve surfaces.

$$\dot{H}_0 = \frac{S}{4\pi r^2}\left(\frac{\mu_{en}}{\rho}\right) EY,$$

where S is the source strength (3.7×10^5 MBq), μ_{en}/ρ is the energy absorption coefficient for muscle tissue (0.0258 cm^2/g), E is the gamma energy of radionuclide A (2.0 MeV), and Y is its yield (1.0). Using these values, the effective dose rate is determined

$$\dot{H}_0 = \frac{(3.7 \times 10^5 \text{ MBq})(1.0 \times 10^6 \text{ Bq/MBq})(1 \text{ dis/Bq s})}{4\pi(200 \text{ cm})^2}$$
$$\times (0.0258 \text{ cm}^2/\text{g})(1000 \text{ g/kg})(2.0 \text{ MeV/dis})$$
$$\times (1.6 \times 10^{-13} \text{ J/MeV})(1 \text{ Sv kg/J})(3600 \text{ s/h}) = 0.0219 \text{ Sv/h}.$$

(b) The shielded effective dose is related to the unshielded effective dose by the relationship:

$$E = \dot{E}_0 t B \exp(-\mu z),$$

where B is the buildup factor, t is the task duration (4 h), μ/ρ is the attenuation coefficient for lead (0.0461 cm^2/g), and z is the lead thickness. With the effective dose goal for this task being 0.001 Sv, the required transmission factor T is

$$T = \frac{E}{E_0 t} = B\exp(-\mu z) = \frac{0.001 \text{ Sv}}{(0.0219 \text{ Sv/h})(4 \text{ h})} = 0.0114.$$

The desired μz value is obtained by interpolating available transmission factor values. These values are presented in the following table:

μz	$B(\mu z)$	$\exp(-\mu z)$	$T = B\exp(-\mu z)$
1	1.40	0.368	0.515
2	1.76	0.135	0.238
3	2.14	0.0498	0.107
4	2.52	0.0183	0.0461
5	2.91	0.00 674	0.0196
Desired	—	—	0.0114
6	3.32	0.00 248	0.00 823

From the table, the desired shielding thickness corresponds to a μz value between 5 and 6. The 0.001 Sv value is achieved by selecting $\mu z = 6$ for a conservative basis of the shield thickness. Interpolation would provide a more accurate shield thickness result. The shield thickness (z) is obtained from the relationship

$$z = \frac{6}{\mu} = \frac{6}{\frac{\mu}{\rho}\rho} = \frac{6}{(0.0461 \text{ cm}^2/\text{g})(11.35 \text{ g/cm}^3)} = 11.5 \text{ cm}.$$

02-04

(a) The production mechanisms for the five isotopes of stainless steel are as follows:

Isotope	Production mechanism	Neutron activation type
^{60}Co	^{59}Co(n,γ)^{60}Co	Thermal
^{58}Co	^{58}Ni(n,p)^{58}Co	Fast
^{54}Mn	^{54}Fe(n,p)^{54}Mn	Fast
^{56}Mn	^{55}Mn(n,γ)^{56}Mn	Thermal
^{59}Fe	^{58}Fe(n,γ)^{59}Fe	Thermal

(b) Neglecting target and product burnup, the ^{60}Co activity (A) at 30 days after shutdown is determined from the activation relationship:

$$A = N\sigma\phi[1-\exp(-\lambda t_{\text{irradiation}})]\exp(-\lambda t_{\text{decay}}),$$

where N is the number of ^{59}Co atoms activated, σ is the ^{59}Co(n,γ)^{60}Co activation cross section (37b), ϕ is the thermal neutron flux (2×10^{13} n/cm^2 s), λ is the ^{60}Co disintegration constant, $t_{\text{irradiation}}$ is the irradiation time (24 months), and t_{decay} is the decay time (30 days). The number of ^{59}Co atoms in the detector is

$$N = \frac{mfA_v}{M},$$

where m is the detector mass (10 g), f is the fraction of ^{59}Co in stainless steel (1.4×10^{-4}), A_v is Avogadro's number (6.02×10^{23} atoms), and M is the gram atomic weight of ^{59}Co (59 g). Using these values, the number of ^{59}Co target atoms can be calculated:

$$N = \frac{(10\,\text{g})(1.4 \times 10^{-4})(6.02 \times 10^{23}\,\text{atoms})}{(59\,\text{g})} = 1.43 \times 10^{19}\,\text{atoms}.$$

The last parameter needed to define the activity is the ^{60}Co disintegration constant, which is inversely proportional to the physical half-life ($T_{1/2}$) of 1923 days:

$$\lambda = \frac{\ln 2}{T_{1/2}} = \frac{0.693}{1923\,\text{days}} = 3.60 \times 10^{-4}/\text{day}.$$

With these values, the ^{60}Co activity is determined:

$$A = (1.43 \times 10^{19}\,\text{atoms})(37\,b/\text{atom})(10^{-24}\,\text{cm}^2/b)(2 \times 10^{13}\,\text{n/cm}^2\,\text{s})$$
$$\times (1\,\text{dis}/n)(1-\exp[(3.6 \times 10^{-4}/\text{days})(24\,\text{months})$$
$$\times (1\,\text{year}/12\,\text{months})(365\,\text{days/year})]\exp[(3.6 \times 10^{-4}/\text{days})$$
$$\times (30\,\text{days})],$$

$$A = (1.06 \times 10^{10}\,\text{dis/s})(1\,\text{Bq s/dis})(1\,\text{MBq}/10^6\,\text{Bq})(1-0.769)(0.989)$$
$$= 2.42 \times 10^3\,\text{MBq}.$$

(c) The effective dose rate ($\dot{E}$) from ^{60}Co activation at 30 cm (r) from one of the failed low power range detectors at 30 days after shutdown is obtained from the activity ($A = 2.42 \times 10^3$ MBq) determined in part (b). The size of the detector and the location of interest justify the use of a point source approximation:

$$\dot{E} = \frac{A\Gamma}{r^2}$$

where Γ is the ^{60}Co dose factor (3.56×10^{-7} Sv m^2/MBq h). Using these values, the desired effective dose rate can be calculated:

$$\dot{E} = \frac{(2.42 \times 10^3\,\text{MBq})(3.56 \times 10^{-7}\,\text{Sv m}^2/\text{MBq h})}{(0.3\,\text{m})^2} = 9.57 \times 10^{-3}\,\text{Sv/h}.$$

02-05

The most likely production modes in an LMFBR for the requested isotopes are

(a) $^{238}U + n \rightarrow {}^{239}U \xrightarrow{\beta} {}^{239}Np \xrightarrow{\beta} {}^{239}Pu$,
(b) $^{239}Pu + n \rightarrow {}^{240}Pu + n \rightarrow {}^{241}Pu$,
(c) $^{241}Pu \xrightarrow{\beta} {}^{241}Am$.

02-06

(a) In this question, you are requested to list considerations when estimating the ^{131}I airborne concentration in containment 24 h after reactor shutdown. The ^{131}I concentration in containment (C) is the activity (A) per unit volume (V) of containment atmosphere:

$$C = \frac{A}{V}.$$

The rate of change of activity in containment ($\dot{A}$) is given by the relationship:

$$\dot{A} = P\exp(-kt),$$

where P is the rate of ^{131}I input into containment and k is the total removal rate of the ^{131}I from containment. P is derived from the RCS leakage to the containment atmosphere and is written as

$$P = \xi L C'.$$

In this equation, L is the primary coolant system leak rate to the containment atmosphere, ξ is the partition fraction that quantifies the fraction of ^{131}I that evolves from the primary coolant and enters the containment atmosphere, and C' is the ^{131}I concentration in the primary coolant system. C' varies with time because the primary coolant cleanup system is operating:

$$C'(t) = C'(0)\exp(-Kt),$$

where t is the time after shutdown, $C'(0)$ is the shutdown primary coolant ^{131}I concentration, and K is the primary coolant system ^{131}I removal rate:

$$K = \lambda + e'\frac{f}{v}.$$

In defining the primary coolant system removal rate, λ is the ^{131}I physical disintegration constant, e' is the ^{131}I removal efficiency from the primary coolant by the cleanup system, f is the primary cleanup system flow rate, and v is the volume of the primary coolant system.

Since there is also a containment ^{131}I removal system, the total removal rate of ^{131}I is defined as

$$k = K + \frac{F'}{V} + e\frac{F''}{V},$$

where F' is the containment ventilation flow rate, V is the containment volume, e is the containment atmosphere charcoal filter efficiency, and F'' is the containment atmosphere charcoal filter cleanup flow rate.

Given these quantities, the containment air activity rate equation is integrated to obtain the activity concentration within containment:

$$\dot{A} = P\exp(-kt),$$

$$A(T) = \int_0^T P\exp(-kt)dt,$$

where T is the period of interest (24 h).

With his information, the containment air ^{131}I concentration is determined:

$$C(T) = \frac{A(T)}{V} = \frac{\xi LC'(0)}{V}\int_0^T \exp\left[-\left(\lambda + \frac{e'f}{v} + \frac{F'}{V} + \frac{eF''}{V}\right)t\right]dt,$$

$$C(T) = \frac{\xi LC(0)}{V\left(\lambda + \frac{e'f}{v} + \frac{F'}{V} + e\frac{F''}{V}\right)}\left(1-\exp\left[-\left(\lambda + \frac{e'f}{v} + \frac{F'}{V} + e\frac{F''}{V}\right)T\right]\right).$$

Using this relationship, the following are considered when estimating the ^{131}I concentration 24 h after the shutdown:
1. Primary coolant volume.
2. The shutdown primary coolant ^{131}I concentration.
3. Primary coolant leak rate to the containment atmosphere.
4. Primary coolant to containment atmosphere partition fraction.
5. Primary coolant system cleanup rate.
6. Primary coolant system ^{131}I cleanup efficiency.
7. Containment ^{131}I concentration.
8. Containment free air volume.
9. Containment atmosphere ventilation rate.
10. Containment atmosphere charcoal filter cleanup flow rate.
11. Containment atmosphere charcoal filter efficiency.
12. Physical half-life of ^{131}I.
13. Time after reactor shutdown.

(b) In this question, you are requested to determine the committed dose equivalent (CDE) to the worker's thyroid from a 10-h exposure to an ^{131}I concentration of 2.96×10^{-4} MBq/m^3. The worker did not use respiratory protection. Also, you are to determine the worker's committed effective dose equivalent (CEDE). Assumptions:
1. There is no plate-out of ^{131}I within the containment.
2. The ^{131}I concentration is constant during the 10-h period.
3. The worker's breathing rate is constant.
4. No ventilation is operating to remove the ^{131}I.
5. Radioactive decay is ignored.
6. No other removal mechanisms are present.

The CDE and the CEDE are readily determined from the ^{131}I DAC value (7.4×10^{-4} MBq/m^3). For the stated assumptions, the CDE to the thyroid is given by the relationship:

$$\text{CDE} = \frac{kCt}{\text{DAC}},$$

where
C = concentration of ^{131}I in the containment = 2.96×10^{-4} MBq/m^3,
DAC = ^{131}I derived air concentration (for iodine, the DAC is based on nonstochastic effects) = 7.4×10^{-4} MBq/m^3,
t = exposure time = 10 h,
k = conversion factor = 0.50 Sv/2000 DAC h.

$$\text{CDE} = \left(\frac{0.5 \text{ Sv}}{2000 \text{ DAC h}}\right)\left(\frac{(2.96 \times 10^{-4} \text{ MBq/m}^3)(10 \text{ h})}{7.4 \times 10^{-4} \text{ MBq/m}^3 \text{ DAC}}\right) = 0.001 \text{ Sv} = H_{50,T}.$$

The CEDE is readily obtained from the CDE ($H_{50,T}$) and the ICRP 26 thyroid-weighting factor w_T (0.03):

$$\text{CEDE} = w_T H_{50,T} = (0.03)(0.001 \text{ Sv})(1000 \text{ mSv/Sv}) = 0.03 \text{ mSv}.$$

(c) The ALARA evaluation considers the total dose that includes the internal dose from airborne radioactive material and the external dose from activation products. Factors that should be considered in the prejob analysis for a containment entry after reactor shutdown to keep the worker's total effective dose equivalent ALARA include the following:
1. The time after reactor shutdown. By delaying the entry, the short-lived isotopes have time to decay.
2. The ventilation rates for the containment ventilation and charcoal filter systems. These flow rates should be maximized to reduce the ^{131}I air concentration.
3. The containment airborne radionuclide concentrations.
4. The charcoal filter efficiency.

5. Operational status of the charcoal filters.
6. Operational status of other ventilation systems (e.g., cleanup fans, pool fans, reactor cavity fans, and reactor vessel head fans).
7. Use of respiratory protection.
8. The estimated entry duration.
9. The letdown flow rate (primary coolant system cleanup rate). This flow rate should be maximized to reduce the primary coolant ^{131}I concentration.
10. Primary coolant cleanup efficiency.
11. The containment locations to be entered and their dose rates.
12. The location of low dose areas.
13. Feasibility of installing temporary shielding.
14. Contamination levels in areas of interest.
15. Protective clothing to be utilized.
16. Status of the fuel fission product barrier.
17. Status of the primary coolant system fission product barrier.
18. Task duration in the various containment areas.
19. The doses of workers qualified to perform the task. Dose leveling should be utilized to equalize qualified workers' doses.

(d) In this question, you are requested to state methods for reducing the primary coolant ^{58}Co cleanup time. To answer this question, consider the activity of a given isotope (e.g., ^{58}Co) that builds up on a demineralizer as a function of time (t):

$$A(t) = \frac{CFe}{\lambda}(1-\exp(-\lambda t_{\text{online}}))\exp(-\lambda t_{\text{decay}}),$$

where C is the radionuclide concentration entering the demineralizer, F is the flow rate of radioactive fluid entering the demineralizer, e is the demineralizer removal efficiency for a given isotope, λ is the radioactive decay constant, t_{online} is the time the demineralizer is in service, and t_{decay} is the time following isolation of the demineralizer from the influent flow. The cleanup time is reduced as follows:

1. Increasing the demineralizer flow rate (F).
2. Improving the removal efficiency (e) of the demineralizer for ^{58}Co.
3. Ensuring the plant chemistry meets the hydrogen peroxide addition requirements to maximize ^{58}Co solubilization.
4. Increasing the number of demineralizers online.
5. Performing a feed and bleed on the primary coolant system. This entails adding clean water to the primary coolant system while draining water containing dissolved or suspended activity. The bleed volume is stored in available tank space and processed at a later time.
6. Ensuring the chemical addition occurs within the prescribed temperature and pH ranges to maximize the solubility of ^{58}Co.
7. The demineralizer resin is fresh and has the maximum number of ion-exchange sites.

(e) Benefits of adding hydrogen peroxide to the primary coolant system at the onset of refueling include the following:
   1. Increasing the solubility of fission and activation products early in the outage facilitates their removal. Following removal, the primary coolant system source term is reduced and subsequent outage doses are lowered.
   2. Increasing the solubility of radioactive material earlier provides sufficient time for their removal by demineralization or feed and bleed operations.
   3. Primary coolant system component dose rates (e.g., pumps, valves, steam generator channel heads) are lowered with a subsequent reduction in personnel doses during surveillance and maintenance operations.
   4. If the chemistry is properly balanced, the solubilized material is preferentially deposited on the fuel and not on primary coolant system surfaces during plant operations. This reduces the source term for subsequent outages.
   5. Early addition effectively utilizes the time period for meeting the conditions for transition to the decay heat removal system (e.g., primary coolant system pressure and heat removal capability of the decay heat system). During this period, only limited primary system maintenance is practical.

(f) The total ^{58}Co activity contained within a pipe is to be determined given an effective dose rate of 2.5×10^{-3} mSv/h at a distance of 2 m from the midpoint of a 2.0-m long pipe containing a uniform concentration of ^{58}Co.

This source is credibly approximated by the line source approximation relationship:

$$\dot{E} = \frac{C_L \Gamma \theta}{w},$$

where
  $\dot{E}$ = effective dose rate at a distance of 2.0 m from the center of the pipe
     $= 2.5 \times 10^{-3}$ mSv/h,
  C_L = concentration per unit length of the pipe
     $= A/L$,
  A = total activity contained within the pipe,
  L = pipe length = 2.0 m,
  θ = included angle between the point of interest and the ends of the pipe,
     $\tan \theta/2 = 1$ m/2 m $= 0.5$,
     $\theta/2 = \tan^{-1}(0.5) = 26.57° \times \pi$ rad/$180°$,
     $\theta = 0.927$ rad,
  $\Gamma = {}^{58}$Co effective dose factor (gamma constant) $= 1.66 \times 10^{-7}$ Sv m^2/MBq h,
  w = perpendicular distance from the pipe $= 2.0$ m.

The effective dose rate equation is solved for the desired activity (A):

$$\dot{E} = \frac{C_L \Gamma \theta}{w} = \frac{A \Gamma \theta}{L w},$$

$$A = \frac{\dot{E} w L}{\theta \Gamma} = \frac{(2.5 \times 10^{-3} \text{ mSv/h})\left(\frac{\text{Sv}}{1000 \text{ mSv}}\right)(2 \text{ m})(2 \text{ m})}{(0.927)\left(1.66 \times 10^{-7} \frac{\text{Sv m}^2}{\text{MBq h}}\right)} = 65.0 \text{ MBq}.$$

In determining this activity, the following assumptions were utilized:

1. The pipe fluid provides insignificant attenuation of the ^{58}Co photons.
2. The source atoms radiate isotropically.
3. The pipe is reasonably approximated by a line source.
4. Significant gamma-ray buildup does not occur within the fluid or pipe wall.
5. Attenuation by the pipe wall is ignored.

02-07

(a) The dose equivalent that a worker receives from a room air concentration (C) of 185 MBq/m³ as measured by workplace air monitoring is to be determined. Assuming the tritium intake occurs in an HTO form, the committed effective dose equivalent is calculated from the exposure time (t) and the given concentration:

$$H = \frac{C \times t \times 50\,\text{mSv}}{2000\,\text{DAC}\,h \times C'},$$

where
H = committed effective dose equivalent from the HTO intake,
C = HTO concentration = 185 MBq/m³,
t = exposure duration = 1 min,
C' = DAC (HTO) = 0.74 MBq/m³ DAC.

$$H = \frac{(185\,\text{MBq/m}^3)(1\,\text{min})(1\,\text{h}/60\,\text{min})}{(2000\,\text{DAC}\,h)(0.74\,\text{MBq/m}^3\,\text{DAC})} 50\,\text{mSv},$$
$$= 0.104\,\text{mSv}.$$

(b) The dose received in mSv is obtained from the acute intake DCF. This assumes that the urine tritium concentration is because of the acute exposure and not from previous chronic occupational exposure.

$$H = (1850\,\text{Bq/l})(7.57 \times 10^{-4}\,\mu\text{Sv}\,\text{l/Bq}) = 1.4\,\text{mSv}.$$

(c) The dose equivalent calculated from the urine concentration differs from the dose equivalent calculated from the room air concentration. Assuming that the measurements and calculations were done correctly, likely sources of this discrepancy include the following:
1. The workplace air monitor location was not representative of the concentration the worker experienced.
2. The urine concentration from the event is masked by previous chronic tritium occupational exposure that was experienced by the worker before the event.
3. The worker's tritium metabolism is not equivalent to the model assumed in formulating the dose conversion factor.
4. The tritium gas did not fully oxidize and includes both HT and HTO components.
5. The worker's intake pathways (inhalation, ingestion, and skin absorption) are not well represented by the dosimetric model used to calculate the tritium dose.

6. The room residence time estimate is in error.
7. The instrumentation used to estimate the volume of air passing through the ion chamber is inaccurate.
8. Noble gas entered the ion chamber and interfered with the tritium estimate.
9. Gamma-ray fields (larger than about 100 μGy/h) are interfering with the instrumentation.

(d) In this question, you have to identify two techniques that can be used for tritium air monitoring and to specify one advantage and one disadvantage of each technique.

1. Ion chamber tritium-in-air monitors. The measurement of tritium in air presents special problems because the average energy of the beta particles is so low (about 6 keV) that it is difficult to design a detector whose walls can be penetrated by the beta particles. Accordingly, the tritium-contaminated air is pumped through the detector so that all the beta particles energy produces ion pairs inside the detector. External radiation also creates ion pairs in the detector. Accordingly, a second, sealed detector is used to compensate for external radiation.

   The detector has the advantage of being convenient and the tritium concentration is determined in real time, particularly in low gamma background environments.

   The instrument has a number of limitations. For example, any radioactive gas present in the air is measured as tritium and leads to a higher than actual reading. In addition, the gamma compensation is adequate only in relatively low gamma fields of about 100 μGy/h or less.

2. Tritium bubbler. The tritium bubbler is simple, accurate, and not affected by any gamma-ray background or the presence of noble gases. This technique consists of bubbling tritiated air through clean water that traps the tritiated water vapor. The tritium content of the water is then analyzed using liquid scintillation counting. The bubbler consists of a pump, a timer, a flow gauge, and a removable water jar containing about 100 ml of clean water. The bubbler fluid is then counted using liquid scintillation techniques to obtain the tritium air concentration.

The tritium bubbler yields more accurate results than the ion chamber, but it is not as convenient. Although the technique is accurate, time must be allowed for sample preparation, counting, and processing. Therefore, the technique does not provide real time tritium air concentration information.

02-08

(a) The following documents are needed to perform this evaluation:
   1. Current and planned system design descriptions for the demineralizer modification that includes flow rates, materials of construction, performance characteristics, and connections to other systems.

2. Current and planned piping and instrumentation drawings for the demineralizer system.
3. Demineralizer procedures for normal operations, sluicing, and resin addition.
4. Government radiation regulations (e.g., Title 10, Code of Federal Regulations, Part 20, Standards for Protection Against Radiation).
5. Final Safety Analysis Report for the BWR.
6. Technical Specifications for the BWR.
7. Radiation survey records for the existing demineralizer and adjacent areas.
8. Previous ALARA evaluations for demineralizer systems and components.
9. Radiation Work Permit results for previous demineralizer system activities (e.g., surveillance, maintenance, and testing activities).
10. Work packages for the proposed activity.
11. Vendor estimates of the new demineralizer's radiological characteristics including radionuclide retention efficiency, resin lifetime, and dose rates as a function of radionuclide loading.
12. Safety analyses for the new demineralizer system.
13. Design reviews for the new demineralizer system.

(b) The following items should be considered when evaluating the demineralizer from an ALARA perspective:
1. Estimated contamination levels both inside and outside the demineralizer systems. These levels govern the use of protective clothing and respiratory protection that impact the time required to perform the individual tasks supporting the demineralizer modification.
2. Estimated times for the various work package activities.
3. Estimated number of personnel by work group to perform the various work package activities. The number of personnel is used to establish the collective dose for the task and to ensure that the incurred dose is distributed in an ALARA manner.
4. Estimated dose rates (by radiation type) for the various work-related activities. The dose rates when multiplied by time (item 2) give the dose for completion of the various tasks supporting the demineralizer job.
5. Location of proposed shielding installations. These locations permit an assessment of the anticipated dose to install the shielding versus the dose savings once the shielding is installed.
6. Accessibility to the demineralizer cubicle and associated systems. The installation dose as well as the expected dose during the various routine, abnormal, and emergency operations should be evaluated.
7. Post installation dose rates. Estimated dose rates for new demineralizer operations should be determined to assess the long-term dose impacts of the new system.
8. System valve types. The type of valves has a significant impact on operational exposures. The use of motor operators, air-operated valves, valve extension handles/reach rods should be considered as a means of reducing operator doses.

9. Monitors. Installed radiation and contamination monitors minimize the need for routine health physics surveys in elevated radiation areas.
10. Remote viewing devices. The use of remote viewing devices (e.g., closed-circuit TV, periscopes, and viewing windows) minimizes the need for operator entries into radiation areas for routine rounds or surveillance activities.

(c) The total ^{60}Co activity in MBq present in the demineralizer at the end of its run time and at the end of its down time is to be determined. The activity buildup on the demineralizer at the end of its run is given by the production relationship:

$$A(t) = \frac{CFe}{\lambda}(1-e^{-\lambda t}),$$

where

$A(t)$ = demineralizer activity as a function of time,
C = influent concentration of ^{60}Co = 70.3 MBq/m^3,
F = demineralizer flow rate = 1000 l/min,
e = demineralizer ^{60}Co removal efficiency = 0.95,
λ = ^{60}Co disintegration constant = $\ln(2)/T_{1/2}$,
$T_{1/2}$ = ^{60}Co half-life = 5.27 years,
λ = 0.693/5.27 years = 0.131/years,
t = demineralizer run time = 100 days.

Using these values, the total ^{60}Co activity present in the demineralizer at the end of its run time is

$$A(t) = \frac{(70.3\,\text{MBq/m}^3)(1000\,\text{l/min})(0.95)(1\,\text{m}^3/1000\,\text{l})}{(0.131/\text{years})(1\,\text{year}/365\,\text{days})(1\,\text{day}/24\,\text{hours})(1\,\text{hour}/60\,\text{min})}$$
$$\times (1-e^{-(0.131/\text{years})(100\,\text{days})(1\,\text{year}/365\,\text{days})})$$
$$= (2.68 \times 10^8\,\text{MBq})(1-0.965) = 9.38 \times 10^6\,\text{MBq}.$$

The total activity in the demineralizer at the end of its down time is given by the relationship:

$$A(t) = \frac{CFe}{\lambda}(1-e^{-\lambda t})e^{-\lambda T} = A(\text{previous problem})e^{-\lambda T},$$

where T = decay time = 60 days.

$$A(t) = (9.38 \times 10^6\,\text{MBq})e^{-(0.131/\text{years})(60\,\text{days})(1\,\text{years}/365\,\text{days})}$$
$$= 9.38 \times 10^6\,\text{MBq} \times 0.979 = 9.18 \times 10^6\,\text{MBq}.$$

(d) The decontamination factor (DF) is defined in terms of the efficiency (ε = 0.95):

$$\text{DF} = \frac{1}{1-\varepsilon} = \frac{1}{1-0.95} = 20.$$

(e) The calculation of the effective dose rate 20 m from the demineralizer at the end of its down time is desired. Any shielding from the demineralizer water, resin, and demineralizer shell is to be ignored. Since the distance from the demineralizer vessel is greater than three times the largest source dimension, a point source approximation is justified. The effective dose rate is

$$\dot{E}(r) = A\Gamma/r^2,$$

where

$\dot{E}(r)$ = effective dose rate from the demineralizer at the end of its down time,
r = distance from the demineralizer at the point of interest = 20 m,
A = demineralizer activity = 9.18×10^6 MBq,
Γ = gamma constant for ^{60}Co = 3.57×10^{-7} Sv m^2/MBq h,
$\dot{E}(r) = (9.18 \times 10^6 \text{ MBq})(3.57 \times 10^{-7} \text{ Sv m}^2/\text{MBq h})/(20 \text{ m})^2$
   $= (0.00819 \text{ Sv/h})(1000 \text{ mSv/Sv}) = 8.19 \text{ mSv/h}.$

(f) The following methods minimize exposure to plant personnel during maintenance of the demineralizer:
  1. Use shielding in high dose rate areas when justified by the ALARA evaluation.
  2. Limit the demineralizer activity to keep dose rates low.
  3. Train workers and practice the various demineralizer tasks using mock-ups.
  4. Use low-cobalt alloys to minimize ^{60}Co deposition on demineralizer piping and in the demineralizer.
  5. Utilize low dose rate waiting areas during maintenance tasks.
  6. Use glove bags for maintenance tasks to minimize internal depositions.
  7. Decontaminate system components (pumps and valves) prior to maintenance.
  8. Flush system components to minimize their contamination levels.
  9. Use remote tools or robotics wherever practical.
  10. Maintain primary system chemistry to ensure activity plates out on core surfaces instead of primary system and interfacing piping.
  11. Plan demineralizer maintenance to maximize the decay of short-lived radionuclides, that is, at the end of the down time period.
  12. Planned maintenance should be performed after sluicing old resin from the demineralizer. After sluicing, the demineralizer should be flushed with high-pressure water and decontaminated to reduce the radiation levels.
  13. When practical, remove pumps and valves to low-dose areas for performance of the required maintenance.

02-09

(a) The iodine (I) and noble gas (NG) release rates through the main steam line relief valve are the steam generator (SG) release rates divided by the appropriate decontamination factors (DF) applicable to the relief valve (RV). The DF represents the reduction in the concentration of radioactive material as a result of the scrubbing of this material by the steam system piping and two-phase steam-water flow:

$$Q^{RV}_{NG} = Q^{SG}_{NG}/DF_{NG} = \frac{3 \text{ MBq/s}}{1} = 3 \text{ MBq/s},$$

$$Q_I^{RV} = Q_I^{SG}/DF_I = \frac{2\,\text{MBq/s}}{100} = 0.02\,\text{MBq/s}.$$

The DF values derived from Generation II operating experience have been assumed to be applicable to the Generation IV reactor considered in this problem.

(b) The release rates of these radioactive materials through the condenser (C) are

$$Q_{NG}^C = Q_{NG}^{SG}/DF_{NG} = \frac{3\,\text{MBq/s}}{1} = 3\,\text{MBq/s},$$

$$Q_I^C = Q_I^{SG}/DF_I = \frac{2\,\text{MBq/s}}{10\,000} = 0.0002\,\text{MBq/s}.$$

(c) From an ALARA perspective, release through the condenser is preferable because it results in a reduced iodine source term. The noble gas source term is not reduced by either pathway. The reduction in the iodine source term is attributable to the increased scrubbing action of the piping and fluid between the relief valve and the condenser/air ejector.

02-10

The air ejector (AE) noble gas activity is obtained from an activity balance that assumes that no noble gas activity is lost during transit to the air ejector:

$$C_{NG}^P L_{P \to S} = C_{NG}^{AE} F_{AE},$$

where C_{NG}^P is the ^{85}Kr noble gas concentration in the primary system (1000 MBq/cm^3), $L_{P \to S}$ is the primary-to-secondary leak rate (500 l/day), C_{NG}^{AE} is the ^{85}Kr noble gas concentration in the air ejector effluent, and F_{AE} is the air ejector flow rate (5000 l/min). Using these values, the air ejector ^{85}Kr noble gas activity is

$$C_{NG}^{AE} = C_{NG}^P \frac{L_{P \to S}}{F_{AE}} = (1000\,\text{MBq/cm}^3)\frac{(500\,\text{l/day})(1\,\text{h/60 min})(1\,\text{day/24 h})}{5000\,\text{l/min}}$$
$$= 0.0694\,\text{MBq/cm}^3.$$

Solutions for Chapter 3

03-01

(a) The binary reaction channels in the ^{4}He system are p + ^{3}H, n + ^{3}He, and D + D.

The threshold energies relative to the ^{4}He ground state are determined from the relationship:

$$Q(a+b) = \Delta(a) + \Delta(b) - \Delta(^4\text{He}),$$

where a and b are the members of the binary channel.

$$Q(p + {}^3H) = \Delta(p) + \Delta({}^3H) - \Delta({}^4He),$$

$$Q(p + {}^3H) = 7.289 \text{ MeV} + 14.950 \text{ MeV} - 2.425 \text{ MeV} = 19.814 \text{ MeV},$$

$$Q(n + {}^3He) = \Delta(n) + \Delta({}^3He) - \Delta({}^4He),$$

$$Q(n + {}^3He) = 8.071 \text{ MeV} + 14.931 \text{ MeV} - 2.425 \text{ MeV} = 20.577 \text{ MeV},$$

$$Q(D + D) = \Delta(D) + \Delta(D) - \Delta({}^4He),$$

$$Q(D + D) = 13.136 \text{ MeV} + 13.136 \text{ MeV} - 2.425 \text{ MeV} = 23.847 \text{ MeV}.$$

(b) For DD fusion, the following reactions are predominant: $D + D \to p + {}^3H$ and $D + D \to n + {}^3He$. The total energy available from each reaction is determined by the difference in binary channel energies (Q). For a given Q value and binary channel $a + b$, the energies of the exit channel particles a and b are

$$E_a = \frac{m_b}{m_D + m_D} Q,$$

$$E_b = \frac{m_a}{m_D + m_D} Q.$$

If the mass of a nucleon is defined as m, then the exit channel particle energy relationships simplify as noted below. The total reaction energy available and the exit channel particle energies are

$$Q(D + D \to p + {}^3H) = Q(D + D) - Q(p + {}^3H) = 23.847 \text{ MeV} - 19.814 \text{ MeV}$$
$$= 4.033 \text{ MeV},$$

$$E_p = \left(\frac{3\,m}{4\,m}\right) 4.033 \text{ MeV} = 3.02 \text{ MeV},$$

$$E_{{}^3H} = \left(\frac{1\,m}{4\,m}\right) 4.033 \text{ MeV} = 1.01 \text{ MeV},$$

$$Q(D + D \to n + {}^3He) = Q(D + D) - Q(n + {}^3He)$$
$$= 23.847 \text{ MeV} - 20.577 \text{ MeV} = 3.270 \text{ MeV},$$

$$E_n = \left(\frac{3\,m}{4\,m}\right) 3.270 \text{ MeV} = 2.45 \text{ MeV},$$

$$E_{{}^3He} = \left(\frac{1\,m}{4\,m}\right) 3.270 \text{ MeV} = 0.82 \text{ MeV}.$$

These results verify Equations 3.1 and 3.2.

03-02

An alpha particle formed from DT fusion has a velocity $\vec{v} = v_0 \hat{i}$ at a particular location in a tokamak vessel. If the magnetic field at this location is $\vec{B} = B_0 \hat{j}$, the instantaneous magnetic force $(\vec{F})$ on the alpha particle is given by the force relationship:

$$\vec{F} = q\vec{v} \times \vec{B} = q|v||B|\sin\theta = qvB = 2ev_0 B_0,$$

where θ is the angle between v and B. Since the coordinate system is cartesian, this angle is $90°$. In this result, we use the fact that the alpha particle's charge is $2e$, where e is the electron charge. The instantaneous force is in the z-direction.

The results are also written in a more explicit form using the definition of the cross product to obtain the same result:

$$\vec{F} = q\vec{v} \times \vec{B} = 2e \begin{vmatrix} \hat{i} & \hat{j} & \hat{k} \\ v_0 & 0 & 0 \\ 0 & B_0 & 0 \end{vmatrix} = 2ev_0 B_0 \hat{k}.$$

03-03

(a) ^{56}Mn is produced from the following two neutron-induced reactions:

- thermal: ^{55}Mn(n,γ)^{56}Mn;
- fast: ^{56}Fe(n,p)^{56}Mn.

The ratio of the ^{56}Mn activities (A) produced from thermal and fast capture reactions is determined from the activation relationship:

$$A = N\sigma\phi[1-\exp(-\lambda t_{\text{irradiation}})]\exp(-\lambda t_{\text{decay}}),$$

where N is the number of atoms activated, σ is the energy-dependent activation cross section, ϕ is the energy-dependent flux, λ is the ^{56}Mn disintegration constant, $t_{\text{irradiation}}$ is the irradiation time, and t_{decay} is the decay time. The ratio (ξ) of ^{56}Mn activities due to thermal and fast capture is

$$\xi = \frac{A_{\text{thermal}}}{A_{\text{fast}}} = \frac{N(^{55}\text{Mn})\sigma_{\text{thermal}}\phi_{\text{thermal}}[1-\exp(-\lambda t_{\text{irradiation}})]\exp(-\lambda t_{\text{decay}})}{N(^{56}\text{Fe})\sigma_{\text{fast}}\phi_{\text{fast}}[1-\exp(-\lambda t_{\text{irradiation}})]\exp(-\lambda t_{\text{decay}})},$$

$$\xi = \frac{N(^{55}\text{Mn})\sigma_{\text{thermal}}\phi_{\text{thermal}}}{N(^{56}\text{Fe})\sigma_{\text{fast}}\phi_{\text{fast}}}.$$

In calculating this ratio, a number of quantities must be determined including N (^{55}Mn), which is the number of ^{55}Mn atoms in the coupon. $N(^{55}$Mn) is given by the relationship:

$$N(^{55}\text{Mn}) = \frac{m(^{55}\text{Mn})A_v}{\text{GAW}(^{55}\text{Mn})},$$

and m is the ^{55}Mn coupon mass (0.1 g), A_v is Avogadro's number (6.02×10^{23} atoms/mole), and GAW is the ^{55}Mn gram atomic weight (55 g). Using these values, the number of ^{55}Mn atoms is determined:

$$N(^{55}\text{Mn}) = \frac{(0.1\,\text{g})(6.02 \times 10^{23}\,\text{atoms/mole})}{55\,\text{g/mole}} = 1.09 \times 10^{21}\,\text{atoms}.$$

In a similar manner, the number of ^{56}Fe atoms is

$$N(^{56}\text{Fe}) = \frac{m(^{56}\text{Fe})A_v}{\text{GAW}(^{56}\text{Fe})},$$

$$N(^{56}\text{Fe}) = \frac{(100\,\text{g})(6.02 \times 10^{23}\,\text{atoms/mole})}{56\,\text{g/mole}} = 1.08 \times 10^{24}\,\text{atoms}.$$

The values of the cross sections and fluence rates are obtained from the problem statement:

$$\sigma_{\text{thermal}} = 13.3\,b/\text{atom},$$

$$\sigma_{\text{fast}} = 0.001\,b/\text{atom},$$

$$\phi_{\text{thermal}} = 1 \times 10^{13}\,\text{n/cm}^2\,\text{s},$$

$$\phi_{\text{fast}} = 8 \times 10^{13}\,\text{n/cm}^2\,\text{s}.$$

With the specification of the number of target atoms and the fast and thermal cross sections and fluence rates, the activity ratio (ξ) is determined by

$$\xi = \frac{A_{\text{thermal}}}{A_{\text{fast}}} = \frac{(1.09 \times 10^{21}\,\text{atoms})(13.3\,b/\text{atom})(1 \times 10^{13}\,\text{n/cm}^2\,\text{s})}{(1.08 \times 10^{24}\,\text{atoms})(0.001\,b/\text{atom})(8 \times 10^{13}\,\text{n/cm}^2\,\text{s})} = 1.68.$$

(b) The effective dose rate ($\dot{E}$) from these sources can be written as

$$\dot{E} = \dot{E}_{\text{fast}} + \dot{E}_{\text{thermal}} = \dot{E}_{\text{fast}}(1 + \xi),$$

where $\dot{E}_{\text{fast}}$ is the effective dose rate contribution from the activation reaction produced from fast neutrons (^{56}Fe(n,p)^{56}Mn) and $\dot{E}_{\text{thermal}}$ is the effective dose rate contribution from the activation reaction produced from thermal neutrons (^{55}Mn(n,γ)^{56}Mn). The effective dose rate contributions from fast and thermal neutrons is written as

$$\dot{E}_{\text{fast}} = \frac{A_{\text{fast}}\Gamma}{r^2},$$

$$\dot{E}_{\text{thermal}} = \frac{A_{\text{thermal}}\Gamma}{r^2} = \frac{\xi A_{\text{fast}}\Gamma}{r^2} = \xi \dot{E}_{\text{fast}}.$$

where A_{fast} is the ^{56}Mn activity produced by fast neutrons, Γ is the ^{56}Mn gamma constant, and r is the distance from the source. A point source relationship is justified on the basis of the size of the sources and the distance from the source. The activity from fast neutrons is determined from the activation relationship:

$$A_{\text{fast}} = N(^{56}\text{Fe})\sigma_{\text{fast}}\phi_{\text{fast}}[1-\exp(-\lambda t_{\text{irradiation}})]\exp(-\lambda t_{\text{decay}}).$$

All parameters were previously defined in this equation except for the time terms. For specificity,

$$t_{\text{irradiation}} = 100 \, \text{days},$$

$$t_{\text{decay}} = 10 \, \text{min},$$

$$\lambda = \frac{\ln 2}{2.58 \, \text{h}} = 0.269/\text{h},$$

$$\Gamma = 2.5 \times 10^{-4} \, \text{mSv m}^2/\text{MBq h},$$

$$r = 5.0 \, \text{m}.$$

Using these values, the fast activity is determined:

$$A_{\text{fast}} = (1.08 \times 10^{24} \, \text{atoms})(0.001 b/\text{atom})(1.0 \times 10^{-24} \, \text{cm}^2/b)$$
$$\times (8 \times 10^{13} \, \text{n/cm}^2 \, \text{s}) \times [1-\exp(-(0.269/\text{h})(24 \, \text{h/day})(100 \, \text{days}))]$$
$$\times \exp(-(0.269/\text{h})(1 \, \text{h}/60 \, \text{min})(10 \, \text{min}))(1 \, \text{dis/n}),$$

$$A_{\text{fast}} = (8.64 \times 10^{10} \, \text{dis/s})(1-0)(0.956)(1 \, \text{MBq}/1.0 \times 10^6 \, \text{Bq})(\text{Bq s/dis})$$
$$= 8.26 \times 10^4 \, \text{MBq}.$$

With these values, the effective dose from the fast neutron reaction is determined:

$$\dot{E}_{\text{fast}} = \frac{(8.26 \times 10^4 \, \text{MBq})(2.5 \times 10^{-4} \, \text{mSv m}^2/\text{MBq h})}{(5.0 \, \text{m})^2} = 0.826 \, \text{mSv/h}.$$

Given the fast neutron effective dose rate, the total effective dose rate is determined from the relationship:

$$\dot{E} = \dot{E}_{\text{fast}}(1 + \xi) = (0.826 \text{ mSv/h})(1 + 1.68) = 2.21 \text{ mSv/h}.$$

03-04

(a) The maximum effective dose (E) from the tritium intake that occurred during the worker's investigation in the torus containment is determined from the relationship:

$$E = I(\text{DCF}),$$

where I is the intake and DCF is the tritium dose conversion factor (4.27×10^{-2} mSv/MBq), which includes absorption through the skin. The maximum intake is determined from the relationship:

$$I = C(\text{BR})(t),$$

where C is the tritium concentration following the torus breach, BR is the worker's breathing rate (3.5×10^{-4} m³/s), and t is the exposure time (30 min). The intake equation is based on a number of assumptions, including

- no removal of tritium from the torus containment through ventilation, plate-out, or radioactive decay;
- tritium is at an equilibrium concentration during the worker's residence in the torus containment;
- tritium concentration does not vary with position within the torus containment.

Given these assumptions, the concentration is

$$C = \frac{A}{V},$$

where
A = activity of tritium released into the torus containment = $(M)(\text{SA})(\text{RF})$,
M = mass of tritium available for release = 10 g,
SA = tritium-specific activity = 3.6×10^8 MBq/g,
RF = release fraction into containment = 1 (for the assumed maximum dose),
V = volume of containment = 50 m × 50 m × 50 m = 1.25×10^5 m³.

Using these values, the total tritium activity is

$$A = (10 \text{ g})(3.6 \times 10^8 \text{ MBq/g})(1) = 3.6 \times 10^9 \text{ MBq}.$$

The concentration is determined using these values:

$$C = \frac{3.6 \times 10^9 \text{ MBq}}{(50 \text{ m})^3} = 2.88 \times 10^4 \text{ MBq/m}^3.$$

The intake is now determined from the relationship:

$$I = C(BR)t,$$

$$I = (2.88 \times 10^4 \text{ MBq/m}^3)(3.5 \times 10^{-4} \text{m}^3/\text{s})(30 \min)(60 \text{ s/min})$$
$$= 1.81 \times 10^4 \text{ MBq}.$$

With the determination of the intake, the effective dose is determined:

$$E = I(DCF),$$

$$E = (1.81 \times 10^4 \text{ MBq})(4.27 \times 10^{-2} \text{ mSv/MBq}) = 773 \text{ mSv}.$$

(b) The external gamma effective dose from the walls, floor, and ceiling is determined by summing the contribution from these structures. Given their size and the location of interest, these plane sources are reasonably approximated as a thin-disk source. By computing the area of a 50 m × 50 m face, the radius of the equivalent disk can be determined:

$$\text{area} = (50 \text{ m})(50 \text{ m}) = \pi r^2,$$

$$r = 28.2 \text{ m}.$$

The gamma effective dose rate $\dot{E}$ is obtained from the thin-disk relationship:

$$\dot{E} = \pi C_a \Gamma \ln\left(\frac{r^2 + h^2}{h^2}\right),$$

where C_a is the activity per unit area on each wall, floor, and ceiling, Γ is the gamma constant, r is the radius of the thin-disk source, and h is the perpendicular distance from the thin-disk source on its axis. Given the information in the problem statement, the following values are derived:

$$C_a = \frac{\text{activity}}{\text{area}} = \frac{A}{\pi r^2} = \frac{1.0 \times 10^8 \text{ MBq}}{\pi (28.2 \text{ m})^2} = 4.0 \times 10^4 \text{ MBq/m}^2,$$

$$\Gamma = 0.0005 \text{ mSv m}^2/\text{MBq h},$$

$h = 25$ m (the point of interest is the center of the torus containment). Using these values, the effective dose rate from one 50 m × 50 m face is

$$\dot{E} = \pi(4.0 \times 10^4 \text{ MBq/m}^2)(0.0005 \text{ mSv m}^2/\text{MBq h}) \ln \frac{(28.2 \text{ m})^2 + (25 \text{ m})^2}{(25 \text{ m})^2},$$

$$\dot{E} = (62.8 \text{ mSv/h})(0.821) = 51.6 \text{ mSv/h}.$$

Since there are six faces in the torus cubicle with equal activity and the residence time is 30 min, the worker's gamma effective dose is

$$E = \dot{E}t = (6)(51.6\,\text{mSv/h})(30\,\text{min})(1\,\text{h}/60\,\text{min}) = 155\,\text{mSv}.$$

This result assumes the worker resides at the center of the torus containment for the entire 30-min period.

(c) To determine the effective dose from tritium received by a member of the public located 1 mile from the center of the torus containment, the following assumptions are made:
- The individual's breathing rate is constant.
- The meteorology (χ/Q) is constant during the event.
- The individual resides at the same location until the entire tritium plume passes.
- Ten percent of the available tritium is released to the environment.

The total activity (A) available for release is

$$A = (0.1)(10\,\text{g})(3.6 \times 10^8\,\text{MBq/g}) = 3.6 \times 10^8\,\text{MBq}.$$

The effective dose (E) to the member of the public from the tritium release is determined from the relationship:

$$E = A\frac{\chi}{Q}(\text{BR})(\text{DCF}),$$

χ/Q = atmospheric dispersion factor at 1 mile = $1.0 \times 10^{-4}\,\text{s/m}^3$,
BR = breathing rate = $3.5 \times 10^{-4}\,\text{m}^3/\text{s}$,
DCF = tritium dose factor = $4.27 \times 10^{-2}\,\text{mSv/MBq}$ inhaled including tritium absorption through the skin,
$E = (3.6 \times 10^8\,\text{MBq})(1.0 \times 10^{-4}\,\text{s/m}^3)$
$\quad \times (3.5 \times 10^{-4}\,\text{m}^3/\text{s})(4.27 \times 10^{-2}\,\text{mSv/MBq})$,
$E = 0.538\,\text{mSv}$.

(d) There is minimal potential for an alpha intake assuming the worker enters the torus containment without respiratory protection. In a fusion reactor, the production of alpha emitters is considerably smaller than in a fission reactor. Although alpha emitters are produced (e.g., ^{5}He (7.6×10^{-22} s) and ^{16}N (7.13 s)), their health physics impact is minimal. Helium nuclei are formed in the fusion process. Once the plasma is quenched, the helium nucleus captures two electrons and forms helium gas.

03-05

The fact that a distributed source has a lower effective dose rate than an equivalent point source is illustrated by considering a point source and thin-disk source of equal activity (A). In this example, consider a pure photon emitter characterized by a gamma constant (Γ). A pure photon emitter is selected to simplify the calculation, but the result is valid for any radiation type that can traverse distances on the scale of a fusion reactor.

The effective dose rate ($\dot{E}$) from a point (P) source and thin-disk (TD) source are

$$\dot{E}_P = \frac{A\Gamma}{h^2},$$

$$\dot{E}_{TD} = \pi C_a \Gamma \ln \frac{R^2 + h^2}{h^2},$$

where h is the distance from the point source or the perpendicular distance from the thin disk along its axis, R is the radius of the thin-disk source, and C_a is the uniform activity per unit area of the thin disk:

$$C_a = \frac{A}{\pi R^2}.$$

The ratio of the thin disk and point source effective dose rates is

$$\frac{\dot{E}_{TD}}{\dot{E}_P} = \frac{\pi C_a \Gamma \ln \frac{R^2 + h^2}{h^2}}{\frac{A\Gamma}{h^2}} = \frac{\pi \frac{A}{\pi R^2} \Gamma \ln \frac{R^2 + h^2}{h^2}}{\frac{A\Gamma}{h^2}},$$

which simplifies to

$$\eta = \frac{\dot{E}_{TD}}{\dot{E}_P} = \frac{h^2}{R^2} \ln \frac{R^2 + h^2}{h^2}.$$

This effective dose rate ratio η is simplified by defining a scale factor ξ:

$$h = \xi R.$$

Using the scale factor simplifies the expression for η:
Substituting $h = \xi R$ in the above equation leads to

$$\eta = \frac{\dot{E}_{TD}}{\dot{E}_P} = \frac{\xi^2 R^2}{R^2} \ln \frac{R^2 + \xi^2 R^2}{\xi^2 R^2} = \xi^2 \ln \frac{1 + \xi^2}{\xi^2}.$$

An examination of this equation suggests that $\dot{E}_P \geq \dot{E}_{TD}$ for all values of ξ. The following conclusions are derived from an evaluation of this equation:

- Near the source ($\xi \ll 1$), $\eta \to 0$ as $\xi \to 0$.
- At $\xi = 1$, $\eta \approx 0.7$.
- As the point of interest moves away from the source ($\xi > 1$), $\eta \to 1$. The $\eta \to 1$ result holds to an accuracy of about 1% as the distance from the source approaches three times the maximum source dimension (i.e., $h \approx 6R$ in the thin-disk example).

03-06

A review of dosimetry records at a twenty-first century 1500 MW$_e$ fusion reactor indicates the following results:

Bioassay and dosimetry results by work group

Work group	Tritium bioassay	Measurable (β, γ) dose	Measurable neutron dose
Operations	Positive	Yes	Yes
Maintenance	Some positive but most are negative	Yes	Most personnel have none

These results are credible. Operations personnel enter areas adjacent to the vacuum vessel/torus as part of their duties including routine inspections, performing valve lineups, surveillances, and operations testing. For example, entries into tritium recovery and torus cooling systems expose operators to tritium and neutron radiation from the fusion process. Therefore, operators receive neutron and tritium doses as part of their routine duties.

Beta-gamma radionuclides are deposited throughout the facility as a result of activation reactions. Accordingly, both operations and maintenance personnel are exposed to beta-gamma activity.

Maintenance personnel do not normally perform work in tritium contaminated areas or neutron radiation fields. Work activities are planned and include an ALARA review to minimize the effective dose. Part of this review involves timing the work to avoid neutron radiation or installing shielding to minimize the neutron dose. Maintenance work is often delayed until a fission reactor is shutdown, and a similar approach may be utilized at a fusion facility. In addition, components are often decontaminated or enclosed in a containment structure prior to the performance of maintenance. The component (e.g., valve operators, motors, motor shafts, and valve seats) may also be removed and the maintenance performed in low dose and low tritium contamination areas. These actions would minimize both neutron and tritium doses received by maintenance personnel.

03-07

(a) The DD fusion neutron spectrum has a threshold energy of 2.45 MeV and the DT fusion neutron spectrum has a higher threshold of 14.1 MeV:

$$D + D \xrightarrow{50\%} T(1.01 \text{ MeV}) + p(3.02 \text{ MeV})$$
$$\xrightarrow{50\%} {}^3He(0.82 \text{ MeV}) + n(2.45 \text{ MeV}),$$
$$D + T \rightarrow {}^4He(3.50 \text{ MeV}) + n(14.1 \text{ MeV}).$$

The neutron spectrum of a fission reactor (PWR, BWR, or CANDU) involves the thermal fission of ^{235}U and ^{239}Pu or the fast fission of ^{238}U. The fission neutron spectrum in these reactors has a most probable energy of about 0.7 MeV and an average energy of about 2 MeV. Only about 5% of the neutrons have energy above 5 MeV. In general, the fission neutron spectrum has a lower energy than the fusion neutron spectrum. Accordingly, more reaction channels are accessible with the higher neutron fusion energy process.

(b) The likelihood of the production of tritium, noble gas, iodine, actinides, and other beta-gamma emitters and their principle means of production in fission and fusion reactors are summarized in the following table:

Comparison of selected fission and fusion reactor properties

Species produced	Fission reactor	Fusion reactor
Tritium	The primary tritium production reactions are: 1. $^2H(n,\gamma)$ 2. $^6Li(n,\alpha)$ 3. $^{10}B(n,2\alpha)$ 4. Tertiary fission BWRs: limited tritium is produced via reactions 1 and 4. Control blades are composed of B_4C. If the cladding is breached, tritium production occurs by reaction 3 PWRs: All four reactions contribute. A 1000 MW$_e$ reactor produces about 3.7×10^7 MBq per year CANDUs: copious quantities of tritium are produced through the deuterium neutron capture reaction. At a Generation II CANDU, 30–40% of the occupational dose is from tritium intakes	In a DT process, tritium is the fuel source In a DD process, tritium is produced as part of the fusion reaction
Noble gas	Noble gases are produced from the fission process with isotopes of krypton and xenon being the dominant species. Argon activation also occurs from neutron irradiation of air	Other than the activation of argon in the air $[^{40}Ar(n,\gamma)^{41}Ar]$, significant noble gas activity is not produced

Comparison of selected fission and fusion reactor properties

Species produced	Fission reactor	Fusion reactor
Noble gas(*continued*)	Noble gas is a dominant source term for off-site accidents at a fission reactor	
Iodine	Radioiodine isotopes are produced from the fission process. Radioiodine is a dominant source term for off-site accidents at a fission reactor	Radioiodine is not produced in a fusion reactor
Actinides	Actinides are produced from the sequential neutron capture in uranium isotopes that produce neptunium, plutonium, americium, and curium	Actinides are not produced in a fusion reactor
Other beta–gamma emitters	Activation reactions are primarily produced by neutron-induced reactions such as (n,γ), (n,p), and (n,α). These reactions produce beta–gamma activation products in fuel, coolant, and structural members. The lower energy neutron spectrum allows only limited reaction channels for production of beta–gamma emitters	Activation products are produced by fusion generated particles (e.g., p, n, d, ^{3}He, ^{3}H, and ^{4}He). The higher energy fusion neutron spectrum opens a wide variety of reaction channels. Although many of the isotopes produced are similar to the fission isotopes, their modes of production can be quite different. Examples of fusion neutron reactions include (n,2n), (n,γ), (n,p), (n,α), (n,d), (n,nα), and (n,^{3}H)

03-08

(a) The neutron dose to the workers located at 200 m from the vacuum vessel/torus is approximated using a point-source relationship. This is reasonable since the distance from the torus is more than three times the ITER torus diameter. Using this approximation, the total fluence at 200 m from the torus is

$$\Phi_{total} = \frac{S}{4\pi r^2},$$

where S is the total number of fusions (5×10^{19}) and r is the distance from the torus (200 m). Using these values:

$$\Phi_{total} = \frac{(5 \times 10^{19} \text{ fusions})(1 \text{ n/fusion})}{4\pi(200 \text{ m})^2(100 \text{ cm/m})^2} = 9.95 \times 10^9 \text{ n/cm}^2.$$

The fluences for each of the three measured energy groups is given by

$$\Phi_j = f_j \Phi_{total},$$

where f_j is the energy-dependent fusion yield for the jth neutron energy group. The total absorbed dose is the sum over the individual energy groups:

$$D_j = \sum_{j=1}^{3} \Phi_j k_j = \Phi_{total} \sum_{j=1}^{3} f_j k_j,$$

where k_j is the flux-to-dose conversion factor for the jth neutron energy group, and $j = 1, 2,$ and 3 correspond to thermal, 1 MeV, and 14 MeV neutrons. Applicable values of f_j and k_j are provided in the following table:

Neutron energy (MeV)	f_j neutron yield/fusion	k_j flux-to-dose conversion factor (Gy cm²/n)
($j=1$) 2.5×10^{-8}	0.1	5.1×10^{-12}
($j=2$) 1	0.2	3.3×10^{-11}
($j=3$) 14	0.7	7.7×10^{-11}

Using these values, we obtain the dose of absorbed neutron:

$$D = (9.95 \times 10^9 \text{ n/cm}^2)[(0.1)(5.1 \times 10^{-12} \text{Gy cm}^2/\text{n}) + (0.2) \\ \times (3.3 \times 10^{-11} \text{Gy cm}^2/\text{n}) + (0.7)(7.7 \times 10^{-11} \text{Gy cm}^2/\text{n})],$$

$$D = 0.00507 \text{ Gy} + 0.0657 \text{ Gy} + 0.536 \text{ Gy} = 0.607 \text{ Gy}.$$

(b) The ^{60}Co activity (A) is obtained from the production equation:

$$A = N\sigma\phi(1 - e^{-\lambda t_{irr}}),$$

where

N = number of ^{59}Co atoms

$$= \frac{(50 \text{ kg})(1000 \text{ g/kg})(6.02 \times 10^{23} \text{ atoms})}{(59 \text{ g})}$$

$= 5.10 \times 10^{26}$ atoms,

σ = activation cross section for the ^{59}Co$(n,\gamma)^{60}$Co thermal neutron capture reaction = 37 b,

ϕ = average total fluence rate during the inadvertent fusion event of duration t (0.001 s).

$$\phi = \frac{\Phi_{\text{total}}}{t} = \frac{9.95 \times 10^9 \text{ n/cm}^2}{0.001 \text{ s}} = 9.95 \times 10^{12} \text{ n/cm}^2 \text{ s},$$

$$\lambda = \frac{\ln 2}{T_{1/2}} = \frac{\ln 2}{5.27 \text{ years}} \frac{1 \text{ years}}{365 \text{ days}} \frac{1 \text{ day}}{24 \text{ h}} \frac{1 \text{ h}}{3600 \text{ s}} = 4.17 \times 10^{-9}/\text{s}.$$

The desired ^{60}Co activity is obtained from these values. However, only 10% of the total fluence is in the thermal range and contributes to the production of ^{60}Co:

$$A = (5.10 \times 10^{26} \text{ atoms})(37 \times 10^{-24} \text{cm}^2/\text{atom}) \left(\frac{0.1 \text{ thermal n}}{\text{total n}}\right)$$

$$\times (9.95 \times 10^{12} \text{ total n/cm}^2 \text{ s}) \frac{1 \text{ dis}}{\text{thermal n}} \frac{1 \text{ Bq s}}{\text{dis}}$$

$$\times [1-\exp(-(4.17 \times 10^{-9}/\text{s})((0.001 \text{ s}))].$$

As the argument of the exponential is small, it can be simplified using the power series expansion:

$$1-e^{-x} = 1-(1-x) = x.$$

Using the expansion of the exponential simplifies the activity relationship and determines the ^{60}Co activity:

$$A = (1.88 \times 10^{16} \text{ Bq})(4.17 \times 10^{-12}) = 7.84 \times 10^4 \text{ Bq}.$$

(c) The effective dose rate 1 m (h) above the disk source on its axis is

$$\dot{E} = \pi C_a \Gamma \ln \frac{r^2 + h^2}{h^2},$$

$$\dot{E} = (\pi) \left(\frac{(7.84 \times 10^4 \text{ Bq})(10^5)}{\pi (5 \text{ m})^2} \frac{1 \text{ MBq}}{10^6 \text{ Bq}}\right) (3.5 \times 10^{-4} \text{ mSv m}^2/\text{MBq h})$$

$$\times \ln\left(\frac{(5 \text{ m})^2 + (1 \text{ m})^2}{(1 \text{ m})^2}\right) = (0.110 \text{ mSv/h})(3.26) = 0.359 \text{ mSv/h}.$$

(d) Initially, the operational testing was to be performed in an area having no measurable radiation. However, the fusion event created a radiation environment that affected this testing. If the task is split equally between the two operators, the required effective dose per operator to complete the testing is

$$\dot{E} = (0.359\,\text{mSv/h})(30\,\text{h}) = 10.8\,\text{mSv}.$$

Although this effective dose is higher than anticipated, it is less than the ICRP 60's recommended 20 mSv/year. Careful planning should permit dose leveling among all operators over a period of a year or two. The fusion pulse will have a significant impact on operational testing. The event will be thoroughly reviewed. Affected equipment will be reevaluated and personnel retrained. After the root cause is determined and associated corrective actions taken, start-up testing will likely be revised and significantly delayed.

03-09

(a) Hot particle absorbed doses are usually dominated by the beta absorbed dose. However, the gamma contribution can contribute up to 30% of the total dose and must be included in the dose calculation.

(b) The skin dose is obtained from the relationship:

$$D = \sum_{i=1}^{3} \frac{A_i(DF)_i t}{a} = \frac{t}{a}\sum_{i=1}^{3} A_i(DF)_i,$$

where A_i is the activity for the ith isotope, DF_i is the isotope-specific dose factor, t is the residence time of the particle on the skin (4 h), and a is the area over which the dose is evaluated (10 cm^2 following the guidance of NCRP 130). Using the activity values and dose factors provided in the problem statement, the absorbed dose is

$$D = \left(\frac{4\,\text{h}}{10\,\text{cm}^2}\right)\left((0.15\,\text{MBq})\left(1.51\,\frac{\text{Gy cm}^2}{\text{MBq h}}\right) + (2.7\,\text{MBq})\left(0.07\,\frac{\text{Gy cm}^2}{\text{MBq h}}\right) + (0.5\,\text{MBq})\left(1.12\,\frac{\text{Gy cm}^2}{\text{MBq h}}\right) \right) = 0.39\,\text{Gy}.$$

(c) The following actions should be taken after this event:
1. Perform a reconstruction to verify the sequence of events and to verify the accuracy of the assumed information including the event duration, particle isotopic and activity composition, and plant area where the contamination occurred.
2. Perform a dose calculation to determine the worker's skin dose.
3. Discuss the dose calculation and any associated health effects with the worker. Given the calculated dose, a physician should be available to answer any

medical questions posed by the worker. Periodic monitoring of the irradiated area by a physician should also be scheduled. Although the calculated absorbed dose is less than the anticipated epilation threshold of 2–3 Gy, monitoring is a prudent action and is also recommended by NCRP 130.
4. Survey/decontaminate the valve repair area to ensure no other hot particles are present.
5. Evaluate the health physics coverage of similar tasks. Having 4 h pass without detecting the hot particle is not acceptable, and the task survey frequency should be reviewed.

03-10

(a) The plume height of the released tritium is related to the effective release height (H). The effective release height is a function of the physical stack height (h), stack exit diameter at the release point (d), release velocity of the gas (v), the mean wind speed (u), absolute temperature of the released gas (T), and absolute ambient temperature (T_0) according to the relationship:

$$H = h + d\left(\frac{v}{u}\right)^{1.4}\left(1 + \frac{T-T_0}{T}\right).$$

(b) Using the Gaussian plume model, the ground-level tritium activity concentration at the plume centerline (χ) is given by the relationship:

$$\chi = \frac{Q}{\pi \sigma_y \sigma_z u} e^{-\left[(y^2/2\sigma_y^2) + (H^2/2\sigma_z^2)\right]},$$

where
Q = release rate
$= \dfrac{(10 \text{ blankets})(5 \times 10^{11} \text{ MBq/blanket})}{(10 \text{ min})(60 \text{ s/min})}$
$= 8.33 \times 10^9$ MBq/s,
σ_y = horizontal standard deviation = 40 m,
σ_z = vertical standard deviation = 20 m,
y = plume centerline = 0,
u = mean wind speed = 2 m/s,
H = effective release height = 65 m.

Using these values, the ground-level tritium concentration is determined:

$$\chi = \left(\frac{8.33 \times 10^9 \text{ MBq/s}}{\pi(40 \text{ m})(20 \text{ m})(2 \text{ m/s})}\right) e^{-(65 \text{ m})^2/\left[2(20 \text{ m})^2\right]}$$

$= (1.66 \times 10^6 \text{ MBq/m}^3)(5.09 \times 10^{-3}),$

$\chi = 8.45 \times 10^3$ MBq/m^3.

(c) Assumptions that may contribute to the inaccuracy of the Gaussian plume model include the following:

1. The site topography may not be adequately represented by the assumed horizontal and vertical standard deviation values that were derived from flat, featureless topography. Hills, valleys, and facility structures alter the assumed plume standard deviation values.
2. Constant meteorology is assumed.
3. The presence of lakes or bodies of water affects the release characteristics. Since the facility is on Lake Michigan, sea breeze effects should be evaluated.
4. The effective stack height is not likely constant. The factors noted in (a) are likely to change during a release scenario.
5. The Gaussian plume model is essentially a quasi-steady-state model that requires average periods of at least 15 min. Values over a shorter period or instantaneous values have additional uncertainty.
6. Plume reflection by the ground is not considered.
7. The model provides no provision for the removal of tritium from the plume.
8. Dispersion is only included in the horizontal and vertical directions, and these parameters are constant during the release.
9. No change in tritium state is allowed. For example, a T_2 release does not weather to HTO as expected during the plume transit through the atmosphere.
10. The release rate and wind speed are assumed to be constant.

(d) The Gaussian plume model tends to overestimate the ground-level concentration. The items noted in problem (c) contribute to this overestimate. In addition, the Gaussian plume model parameters are inherently conservative and tend to artificially increase the ground-level concentration. The author's experience from routine releases at Generation II fission power reactors suggests a factor of 100–1000 overestimation in the ground-level concentration when the model is compared to measured data.

Solutions for Chapter 4

04-01

(a) The Coulomb barrier (E_C) for the ^{208}Pb + ^{238}U interaction is given by Adler's relationship:

$$E_C = \frac{Z_1 Z_2 (1 + A_1/A_2)}{A_1^{1/3} + A_2^{1/3} + 2} \text{ MeV},$$

where Z_1 and Z_2 are the charge of the heavy ion (82) and target nucleus (92), respectively. A_1 (208) and A_2 (238) are their respective mass numbers. Using these values:

$$E_C = \frac{(82)(92)(1 + 208/238)}{208^{1/3} + 238^{1/3} + 2} \text{ MeV} = 1000 \text{ MeV}.$$

(b) If the beam energy is 1200 GeV, a variety of radiation types are produced including neutrons, high-energy particles (e.g., pions, muons, and other mesons) gamma rays, and heavy fragments. Depending on the location, either neutrons or muons dominate the health physics considerations.

(c) For a beam current of 1 mA and accelerated ions with a charge of +20 e, the number of lead ions striking the target per second (n) is

$$n = (0.001 \text{ A})\text{C/A s}\left(\frac{1 \text{ Pb ion}}{(20)(1.6 \times 19^{-19} \text{ C})}\right) = 3.13 \times 10^{14} \text{ Pb ions/s}.$$

(d) For the conditions of part (c), 10 neutrons are produced for every lead ion striking the target. If the neutrons are produced in an isotropic manner, the neutron fluence at 5 m from the target is

$$\phi = \frac{(3.13 \times 10^{14} \text{ Pb ions/s})(10 \text{ n/Pb ion})}{(4\pi)(500 \text{ cm})^2} = 9.97 \times 10^{8} \text{n/cm}^2 \text{ s}.$$

04-02

(a) The neutron effective dose rate is determined from the fluence rate. Using an isotropic point source relationship, the fluence rate is

$$\phi = \frac{S}{4\pi r^2},$$

where S is the neutron emission rate from the target as a result of the ^{3}H(d,n) reaction and r is the distance from the target (1.2 m). The neutron emission rate (S) produced in the target is

$$S = N(^3\text{H})\sigma(d,n)\phi_d,$$

where $N(^3\text{H})$ is the number of ^{3}H atoms in the target, $\sigma(d,n)$ is the total reaction cross section for the ^{3}H(d,n)^{4}He reaction (5b), and ϕ_d is the deuteron fluence rate striking the target (6.25 × 10^{13} d/cm^2 s). The number of ^{3}H atoms in the target is determined from the activity (A) relationship and the tritium activity per unit area in the target (3.7 × 10^5 MBq/cm^2):

$$N(^3\text{H}) = \frac{A}{\lambda}$$

$$= \frac{(3.7 \times 10^5 \text{ MBq/cm}^2)(1 \text{ cm}^2)(10^6 \text{ Bq/MBq})(1 \text{ dis/Bq s})(1 \text{ atom/dis})}{\frac{\ln 2}{12.3 \text{ years}} \frac{1 \text{ year}}{365 \text{ day}} \frac{1 \text{ day}}{24 \text{ h}} \frac{1 \text{ h}}{3600 \text{ s}}},$$

$$N(^3\text{H}) = 2.07 \times 10^{20} \text{ atoms}.$$

With these values, the neutron emission rate from the target due to the incident deuterons is

$$S = (2.07 \times 10^{20} \text{atoms})(5b/\text{atom})(10^{-24} \text{cm}^2/b)$$
$$\times (6.25 \times 10^{13} \text{d/cm}^2\text{s})(1\text{n/d}),$$

$S = 6.47 \times 10^{10}$ n/s.

The neutron fluence rate is now determined:

$$\phi = \frac{6.47 \times 10^{10}\ \text{n/s}}{4\pi[(1.2\ \text{m})(100\ \text{cm/m})]^2} = 3.58 \times 10^5\ \text{n/cm}^2\ \text{s}.$$

The neutron effective dose rate ($\dot{E}$) is determined from the flux since the dose factor (K) is known (5×10^{-10} Sv cm^2/n):

$$\dot{E} = K\phi = (5 \times 10^{-10}\ \text{Sv cm}^2/\text{n})(3.58 \times 10^5\ \text{n/cm}^2\ \text{s})(3600\ \text{s/h})$$
$$= 0.644\ \text{Sv/h}.$$

(b) The major elements of this accelerator's radiation protection program include the following:
- An ALARA program is implemented to minimize worker doses. Dose control and source term reduction are program priorities because radiation risk is reduced when the effective dose is reduced.
- Contamination controls are implemented with special provisions for tritium. Tritium could be released from the target area and potentially contaminate cooling water systems, vacuum systems, oil and lubricants, beam lines, and other systems interfacing with the target. Surface contamination surveys in the target room and adjacent areas are performed periodically.
- A bioassay program is implemented. Urinalysis should be used to assess the effectiveness of the contamination control program.
- The target area and beam lines are interlocked to minimize direct exposure to the deuteron beam. Warning signs and lights signal that the accelerator is in operation.
- Instrumentation including radiation monitors and beam line vacuum detectors are interlocked to terminate operations if a beam loss event occurs.
- A records program including personnel dosimetry documentation, radiation survey sheets, and radiation work permits is implemented. This program ensures programmatic consistency and litigation assurance.
- An independent review and oversight program is implemented to ensure quality and consistency.
- Administrative and procedural controls are in place to enhance and enforce the requirements of the radiological controls program.
- Environmental sampling and direct dose rate surveys are implemented to ensure the public is protected during machine operation. Soil activation and groundwater contamination are monitored as part of this program.
- Periodic surveys for external radiation are performed to ensure that the accelerator staff is being properly protected and that the shielding controls are effective. Skyshine and scattered radiation should be carefully assessed. Both shutdown and operating surveys are appropriate.
- Personnel dosimetry is provided for the assessment and monitoring of external exposures including neutron, gamma, muon, and beta radiation. Extremity monitoring is appropriate.

(c) Lead and normal polyethylene are available to construct a temporary shield around the target. The order of these shielding materials is dictated by the neutron energy spectrum. For neutron energies below about 5 MeV, these materials should be placed in the following order: polyethylene should be placed closest to the source and be followed by lead. Polyethylene degrades the neutron energy via elastic scattering with the hydrogen nuclei. Thermal neutron captures in hydrogen and inelastic scattering with carbon produce gamma rays. These gamma rays are then attenuated by the outer thickness of lead.

In the ITER, the DT fusion process produces 14.1 MeV neutrons. At 14.1 MeV, elastic scattering with hydrogen nuclei is ineffective in degrading the neutron energy spectrum. It is more efficient to degrade the neutron energy through inelastic scattering with lead [^{208}Pb(n,n′)] until the neutron energies are reduced to about 5 MeV. At this energy, elastic scattering with hydrogen is effective in further degrading the neutron energy spectrum. Therefore, the correct order of materials to shield the 14.1 MeV neutrons is lead (closest to the target), followed by polyethylene, and then lead.

04-03

(a) This question requires the calculation of the effective dose rate from the measured current. Assuming the anode is 100% efficient in collecting the current, the effective dose rate ($\dot{E}$) is related to the current by the relationship:

$$I = \rho V \frac{T_{stp}}{T} \frac{P}{P_{stp}} \dot{E},$$

where
  I = ionization chamber current = 10^{-12} A,
  ρ = density of air at 1 atm absolute pressure and at 273 K temperature = 1.293 g/l,
  V = ionization chamber volume,
  $V = \pi r^2 h$,
    r = radius of the ion chamber = 5 cm,
    h = length of the ion chamber = 20 cm,
  $V = (3.14)(5\,\text{cm})^2(20\,\text{cm}) = 1570\,\text{cm}^3 \times 1\,\text{l}/1000\,\text{cm}^3 = 1.57\,\text{l}$,
  T = room temperature = 20°C = (273 + 20) K = 293 K,
  T_{stp} = standard temperature = 273 K,
  P = room pressure = 1 atm,
  P_{stp} = standard pressure = 1 atm.

A conversion factor is needed to convert C/kg in air into Sv in the current equation:

$$\frac{C}{kg} = \frac{C}{kg} \frac{1\,\text{ion}}{1.6 \times 10^{-19}\,C} \frac{34\,\text{eV}}{\text{ion}} \frac{1.6 \times 10^{-19}\,J}{\text{eV}} \frac{Sv}{J/kg} = 34\,\text{Sv}.$$

This relationship may be solved for the effective dose rate:

$$\dot{E} = I / \left(\rho V \frac{T_{stp}}{T} \frac{P}{P_{stp}} \right).$$

Using the previously defined parameter values, the effective dose is

$$\dot{E} = \frac{(10^{-12}\,\text{A})(1\,\text{C/A s})(34\,\text{Sv kg/C})(3600\,\text{s/h})(1000\,\text{mSv/Sv})}{(1.293\,\text{g/l})(1.57\,\text{l})(273\,\text{K}/293\,\text{K})(1\,\text{atm}/1\,\text{atm})(1\,\text{kg}/1000\,\text{g})}$$
$$= 6.47 \times 10^{-2}\,\text{mSv/h}.$$

(b) Conditions that could affect the accuracy of the ionization chamber measurements include
   1. uniformity of the electric field within the ionization chamber volume;
   2. losses caused by ion volume recombination and diffusion within the chamber;
   3. leakage current losses through insulating structures in the ionization chamber;
   4. orientation of the tube relative to the radiation source;
   5. uniformity of the radiation field;
   6. presence of mixed radiation fields;
   7. the presence of radioactive gases that induce a "memory effect" in the chamber;
   8. anode efficiency less than the assumed 100%;
   9. presence of pulsed radiation fields;
   10. decay of short-lived particles inside the ion chamber;
   11. degradation of components in the detection circuit such as resistors, power supplies (batteries), and short circuits;
   12. electronic noise;
   13. existence of very low or very high dose rates;
   14. electromagnetic interference;
   15. humidity.

(c) Muon creation in an accelerator depends on the accelerator energy and reaction being investigated. At lower energy accelerators encountered in many health physics applications, muons are created in either electron or proton accelerators. When an incident electron or proton beam excites a nucleus in excess of 140 MeV, pions (π) are produced. Muons (μ) are produced from the decay of the pions. The most significant reactions are

$$\pi^+ \rightarrow \mu^+ + \nu_\mu,$$

$$\pi^- \rightarrow \mu^- + \bar{\nu}_\mu,$$

where ν_μ is a muon neutrino.

At higher energies, muons are created from kaon decays

$$K^+ \to \mu^+ + \nu_\mu,$$

$$K^- \to \mu^- + \bar{\nu}_\mu,$$

and the electromagnetic and nuclear cascade sequences. These cascade sequences occur at high-energy electron, proton, and heavy-ion accelerators. Muons also result from the decay of third-generation leptons such as the tau.

As muons are decay products of pions, hadron machines also produce muons. Therefore, most high-energy accelerators lead to muon production.

(d) The charge of a muon can be +1 or −1. The muon has a charge of −1 and its antiparticle's charge is +1.

(e) The muon mass, mean lifetime, and decay mode are compared to other elementary particles in the following table:

Particle	Mass (MeV)	Mean lifetime	Decay mode
Electron (e^-)	0.511	$>4.6 \times 10^{26}$ yr	Stable[a]
Positron (e^+)	0.511	$>4.6 \times 10^{26}$ yr	Stable[a]
Muon (μ^+)	105.7	2.2×10^{-6} s	$\mu^+ \to e^+ + \bar{\nu}_\mu + \nu_e$
Muon (μ^-)	105.7	2.2×10^{-6} s	$\mu^- \to e^- + \nu_\mu + \bar{\nu}_e$
Pion (π^0)	135.0	8.4×10^{-17} s	$\pi^0 \to \gamma + \gamma$
Pion (π^+)	139.6	2.6×10^{-8} s	$\pi^+ \to \mu^+ + \nu_\mu$
Pion (π^-)	139.6	2.6×10^{-8} s	$\pi^- \to \mu^- + \bar{\nu}_\mu$
Proton (p)	938.3	$>2.1 \times 10^{29}$ yr	Stable[a]
Antiproton ($\bar{p}$)	938.3	$>2.1 \times 10^{29}$ yr	Stable[a]
Neutron (n)	939.6	885.7 s	$n \to p + e^- + \bar{\nu}_e$
Antineutron ($\bar{n}$)	939.6	885.7 s	$\bar{n} \to \bar{p} + e^+ + \nu_e$

[a] These particles are stable from a health physics perspective.

(f) The radiation field from a misdirected particle beam can create a temporary muon radiation field. Factors to be considered when determining an ionization chamber's location to measure the muon field include

1. occupancy factors for the area receiving the misdirected beam and adjacent areas;
2. presence of components that produce electromagnetic interference;
3. anticipated direction of the beam as a result of the component failures;
4. the radiation characteristics of the beam particle if the failure occurs on the beam line;
5. the beam energy if the failure occurs on the beam line;
6. the radiation characteristics of the produced particles;
7. the produced particle energies;

8. beam current or flux;
9. produced particle flux;
10. mass stopping power of the beam particle;
11. mass stopping power of the produced particles;
12. time to restore power to the device initiating the beam misdirection or to terminate the beam;
13. location of beam optics components such as electromagnetic field devices including focusing magnets and accelerating sections;
14. location of the beam stop;
15. proximity of the misdirected beam to sensitive equipment;
16. presence of radiation shielding;
17. presence of other sources of radiation; and
18. location of the area relative to the beam line or projected point of beam exit.

(g) Hazards (other than ionizing radiation) associated with high-energy accelerator facilities include
1. fire;
2. electrical shocks;
3. chemical hazards including SF_6 gas in van de Graaff accelerator modules;
4. fall and trip hazards;
5. high-voltage power supplies;
6. lasers;
7. microwave generators;
8. microwave wave guides;
9. high temperatures (e.g., ion sources, target, and beam stop);
10. low temperatures (e.g., cryogenic components);
11. confined spaces;
12. toxic gases including ozone and nitrous oxides;
13. steam;
14. noise;
15. contained energy;
16. heat stress.

04-04

(a) In formulating the reentry plan, your primary considerations include the workers' conditions, facility's status, and radiological conditions. These items are determined by evaluating a number of considerations including the following:

1. Determining if the workers are injured and the extent of their injuries is a priority action. Medical assistance may be required depending on the seriousness of the injuries to the workers.

2. An assessment of the workers' conditions and the ambient radiological conditions are needed to determine the radiological significance of this event. The flux and energy of protons that struck the structural member, the flux and energy of secondary muons, neutrons, and photons, and the flux and energy of other radiological species need to be estimated.
3. The status of the proton beam needs to be determined to ascertain if it has been terminated or is still active. Since interlocks failed, the plant instrumentation status requires a careful review to ensure that it is displaying an accurate representation of the radiological conditions. If the event was terminated, then the dose rates are decreasing. If the event is ongoing, then the dose rates are elevated or increasing.
4. The location and initiating cause of the event and current radiation levels must be ascertained. Planning entry and exit routes depends on this information. All activities are to be performed in an ALARA manner.
5. The areas affected by the event and their current radiological condition need to be determined. Materials in these areas that would lead to an enhanced radiological source term need to be assessed.
6. The current external radiation levels, airborne levels, and contamination levels need to be determined. Installed radiation instrumentation should be monitored from a remote location. The facility's safety analysis report provides bounding values and a conservative radiological assessment of a loss-of-beam event. This information provides a basis for initial reentry planning.
7. Any toxic material in the area and its concentration should be determined. Relevant, installed instrumentation should be monitored from a remote location.
8. The availability of shielding materials should be ascertained.
9. The proximity of personnel to the site of the proton beam impact should be determined. The worker's position as a function of time following the event has a significant impact on the delivered dose.
10. The time and duration of the event must be ascertained. Considerable dose is incurred after the event is terminated from activation reactions and subsequent decays. The total effective dose must be considered in the planning process.

(b) A method that could be used to quickly screen persons is the collection and reading of the individuals' thermoluminescent or self-reading dosimetry. This method is quick if predetermined ratios of neutron and gamma doses are available.

If the dosimetry was off-scale or not worn at the time of the event, gold jewelry or other metal worn by the individual should be counted to assess the neutron dose and spectrum. Counting is expedited if a log of worker's jewelry (e.g., wedding bands) and their characteristics (e.g., mass and composition) are available. If not, this method is not as timely as desired.

Under emergency conditions, a rapid screening procedure is frequently used when exposure to a neutron field is suspected. Capture of neutrons by sodium atoms in the blood via the ^{23}Na(n,γ)^{24}Na reaction results in the formation of ^{24}Na, which decays by photon emission. A Geiger–Mueller detector placed in the vicinity of a large blood volume (i.e., under the arm pit) detects the ^{24}Na photons. As a point of reference, about 1 mrem/h results from the acute exposure of 5 Gy of fast neutrons.

(c) Medical interventions that could positively change the health outcome for an individual exposed to 8 Gy (deep dose) if administered during the first month following the incident include
1. place the victim in a sterile room to minimize infections since the individual's ability to fight infection has been reduced;
2. administer antibiotics to fight infection;
3. administer fluids to minimize dehydration;
4. bone marrow transplant therapy should be considered if a suitable donor or match is available;
5. administer blood transfusions;
6. administer hormones to assist lung tissue regeneration;
7. consider the use of radioprotective chemicals.

These interventions are intended to counter the acute radiation syndrome. The depletion of blood cells and blood-forming organs and damage to lung tissue are immediate concerns. These conditions may manifest themselves as limited ability to fight infection, dehydration, deterioration of blood-forming organs, and deterioration of the lung and its ability to exchange gases.

Doses in the 8 Gy range were received as a result of the 1999 Tokai Mura criticality in Japan. Although many of the interventions noted above were utilized, the exposed workers did not survive. Damage to lung tissue was a major factor in their deaths.

04-05

(a) In this question, the dominant ^{15}O removal mechanism is to be determined. The two available removal mechanisms are room ventilation and radioactive decay.

Room ventilation rate:

$$\lambda_v = F/V,$$

where
λ_v = room ventilation removal rate,
F = room exhaust rate = 30 m^3/min,
V = room volume = 6 m $\times$ 6 m $\times$ 3 m = 108 m^3,
$\lambda_v = (30\,\text{m}^3/\text{min})(1\,\text{min}/60\,\text{s})/(108\,\text{m}^3)$
$= 4.63 \times 10^{-3}\,\text{s}^{-1}$.

Radioactive decay removal rate:

$$\lambda = \ln(2)/T_{1/2}$$

where λ = radioactive decay constant,
$T_{1/2}$ = physical half-life of ^{15}O = 122 s,
$\lambda = 0.693/122 \text{ s} = 5.68 \times 10^{-3} \text{ s}^{-1}$.

Since the radioactive decay constant is larger than the ventilation removal rate, radioactive decay is the dominant removal mechanism.

(b) The ^{15}O air activity concentration after 4 min of release is obtained from the production equation. The total activity $A(t)$ released into the target area as a function of time t is given in terms of the release rate P and effective (total) removal rate k:

$$A(t) = (P/k)(1-e^{-kt}),$$

$$A(t) = (\lambda \dot{N}/k)(1-e^{-kt}),$$

where
$\dot{N} = {}^{15}$O release rate $= 2.6 \times 10^9$ atoms/s,
k = effective (total) removal rate $= \lambda_v + \lambda$,
λ = radioactive decay constant $= 5.68 \times 10^{-3} \text{ s}^{-1}$,
λ_v = room ventilation removal rate $= 4.63 \times 10^{-3} \text{ s}^{-1}$,
$k = 4.63 \times 10^{-3} \text{ s}^{-1} + 5.68 \times 10^{-3} \text{ s}^{-1} = 0.0103 \text{ s}^{-1}$,
t = release time = 4 min.

With these values, the ^{15}O concentration at 4 min after the release is

$$C(t) = A(t)/V,$$

where V = room volume = 6 m × 6 m × 3 m = 108 m³.

$$\begin{aligned}
A(t) &= [(2.6 \times 10^9 \,{}^{15}\text{O atoms/s})(5.68 \times 10^{-3}/\text{s})(1 \text{ dis}/{}^{15}\text{O atom})/(0.0103/\text{s})] \\
&\quad \times [1 - e^{-(0.0103/\text{s})(4 \text{ min})(60 \text{ s/min})}] \\
&= (1.43 \times 10^9 \text{ dis/s})(1 - 0.0844) \\
&= (1.31 \times 10^9 \text{ dis/s})(1 \text{ MBq}/1.0 \times 10^6 \text{ dis/s}) = 1.31 \times 10^3 \text{ MBq}, \\
C(t) &= A(t)/V \\
&= (1.31 \times 10^3 \text{ MBq})/(108 \text{ m}^3) = 12.1 \text{ MBq/m}^3.
\end{aligned}$$

(c) If ventilation flow was terminated after 6 min, the activity released into the room is obtained from the relationships derived in (b). Using this methodology, the room air concentration at 6 min after the release is

$$C(t) = A(t)/V,$$

where
V = room volume = 6 m × 6 m × 3 m = 108 m³,
$A(t)$ = activity released into the room after 6 min,

$$\begin{aligned}
A(t) &= [(2.6 \times 10^9 \,{}^{15}\text{O atoms/s})(5.68 \times 10^{-3}/\text{s})(1 \text{ dis}/{}^{15}\text{O atom})/(0.0103/\text{s})] \\
&\quad \times [1 - e^{-(0.0103/\text{s})(6 \text{ min})(60 \text{ s/min})}] \\
&= (1.43 \times 10^9 \text{ dis/s})(1 - 0.0245) \\
&= (1.39 \times 10^9 \text{ dis/s})(1 \text{ MBq}/1.0 \times 10^6 \text{ dis/s}) = 1.39 \times 10^3 \text{ MBq}.
\end{aligned}$$

With these values, the room concentration calculated by the experimenter is verified:

$$C(t) = A(t)/V = (1.39 \times 10^3 \text{ MBq})/(108 \text{ m}^3) = 12.9 \text{ MBq/m}^3.$$

Although the calculated air concentration exceeds the DAC, it does not lead to an overexposure. Two reasons why an overexposure did not occur are
1. The room size (6 m × 6 m × 3 m) is smaller than the size of the semi-infinite cloud used in deriving the ^{15}O DAC.
2. The radiation dose is determined by the number (N) of DAC h and not just the air concentration. The ALI corresponds to 2000 DAC h. In this question, the dose equivalent (H) is given by the relationship:

$$H = (50 \text{ mSv})(N/2000 \text{ DAC h}),$$

where
N = number of DAC h received,
$N = (1.29 \times 10^7 \text{ Bq/m}^3)(6 \text{ min})$
$\quad \times (1 \text{ h}/60 \text{ min}) (1 \text{ DAC}/4000 \text{ Bq/m}^3) = 323 \text{ DAC h},$
$H = 50 \text{ mSv} \times (323 \text{ DAC h}/2000 \text{ DAC h}) = 8.08 \text{ mSv}.$

Clearly, no overexposure occurred.

04-06

Given the time constraints, the following assumptions are made to simplify the calculation:

1. All low-energy protons and oxygen nuclei are absorbed in the hand. This is reasonable given the range of these particles.
2. Although a small fraction of the neutrons and high-energy photons are absorbed in the hand, no energy absorption from these radiation types is assumed. This is reasonable given the small fraction of energy deposited by neutrons and high-energy photons.
3. No neutrino energy is deposited in the hand. This is a valid assumption given the energy range considered in this problem.

Using these assumptions, the absorbed dose (D) is the total energy (E) deposited in the tissue divided by the tissue mass (m):

$$D = \frac{E}{m} = \left(\frac{1 \times 10^8 \pi^-}{\text{cm}^3}\right) \frac{(100 \text{ MeV} + 150 \text{ MeV})/\pi^-}{1 \text{ g/cm}^3} \left(\frac{1000 \text{ g}}{\text{kg}}\right)$$
$$\times \left(\frac{1.6 \times 10^{-13} \text{ J}}{\text{MeV}}\right) \left(\frac{\text{Gy kg}}{\text{J}}\right) = 4 \text{ Gy}.$$

04-07

The annual neutrino effective dose (H) for a 1000 TeV muon linear collider is determined from Equation 4.44:

$$H = \frac{K}{2}\frac{N}{g}X(E_0/2)E_0^2.$$

For the 1000 TeV muon linear accelerator, E_0 is 500 TeV (i.e., two, 500 TeV linear muon accelerators). The other parameters used in developing Equation 4.45 are $N = 6.4 \times 10^{18}$ muon decays per year, $g = 1$ GeV/m, and $K = 6.7 \times 10^{-21}$ mSv GeV/m TeV^2. The remaining parameter in Equation 4.45 is the cross section factor X:

$$X(E) = (0.512(3-\alpha) + 0.175(\alpha-2))/1.453,$$

where $\alpha = \log_{10}(E)$, where E is the muon energy expressed in TeV.

For the 500 TeV muon beam:

$$\alpha = \log_{10}(E_0/2) = \log_{10}(250) = 2.4,$$
$$X(E) = (0.512(3-\alpha) + 0.175(\alpha-2))/1.453$$
$$= (0.512(3-2.4) + 0.175(2.4-2))/1.453 = 0.26.$$

Using these values in Equation 4.44, we obtain the desired result:

$$H = \left(\frac{6.7 \times 10^{-21} \text{ mSv GeV/m TeV}^2}{2}\right)\left(\frac{6.4 \times 10^{18}/\text{year}}{1 \text{ GeV/m}}\right)(0.26)(500 \text{ TeV})^2,$$

$$H = 1394 \text{ mSv/year} = 1.4 \text{ Sv/year}.$$

04-08

(a) Sproton–antiproton annihilation events occur uniformly along a 100 m (L) zone. This zone resembles a line source of radioactivity. Therefore, the effective dose rate ($\dot{H}$) is obtained from the line source approximation:

$$\dot{H} = \frac{C_L \Gamma \theta}{w},$$

where C_L is the activity (A) per unit source length (L), θ is the included angle that the point of interest makes with the ends of the line source, Γ is the dose factor, and w is the perpendicular distance from the line source to the point of interest (10 m).

The values of C_L, Γ, and θ need to be determined to obtain the effective dose rate ($\dot{H}$):

$$C_L = \frac{A}{L},$$

$$A = \frac{(10^5 \text{ collision events})}{(10^{-6} \text{ s})} \left(\frac{1 \text{ dis}}{\text{collision event}}\right) \left(\frac{\text{Bq s}}{\text{dis}}\right) \left(\frac{\text{MBq}}{10^6 \text{ Bq}}\right) = 1 \times 10^5 \text{ MBq},$$

$$C_L = \frac{\left(\frac{1 \times 10^5 \text{ dis}}{1 \times 10^{-6} \text{ s}}\right)\left(\frac{\text{Bq s}}{\text{dis}}\right)\left(\frac{1 \text{ MBq}}{10^6 \text{ Bq}}\right)}{100 \text{ m}} = \frac{1 \times 10^5 \text{ MBq}}{100 \text{ m}} = 1 \times 10^3 \frac{\text{MBq}}{\text{m}}.$$

In calculating the gamma constant, the energy $0.1 \text{ TeV} = 10^5 \text{ MeV}$ and yield of 2 are used:

$$\begin{aligned}\Gamma &= (1.35 \times 10^{-7} \text{ Sv m}^2/\text{MBq h}) E(\text{MeV}) Y \\ &= (1.35 \times 10^{-7} \text{ Sv m}^2/\text{MBq h})(10^5)(2) \\ &= 0.027 \text{ Sv m}^2/\text{MBq h},\end{aligned}$$

$$\tan \theta = \frac{100 \text{ m}}{10 \text{ m}} = 10,$$

$$\theta = \tan^{-1}(10) = (84.3°)\left(\frac{\pi}{180°}\right) = 1.47.$$

Using these values leads to the effective dose rate:

$$\dot{H} = \left(\frac{(1 \times 10^3 \text{ MBq/m})(0.027 \text{ Sv m}^2/\text{MBq h})(1.47)}{10 \text{ m}}\right) = 3.97 \text{Sv/h}.$$

(b) The shielded ($\dot{H}$) and unshielded effective ($\dot{H}_0$) dose rates are related through the relationship involving the number of half-value layers (N):

$$\dot{H} = \dot{H}_0 \left(\frac{1}{2}\right)^N,$$

where $\dot{H}$ is the design goal 0.2 mSv/h and $\dot{H}_0$ was provided in question (a). Using these values:

$$N = \frac{\ln \frac{\dot{H}}{\dot{H}_0}}{\ln(1/2)} = \frac{\ln \frac{(0.2 \text{ mSv/h})(1 \text{ Sv}/1000 \text{ mSv})}{(3.97 \text{ Sv/h})}}{\ln(1/2)} = 14.3.$$

The required shielding thickness (t) is N times the half-value layer (t_{HVL}) value (5 cm):

$$t = nt_{HVL} = (14.3)(5 \text{ cm}) = 71.5 \text{ cm}.$$

(c) Since the sproton decays into heavy charged hadrons, these decay products are best detected using large, multicomponent detectors similar to those used at the Large Hadron Collider. Large, massive detectors are needed because the supersymmetry particles will likely have high energies and a diverse character. It is likely that more than one detector type will be needed since the sproton properties are not known.

As an additional example of these complex detectors consider the Collider Detector at the Fermi (CDF) National Accelerator Laboratory or Fermilab. CDF is a 100-ton detector that measures most of the particles originating in proton–antiproton collisions. Particles enter the interior of the roughly cylindrical CDF and propagate from the interior to the exterior of the device. As the particles move outward, they encounter various detector types used in analyzing the products of the proton–antiproton annihilation. Progressing radially outward, these detectors/components include a Silicon Vertex Tracker (SVT), a Central Tracker (CT), an Electromagnetic Calorimeter (EMC), Hadron Calorimeter (HC), an iron absorber, and a Muon Chamber (MC).

Particles first encounter the SVT detector that measures charged particles. The SVT is an accurate device for measuring the position of charged particles and it has an accuracy of a few tens of microns. Moving radially outward, particles next encounter the CT that also detects charged particles. The CT measures the particle's trajectory, which defines its momentum.

An EMC consists of lead sheets sandwiched with scintillation material that measures the ionization from tracks in the electromagnetic shower. The detected light energy is proportional to the energy of the electron or photon. With suitable calibration, the energy of the electron or photon is determined with the EMC.

An HC includes iron plates sandwiched with scintillation material. Almost all EM showers are terminated by the time the remaining particles traverse the HC. For hadrons (e.g., charged pions, kaons, and protons), most of the energy is deposited in the HC with 10–30% in the EMC.

An iron absorber is located after the calorimeters to absorb the remaining shower products. The final detector, a Muon Chamber, is the outermost device that follows the iron absorber. If a charged particle reaches the MC, it is unlikely to be a hadron. Therefore, the particle is a good muon candidate.

04-09

(a) Since the photon source is not polarized and no electromagnetic fields are specified, the photon emission should be isotropic. The photon release can also be modeled as a point source since the size of a 100 g sample is small compared to the 10 km distance. The gamma-ray absorbed dose (D) is

$$D = \frac{kS}{4\pi r^2},$$

where k = dose factor = 1 Gy/h = $5.5 \times 10^7 \, \gamma/cm^2 \, s$,

$S =$ gamma-ray source strength $= NY$,
$N =$ number of $^{472\mathrm{m}}$X metastable state nuclei $= (100\,\mathrm{g}/472\,\mathrm{g})$
$\times\; 6.02 \times 10^{23}$ nuclei $= 1.28 \times 10^{23}$ nuclei,
$Y =$ gamma-ray yield $= 10$,
$r =$ distance from the source $= 10\,\mathrm{km} = 10^4\,\mathrm{m}$.
Using these values determines S:

$$S = (1.28 \times 10^{23}\,\text{nuclei})\left(10\,\frac{\gamma}{\text{nucleus}}\right) = 1.28 \times 10^{24}\gamma.$$

These parameters determine the absorbed dose:

$$D = \left(\frac{1.28 \times 10^{24}\gamma}{(4\pi)(10^4\,\mathrm{m})^2}\right)\left(\frac{1\,\mathrm{m}}{100\,\mathrm{cm}}\right)^2\left(\frac{1\,\mathrm{Gy/h}}{5.5 \times 10^7\,\frac{\gamma}{\mathrm{cm}^2\,\mathrm{s}}}\right)\left(\frac{1\,\mathrm{h}}{3600\,\mathrm{s}}\right) = 0.515\,\mathrm{Gy}.$$

(b) The temperature increase (ΔT) in the calorimeter is obtained from the relationship:

$$Q = mc\Delta T,$$

$$\Delta T = \frac{1}{c}\left(\frac{Q}{m}\right),$$

where Q/m is the absorbed dose (D) or the energy absorbed (Q) per unit calorimeter mass (m) and c is the specific heat of water ($1\,\mathrm{cal/g\,^\circ C}$).
The absorbed dose at 100 m is obtained as outlined in part (a):

$$D = \left(\frac{1.28 \times 10^{24}\gamma}{(4\pi)(100\,\mathrm{m})^2}\right)\left(\frac{1\,\mathrm{m}}{100\,\mathrm{cm}}\right)^2\left(\frac{1\,\mathrm{Gy/h}}{5.5 \times 10^7\,\frac{\gamma}{\mathrm{cm}^2\,\mathrm{s}}}\right)\left(\frac{1\,\mathrm{h}}{3600\,\mathrm{s}}\right) = 5.15 \times 10^3\,\mathrm{Gy}.$$

The information needed to determine the temperature is now determined:

$$\Delta T = \frac{(5.15 \times 10^3\,\mathrm{Gy})\left(\frac{1\,\mathrm{J}}{\mathrm{kg\,Gy}}\right)\left(\frac{1\,\mathrm{kg}}{1000\,\mathrm{g}}\right)\left(\frac{1\,\mathrm{cal}}{4.186\,\mathrm{J}}\right)}{\left(\frac{1\,\mathrm{cal}}{\mathrm{g\,^\circ C}}\right)} = 1.23\,^\circ\mathrm{C}.$$

(c) For an isotropic, unattenuated point source, the distance from the source (r) and absorbed dose (D) have a well-defined relationship:

$$Dr^2 = k,$$

where k is a constant. The location where the absorbed dose decreases to 10 mSv is determined using the relationship:

$$D_1 r_1^2 = D_2 r_2^2,$$

or

$$r_2 = r_1 \left(\frac{D_1}{D_2}\right)^{1/2} = (10\,\text{km})\left(\frac{0.515\,\text{Gy}}{10\,\text{mGy}}\frac{1000\,\text{mGy}}{\text{Gy}}\right)^{1/2} = 71.8\,\text{km}.$$

04-10

(a) The average absorbed dose rate ($\dot{D}$), during the time t (1 h), at location A is obtained from the energy deposited (E) in the 1.0 cm diameter water sphere and the mass of this volume (m). The energy deposited is the product of the fraction (f_i) of each radiation type (i) deposited at a location and the total energy available for deposition for each radiation type (E_i):

$$E = \sum_{i=1}^{4} f_i E_i,$$

where $i = 1, 2, 3,$ and 4 for neutrons, protons, heavy nuclear fragments, and gamma photons, respectively. With these values, the average absorbed dose is

$$\dot{D} = \frac{E}{mt} = \frac{(750\,\text{MeV})(0.05) + (400\,\text{MeV})(0.99) + (350\,\text{MeV})(1.0) + (75\,\text{MeV})(0.1)}{\left[\frac{4}{3}\pi(0.5\,\text{cm})^3\right](1\,\text{g/cm}^3)(s^0\,\text{capture})}$$

$$\times \left(\frac{10^7\,s^0\,\text{captures}}{1\,\text{h}}\right)\left(\frac{1.6\times 10^{-13}\,\text{J}}{\text{MeV}}\right)\left(\frac{1000\,\text{g}}{\text{kg}}\right)\left(\frac{\text{Gy\,kg}}{\text{J}}\right) = 2.42\,\text{Gy/h}.$$

(b) The ratio (ξ) of average absorbed dose rates at locations A and B is

$$\xi = \frac{\dot{D}_A}{\dot{D}_B} = \frac{\frac{E_A}{mt}}{\frac{E_B}{mt}} = \frac{E_A}{E_B},$$

$$\xi = \frac{(750\,\text{MeV})(0.05) + (400\,\text{MeV})(0.99) + (350\,\text{MeV})(1.0) + (75\,\text{MeV})(0.1)}{(750\,\text{MeV})(0.95) + (400\,\text{MeV})(0.01) + (350\,\text{MeV})(0.0) + (75\,\text{MeV})(0.9)},$$

$$\xi = \frac{791\,\text{MeV}}{784\,\text{MeV}} = 1.01.$$

(c) The tabulated data suggest a complex interaction behavior for the s^0-particles. Proton and heavy fragment (HF) interactions resemble the Bethe stopping power curve for electrically charged particles with the Bragg peak occurring at location A. However, since the s^0-particle is uncharged, this behavior is peculiar.

The gamma and neutron components also have a Bragg type peak at location B. This character is atypical of the normal energy deposition for gamma and neutron radiations.

The energy deposition profile of the neutron, proton, heavy ion, and gamma radiation types does not follow the expected behavior. Therefore, it appears that the s^0-particles have an unusual character that is not well represented by conventional interaction properties.

(d) The ICRP 60 recommendation for an effective dose of 20 mSv/year is the basis for a stay-time limit (T). The effective dose is written in terms of the absorbed dose (D) and the radiation-weighting factor (w_R) for each particle type:

$$E = \sum_{i=1}^{4} w_R D_R.$$

Given the energies specified in the problem, the following ICRP 60 radiation-weighting factors are used:
- a value of 5 for neutrons with energies <10 keV;
- a value of 5 for protons with an energy of 7 MeV;
- a value of 20 for heavy fragments with an energy of 20 MeV/amu;
- a value of 1 for gamma rays with an energy of 0.25 MeV.

The annual effective dose rate at point A ($\dot{E}_A$) is

$$\dot{E}_A = \frac{E}{mt} = \frac{\begin{array}{c}[(750\,\text{MeV})(0.05)(5\,\text{Sv/Gy}) + (400\,\text{MeV})(0.99)(5\,\text{Sv/Gy}) \\ +(350\,\text{MeV})(1.0)(20\,\text{Sv/Gy}) + (75\,\text{MeV})(0.1)(1\,\text{Sv/Gy})]\end{array}}{\left[\frac{4}{3}\pi(0.5\,\text{cm})^3\right]\left(\frac{1\,\text{g}}{\text{cm}^3}\right)(s^0\,\text{capture})}$$

$$\times \left(\frac{10^7\,s^0\,\text{captures}}{1\,\text{month}}\right)\left(\frac{1.6 \times 10^{-13}\,\text{J}}{\text{MeV}}\right)\left(\frac{1000\,\text{g}}{\text{kg}}\right)\left(\frac{\text{Gy}\,\text{kg}}{\text{J}}\right)\frac{(12\,\text{months})}{\text{year}}$$

$$= 337\,\text{Sv/year}.$$

The effective dose and ICRP 60 effective dose recommendation determines the point A stay time:

$$T = \left(\frac{20\,\text{mSv}}{\left(\frac{337\,\text{Sv}}{\text{year}}\right)\left(\frac{1000\,\text{mSv}}{\text{Sv}}\right)\left(\frac{1\,\text{year}}{365\,\text{days}}\right)\left(\frac{1\,\text{days}}{24\,\text{h}}\right)\left(\frac{1\,\text{h}}{60\,\text{min}}\right)}\right) = 31.2\,\text{min}.$$

Since the interaction character of the s^0-particles is not fully understood, an exclusion zone beyond the 25 m point B location should be established. Entry time restrictions are imposed if an entry into the region penetrated by the s^0-particles occurs. The point B access times are calculated in a similar manner to the point A calculations noted above.

04-11

(a) The average absorbed dose (D) received by the reaction chamber is determined from its gamma (γ), beta (β), neutron (n), proton (p), and HF components:

$$D = D_\gamma + D_\beta + D_n + D_p + D_{HF},$$

and

$$D_i = f_i E_i,$$

where f_i is the fraction of the fission energy (E_i) of radiation type i deposited in the chamber.

$$D = \frac{E}{m} = (10^{20} \text{ fissions}) \left(\frac{1.6 \times 10^{-13} \text{ J}}{\text{MeV}}\right) \left(\frac{\text{Gy kg}}{\text{J}}\right)$$

$$\times \frac{\begin{bmatrix}(0.05)(70 \text{ MeV/fission}) + (0.62)(100 \text{ MeV/fission}) + \\ (0.12)(120 \text{ MeV/fission}) + (0.82)(50 \text{ MeV/fission}) + \\ (0.97)(100 \text{ MeV/fission})\end{bmatrix}}{5 \text{ kg}}$$

$$= 6.97 \times 10^8 \text{ Gy}.$$

(b) Heavy element ^{450}Bv and iron shielding attenuate beta particles, protons, photons, and heavier fragments. The dominant contributor to the effective dose inside the control room is derived from the neutron component and bremsstrahlung photons from beta particle attenuation in the iron and high Z shielding. Electromagnetic and hadron-initiated cascade reactions in the iron and ^{450}Bv shielding require evaluation to determine their impact in the control room.

(c) The effective dose (E) is the sum of the gamma-ray (E_γ) and neutron (E_n) components:

$$E = E_\gamma + E_n.$$

This fission reaction has no preferred direction, and it is reasonable to assume that the primary source photons and neutrons are produced in an isotropic manner. The fission source is well approximated by a point source since it is small relative to the 500 m distance to the location of interest. Using a point source relationship, the gamma effective dose is written in terms of the relationship:

$$E_\gamma = (1 - f_\gamma)(\text{AF}_\gamma) \frac{kS_\gamma}{4\pi r^2},$$

where AF_γ = shielding attenuation factor for gamma rays = 0.72; f_γ = fraction of gamma-ray energy absorbed in the reaction chamber = 0.05; k = gamma-ray dose factor = 1 Sv/h = 5.5×10^7 γ/cm² s; r = distance between the control room and the reaction chamber = 500 m; S_γ = gamma-ray source strength = (10^{20} fissions)(8 γ/fission) = $8 \times 10^{20} \gamma$.

Using these values, the gamma effective dose is

$$E_\gamma = (1-0.05)(0.72) \left(\frac{1 \text{ Sv/h}}{5.5 \times 10^7 \gamma/\text{cm}^2 \text{ s}}\right) \left(\frac{8 \times 10^{20} \gamma}{(4\pi)(500 \text{ m})^2}\right) \left(\frac{1 \text{ m}}{100 \text{ cm}}\right)^2$$

$$\times \left(\frac{1 \text{ h}}{3600 \text{ s}}\right) = 0.088 \text{ Sv}.$$

In a similar manner, the neutron effective dose is

$$E_n = (1 - f_n)(\text{AF}_n) \frac{kS_n}{4\pi r^2},$$

where AF_n = shielding attenuation factor for neutrons = 0.036; f_n = fraction of neutron energy absorbed in the reaction chamber = 0.12; k = neutron dose factor = 1 Sv/h = 8.0×10^5 n/cm² s; S_n = neutron source strength = (10^{20} fissions)(5 n/fission) = 5.0×10^{20} n.

Using these values, the neutron effective dose is

$$E_n = (1-0.12)(0.036)\left(\frac{1\text{ Sv/h}}{8.0 \times 10^5 \text{n/cm}^2\text{ s}}\right)\left(\frac{5.0 \times 10^{20}\text{ n}}{(4\pi)(500\text{ m})^2}\right)$$

$$\times \left(\frac{1\text{ m}}{100\text{ cm}}\right)^2 \left(\frac{1\text{ h}}{3600\text{ s}}\right) = 0.175 \text{ Sv}.$$

The gamma and neutron effective dose values lead to the total effective dose in the control room:

$$E = 0.088\text{ Sv} + 0.175\text{ Sv} = 0.263\text{ Sv}.$$

(d) Using the ^{450}Bv shielding the effective dose is

$$E = (0.088\text{ Sv})B(\mu_\gamma x)e^{-\mu_\gamma x} + (0.175\text{ Sv})e^{-\mu_n x},$$

where $B(\mu_\gamma x) = 1.6 + 3\mu_\gamma x = 1.6 + (3)(0.8\text{ cm}^{-1})x$ and x is the shield thickness.

$$E = (0.088\text{ Sv})(1.6 + (3)(0.8/\text{cm})x)e^{-\left(\frac{0.8}{\text{cm}}\right)x} + (0.175\text{ Sv})e^{-\frac{2.6}{\text{cm}}x},$$

$$E = (0.088\text{ Sv})\left(1.6 + \frac{2.4x}{\text{cm}}\right)e^{-\left(\frac{0.8}{\text{cm}}\right)x} + (0.175\text{ Sv})e^{-\frac{2.6}{\text{cm}}x}.$$

The required shielding to reduce the dose to 0.1 mSv inside the control room is obtained by solving the equation:

$$0.0001\text{ Sv} = (0.088\text{ Sv})\left(1.6 + \frac{2.4x}{\text{cm}}\right)e^{-\left(\frac{0.8}{\text{cm}}\right)x} + (0.175\text{ Sv})e^{-\left(\frac{2.6}{\text{cm}}\right)x}.$$

An examination of the equation suggests that the first term controls the shielding calculation. The following table summarizes the effective dose as a function of added shielding thickness x:

x (cm)	E_γ (Sv)	E_n (Sv)	E (Sv)
5	0.0219	3.96×10^{-7}	0.0219
10	7.56×10^{-4}	$<10^{-12}$	7.56×10^{-4}
11	3.71×10^{-4}	$<10^{-12}$	3.71×10^{-4}
12	1.81×10^{-4}	$<10^{-12}$	1.81×10^{-4}
12.82	1.00×10^{-4}	$<10^{-12}$	1.00×10^{-4}
13	8.78×10^{-5}	$<10^{-12}$	8.78×10^{-5}

The effective dose limit of 0.1 mSv is achieved by adding 12.82 cm of ^{450}Bv to the control room.

Solutions for Chapter 5

05-01

(a) The total electron energy (W) is the sum of the kinetic energy (T) of the electron and its rest mass equivalent:

$$W = m_o c^2 + T = 0.511 \text{ MeV} + 2000 \text{ MeV} = 2000.511 \text{ MeV}.$$

(b) The velocity β of the electron is

$$\beta = \frac{v}{c} = \left[1 - \left(\frac{m_o c^2}{E + m_o c^2}\right)^2\right]^{1/2} = \left[1 - \left(\frac{E_o}{W}\right)^2\right]^{1/2}$$

$$= \left[1 - \left(\frac{0.511 \text{ MeV}}{0.511 \text{ MeV} + 2000 \text{ MeV}}\right)^2\right]^{1/2}.$$

This expression is most readily evaluated using the power series expansion

$$(1-x)^{1/2} = 1 - \frac{1}{2}x + \cdots \quad \text{for small } x,$$

where

$$x = \left(\frac{0.511 \text{ MeV}}{0.511 \text{ MeV} + 2000 \text{ MeV}}\right)^2 = 6.524\,690\,457 \times 10^{-8}.$$

$$\beta = 1 - \frac{1}{2}x = 1 - \frac{6.524\,690\,457 \times 10^{-8}}{2} = 0.999\,999\,967.$$

(c) The Lorentz factor (γ) is determined from the relationship

$$\gamma = \frac{1}{\sqrt{1-\beta^2}} = \frac{1}{\sqrt{1-(0.999\,999\,967)^2}} = 3915.1.$$

For electrons and positrons, γ is reasonably well approximated by

$$\gamma \approx 1957 E \text{ [GeV]} = (1957)(2) = 3914.$$

(d) The average electron velocity ($\bar{\beta}$) along the direction of motion is written in terms of the undulator parameter ($K=2$):

$$\bar{\beta}_z = \bar{\beta} \cong \beta\left(1 - \frac{K^2}{4\gamma^2}\right) \cong 1 - \frac{1}{2\gamma^2} - \frac{K^2}{4\gamma^2} = 1 - \frac{1}{2(3915.1)^2} - \frac{(2)^2}{4(3915.1)^2} \approx 1.$$

(e) The electron's maximum deflection angle (ξ) in milliradians (mr) is determined from the relationship

$$\xi = \frac{K}{\gamma} = \frac{2}{3915.1} = 0.511 \text{ mr}.$$

(f) The maximum oscillation amplitude of the electron is obtained from the relationship

$$x = \frac{Kc}{\gamma\Omega}\sin\Omega t,$$

where the oscillation frequency Ω is obtained from the relationships

$$kz = \Omega t,$$

$$k = \frac{2\pi}{\lambda_o},$$

$$z = \bar{\beta}ct,$$

where k is the wave number, t is the time, and λ_o is the undulator period (50 mm). Using these relationships,

$$\Omega = \frac{2\pi\bar{\beta}c}{\lambda_o} = \frac{2\pi(1)(3 \times 10^8 \text{ m/s})}{0.05 \text{ m}} = 3.77 \times 10^{10} \text{ s}^{-1}.$$

The maximum amplitude (A) occurs when $\sin\Omega t = 1$:

$$A = \frac{Kc}{\gamma\Omega} = \frac{(2)(3 \times 10^8 \text{ m/s})}{(3915.1)(3.77 \times 10^{10}/\text{s})} = 4.07 \times 10^{-6} \text{ m} = 4.1 \text{ μm}.$$

(g) The electron's maximum displacement in the z-direction (η) is obtained from the relationship

$$\eta = -\frac{K^2 c}{8\gamma^2\Omega}\sin 2\Omega t.$$

When sin$2\Omega t = 1$, the maximum displacement is determined:

$$\eta = \frac{K^2 c}{8\gamma^2 \Omega} = \frac{(2)^2(3 \times 10^8 \text{ m/s})}{(8)(3915.1)^2(3.77 \times 10^{10}/\text{s})} = 2.6 \times 10^{-10} \text{ m} = 2.6 \text{ Å}.$$

(h) The emitted wavelength of the fundamental ($n = 1$) of this device along the beam direction ($\theta = 0$) is

$$\lambda = \frac{\lambda_o}{2\gamma^2} \frac{1}{n} \left(1 + \frac{K^2}{2} + \gamma^2 \theta^2\right) = \frac{\lambda_o}{2\gamma^2} \frac{1}{n} \left(1 + \frac{K^2}{2}\right),$$

$$\lambda = \frac{0.05 \text{ m}}{(2)(3915.1)^2} \left(1 + \frac{(2)^2}{2}\right) = 16.3 \times 10^{-10} \text{ m}(1 + 2 + 0)$$
$$= 48.9 \times 10^{-10} \text{ m} = 48.9 \text{ Å}.$$

05-02

The FEL output is a beam not a broad source of photons. As such, no buildup factor is required. However, the design calculations involve some photon scatter that should be evaluated using codes summarized in Appendix J (e.g., Electron Gamma Shower (EGS)).

05-03

The shielding for the X-ray tube is minimal in comparison to the shielding associated with an XFEL. As the output energies of the X-ray tube and the XFEL are the same and each is to be designed to meet the same design kerma rate $\dot{K}_{STD}$, expressions can be written as follows for the shielding requirements for each device:

$$\frac{\dot{K}_{STD}}{\dot{K}_T} = \left(\frac{1}{2}\right)^{N_T},$$

$$\frac{\dot{K}_{STD}}{\dot{K}_{FEL}} = \left(\frac{1}{2}\right)^{N_{FEL}},$$

where $\dot{K}_T$ is the unshielded kerma rate from the X-ray tube, N_T is the number of half-value layers required to bring the tube kerma rate to the design value, $\dot{K}_{FEL}$ is the unshielded kerma rate from the XFEL, and N_{FEL} is the number of half-value layers required to bring the XFEL kerma rate to the standard value. Using the previous equations, the X-ray tube and XFEL outputs may be compared:

$$\dot{K}_{STD} = \dot{K}_T \left(\frac{1}{2}\right)^{N_T} = \dot{K}_{FEL}\left(\frac{1}{2}\right)^{N_{FEL}},$$

or

$$\frac{\dot{K}_T}{\dot{K}_{FEL}} = \left(\frac{1}{2}\right)^{N_{FEL}-N_T}.$$

As the kerma rate is proportional to the brightness values, the difference in shielding requirements between X-ray tubes and XFELs is determined as

$$\frac{10^8}{10^{35}} = \left(\frac{1}{2}\right)^N,$$

where

$$N = N_{FEL} - N_T.$$

The solution of this equation leads to the result

$$10^{-27} = \left(\frac{1}{2}\right)^N,$$

$$\ln(10^{-27}) = N \ln(1/2),$$

$$N = \frac{\ln(10^{-27})}{\ln(1/2)} = 89.7,$$

or about 90 additional half-value layers are needed to shield the XFEL.

05-04

(a) The spectral flux density $(dF/d\theta)$ at the critical wavelength (λ_c) is

$$\frac{dF}{d\theta} = \frac{2.46 \times 10^{13} \gamma}{\text{s mr } 0.1\%\text{bandwidth A GeV}} I_o \,[\text{A}]\, E\,[\text{GeV}] \left(\frac{\lambda}{\lambda_c}\right)^2 G\left(\frac{\lambda}{\lambda_c}\right),$$

where I_o is the electron beam current (0.2 A), the wavelength (λ) is equal to the critical wavelength (λ_c), and the beam energy (E) is 6 GeV. Using these values, the spectral flux density is

$$\frac{dF}{d\theta} = \frac{2.46 \times 10^{13} \gamma}{\text{s mr 0.1\%bandwidth A GeV}} (0.2\,\text{A})(6\,\text{GeV})(1)^2(0.65).$$

$$\frac{dF}{d\theta} = 1.92 \times 10^{13} \frac{\gamma}{\text{s mr 0.1\%bandwidth}}.$$

(b) The electron beam standard deviations ($\sigma_x = 0.07$ mm, $\sigma_z = 0.032$ mm, and $\sigma_z' = 0.055$ mr) lead to the average dipole brightness (AB). Before calculating the AB, the σ_γ' parameter used in its definition is determined:

$$\gamma = 1957 E\,[\text{GeV}] = (1957)(6) = 1.17 \times 10^4,$$

$$\sigma_\gamma' = \sqrt{(\sigma_z')^2 + 0.41 \frac{\lambda}{\lambda_c} \frac{1}{\gamma^2}} = \sqrt{(0.055)^2 + (1)(1.17 \times 10^4)^{-2}}\,\text{mr} = 0.055\,\text{mr},$$

$$\text{AB} = \frac{\frac{dF}{d\theta}}{(2.36\sigma_x)(2.36\sigma_z)(2.36\sigma_\gamma')},$$

$$\text{AB} = \frac{1.92 \times 10^{13} \frac{\gamma}{\text{s mr 0.1\%bandwidth}}}{(2.36)^3 (0.07\,\text{mm})(0.032\,\text{mm})(0.055\,\text{mr})},$$

$$\text{AB} = 1.19 \times 10^{16} \frac{\gamma}{\text{s mm}^2\,\text{mr}^2\,0.1\%\text{bandwidth}}.$$

05-05

(a) The total power in the synchrotron spectrum is

$$P\,[\text{kW}] = \left(88.5 \frac{\text{kW m}}{\text{GeV}^4\,\text{A}}\right) \frac{E^4\,[\text{GeV}]\,I_o\,[\text{A}]}{R\,[\text{m}]},$$

where the electron beam energy E is 50 GeV, the average beam current I_o is 6 mA, and the bending radius r is 3096 m. Using these values, the total power is

$$P = \frac{\left(88.5 \frac{\text{kW m}}{\text{GeV}^4\,\text{A}}\right)(50\,\text{GeV})^4(0.006\,\text{A})}{(3096\,\text{m})} = 1.07 \times 10^3\,\text{kW}.$$

(b) The critical wavelength is

$$\lambda_c [\text{Å}] = \frac{5.59\,\text{Å}\,\text{GeV}^3}{\text{m}} \frac{R[\text{m}]}{E^3[\text{GeV}]},$$

$$\lambda_c = \frac{5.59\,\text{Å}\,\text{GeV}^3}{\text{m}} \frac{3096\,\text{m}}{(50\,\text{GeV})^3} = 0.138\,\text{Å}.$$

(c) The critical energy is

$$\varepsilon_c [\text{keV}] = \frac{12.39\,\text{keV}\,\text{Å}}{\lambda_c[\text{Å}]},$$

$$\varepsilon_c [\text{keV}] = \frac{12.39\,\text{keV}\,\text{Å}}{0.138\,\text{Å}} = 89.8\,\text{keV}.$$

(d) As the dose rate scales with the synchrotron power output, any change in parameters affects the dose rate according to the relationship

$$\dot{D} = \frac{cE^4 I_o}{R},$$

where c is a constant.
Therefore, the ratio of dose rates is

$$\frac{\dot{D}_2}{\dot{D}_1} = \frac{E_2^4\,I_{o,2}\,R_1}{E_1^4\,I_{o,1}\,R_2} = \frac{(500\,\text{GeV})^4}{(50\,\text{GeV})^4}\frac{(75\,\text{mA})}{(6\,\text{mA})}\frac{(3.096\,\text{km})}{(10\,\text{km})} = 3.87 \times 10^4.$$

05-06

(a) The wavelength (λ) of the fundamental ($n=1$) is determined from the relationship

$$\lambda_n = \frac{\lambda_o}{n}\frac{1}{2\gamma^2}\left(1+\frac{K^2}{2}\right),$$

where the electron beam energy (E) is 6 GeV, the undulator period (λ_o) is 46 mm, and the effective magnetic field (B_o) is 0.233 T.
The deflection parameter K and Lorentz factor (γ) are calculated using the relationships

$$K = \frac{93.4}{\text{m}\,\text{T}}\lambda_o[\text{m}]\,B_o[\text{T}] = \frac{93.4}{\text{m}\,\text{T}}(0.046\,\text{m})(0.233\,\text{T}) = 1,$$

$\gamma = 1957 E\,[\text{GeV}] = (1957)(6) = 1.17 \times 10^4.$

These values determine the fundamental wavelength:

$$\lambda_1 = \frac{(0.046\,\text{m})}{(1)} \frac{1}{2(1.17 \times 10^4)^2}\left(1 + \frac{1^2}{2}\right) = 2.52 \times 10^{-10}\,\text{m} = 2.52\,\text{Å}.$$

(b) The fundamental flux in the central radiation cone (F_1) is

$$F_1 = \frac{1.43 \times 10^{14}\gamma}{\text{s A 0.1\%bandwidth}} \frac{L\,[\text{m}]}{\lambda_o\,[\text{m}]} I_o\,[\text{A}] Q_1(K),$$

where L is the undulator length (1.66 m), the average beam current I_o is 0.2 A, and the spectral harmonic factor for the fundamental (f_1) is 0.37. The final factor needed to determine the fundamental flux is the parameter Q_1 defined by the relationship

$$Q_1(k) = \left(1 + \frac{K^2}{2}\right)\frac{f_1}{n} = \left(1 + \frac{(1)^2}{2}\right)\frac{0.37}{1} = 0.555.$$

These values determine the flux in the central radiation cone:

$$F_1 = \left(1.43 \times 10^{14} \frac{\gamma}{\text{s A 0.1\%bandwidth}}\right)\left(\frac{1.66\,\text{m}}{0.046\,\text{m}}\right)(0.2\,\text{A})(0.555),$$

$$F_1 = 5.73 \times 10^{14} \frac{\gamma}{\text{s 0.1\% bandwidth}}.$$

(c) The average on-axis brightness of the fundamental is

$$\text{AOAB} = \frac{F_1}{(2.36)^4 \sigma_{\gamma x}\sigma'_{\gamma x}\sigma_{\gamma z}\sigma'_{\gamma z}},$$

where the photon source size $\sigma_{\gamma x}(\sigma_{\gamma z})$ in the x (z)-direction is 0.06 mm (0.013 mm) and the photon source divergence $\sigma'_{\gamma x}(\sigma'_{\gamma z})$ in the x (z)-direction is 0.12 mr (0.012 mr). Using these values, the AOAB is

$$\text{AOAB} = \frac{5.73 \times 10^{14}\,\dfrac{\gamma}{\text{s 0.1\%bandwidth}}}{(2.36)^4 (0.06\,\text{mm})(0.12\,\text{mr})(0.013\,\text{mm})(0.012\,\text{mr})},$$

$$\text{AOAB} = 1.64 \times 10^{19} \frac{\gamma}{\text{s mm}^2\,\text{mr}^2\,0.1\%\text{bandwidth}}.$$

(d) Given that the fundamental flux in the central cone expands to fully illuminate 1.0 sr at a distance (r) of 500 m, the flux-to-dose-conversion factor (DCF) is 2.2×10^{-10} (Gy/h)/(γ/m² s), and that the bandwidth multiplier (p) is $2.0 \times 0.1\%$bandwidth, the absorbed dose rate ($\dot{D}$) at 500 m is

$$\dot{D} = \frac{F_1 p}{g r^2} (\mathrm{DCF}),$$

where g is the opening angle for flux propagation (1 sr). Using these values, the absorbed dose rate is

$$\dot{D} = \frac{\left(5.73 \times 10^{14} \frac{\gamma}{\mathrm{s}\, 0.1\%\mathrm{bandwidth}}\right)(2)(0.1\%\mathrm{bandwidth})\left(2.2 \times 10^{-10} \frac{\mathrm{Gy/h}}{\gamma/\mathrm{m}^2\mathrm{s}}\right)}{(1)(500\,\mathrm{m})^2}$$

$$= 1.01\,\mathrm{Gy/h}.$$

(e) As all three options meet the scientific objectives of the USEA, the preferred option from a radiological perspective is the one with the lowest absorbed dose rate. The absorbed dose rate is written as

$$\dot{D} = \frac{F_1 p}{g r^2}(\mathrm{DCF}) = \frac{\zeta}{g r^2} L I_\mathrm{o},$$

where ζ is a constant and g is the number of steradians illuminated at a distance r. Using this relationship and the values provided in the problem statement for the number of steradians illuminated at a distance r, average beam current (I_o), and undulator length (L), the absorbed dose rates for the three options are

$$\dot{D}_1 = \frac{\zeta (0.5\,\mathrm{A})(2.5\,\mathrm{m})}{(0.5)(600\,\mathrm{m})^2} = 6.94 \times 10^{-6}\zeta,$$

$$\dot{D}_2 = \frac{\zeta (1.0\,\mathrm{A})(3.0\,\mathrm{m})}{(2.0)(400\,\mathrm{m})^2} = 9.38 \times 10^{-6}\zeta,$$

$$\dot{D}_3 = \frac{\zeta (0.3\,\mathrm{A})(5.0\,\mathrm{m})}{(\pi)(500\,\mathrm{m})^2} = 1.91 \times 10^{-6}\zeta.$$

From a radiological perspective, the preferred option is the one with the lowest dose rate. Therefore, one should select Option 3.

05-07

(a) The output wavelength of the first harmonic ($n = 2$) of the FEL is obtained from the relationship

$$\lambda = \left(\frac{1 + K^2}{2\gamma^2}\right)\frac{\lambda_\mathrm{o}}{n},$$

where the wiggler period (λ_o) is 2.5 cm, the wiggler parameter (K) is 1.0, and the Lorentz factor (γ) is given by the relationship

$$\gamma = 1957 E\,[\text{GeV}] = 1957(10) = 1.96 \times 10^4.$$

Using these values, the output wavelength of the first harmonic of the FEL is

$$\lambda = \left(\frac{1+(1.0)^2}{2(1.96 \times 10^4)^2}\right)\frac{0.025\,\text{m}}{2} = 3.25 \times 10^{-11}\,\text{m} = 0.325\,\text{Å}.$$

(b) The electron beam energy corresponding to an output wavelength of 0.1 Å for the first harmonic is obtained from the relationships presented in the previous question:

$$\lambda = \left(\frac{1+K^2}{2\left(\frac{1957E}{\text{GeV}}\right)^2}\right)\frac{\lambda_o}{2},$$

$$E = \frac{\text{GeV}}{1957(2)}\sqrt{(1+K^2)\frac{\lambda_o}{\lambda}} = \frac{\text{GeV}}{1957(2)}\sqrt{(1+1^2)\frac{0.025\,\text{m}}{0.1\times 10^{-10}\,\text{m}}} = 18.1\,\text{GeV}.$$

05-08

(a) The required optical density (OD) of the goggles must reduce the radiant exposure (H) at 1 m to the maximum permissible exposure (MPE) value. The MPE is determined from the requirements of ANSI Z136.1-2000. This standard requires the computation of three values and the MPE is the smallest value in this set.

MPE-1: Single-pulse MPE

MPE-1 uses the pulse width as the exposure time. MPE-1 is determined from the wavelength (694.3 nm) and pulse duration (exposure time) of 10 μs. Using the table provided in the problem and these values leads to MPE-1 = $5 \times 10^{-7}\,\text{J/cm}^2$.

MPE-2: Average power MPE for thermal and photochemical hazards

MPE-2 is determined from the relationship

$$\text{MPE-2} = \frac{\text{Continuous wave(CW) MPE for the same wavelength}}{\text{PRF}}.$$

To determine MPE-2, the following values are required:

Pulse repetition frequency(PRF) = $2\,\text{min}^{-1}$,

CW-MPE@694.3 nm = $C_B \times 10^{-6}\,\text{W/cm}^2$,

where

$$C_B = 10^{15(0.6943-0.550)} = 146.$$

Using these values MPE-2 is obtained:

$$\text{MPE-2} = \frac{(146 \times 10^{-6}\,\text{W/cm}^2)(60\,\text{s/min})(\text{J/W s})}{2\,\text{min}^{-1}} = 4.38 \times 10^{-3}\,\text{J/cm}^2.$$

MPE-3: Multiple-pulse MPE for thermal hazards

$$\text{MPE-3} = \text{Single-pulse MPE} \times n^{-1/4} \text{ with } n^{-1/4} \leq 1,$$

where n is the number of pulses over the exposure duration, which is given by the product of the PRF and exposure duration. Following ANSI Z136.1-2000, the exposure duration is taken to be 0.25 s for visible lasers unless purposeful staring is intended. Therefore,

$$n = (2\,\text{min}^{-1})(1\,\text{min}/60\,\text{s})(0.25\,\text{s}) = 0.00833,$$

$$\text{MPE-3} = (5 \times 10^{-7}\,\text{J/cm}^2)(1)$$
$$= 5 \times 10^{-7}\,\text{J/cm}^2$$

Comparing the three calculated values [MPE-1 $= 5 \times 10^{-7}$ J/cm^2, MPE-2 $= 4.38 \times 10^{-3}$ J/cm^2, and MPE-3 $= 5 \times 10^{-7}$ J/cm^2] leads to the limiting MPE value of 5×10^{-7} J/cm^2. With the determination of the MPE, the radiant exposure (H) at 1 m is determined:

$$H = \frac{e}{\pi r^2},$$

where e is the laser pulse energy (20 J) and r is the laser beam radius at 1 m. The laser beam radius is obtained from the hyperbolic beam expansion criterion of ANSI Z 136.1-2000:

$$r = \left(a^2 + \frac{d^2 D^2}{4}\right)^{1/2},$$

where a is the aperture radius (1 mm), d is the distance from the aperture (1 m), and D is the divergence angle (15 mr). Using these values,

$$H = \frac{20\,\text{J}}{\pi[(0.1\,\text{cm})^2 + (100\,\text{cm})^2(0.015)^2/4]^{1/2}} = \frac{20\,\text{J}}{\pi(0.757\,\text{cm}^2)} = 8.41\,\text{J/cm}^2.$$

The laser beam area should be compared with the limiting aperture defined in ANSI Z 136.1-2000. At 100 cm, the laser beam area is $\pi(0.757\,\text{cm}^2)$ or 2.38 cm^2, which corresponds to a radius of 0.870 cm or a diameter of 1.74 cm. As the beam diameter (area) is larger than that corresponding to the limiting aperture diameter (0.7 cm), the actual area is used in the radiant exposure calculation.

Specification of the MPE and the radiant exposure permits the OD to be determined:

$$OD = \log_{10}[H/MPE] = \log_{10}\frac{8.41 \text{ J/cm}^2}{5 \times 10^{-7} \text{ J/cm}^2} = 7.23.$$

(b) The 633-nm beam has a 50 mW beam power (P), an aperture diameter of 3 mm, and a beam divergence of 0.3 mr. The emergent irradiance (E) is the radiant flux density leaving the aperture's surface and is usually expressed in W/cm^2:

$$E = \frac{P}{A} = \frac{50 \text{ mW}}{\pi(0.15 \text{ cm})^2} = 708 \text{ mW/cm}^2.$$

(c) Using the conditions of part (b), the hazardous intrabeam viewing distance is to be determined. The hazardous intrabeam viewing distance or the nominal ocular hazard distance (NOHD) is given by the ANSI Z 136.1-2000 relationship:

$$NOHD = \frac{2}{D}\sqrt{\frac{P}{\pi(MPE)} - a^2},$$

where D is the beam divergence (0.3 mr = 3×10^{-4} rad), P is the beam power (50 mW = 0.05 W), a is the aperture radius (1.5 mm), and MPE is the maximum permissible exposure for continuous viewing. The MPE is derived from the values provided in the problem table:

$$MPE = C_B \times 10^{-6} \text{ W/cm}^2 = 10^{15(\lambda - 0.550)} \times 10^{-6} \text{ W/cm}^2,$$
$$= 10^{15(0.633 - 0.550)} \times 10^{-6} \text{ W/cm}^2 = 1.76 \times 10^{-5} \text{ W/cm}^2.$$

Using the MPE and the previously defined values leads to the NOHD:

$$NOHD = \frac{2}{3 \times 10^{-4}}\sqrt{\frac{(0.05 \text{ W})}{\pi(1.76 \times 10^{-5} \text{ W/cm}^2)} - (0.15 \text{ cm})^2}$$
$$= 2.01 \times 10^5 \text{ cm} = 2.01 \text{ km}.$$

05-09

(a) The expected activation products of air are ^{11}C, ^{13}N, and ^{15}O. The production mechanisms for these radionuclides include ^{12}C(γ,n)^{11}C, ^{12}C(n,2n)^{11}C, ^{14}N(p,α)^{11}C, ^{14}N(γ,n)^{13}N, ^{14}N(n,2n)^{13}N, ^{16}O(p,α)^{13}N, ^{16}O(γ,n)^{15}O, ^{16}O(n,2n)^{15}O, and ^{14}N(p,γ)^{15}O. The photons and neutrons initiating these reactions are produced from a variety of processes including bremsstrahlung and (γ, n) and (e, e'n) reactions.

(b) The activation products of water include ^{3}H, ^{7}Be, ^{11}C, ^{13}N, and ^{15}O. ^{3}H and ^{7}Be are produced from spallation reactions with oxygen. Tritium is also produced from the ^{2}H(n, γ)^{3}H capture reaction. The production mechanisms for ^{11}C, ^{13}N, and ^{15}O and the mechanisms for photon and neutron production are the same as in part (a). Other water activation products include ^{16}N[^{16}O(n, p)^{16}N] and ^{17}N[^{17}O(n, p)^{17}N].

(c) The activation products of soil include ^{3}H and ^{22}Na. Tritium is produced from spallation reactions with soil constituents and the ^{2}H(n, γ)^{3}H reaction. ^{22}Na is generated by the ^{23}Na(γ, n)^{22}Na and ^{23}Na(n, 2n)^{22}Na reactions.

(d) There are a variety of reactions that occur as a result of the 6-GeV electrons that strike ferrous structural members. The ^{56}Fe(γ, n)^{55}Fe reaction is the most common reaction with the iron constituents. Other activation products are also produced from neutron and gamma induced reactions. From a radiological perspective, most of the personal dose is attributed to the following isotopes: ^{56}Co, ^{57}Co, ^{58}Co, ^{60}Co, ^{55}Fe, ^{59}Fe, ^{54}Mn, and ^{56}Mn.

05-10

The 6-GeV electron beam produces a variety of radiation types upon interacting with FEL structures. These reaction types include neutrons, pions (π^+, π^-, π^0), muons (μ^+ and μ^-), neutrinos (v_e and v_μ and their associated antiparticles), scattered electrons, and bremsstrahlung. Mesons heavier than pions and leptons heavier than muons will also be produced. The activation products decay primarily by beta, positron, and gamma emission. Alpha particles are also emitted in the decay of ^{16}N and neutrons are released when ^{17}N decays. In terms of effective dose, the dominant radiation types are neutrons, muons, and photons.

05-11

(a) Assuming the isotropic emission of neutrons, the direct thermal neutron fluence rate (ϕ) at a distance (r) of 3 m from the exit window is given by the relationship

$$\phi = \frac{S}{4\pi r^2},$$

where S is the thermal neutron emission source strength

$$S = IYk,$$

I is the average beam current (400 μA), Y is the thermal neutron yield (0.001 n/e), and k is a conversion factor.

$$S = [(400\,\mu A)(1.0 \times 10^{-6}\,A/\mu A)](0.001\,n/e)(C/A\,s)(1e/1.6 \times 10^{-19}\,C)$$
$$= 2.5 \times 10^{12}\,n/s.$$

Using these values, the thermal neutron fluence rate is determined:

$$\phi = \frac{2.5 \times 10^{12} \text{ n/s}}{(4\pi(300\text{ cm})^2)} = 2.21 \times 10^6 \text{ n/cm}^2 \text{ s}.$$

(b) For a thermal neutron fluence rate (ϕ) of 2.0×10^7 n/cm^2 s, the saturation activity (A_{sat}) of ^{24}Na in 1 cm^3 of concrete is given by

$$A_{sat} = N\sigma\phi = \mu\phi = \rho\frac{\mu}{\rho}\phi,$$

where ρ is the number of grams of ^{23}Na per cm^3 of concrete (0.012 g/cm^3) and μ/ρ is the cross section for the ^{23}Na(n, γ)^{24}Na reaction (0.0139 cm^2/g). Using these values, the saturation activity of ^{24}Na is

$$A_{sat} = (0.012 \text{ g/cm}^3)(0.0139 \text{ cm}^2/\text{g})(2.0 \times 10^7 \text{ n/cm}^2 \text{ s})\left(\frac{1 \text{ dis}}{n}\right)\left(\frac{1 \text{ Bq s}}{\text{dis}}\right),$$

$$A_{sat} = 3.34 \times 10^3 \text{ Bq/cm}^3.$$

(c) The ratio of saturation activities of ^{42}K and ^{24}Na is

$$\frac{A_{sat}(^{42}\text{K})}{A_{sat}(^{24}\text{Na})} = \frac{\rho(^{42}\text{K})a(^{42}\text{K})\frac{\mu}{\rho}(^{42}\text{K})\varphi}{\rho(^{24}\text{Na})a(^{24}\text{Na})\frac{\mu}{\rho}(^{24}\text{Na})\varphi},$$

where a is the abundance of the isotope. Using the values in the problem statement leads to the desired ratio

$$\frac{A_{sat}(^{42}\text{K})}{A_{sat}(^{24}\text{Na})} = \frac{(0.008 \text{ g/cm}^3)(0.0677)(1.22 \times 10^{-3} \text{ cm}^2/\text{g})(2.0 \times 10^7 \text{ n/cm}^2 \text{ s})}{(0.012 \text{ g/cm}^3)(1.0)(0.0139 \text{ cm}^2/\text{g})(2.0 \times 10^7 \text{ n/cm}^2 \text{ s})},$$

$$\frac{A_{sat}(^{42}\text{K})}{A_{sat}(^{24}\text{Na})} = 3.96 \times 10^{-3}.$$

05-12

(a) The beam dump resides in a cubicle having a volume (V) of 500 m^3, and the cubicle has an exhaust velocity (F) of 4 m^3/s. The time following shutdown of the electron beam for the radioactive gas concentration to be reduced to 2 Bq/cm^3 is obtained most expeditiously by noting that for each 10-min period ^{13}N undergoes one physical half-life while ^{15}O undergoes five. As the initial activities are similar, the problem is solved by only considering the ^{13}N half-life. Similar logic justifies the exclusion of ^{16}N from consideration.

Given the short half-lives of the radioactive gas, it is reasonable to assume that the accelerator reaches a saturation gas concentration $C(0)$. Given this condition, the radioactive gas concentration as a function of time $C(t)$ following beam shutdown is

$$C(t) = C(0)\exp\left[-\left(\frac{F}{V}+\lambda\right)t\right].$$

This equation is solved for the desired time for ^{13}N to reach $2\,\text{Bq/cm}^3$:

$$t = \frac{\ln\left(\frac{C(t)}{C(0)}\right)}{-\left(\frac{F}{V}+\lambda\right)} = \frac{\ln\left(\frac{2\,\text{Bq/cm}^3}{7.3\times 10^4\,\text{Bq/cm}^3}\right)}{-\left(\frac{4\,\text{m}^3/\text{s}}{500\,\text{m}^3}\times\frac{60\,\text{s}}{\text{min}}+\frac{0.693}{10\,\text{min}}\right)} = \frac{-10.5}{-0.549\,\text{min}^{-1}} = 19.1\,\text{min}.$$

The fact that the ^{15}O concentration is insignificant is demonstrated by calculating the concentration at 19.1 min:

$$C(t) = C(0)\exp\left[-\left(\frac{F}{V}+\lambda\right)t\right],$$

$$C(t) = (3.9\times 10^4\,\text{Bq/m}^3)\exp\left[-\left(\frac{4\,\text{m}^3/\text{s}}{500\,\text{m}^3}\times\frac{60\,\text{s}}{\text{min}}+\frac{0.693}{2\,\text{min}}\right)(19.1\,\text{min})\right]$$

$$= 0.005\,\text{Bq/cm}^3.$$

(b) The equilibrium toxic gas concentration $Z(0)$ in the beam dump cubicle is 5.5 ppm. If the mean lifetime (T) of the toxic gas is 30 min, the time for the concentration $Z(t)$ to be reduced to 0.1 ppm is obtained from the relationship

$$Z(t) = Z(0)\exp\left[-\left(\frac{F}{V}+\frac{1}{T}\right)t\right],$$

$$t = \frac{\ln\left(\frac{Z(t)}{Z(0)}\right)}{-\left(\frac{F}{V}+\frac{1}{T}\right)} = \frac{\ln\left(\frac{0.1\,\text{ppm}}{5.5\,\text{ppm}}\right)}{-\left(\frac{4\,\text{m}^3/\text{s}}{500\,\text{m}^3}\times\frac{60\,\text{s}}{\text{min}}+\frac{1}{30\,\text{min}}\right)} = \frac{-4.01}{-0.513\,\text{min}^{-1}} = 7.82\,\text{min}.$$

05-13

(a) The ozone production rate in molecules/cm^3 s is obtained from the empirical relationship

$$P(\text{molecules/cm}^3\,\text{s}) = (600\,\text{eV/cm}^4\,\text{A s})\times G I d,$$

where G has the value 10.3 molecules/100 eV for ozone, I is the average beam current (A), and d is the path length (10 m) traveled by the electron beam in air (cm).

The average beam current is obtained from the relationship

$$P = IV.$$

In this equation, V is the accelerating potential for the electron beam and P is the average beam power. The average power is derived from the duty factor (DF = 0.12) and peak power (P_{peak} = 50 MW):

$$P = (DF)P_{peak} = (0.12)(50\,\text{MW}) = 6.0\,\text{MW}.$$

The electron beam terminal voltage (V) is readily obtained from the beam energy (E) and electron charge (e):

$$E = eV,$$
$$V = E/e,$$
$$V = 12\,\text{GeV}/e = 1.2 \times 10^{10}\,\text{eV}/e = 1.2 \times 10^{10}\,\text{V}.$$

These values determine the average beam current:

$$I = \frac{P}{V} = \frac{6 \times 10^6\,\text{W}}{1.2 \times 10^{10}\,\text{V}} = 5 \times 10^{-4}\,\text{A}.$$

With these values, the ozone production rate is determined:

$$\begin{aligned}P(\text{molecules/cm}^3\,\text{s}) &= (600\,\text{eV/cm}^4\,\text{A s}) \times GId, \\ &= (600\,\text{eV/cm}^4\,\text{A s})(10.3\,\text{molecules/100 eV}) \\ &\quad \times (5 \times 10^{-4}\,\text{A})(1000\,\text{cm}), \\ &= 30.9\,\text{molecules/cm}^3\,\text{s}.\end{aligned}$$

(b) Given an NO_x production rate [$\dot{Z}(0)$] of 150 molecules/cm^3 s and the mean NO_x lifetime (T) of 1800 s, the steady-state concentration in the support area is obtained by integrating the concentration rate expression from $t = 0$ to time τ:

$$\dot{Z}(t) = \dot{Z}(0)\exp\left[-\left(\frac{F}{V} + \frac{1}{T}\right)t\right],$$

$$\int_0^\tau \dot{Z}(t)dt = \int_0^\tau \dot{Z}(0)\exp\left[-\left(\frac{F}{V} + \frac{1}{T}\right)t\right]dt,$$

where the support area volume (V) is 95 m³ and the support area ventilation rate (F) is 5 m/s.

As the initial concentration at the start of operations $Z(0) = 0$ and the production rate $\dot{Z}(0)$ is a constant, the concentration at time τ, $Z(\tau)$, is determined:

$$Z(\tau) = \frac{\dot{Z}(0)}{\left(\dfrac{F}{V} + \dfrac{1}{T}\right)} \left\{ 1 - \exp\left[-\left(\dfrac{F}{V} + \dfrac{1}{T}\right)\tau\right]\right\}.$$

The steady-state concentration $Z(\infty)$ occurs at large times relative to the removal term:

$$Z(\infty) = \frac{\dot{Z}(0)}{\left(\dfrac{F}{V} + \dfrac{1}{T}\right)} = \frac{150 \,\text{molecules}/\text{cm}^3\,\text{s}}{\left(\dfrac{5\,\text{m}^3/\text{s}}{95\,\text{m}^3} + \dfrac{1}{1800\,\text{s}}\right)} = 2.82 \times 10^3 \,\text{molecules}/\text{cm}^3.$$

05-14

(a) For a helical wiggler and the TQFELL parameters, the j-value is given by the relationship

$$j = \left[\frac{\pi Z e}{mc^2}\right] \frac{IL^3 K^2 (1+K^2)}{A\lambda \gamma^5},$$

where the beam current (I) is 10 A, the interaction length (L) is 1 m, the undulator parameter (K) is 0.7, the Lorentz factor γ is 100, the beam plus optical mode area (A) is $2\times 10^{-6}\,\text{m}^2$, the output wavelength (λ) is 1 μm, and the impedance of free space (Z) is 377 Ω. Given these parameters, the j-value is

$$j = \frac{\pi(377\,\Omega)(1.6\times 10^{-19}\,\text{C})}{(0.511\,\text{MeV})(1.6\times 10^{-13}\,\text{J}/\text{MeV})} \frac{(10\,\text{A})(1\,\text{m})^3(0.7)^2(1+(0.7)^2)}{(2\times 10^{-6}\,\text{m}^2)(1\times 10^{-6}\,\text{m})(100)^5} = 0.846.$$

(b) The maximum gain is obtained from the low gain relationship

$$G = \frac{j}{4} \frac{d}{d\theta}\left(\frac{\sin\theta}{\theta}\right)^2,$$

where the parameter θ is one-half the phase slip from synchronism over the interaction length. As the j-value is a constant for a given configuration, the maximum gain value is obtained by evaluating the function ξ

$$\xi = \frac{d}{d\theta}\left(\frac{\sin\theta}{\theta}\right)^2.$$

This maximum occurs when $\theta = -1.3$ rad (r). The reader should verify this by plotting the derivative or numerically evaluating ξ. The maximum gain is obtained by evaluating ξ at $\theta = -1.3\, r$.

$$G_{max} = \frac{j}{4}\left(\frac{2\theta^2 \sin\theta\cos\theta - 2\theta\sin^2\theta}{\theta^4}\right)_{\theta=-1.3\,r},$$

$$G_{max} = \frac{j}{2}\left(\frac{\theta\sin\theta\cos\theta - \sin^2\theta}{\theta^3}\right)_{\theta=-1.3\,r},$$

$$G_{max} = \left(\frac{0.846}{2}\right)\left(\frac{(-1.3)(-0.964)(0.267)-(-0.964)^2}{(-1.3)^3}\right) = 0.114.$$

(c) The second configuration only differs from the first in the current, interaction length, and undulator parameter. These differences in the second configuration include a beam current of 100 A, an interaction length of 5 m, and an undulator parameter of 1.2.

The absorbed dose is proportional to the gain:

$$D = CG,$$

where C is a constant. Using the gain relationship and the configuration differences, the absorbed dose ratio is

$$\frac{D_2}{D_1} = \frac{I_2 L_2^3 K_2^2 (1+K_2^2)}{I_1 L_1^3 K_1^2 (1+K_1^2)},$$

$$\frac{D_2}{D_1} = \frac{(100\,\text{A})(5\,\text{m})^3 (1.2)^2 (1+(1.2)^2)}{(10\,\text{A})(1\,\text{m})^3 (0.7)^2 (1+(0.7)^2)} = 6.02 \times 10^3.$$

The configuration change increases the absorbed dose by a factor of about 6000.

05-15

(a) The buildup time (t_b) is determined from the relationship

$$t_b = \frac{2L_c}{c} \frac{\log(P/P_o)}{\log(1+G_n)},$$

where the output (P) to input (P_o) power ratio is 10^{10}, the net gain (G_n) is 0.2, and the cavity length (L_c) is 2 m. These values determine the buildup time

$$t_b = \frac{2(2\,\text{m})}{3\times 10^8\,\text{m/s}} \frac{\log(10^{10})}{\log(1+0.2)} = 1.68 \times 10^{-6}\,\text{s}.$$

420 | Solution

(b) A buildup time of 7×10^{-7} s is proposed as a modification to the configuration in Question (a). If all other parameters remain unchanged, the power ratio (P/P_o) is obtained from the buildup time equation

$$t_b = \frac{2L_c}{c} \frac{\log(P/P_o)}{\log(1+G_n)},$$

$$\log(P/P_o) = \frac{ct_b \log(1+G_n)}{2L_c},$$

$$P/P_o = (1+G_n)^{ct_b/2L_c},$$

$$P/P_o = (1+0.2)^{(3\times 10^8 \text{ m/s})(4\times 10^{-7} \text{ s})/(2(2 \text{ m}))} = 1.2^{30} = 237.$$

(c) The half-divergence angle (Ψ) is determined from the output wavelength (λ) of 1 μm and interaction length (L) of 2 m. Using these values, the half-divergence angle is

$$\Psi = \sqrt{\frac{\lambda}{2L}} = \sqrt{\frac{1\times 10^{-6} \text{ m}}{2(2 \text{ m})}} = 5\times 10^{-4} \text{ r} = 0.5 \text{ mr}.$$

(d) For the parameters of Question (c), the beam emittance (ε) is determined from the relationship

$$\varepsilon \leq \frac{\lambda}{2\pi^{1/2}} \leq \frac{1 \text{ μm}}{2\pi^{1/2}} \leq 0.282 \text{ μm}.$$

05-16

(a) The ratio of FEL gain values with and without the optical klystron (OK) are obtained from the individual gain relationships

$$G_{OK} = \frac{16.5}{T^2 \text{ A m}^2} \delta L^2 S^3 B^2,$$

$$G_{FEL} = \frac{3.12 \times 10^{-4}}{A} \delta (1+K^2) L_{FEL}^3,$$

where δ is a configuration independent parameter having units of A/m³, the dispersive magnet length (S) is 0.2 m, the length (L) of the modulation and output sections is 0.4 m, the dispersive magnet's strength is 0.3 T, the wiggler parameter (K) is 1.0, and the length of the free-electron laser (L_{FEL}) is 1 m. Using these values, the desired ratio is determined:

$$\frac{G_{OK}}{G_{FEL}} = \frac{\frac{16.5}{T^2 \, A \, m^2} \delta L^2 S^3 B^2}{\frac{3.12 \times 10^{-4}}{A} \delta(1+K^2) L_{FEL}^3} = \frac{\frac{16.5}{T^2 \, A \, m^2} L^2 S^3 B^2}{\frac{3.12 \times 10^{-4}}{A} (1+K^2) L_{FEL}^3},$$

$$\frac{G_{OK}}{G_{FEL}} = \frac{\left(\frac{16.5}{T^2 \, A \, m^2}\right)(0.4 \, m)^2 (0.2 \, m)^3 (0.3 \, T)^2}{\left(\frac{3.12 \times 10^{-4}}{A}\right)(1+(1)^2)(1 \, m)^3} = 3.05.$$

(b) The ratio of energy acceptance values ($\Delta\gamma/\gamma$) of the OK modification relative to that of the FEL is determined from the gain-energy acceptance relationship:

$$\frac{\left(G\frac{\Delta\gamma}{\gamma}\right)_{OK}}{\left(G\frac{\Delta\gamma}{\gamma}\right)_{FEL}} = \left(\frac{2L}{L_{FEL}}\right)^2,$$

$$\frac{\left(\frac{\Delta\gamma}{\gamma}\right)_{OK}}{\left(\frac{\Delta\gamma}{\gamma}\right)_{FEL}} = \left(\frac{G_{FEL}}{G_{OK}}\right)\left(\frac{2L}{L_{FEL}}\right)^2 = \left(\frac{1}{3.05}\right)\left(\frac{(2)(0.4 \, m)}{(1 \, m)}\right)^2 = 0.210.$$

Solutions for Chapter 6

06-01

(a) The ion's trajectory is a superposition of the motion in the z-direction and the (x, y) plane. In the z-direction (parallel to the magnetic induction), the ion moves with a velocity of magnitude a. In the (x, y) plane, the ion moves in a circle of radius r. The superposition of these two motions is a helical path with the particle moving in the positive z-direction.

(b) The ion's radius (r) in the (x, y) plane is determined by equating the magnitude's of the magnetic force (F_{mag}) and centripetal force (F_C):

$$\vec{F}_{mag} = q\vec{v} \times \vec{B},$$

$$|F_C| = \frac{mb^2}{r},$$

where q is the ion's charge, b is the ion's velocity in the (x, y) plane, B is the magnetic induction, and m is the ion's mass.

For simplicity, the radius is determined by considering the specific instant when the velocity and magnetic induction have the configurations

$$\vec{v} = b\,\hat{j},$$

$$\vec{B} = B_o\hat{k}.$$

Using these values, the magnetic force is

$$\vec{F}_{\text{mag}} = q \begin{vmatrix} \hat{i} & \hat{j} & \hat{k} \\ 0 & b & 0 \\ 0 & 0 & B_o \end{vmatrix} = qbB_o\hat{i},$$

$$|F_{\text{mag}}| = qbB_o.$$

Using these relationships, the ion's radius in the (x, y) plane is determined by equating the forces' magnitude:

$$qbB_o = \frac{mb^2}{r},$$

$$r = \frac{mb}{qB_o}.$$

06-02

Given a cosmic ray flux (ϕ) of 2 protons/cm^2 s, an atmospheric density (ρ) of 5×10^5 atoms/cm^3 at 1000 km altitude, and a mean cross section (σ) describing the interaction of protons with atoms in the atmosphere of 0.2 b, the source strength (S) of cosmic ray generated trapped particles (tp) is

$$S = \rho\sigma\phi,$$

$$S = \left(5 \times 10^5 \frac{\text{atoms}}{\text{cm}^3}\right)\left(\frac{0.2\,\text{b}\ 10^{-24}\,\text{cm}^2}{\text{atom}\ \ \text{b}}\right)\left(2\frac{\text{protons}}{\text{cm}^2\,\text{s}}\right)\left(\frac{1\,\text{tp}}{\text{proton}}\right)$$

$$= 2 \times 10^{-19} \frac{\text{tp}}{\text{cm}^3\,\text{s}},$$

This trapped particle cosmic ray source is negligible compared to the measured trapped particle density. Review the calculation with the student and have him reconsider his conclusion.

06-03

(a) The energy of solar neutrons is 1 MeV and the neutrons have a mean lifetime of 886 s. To determine the surviving neutron fraction, the neutron velocity is determined from the relativistic relationship for the total energy (W):

$$W = T + mc^2 = \frac{mc^2}{\sqrt{\left(1 - \frac{v^2}{c^2}\right)}},$$

where T is the neutron's kinetic energy (1 MeV), m is the neutron rest mass (939.6 MeV), c is the speed of light (3×10^8 m/s), and v is the neutron's velocity.

The total energy relationship is solved for the neutron's velocity:

$$v = c\left(1 - \left(\frac{mc^2}{mc^2 + T}\right)^2\right)^{1/2} = c\left(1 - \left(\frac{939.6 \text{ MeV}}{939.6 \text{ MeV} + 1 \text{ MeV}}\right)^2\right)^{1/2} = 0.0461\, c.$$

The time (t) for a 1-MeV solar neutron to reach the Earth is given by the relationship

$$s = vt,$$

where s is the distance between the Earth and the Sun (1.5×10^8 km). Given these values, the time is

$$t = \frac{s}{v} = \frac{(1.5 \times 10^8 \text{ km})(1000 \text{ m/km})}{(0.0461)(3 \times 10^8 \text{ m/s})} = 1.08 \times 10^4 \text{ s}.$$

The neutron's mean lifetime (τ) is 886 s, which corresponds to a disintegration constant of

$$\lambda = \frac{1}{\tau} = \frac{1}{886 \text{ s}} = 1.13 \times 10^{-3}/\text{s}.$$

The fraction (f) of neutrons reaching the earth is

$$f = \frac{N(t)}{N(0)} = \frac{N(0)\exp(-\lambda t)}{N(0)} = \exp(-\lambda t) = \exp[(-1.13 \times 10^{-3}/\text{s})(1.08 \times 10^4 \text{ s})],$$

$$f = \exp(-12.2) = 5.03 \times 10^{-6}.$$

Therefore, few neutrons reach the Earth from the Sun. Neutrons produced from cosmic ray interactions dominate the solar neutron source. Therefore, the solar neutrons are of negligible importance to the neutron source term.

(b) The neutron decay products are the proton, electron, and antielectron neutrino:

$$n \rightarrow p + e^- + \bar{\nu}_e.$$

(c) Neutrons produced in the upper atmosphere can decay, interact with atmospheric nuclei, or reach the Earth and contribute to its radiation environment. The neutrons generated from cosmic ray interactions decay as noted in Question (b). For the energies considered in this problem, the antineutrino has no radiological consequences. However, the proton and electron are charged particles that are trapped by the magnetic field of the Earth and become part of the van Allen belts. Neutrons and electrons created by cosmic rays also reach the Earth's surface where they contribute to the background radiation environment.

06-04

(a) The contribution of cosmic rays to the background radiation level at sea level is about 0.26 mSv.

(b) Given the ground-level contribution $H(0)$ from part (a), the cosmic ray dose contribution at 10 000 m is

$$H(10\,000\,\text{m}) = H(0)2^{h/2000\,\text{m}} = (0.26\,\text{mSv})2^{10\,000\,\text{m}/2000\,\text{m}} = 8.32\,\text{mSv}.$$

This relationship is based on the fact that the cosmic ray dose contribution doubles for every 2000 m increase in altitude.

06-05

(a) Artificial radiation belts result from the high altitude detonation of a nuclear weapon and the associated release of energetic charged particles. Electrons produced from the beta decay of fission fragments dominate the artificial radiation belts.

(b) The "Argus I" test (1 kT) was conducted to study the trapping of energetic particles by the Earth's magnetic field. The electrons from the "Argus I" detonation dispersed to form a 100-km thick shell at an altitude of about 2 R_e. The "Starfish" test was of considerably higher yield (1.4 MT) and produced a larger and more intense belt with a large electron flux value extending to 4 R_e and beyond. A maximum flux of 10^9 e/cm^{2}s occurred at about 1.3 R_e.

06-06

(a) The primary constituents of cosmic rays that strike the upper atmosphere are protons (87%), alpha particles (12%), and light nuclei (1%).

(b) The constituents of cosmic rays at sea level are muons (63%), electrons (15%), and neutrons (21%).

(c) Upon entering the upper atmosphere, the primary cosmic rays interact with oxygen and nitrogen atoms. In these collisions, protons, neutrons, and charged and neutral pions are produced. These particles penetrate the atmosphere and participate in additional nuclear interactions producing lower energy particles. A portion of these particles (primarily neutrons, electrons, and muons) reaches the Earth's surface.

The charged pions decay into muons and neutrinos. The muons interact through the weak interaction and reach the Earth's surface where they are the most abundant secondary particle. The neutral pions decay into two photons. These high-energy photons interact with the electric field of the oxygen and nitrogen nuclei and produce electron–positron pairs. These particles interact with the nuclear electric field and yield bremsstrahlung photons. The pair production/bremsstrahlung process continues through the atmosphere until the photon energies fall below the pair production threshold. This pair production/bremsstrahlung process is the electromagnetic cascade discussed in Chapter 4.

06-07

(a) The unattenuated proton effective dose is

$$E = \left(8 \times 10^5 \frac{\text{p}}{\text{cm}^2 \text{ s}}\right) \left(\frac{3000 \text{ pSv cm}^2}{\text{p}}\right) \left(\frac{1 \text{ Sv}}{10^{12} \text{ pSv}}\right) (60 \text{ min}) \frac{60 \text{ s}}{\text{min}} = 8.64 \text{ Sv}.$$

(b) The effective dose (E) to the spacecraft's occupants is

$$E = \sum_{i=1}^{3} E_i^{\text{p}} + E_i^{\text{ap}},$$

where $i = 1$ refers to the first 15-min period, $i = 2$ is the second 15-min interval, and $i = 3$ is the final 30-min interval. The effective dose component E_i^{p} is the contribution from SPE protons that reach the spacecraft without interacting with the rail-gun's antiprotons. E_i^{ap} is the effective dose resulting from the reaction products of the proton–antiproton collisions that reach the spacecraft. During any interval, the proton dose is given by the relationship

$$\dot{H}_{\text{p}} = \beta k \phi,$$

where ϕ is the proton fluence reaching the spacecraft, k is the dose-conversion factor having the value of 3×10^3 pSv cm^2/proton, and $\beta = 1$ for the SPE source term and 0.01 for the proton–antiproton annihilation products.

Solution

The individual terms for the three interaction intervals in the sum are calculated separately:

$$E_1^p = (8 \times 10^5 - 8 \times 10^5) \frac{p}{cm^2 \, s} \frac{3000 \, pSv \, cm^2}{p} \frac{1 \, Sv}{10^{12} \, pSv} (15 \, min) \frac{60 \, s}{min} = 0,$$

$$E_1^{ap} = \frac{1}{100} 3000 \frac{pSv \, cm^2}{proton} \frac{1 \, Sv}{10^{12} \, pSv} \frac{8 \times 10^5 \, proton}{cm^2 \, s} (15 \, min) \frac{60 \, s}{min} = 0.0216 \, Sv,$$

$$E_2^p = (8 \times 10^5 - 6 \times 10^5) \frac{p}{cm^2 \, s} \frac{3000 \, pSv \, cm^2}{p} \frac{1 \, Sv}{10^{12} \, pSv} (15 \, min) \frac{60 \, s}{min} = 0.54 \, Sv,$$

$$E_2^{ap} = \frac{1}{100} 3000 \frac{pSv \, cm^2}{proton} \frac{1 \, Sv}{10^{12} \, pSv} \frac{6 \times 10^5 \, proton}{cm^2 \, s} (15 \, min) \frac{60 \, s}{min} = 0.0162 \, Sv,$$

$$E_3^p = (8 \times 10^5 - 4 \times 10^5) \frac{p}{cm^2 \, s} \frac{3000 \, pSv \, cm^2}{p} \frac{1 \, Sv}{10^{12} \, pSv} (30 \, min) \frac{60 \, s}{min} = 2.16 \, Sv,$$

$$E_3^{ap} = \frac{1}{100} 3000 \frac{pSv \, cm^2}{proton} \frac{1 \, Sv}{10^{12} \, pSv} \frac{4 \times 10^5 \, proton}{cm^2 \, s} (30 \, min) \frac{60 \, s}{min} = 0.0216 \, Sv.$$

The total proton effective dose is

$$E = (0 + 0.0216 + 0.54 + 0.0162 + 2.16 + 0.0216) \, Sv = 2.76 \, Sv.$$

(c) The effective dose using the antiproton beam is higher than desirable. Other options to reduce the crew's dose include the following:
- Relocate the crew to the shielded shelter.
- Move the crew to the most forward location in the ship to maximize the available shadow shielding.
- Request the crew's physician administer radioprotective agents to minimize the effective dose.
- Accelerate the spacecraft to increase the distance from the Sun and reduce the proton fluence.
- Utilize nearby spatial bodies (e.g., planets, asteroids, and moons) to provide shielding during the SPE.
- Repair the EM deflector. Even a partially functioning deflector will reduce the effective dose.

06-08

(a) The annual effective dose for GSDC's LEO operations is

$$E = (0.06 \, mSv/h)(24 \, h/day)(365 \, day/year)(1 \, Sv/1000 \, mSv) = 0.526 \, Sv.$$

(b) NCRP 132 imposes career whole-body exposure limits for a lifetime excess risk of total cancer of 3%. For a 25-year-old male worker, the limit is 1.5 Sv. This limit would not be exceeded unless the worker served three LEO tours of duty. Company policy should address the number of duty tours that could be served by its personnel.

The 10-year career limit, based on 3% excess lifetime risk of cancer mortality, is 0.7 Sv for the worker. This limit is exceeded if more than one duty tour were served.

Assuming 1 Gy Eq is equivalent to 1 Sv, the 1-year limit for BFO of 0.5 Gy Eq is exceeded for the referenced average effective dose. As the 0.06 mSv/h value is an average, it will vary and larger values are possible. CSDC should review its policy, and consider limiting the duration of the duty to less than a year or justify other limits with supporting methodology. As the policy currently exists, a 1-year duty tour exceeds the NCRP 132 recommendations.

Shielding or an EM deflector could be used to lower the average effective dose. Both options require evaluation to determine their effectiveness.

06-09

(a) The absorbed dose received by the crew with no deflector in operation (D_o) is the product of the total fluence (Φ) and the applicable dose-conversion factor (D_F) for each flare event:

$$D_o = \sum_{i=1}^{6} D_o^i = \sum_{i=1}^{6} \Phi_i D_F^i,$$

$$\begin{aligned}D_o = &[(8 \times 10^7)(2500) + (2 \times 10^8)(3000) + (4 \times 10^8)(3200) \\ &+ (1 \times 10^9)(3100) + (5 \times 10^8)(4500) + (8 \times 10^7)(5800)] \\ &\times \left(\frac{\text{protons}}{\text{cm}^2}\right)\left(\frac{\text{pGy cm}^2}{\text{proton}}\right)\left(\frac{1\,\text{Gy}}{10^{12}\,\text{pGy}}\right),\end{aligned}$$

$$D_o = (0.2 + 0.6 + 1.28 + 3.1 + 2.25 + 0.46)\,\text{Gy} = 7.89\,\text{Gy}.$$

(b) The absorbed dose with the deflector in operation during the flare events is written in terms of the unattenuated absorbed dose (D_o) and the deflector efficiency (ε_i). The total absorbed dose is the sum over each SPE:

$$D = \sum_{i=1}^{6} D_o^i (1-\varepsilon_i),$$

$$D = \begin{bmatrix} (0.2)(1-0.9) + (0.6)(1-0.9) + (1.28)(1-0) \\ +(3.1)(1-0.6) + (2.25)(1-0.6) + (0.46)(1-0.6) \end{bmatrix} \text{Gy} = 3.68\,\text{Gy}.$$

06-10

(a) In terms of the absorbed dose rate per unit volume and unit time function ($\xi(x, t)$) defined in the problem statement, the average absorbed dose rate is

$$\dot{D} = \frac{\iiint \xi(x,t) dx dy dz dt}{\iiint dx dy dz dt} = \frac{\iint \xi(x,t) dx dt}{\iint dx dt}.$$

As $\xi(x, t)$ is separable, the distance and time coordinates are:

$$\xi(x,t) = \alpha(x)\beta(t) = (a(1-bx+ce^{-dx}))(e^{-\lambda t}).$$

The average absorbed dose is

$$\dot{D} = \frac{\int \alpha(x) dx \int \beta(t) dt}{\int dx \int dt}.$$

The average absorbed dose rate averaged over 5 cm of tissue during the first hour (T) of the event ($\dot{D}_{5,1}$) is

$$\dot{D}_{5,1} = \frac{\int_0^{5\,cm} a(1-bx+ce^{-dx}) dx}{\int_0^{5\,cm} dx} \cdot \frac{\int_0^{1\,h} e^{-\lambda t} dt}{\int_0^{1\,h} dt},$$

$$\dot{D}_{5,1} = \frac{\left[a\left(x - \frac{bx^2}{2} - \frac{ce^{-dx}}{d}\right)\right]_0^{5\,cm}}{5\,cm} \cdot \frac{\left[\frac{e^{-\lambda t}}{-\lambda}\right]_0^{1\,h}}{1\,h},$$

where a, b, c, d, and λ are $0.2\,mJ/cm^3\,h$, $0.025\,cm^{-1}$, 0.4, $0.1\,cm^{-1}$, and $1\,h^{-1}$, respectively.

$$\dot{D}_{5,1} = \frac{(0.2\,mJ/cm^3\,h)\left(5\,cm - (0.025\,cm^{-1})(5\,cm)^2/2 + \left((0.4)\frac{1-e^{-(0.1\,cm^{-1})(5\,cm)}}{(0.1\,cm^{-1})}\right)\right)}{5\,cm}$$

$$\times \left(\frac{1-e^{-(1h^{-1})(1h)}}{1h^{-1}}\right) = \frac{(0.2\,mJ/cm^3\,h)}{5\,cm}(5\,cm - 0.313\,cm + 1.57\,cm)(0.632),$$

$$= \left(0.25\,\frac{mJ}{cm^3\,h}\right)\left(\frac{1J}{1000\,mJ}\right)\left(\frac{cm^3}{1\,g}\right)\left(\frac{1000\,g}{kg}\right)\left(\frac{Gy\,kg}{J}\right)(0.632) = 0.158\,Gy/h.$$

$$D_{5,1}(total) = \dot{D}_{5,1} T = (0.158\,Gy/h)(1\,h) = 0.158\,Gy.$$

(b) The average absorbed dose averaged over 5 cm of tissue during the first 8 h (T) of the event ($D_{8,1}$) is

$$\dot{D}_{8,1} = \frac{\int_0^{5cm} a(1-bx+ce^{-dx})dx \int_0^{8h} e^{-\lambda t}dt}{\int_0^{5cm} dx \int_0^{8h} dt},$$

$$\dot{D}_{8,1} = \frac{\left[a\left(x - \frac{bx^2}{2} - \frac{ce^{-dx}}{d}\right)\right]_0^{5cm} \left[\frac{e^{-\lambda t}}{-\lambda}\right]_0^{8h}}{5\,cm \quad 1\,h},$$

$$\dot{D}_{8,1} = \frac{(0.2\,\text{mJ/cm}^3\,\text{h})\left(5\,\text{cm} - (0.025\,\text{cm}^{-1})(5\,\text{cm})^2/2 + (0.4)\frac{1-e^{-(0.1\,\text{cm}^{-1})(5\,\text{cm})}}{(0.1\,\text{cm}^{-1})}\right)}{5\,\text{cm}}$$

$$\times \left(\frac{\frac{1-e^{-(1h^{-1})(8h)}}{1h^{-1}}}{8h}\right) = \frac{(0.2\,\text{mJ/cm}^3\,\text{h})}{5\,\text{cm}}(5\,\text{cm} - 0.313\,\text{cm} + 1.57\,\text{cm})(0.125),$$

$$= \left(0.25\frac{\text{mJ}}{\text{cm}^3\,\text{h}}\right)\left(\frac{1\,\text{J}}{1000\,\text{mJ}}\right)\left(\frac{\text{cm}^3}{1\,\text{g}}\right)\left(\frac{1000\,\text{g}}{\text{kg}}\right)\left(\frac{\text{Gy}\,\text{kg}}{\text{J}}\right)(0.125) = 0.0313\,\text{Gy/h}.$$

$D_{8,1}(\text{total}) = \dot{D}_{8,1}T = (0.0313\,\text{Gy/h})(8\,\text{h}) = 0.25\,\text{Gy}.$

(c) You should recommend the acceleration. By leaving the field in 1 h, significant absorbed dose is saved (0.25 Gy − 0.16 Gy = 0.09 Gy).

06-11

(a) The warning time (t) is determined from the distance (s)–velocity (v) relationship:

$$t = \frac{s}{v} = \frac{(4.2\,\text{AU})(1.5 \times 10^{11}\,\text{m/AU})}{(0.1)(3 \times 10^8\,\text{m/s})} = 2.1 \times 10^4\,\text{s}\left(\frac{1\,\text{h}}{3600\,\text{s}}\right) = 5.83\,\text{h}.$$

(b) The absorbed dose (D) received by the crew in the 2-h period prior to the event is determined by the fluence (Φ), the dose-conversion factor (F), the spacecraft shell thickness (t), and the attenuation coefficient (μ):

$$D = \sum_{i=1}^{2} \Phi_i F_i e^{-\mu_i t},$$

where $i=1$ and 2 define the proton and heavy ion source terms, respectively. The absorbed dose is determined from the values provided in the problem statement:

$$D = \left(3 \times 10^5 \frac{p}{cm^2}\right)\left(\frac{3000 \, pGy \, cm^2}{p}\right)\left[\frac{1 \, Gy}{10^{12} \, pGy}\right] e^{-\left(\frac{0.2}{cm}\right)(3cm)}$$

$$+ \left(4 \times 10^4 \frac{HI}{cm^2}\right)\left(\frac{7000 \, pGy \, cm^2}{HI}\right)\left[\frac{1 \, Gy}{10^{12} \, pGy}\right] e^{-\left(\frac{0.35}{cm}\right)(3cm)}$$

$$= 4.94 \times 10^{-4} \, Gy + 0.98 \times 10^{-4} \, Gy = 5.92 \times 10^{-4} \, Gy.$$

(c) The absorbed dose received by the crew if only the spacecraft shell shields them is

$$D = \sum_{i=1}^{2} \Phi_i F_i e^{-\mu_i t},$$

$$D = \left(5 \times 10^9 \frac{p}{cm^2}\right)\left(\frac{4500 \, pGy \, cm^2}{p}\right)\left[\frac{1 \, Gy}{10^{12} \, pGy}\right] e^{-\left(\frac{0.15}{cm}\right)(3cm)}$$

$$+ \left(8 \times 10^9 \frac{HI}{cm^2}\right)\left(\frac{8800 \, pGy \, cm^2}{HI}\right)\left[\frac{1 \, Gy}{10^{12} \, pGy}\right] e^{-\left(\frac{0.30}{cm}\right)(3cm)}$$

$$= 14.3 \, Gy + 28.6 \, Gy = 42.9 \, Gy.$$

(d) The absorbed dose (D) received by the crew in the shelter during the 2-h event is

$$D = \sum_{i=1}^{2} \Phi_i F_i e^{-\mu_i t},$$

$$D = \left(5 \times 10^9 \frac{p}{cm^2}\right)\left(\frac{4500 \, pGy \, cm^2}{p}\right)\left[\frac{1 \, Gy}{10^{12} \, pGy}\right] e^{-\left(\frac{0.15}{cm}\right)(11 cm)}$$

$$+ \left(8 \times 10^9 \frac{HI}{cm^2}\right)\left(\frac{8800 \, pGy \, cm^2}{HI}\right)\left[\frac{1 \, Gy}{10^{12} \, pGy}\right] e^{-\left(\frac{0.30}{cm}\right)(11 cm)}$$

$$= 4.32 \, Gy + 2.60 \, Gy = 6.92 \, Gy.$$

Given these results the crew should be relocated to the shelter. However, the absorbed dose is large enough that other mitigative measures are warranted.

(e) Given the vast distance between the event and the Solar System, the fluence reaching the spacecraft is essentially the same as the fluence reaching the Earth. The unattenuated dose in the vicinity of the Earth is

$$D = \sum_{i=1}^{2} \Phi_i F_i,$$

$$D = \left(5 \times 10^9 \, \frac{p}{cm^2}\right)\left(\frac{4500 \, pGy \, cm^2}{p}\right)\left[\frac{1 \, Gy}{10^{12} \, pGy}\right]$$

$$+ \left(8 \times 10^9 \, \frac{HI}{cm^2}\right)\left(\frac{8800 \, pGy \, cm^2}{HI}\right)\left[\frac{1 \, Gy}{10^{12} \, pGy}\right],$$

$$= 22.5 \, Gy + 70.4 \, Gy = 92.9 \, Gy.$$

However, the radiation is significantly attenuated by the Earth's atmosphere:

$$D = D_{proton} + D_{Heavy \, Ion},$$

$$D = \left[(22.5 \, Gy)2^{\left(-\frac{25 \, km}{2 \, km}\right)} + (70.4 \, Gy)2^{\left(-\frac{25 \, km}{1.5 \, km}\right)}\right]\left(\frac{1000 \, mGy}{Gy}\right),$$

$$= 3.88 \, mGy + 0.68 \, mGy = 4.56 \, mGy.$$

(f) The effective dose is written in terms of the absorbed dose (D_R) from radiation of type (R) and the radiation-weighting factor (w_R). Using the values from Question (e) and the w_R values in the problem statement, the effective dose is

$$E = \sum_{R=1}^{2} w_R D_R = (3.88 \, mGy)\left(5 \, \frac{mSv}{mGy}\right) + (0.68 \, mGy)\left(20 \, \frac{mSv}{mGy}\right) = 33 \, mSv.$$

(g) Depending on the reference source, the effective dose from Question (f) is on the order of 10 times the annual natural background dose. For example, NCRP 93 assigns 3 mSv/year to the natural background dose in the United States.

Any recommendations should consider the diversity of the world's population and their geographic location. In addition, the 6-h warning time (Question (a)) is insufficient to plan and coordinate an effective response. Much of the 3-week exposure period will be consumed with planning, staging, and implementing the response. At best, only a fraction of the population could be relocated. Relocation decisions will likely be addressed on a national or even local basis.

During the 3-week period of the event, the remainder of the world's population should be encouraged to remain indoors except for normal activity. Relocation of all population groups is not warranted based on the magnitude of the effective dose and inherent logistical difficulties. Heroic measures regarding mass population relocation to sea level or underground are also not warranted.

Clear communication with the public is important. The risks from the event should be presented in a clear, logical manner. Dialogue should be encouraged, but care must be taken not to unnecessarily alarm the population. The risk from the radiation dose is relatively low and this fact must be clearly communicated.

06-12

To evaluate the best location for the water shielding, determine the shield thickness for the two possible configurations.

Case I – Shielding added to the outside of the shelter:

The volume (V) occupied by the 2000 kg mass (M) of water is determined from the density relationship

$$V = \frac{M}{\rho} = \frac{2000\,\text{kg}}{1000\,\text{kg/m}^3} = 2\,\text{m}^3.$$

If the water is added outside the shelter, it has a radius (R) given by the relationship

$$V = \frac{4}{3}\pi(R^3 - (2.0\,\text{m})^3),$$

$$R = \left(\frac{3V}{4\pi} + (2.0\,\text{m})^3\right)^{1/3} = \left(\frac{(3)(2\,\text{m}^3)}{4\pi} + (2.0\,\text{m})^3\right)^{1/3} = 2.039\,\text{m}.$$

The water shield would be 3.9 cm thick if added to the outer radius of the shelter.

Case II – Shielding added to the inside of the spacecraft shell:

Adding the water shielding inside the spacecraft's shell results in the water having an inner radius (R) given by the relationship

$$V = \frac{4}{3}\pi((10.0\,\text{m})^3 - R^3),$$

$$R = \left((10.0\,\text{m})^3 - \frac{3V}{4\pi}\right)^{1/3} = \left((10.0\,\text{m})^3 - \frac{(3)(2\,\text{m}^3)}{4\pi}\right)^{1/3} = 9.998\,\text{m}.$$

The shield thickness added to the spacecraft's inner shell is 0.2 cm.

With these values the mission effective dose (E) for the two configurations is determined:

$$E = E_{GCR} + E_{SPE} = \dot{E}_{GCR} t + E_{SPE},$$

where t is the mission time (2 years), $\dot{E}_{GCR}$ is the GCR effective dose rate (0.7 mSv/year), and E_{SPE} is the effective dose from the SPE (12 Sv). From the conditions of the problem, the GCR dose is shielded by the spacecraft shell thickness, and the SPE dose is shielded by the shelter plus shell thickness. A summary of the shielding thickness is provided in the following table:

Case	Shelter shielding (cm)[a]		Spacecraft shielding (cm)[b]	
	Original configuration	Added water shielding	Original configuration	Added water shielding
I	15	3.9	3	0
II	15	0	3	0.2

[a]Only shields the SPE dose.
[b]Shields GCR and SPE dose.

Case I – 3.9-cm water equivalent shielding added to the outer shell of the shelter:

$$E = (2\,\text{years})(0.7\,\text{Sv/year})e^{-(3\,\text{cm})(0.1\,\text{cm}^{-1})}$$
$$+ (12\,\text{Sv})e^{-(3\,\text{cm}+15\,\text{cm}+3.9\,\text{cm})(0.1\,\text{cm}^{-1})},$$
$$E = 1.04\,\text{Sv} + 1.34\,\text{Sv} = 2.38\,\text{Sv}.$$

Case II – 0.2-cm water equivalent shielding added to the inner spacecraft shell:

$$E = (2\,\text{years})(0.7\,\text{Sv/year})e^{-(3\,\text{cm}+0.2\,\text{cm})(0.1/\text{cm})}$$
$$+ (12\,\text{Sv})e^{-(3\,\text{cm}+15\,\text{cm}+0.2\,\text{cm})(0.1/\text{cm})},$$
$$E = 1.02\,\text{Sv} + 1.94\,\text{Sv} = 2.96\,\text{Sv}.$$

On the basis of these results, the water should be stored outside the shelter.

06-13

(a) During stable operating conditions, the magnetic (M) force and centrifugal (C) forces are balanced:

$$F_C = \frac{mv^2}{r},$$

$$F_M = qvB,$$

$$\frac{mv^2}{r} = qvB.$$

The force equation permits the electron momentum to be determined:

$$p = mv = qBr.$$

As the electron is relativistic with a speed approaching the speed of light, its momentum (p) in the collider is related to its energy (E):

$$E = pc.$$

Using this relationship, the electron energy can be rewritten using the momentum equation

$$E = qBrc.$$

For the values specified in the problem, the electron energy is

$$E = \frac{(1.6 \times 10^{-19}\,\text{C})(5 \times 10^{-6}\,\text{T})(6.6 \times 10^6\,\text{m})(3 \times 10^8\,\text{m/s})}{(1.6 \times 10^{-13}\,\text{J/MeV})(1000\,\text{MeV/GeV})} = 9.9\,\text{GeV}.$$

The center of mass energy is twice the electron energy or 19.8 GeV.

(b) The average power (P) radiated by a bunch consisting of 10^{18} electrons is determined from the synchrotron relationship

$$P\,[\text{kW}] = \frac{88.5\,\text{kW m}}{A\,\text{GeV}^4} \frac{E^4\,[\text{GeV}] I_o\,[\text{A}]}{r\,[\text{m}]},$$

where I_o is the beam current and r is the ring radius. The current generated by the electron bunch is

$$I_o = \frac{nQ}{t},$$

where Q is the charge of the electron, n is the number of electrons in the bunch, and t is the time for the bunch to transit the collider. The transit time is the circumference of the ring divided by the electron velocity ($\approx c$):

$$t = \frac{2\pi r}{c}.$$

Combining these two equations provides the current:

$$I_o = \frac{nQc}{2\pi r} = \frac{(1 \times 10^{18}\,\text{e})(1.6 \times 10^{-19}\,\text{C/e})(3 \times 10^8\,\text{m/s})\left(\frac{\text{As}}{\text{C}}\right)}{(2\pi)(6.6 \times 10^6\,\text{m})} = 1.16\,\text{A}.$$

The determination of the current establishes the total power output of the synchrotron radiation spectrum from the 9.9 GeV electron beam:

$$P\,[\text{kW}] = \frac{88.5\,\text{kW m}}{\text{A GeV}^4} \frac{(9.9\,\text{GeV})^4 (1.16\,\text{A})}{(6.6 \times 10^6\,\text{m})} = 0.149\,\text{kW}.$$

Solutions for Chapter 7

07-01

(a) The spectrum of the ^{60}Ar nucleus in both 3D and 4D spaces is calculated using the relations provided in the text and problem statement. In 3D space, the spectrum is given by the relationship

$$E_K = \frac{\hbar^2}{2m}\left(\frac{K\pi}{b}\right)^2 \quad K = 1, 2, 3, \ldots,$$

where $\hbar$ is Planck's constant (1.055×10^{-34} J s), m is the nucleon mass (1.66×10^{-27} kg), and b is a scale factor that was determined to be 36.4 fm. Using these values, the ground state ($K=1$) is

$$E_1 = \frac{(1.055 \times 10^{-34}\,\text{J s})^2}{(2 \times 1.66 \times 10^{-27}\,\text{kg})}\left(\frac{1\,\text{MeV}}{1.6 \times 10^{-13}\,\text{J}}\right)\left(\frac{(1)\pi}{3.64 \times 10^{-14}\,\text{m}}\right)^2\left(\frac{\text{kg m}^2}{\text{s}^2\,\text{J}}\right),$$

$$= 0.156\,\text{MeV}.$$

Using this approach, the other energy levels are calculated in a similar manner, and the photon spectrum is derived from the allowable transitions. The following table summarizes the photon spectrum and resultant energy levels in the 3D space for the allowed transitions.

Quantum number K	3D energy level E_K (MeV)	Photon energy ($K+1 \rightarrow K$ transition) (MeV)
1	0.156	a
2	0.624	$(2 \rightarrow 1)$ 0.468
3	1.404	$(3 \rightarrow 2)$ 0.780
4	2.496	$(4 \rightarrow 3)$ 1.092

^aThe $K=1 \rightarrow 0$ transition is excluded because $K=0$ is not allowed.

The 4D energy levels are based on $\Delta K = \pm 1$ with fixed L and $\Delta L = \pm 1$ with fixed K transitions:

$$E_{K,L} = \frac{\hbar^2}{2m}\left[\left(\frac{K\pi}{b}\right)^2 + \left(\frac{L}{R}\right)^2\right] \quad K = 1, 2, 3, \ldots;\quad L = 0, 1, 2.$$

Solution

This relationship can be rewritten because $R = b/\pi$:

$$E_{K,L} = \frac{\hbar^2}{2m}\left[\left(\frac{K\pi}{b}\right)^2 + \left(\frac{L\pi}{b}\right)^2\right] = \frac{\hbar^2}{2m}\left(\frac{\pi}{b}\right)^2[K^2 + L^2].$$

As an example of the 4D spectrum consider the $E_{1,1}$ energy level:

$$E_{1,1} = \frac{(1.055 \times 10^{-34}\,\text{J s})^2}{(2 \times 1.66 \times 10^{-27}\,\text{kg})}\left(\frac{1\,\text{MeV}}{1.6 \times 10^{-13}\,\text{J}}\right)\left(\frac{\pi}{3.64 \times 10^{-14}\,\text{m}}\right)^2\left(\frac{\text{kg m}^2}{\text{s}^2\,\text{J}}\right)$$
$$\times [(1)^1 + (1)^1] = 0.312\,\text{MeV}.$$

Using this approach, the other 4D energy levels and photon spectra are calculated. The following tables summarize the photon spectrum and resultant energy levels in the 4D space for the allowed transitions.

	4D energy levels allowed transitions with K fixed	
Quantum number K	Allowed transition quantum number L ($L+1 \to L$)	Photon energy ($L+1 \to L$ transition with K fixed) (MeV)
1	$L = 1 \to 0$	0.156
1	$L = 2 \to 1$	0.468
1	$L = 3 \to 2$	0.780
2	$L = 1 \to 0$	0.156
2	$L = 2 \to 1$	0.468
2	$L = 3 \to 2$	0.780
3	$L = 1 \to 0$	0.156
3	$L = 2 \to 1$	0.468

	4D energy levels allowed transitions with L fixed	
Quantum number L	Allowed transition quantum number K ($K+1 \to K$)	Photon energy ($K+1 \to K$ transition with L fixed) (MeV)
0	$K = 2 \to 1$	0.468
0	$K = 3 \to 2$	0.780
0	$K = 4 \to 3$	1.092
1	$K = 2 \to 1$	0.468
1	$K = 3 \to 2$	0.780
2	$K = 2 \to 1$	0.468
2	$K = 3 \to 2$	0.780
3	$K = 2 \to 1$	0.468

The results of the 3D and 4D calculations are summarized in terms of the photon energies and yields. Although individual transitions have a yield of 1.0, if a transition occurs more than once its total yield is greater than 1. The 3D and 4D spectra and total yields are as follows:

Photon energy (MeV)	Total yield	
	3D	4D
0.156	0	3
0.468	1	7
0.780	1	5
1.092	1	1

(b) The 3D and 4D dose factors with units Gy m²/h MBq and Gy m³/h MBq, respectively, are given in terms of a simple relationship. This 3D dose factor relationship was empirically determined in terms of English units (R m²/h Ci) when the gamma-ray energy E is in MeV and Y is the yield of the photon and i is the number of gamma rays:

$$G_3 = \frac{1}{2} \sum_i E_i Y_i.$$

As this relationship has an accuracy of 15–20%, only two significant figures are retained. Converting this relationship to SI units yields

$$G_3 = \frac{1}{2} \sum_i E_i Y_i \frac{\text{R m}^2}{\text{h Ci}} \left[\frac{(0.877 \text{ rad/R})(1 \text{ Gy}/100 \text{ rad})}{3.7 \times 10^4 \text{ MBq/Ci}} \right],$$

$$G_3 = 1.2 \times 10^{-7} \frac{\text{Gy m}^2}{\text{h MBq}} \sum_i E_i Y_i.$$

Using the total yield table from part (a) leads to the 3D result

$$G_3 = 1.2 \times 10^{-7} [(0.468)(1) + (0.780)(1) + (1.092)(1)] \frac{\text{Gy m}^2}{\text{h MBq}}$$

$$= 2.8 \times 10^{-7} \frac{\text{Gy m}^2}{\text{h MBq}}.$$

The 4D result is written in a similar manner:

$$G_4 = 1.2 \times 10^{-7} \frac{\text{Gy m}^3}{\text{h MBq}} \sum_i E_i Y_i,$$

$$G_4 = 1.2 \times 10^{-7} [(0.156)(3) + (0.468)(7) + (0.780)(5) + (1.092)(1)] \frac{\text{Gy m}^3}{\text{h MBq}},$$

$$G_4 = 1.0 \times 10^{-6} \frac{\text{Gy m}^3}{\text{h MBq}}.$$

(c) As the distance of interest (r) is 5 m in both 3D and 4D space, the 3D and 4D absorbed dose rates are calculated using the point, isotropic source activity (A) of 4×10^5 MBq:

$$D_3 = \frac{AG}{r^2} = \frac{(4 \times 10^5 \text{ MBq})\left(2.8 \times 10^{-7} \frac{\text{Gy m}^2}{\text{h MBq}}\right)}{(5\,\text{m})^2} \frac{1000 \text{ mGy}}{\text{Gy}} = 4.5 \text{ mGy/h},$$

$$D_4 = \frac{2}{\sqrt{\pi}} \frac{AG_4}{r^3} = \frac{2}{\sqrt{\pi}} \frac{(4 \times 10^5 \text{ MBq})\left(1.0 \times 10^{-6} \frac{\text{Gy m}^3}{\text{h MBq}}\right)}{(5\,\text{m})^3} \frac{1000 \text{ mGy}}{\text{Gy}} = 3.6 \text{ mGy/h}.$$

07-02

(a) The unshielded neutron effective dose is determined using a point, isotropic source approximation and the relationship

$$D = k(\Phi_S^A + \Phi_S^B),$$

where k is the dose-conversion factor (3.5×10^{-10} Sv cm^2/n) and $\Phi_S^{A(B)}$ is the neutron fluence reaching the spacecraft from the A(B) star system. The fluence at the spacecraft's location is determined from the distance between the spacecraft (S) and probe (P) [D_{SP}], and the distance between the Solar System (SS) and probe (D_{SSP}) assuming a point, isotropic neutron source:

$$\Phi_S (D_{SSP} + D_{SP})^2 = \Phi_P D_{SSP}^2,$$

$$\Phi_S = \Phi_P \left(\frac{D_{SSP}}{D_{SSP} + D_{SP}}\right)^2.$$

Applying this relationship to the two star systems yields

$$\Phi_S^A = \Phi_P^A \left(\frac{D_{SSP}^A}{D_{SP}^A + D_{SSP}^A}\right)^2 = \left(2 \times 10^5 \frac{\text{n}}{\text{cm}^2}\right)\left(\frac{10 \text{ LY}}{2 \text{ LY} + 10 \text{ LY}}\right)^2 = 1.39 \times 10^5 \frac{\text{n}}{\text{cm}^2},$$

$$\Phi_S^B = \Phi_P^B \left(\frac{D_{SSP}^B}{D_{SP}^B + D_{SSP}^B}\right)^2 = \left(3 \times 10^7 \frac{\text{n}}{\text{cm}^2}\right)\left(\frac{50 \text{ LY}}{2 \text{ LY} + 50 \text{ LY}}\right)^2 = 2.77 \times 10^7 \frac{\text{n}}{\text{cm}^2}.$$

Using these values leads to the unshielded neutron effective dose (E):

$$E = 3.5 \times 10^{-10} \frac{\text{Sv cm}^2}{\text{n}} \left(1.39 \times 10^5 \frac{\text{n}}{\text{cm}^2} + 2.77 \times 10^7 \frac{\text{n}}{\text{cm}^2}\right) \frac{1000 \text{ mSv}}{\text{Sv}} = 9.74 \text{ mSv}.$$

(b) The electromagnetic deflector is only effective for charged particles. As the neutron is uncharged, the deflector has a negligible effect.

(c) Assuming that an antineutron source of sufficient intensity is viable, it could be used to reduce the neutron dose. However, the radiation generated in producing the antineutrons and the annihilation photons and pions need to be evaluated to ensure that a net dose savings is obtained. The antineutron source would be ALARA if a dose savings is achieved.

07-03

(a) To derive Equation 7.34, consider two cases. The first case is the unshielded effective dose $E(0, r)$ and the second is the shielded case $E(x, r)$, where x is the shield thickness and r is the distance from the fusion chamber. For these two cases, the effective doses are

$$E(0, r_o) = \frac{kS}{4\pi r_o^2},$$

$$E(x, r) = \frac{kS}{4\pi r^2} e^{-\Sigma x},$$

where k is the dose-conversion factor, S is the DT fusion neutron source strength, $E(0, r_o)$ is the unshielded dose at location r_o, $E(x, r)$ is the shielded dose at distance r because of a shield of thickness x, and Σ is the neutron removal coefficient. The effective dose relationships can be equated using the constant terms

$$\frac{kS}{4\pi} = E(0, r_o) r_o^2 = E(x, r) r^2 e^{-\Sigma x}.$$

Isolating the shield thickness leads to

$$e^{-\Sigma x} = \left(\frac{r^2 E(x, r)}{r_o^2 E(0, r_o)}\right).$$

By taking the natural log of both sides, Equation 7.34 is obtained:

$$x = -\frac{1}{\Sigma} \ln\left(\frac{r^2 E(x, r)}{r_o^2 E(0, r_o)}\right).$$

(b) The shielding required to meet the 0.1-Sv criterion for the DT reactor is obtained from the relationship

$$x = -\frac{1}{\Sigma} \ln\left(\frac{r^2 E(x, r)}{r_o^2 E(0, r_o)}\right),$$

where $E(0, r_o) = 10^{10}$ Sv, $r_o = 10$ m, $E(x, r) = 0.1$ Sv, and $\Sigma = 0.103$ cm^{-1}. As an example calculation, the shielding required for a standoff distance (r) of 100 m is

$$x = \left(\frac{-1}{0.103/\text{cm}}\right) \ln\left(\frac{(100\,\text{m})^2}{(10\,\text{m})^2} \frac{0.1\,\text{Sv}}{10^{10}\,\text{Sv}}\right) = -(9.71\,\text{cm})(-20.7) = 201\,\text{cm}.$$

The other required shielding results are summarized in Table 7.7.

07-04

(a) The major radiation types associated with this propulsion system are 14.1-MeV neutrons from DT fusion, beta particles from tritium decay and from the decay of activation products generated from the 14.1-MeV neutrons, and gamma rays from fusion activation products and fusion events. Other charged particles (e.g., protons, alpha particles) are also produced but are likely retained within the fusion chamber.

(b) The direct neutron and gamma radiation from the fusion event presents an external radiation hazard. The gamma radiation from the activation products is also an external radiation hazard. Internal hazards are derived from the tritium fuel and from the beta-emitting activation products if they become mobilized. External hazards are also presented from the activation of air and water in the vicinity of the reactor.

(c) The various sources of radiation external to the spacecraft and possible ALARA measures to mitigate these exposures include the following:

- Galactic cosmic radiation (GCR)– Within a few light years of our Solar System, GCR would be expected to be relatively constant. The GCR could vary depending on the characteristics of the visited star. The particles composing the GCR should be similar to that observed in the vicinity of the Earth. Protons and alpha particles are the most likely GCR constituents. Other particles were listed in Chapter 6. Shielding and electromagnetic deflection would be appropriate ALARA measures.

- Solar particle events (SPE)– The SPE hazard is significant, but its intensity and periodicity are normally uncertain. This hazard varies significantly and is highly dependent on the visited star. The visited star may have a very different radiation signature than Sol, and its radiation characteristics should be determined prior to the mission to establish appropriate radiation constraints.

    SPE particles primarily comprise protons and alpha particles. This should also be verified as a part of the mission's ALARA evaluation. Shielding and electromagnetic deflection are appropriate ALARA measures. Upon detection of an SPE, the spacecraft could either leave the Solar System until the event ended or position itself behind a nearby planet or moon. This ALARA option depends on the sensitivity of the instrumentation and radiation detection systems and the speed of the spacecraft. Other ALARA options for the SPE event include the use of an electromagnetic deflector and relocating the crew to a shielded shelter.

- Trapped radiation (TR)– The geomagnetic nature of any visited planet has a significant impact on its TR characteristics. These characteristics must be well established prior to an approved ALARA evaluation that supports a planetary mission in a visited star system. Both trapped electrons and protons should be characterized. Shielding and electromagnetic deflection are appropriate ALARA measures. Selecting a low-dose orbit is also an effective ALARA tool.
- Radiation emanating from spatial anomalies – By their nature, radiation from any spatial anomaly is highly uncertain. This uncertainty includes the types of radiation emitted as well as its intensity and spatial distribution. Hopefully, resolution of the pioneer anomaly will begin to reduce this uncertainty.
- Low-probability events – Events such as supernovas and gamma-ray bursts are also possible and involve gamma rays, protons, alpha particles, and heavy charged particles. An electromagnetic deflector is an effective ALARA tool for charged particles. The gamma-ray burst would be mitigated by moving the spacecraft to a lower dose region. This capability depends on the spacecraft's speed and its ability to detect the gamma-ray burst prior to it reaching the spacecraft.
- Dark matter and dark energy (DM/DE) – By their nature DM/DE are uncertain in terms of their radiation characteristics and interaction properties. Accordingly, the radiation consequences of encountering DM/DE are highly uncertain.

(d) A proton–antiproton propulsion system produces pions and photons. The pions decay into muons, electrons, neutrinos, and photons. These radiation types present an external hazard. Unless the protons and antiprotons are collided at extremely high energies, the neutrinos will not represent a significant radiation hazard.

07-05

The stellar mass $m = 250$ kg has a density $\rho = 100$ kg/cm^3. The volume of this matter is determined from the relationship

$$V = \frac{m}{\rho} = \frac{250 \text{ kg}}{100 \text{ kg/cm}^3} = 2.5 \text{ cm}^3.$$

Given the size of the stellar mass, a point-source approximation is justified. Using a point-source relationship, the distance (r) to limit the absorbed dose to 0.01 Gy/h is

$$\dot{D} = \frac{A\Gamma}{r^2},$$

$$r = \left(\frac{A\Gamma}{\dot{D}}\right)^{1/2}.$$

Solution

The activity A of this stellar matter is determined from the specific activity A_{sp} and the stellar mass:

$$A = A_{sp} m = (2.5 \times 10^8 \text{ MBq/g})(250 \text{ kg})(1000 \text{ g/kg}) = 6.25 \times 10^{13} \text{ MBq}.$$

The gamma constant $(0.05 \text{ Gy m}^2/\text{h MBq})$ and limiting absorbed dose rate (0.01 Gy/h) were provided in the problem statement. Using these values, the desired distance is determined:

$$r = \left(\frac{A\Gamma}{\dot{D}}\right)^{1/2} = \left(\frac{(6.25 \times 10^{13} \text{ MBq})(0.05 \text{ Gy m}^2/\text{h MBq})}{0.01 \text{ Gy/h}}\right)^{1/2} = 1.77 \times 10^7 \text{ m},$$
$$= 1.77 \times 10^4 \text{ km}.$$

07-06

The recommended actions should be included in an ALARA evaluation that was a part of the mission planning process. An ALARA evaluation section addressing a large SPE and its radiological consequences to the ship's crew, either on the ship or on a planet, would be an expected component of this report.

The following actions should be recommended to minimize the crew's absorbed dose from the massive Tau Ceti solar particle event:

- The crew should be relocated to the interior of the spacecraft. If a radiation shelter is available, it should be utilized.
- The electromagnetic field generators should be activated and the deflector fields established at their maximum intensity. All power supplies should be online.
- The spacecraft should be oriented to provide maximum shielding to the crew.
- The spacecraft's orbit should be adjusted to have the planet shield the spacecraft from the solar particle event.
- The administration of radioprotective chemicals should be performed if warranted by the projected crew dose.
- If the spacecraft's detection system provides sufficient warning and the propulsion system has sufficient velocity capability, an exit from the Solar System would be the best option to minimize the radiological impact from the SPE.

07-07

Given the conditions of the previous problem, the following actions should be recommended to minimize the absorbed doses to crewmembers on the planet's surface from the massive Tau Ceti event:

- If the crew is in the vicinity of their habitat, they should be relocated to the designated shelter areas that provide maximum shielding from the solar particle event.
- The crew should be relocated to utilize any available surface shielding. If caves or large rock formations are available, they should be utilized.

- Any portable electromagnetic field generators should be activated and the deflector fields should be established at their maximum intensity.
- The administration of radioprotective chemicals should be authorized if warranted by the projected crew dose.
- If justified by ALARA considerations and the event-specific circumstances, the crew should return to the spacecraft.

07-08

The two propulsion systems are defined in terms of a number of key parameters with associated symbols:

Symbol	Power system parameter/characteristic	Propulsion system-1	Propulsion system-2
—	Mean reactor radius (m)	12	20
S_γ	Gamma source strength (γ/s)	4×10^{16}	9×10^{15}
S_n	Neutron source strength (n/s)	5×10^{15}	7×10^{14}
r	Distance from propulsion system to crew quarters (m)	1000	525
E_γ	Mean gamma-ray energy (MeV/γ) and yield	4.5@0.94	6.7@0.65
E_n	Mean neutron energy (MeV/n)	11.0	6.0
k	Neutron flux-to-dose-conversion factor (Sv cm^2/n)	4.3×10^{-8}	4.2×10^{-8}
F_n	Neutron attenuation factor	0.00001	0.00001
F_γ	Gamma-ray attenuation factor (cm^2/g)	0.0005	0.0004
μ_{en}	Gamma-ray mass energy absorption coefficient (cm^2/g)	0.05	0.06

(a) Propulsion system-1:

The mean diameter of propulsion system-1 is 24 m. A comparison of the diameter with the distance from the source (1000 m) is sufficient to justify a point-source approximation. The following relationships are based on an isotropic point source. The effect of shielding is taken into account through the attenuation factor.

Using the aforementioned values, the gamma-ray and neutron-effective dose rates are

Gamma-ray effective dose rate ($\dot{H}_\gamma$):

$$\dot{H}_\gamma = \left(\frac{S_\gamma}{4\pi r^2} \frac{\mu_{en}}{\rho} \sum_i E_i Y_i \right) F_\gamma,$$

$$= \frac{(4 \times 10^{16} \, \gamma/s)(0.05 \, cm^2/g)(1000 \, g/kg)(4.5 \, MeV/\gamma)(0.94)}{(4\pi)(1000 \, m \times 100 \, cm/m)^2}$$

$$\times (1.6 \times 10^{-13} \, J/MeV)(Sv \, kg/J)(3600 \, s/h)(0.0005) = 1.94 \times 10^{-5} \, Sv/h.$$

Neutron-effective dose rate ($\dot{H}_n$):

$$\dot{H}_n = \frac{S_n}{4\pi r^2} kF_n = \frac{(5 \times 10^{15}\,\text{n/s})(4.3 \times 10^{-8}\,\text{Sv cm}^2/\text{n})(3600\,\text{s/h})(0.00001)}{(4\pi)(1000\,\text{m} \times 100\,\text{cm/m})^2},$$
$$= 6.16 \times 10^{-5}\,\text{Sv/h}.$$

The total effective dose rate for propulsion system-1 is the sum of the gamma-ray and neutron components:

$$\dot{H} = \dot{H}_\gamma + \dot{H}_n = 1.94 \times 10^{-5}\,\text{Sv/h} + 6.16 \times 10^{-5}\,\text{Sv/h} = 8.10 \times 10^{-5}\,\text{Sv/h},$$

$$\dot{H} = (8.1 \times 10^{-5}\,\text{Sv/h})(24\,\text{h/day})(365\,\text{day/year}) = 0.710\,\text{Sv/year}.$$

(b) Propulsion system-2:

The mean diameter of propulsion system-2 is 40 m. A comparison of the diameter with the distance from the source (525 m) is sufficient to justify a point-source approximation. The following relationships are based on an isotropic point source. The effect of shielding is taken into account through the attenuation factor.

Using the aforementioned values, the gamma-ray and neutron-effective dose rates are the following:

Gamma-ray effective dose rate ($\dot{H}_\gamma$):

$$\dot{H}_\gamma = \left(\frac{S_\gamma}{4\pi r^2}\frac{\mu_{en}}{\rho}\sum_i E_i Y_i\right) F_\gamma,$$

$$= \frac{(9 \times 10^{15}\,\gamma/\text{s})(0.06\,\text{cm}^2/\text{g})(1000\,\text{g/kg})(6.7\,\text{MeV}/\gamma)(0.65)}{(4\pi)(525\,\text{m} \times 100\,\text{cm/m})^2}$$
$$\times (1.6 \times 10^{-13}\,\text{J/MeV})(\text{Sv kg/J})(3600\,\text{s/h})(0.0004) = 1.57 \times 10^{-5}\,\text{Sv/h}.$$

Neutron-effective dose rate ($\dot{H}_n$):

$$\dot{H}_n = \frac{S_n}{4\pi r^2} kF_n = \frac{(7 \times 10^{14}\,\text{n/s})(4.2 \times 10^{-8}\,\text{Sv cm}^2/\text{n})(3600\,\text{s/h})(0.00001)}{(4\pi)(525\,\text{m} \times 100\,\text{cm/m})^2},$$
$$= 3.06 \times 10^{-5}\,\text{Sv/h}.$$

The total effective dose rate for propulsion system-2 is the sum of the gamma-ray and neutron components:

$$\dot{H} = \dot{H}_\gamma + \dot{H}_n = 1.57 \times 10^{-5}\,\text{Sv/h} + 3.06 \times 10^{-5}\,\text{Sv/h} = 4.63 \times 10^{-5}\,\text{Sv/h},$$

$$\dot{H} = (4.63 \times 10^{-5}\,\text{Sv/h})(24\,\text{h/day})(365\,\text{day/year}) = 0.406\,\text{Sv/year}.$$

(c) Neither system is acceptable with annual effective doses of 0.710 and 0.406 Sv/year for propulsion systems-1 and -2, respectively. These values exceed the project recommendations. Additional shielding or an increased distance between the fusion reactor and occupied areas are needed to reduce these values to the 100 mSv/year Wolf 359 effective dose requirement.

07-09

Shielding the 14.1-MeV DT fusion neutrons is more complex than shielding lower energy (less than about 5 MeV) neutrons. Lower energy neutrons are shielded using elastic scattering with hydrogen nuclei to reduce the neutron's energy to the thermal range. Low-energy elastic scattering with hydrogen degrades the energy and permits neutrons to be effectively captured via the ^{1}H(n, γ) reaction. The residual gamma rays are then removed by high Z material following the hydrogen bearing material. However, at 14.1 MeV, elastic scattering with hydrogen is not efficient in reducing the neutron's energy. The high-energy neutrons should first encounter the lead where Pb (n, n′) inelastic scattering reduces the neutron energy such that elastic scattering with hydrogen becomes effective in degrading the energy to the thermal range. Therefore, the correct shielding arrangement is lead, closest to the neutron source, followed by polyethylene. A final lead layer is added to attenuate the capture gamma rays.

07-10

(a) The annual effective dose (E) for typical years not having a major SPE is

$$E = \sum_{i=1}^{3} E_i = \sum_{i=1}^{3} k_i \Phi_i,$$

where k is the chronic proton dose-conversion factor, Φ is the proton fluence, and i labels the proton energy group. The chronic dose-conversion factor is used because the radiation dose occurs over a prolonged period (e.g., a year). Using the problem statement data, the annual effective dose for each proton energy group is

$$E_1 = \left(200 \frac{\text{pSv cm}^2}{\text{proton}}\right)\left(1.0 \times 10^6 \frac{\text{proton}}{\text{cm}^2}\right)\left(\frac{1 \text{ Sv}}{10^{12} \text{ pSv}}\right) = 2 \times 10^{-4} \text{ Sv},$$

$$E_2 = \left(500 \frac{\text{pSv cm}^2}{\text{proton}}\right)\left(2.0 \times 10^5 \frac{\text{proton}}{\text{cm}^2}\right)\left(\frac{1 \text{ Sv}}{10^{12} \text{ pSv}}\right) = 1 \times 10^{-4} \text{ Sv},$$

$$E_3 = \left(1500 \frac{\text{pSv cm}^2}{\text{proton}}\right)\left(5.0 \times 10^4 \frac{\text{proton}}{\text{cm}^2}\right)\left(\frac{1 \text{ Sv}}{10^{12} \text{ pSv}}\right) = 7.5 \times 10^{-5} \text{ Sv},$$

$$E = (2 + 1 + 0.75) \times 10^{-4} \text{ Sv} = (3.75 \times 10^{-4} \text{ Sv})(1000 \text{ mSv/Sv}) = 0.375 \text{ mSv}.$$

(b) The biological effects of the 0.375-mSv annual effective dose are minimal. This effective dose is a part of the natural background radiation environment of the sphere and the inhabitants experience this effective dose as part of normal life. As the sphere's inhabitants are similar to humans, no health detriment is expected from the 0.375 mSv/year effective dose.

(c) The absorbed dose (D) received during a maximum SPE is

$$D_{SPE} = \sum_{i=1}^{3} D_i = \sum_{i=1}^{3} k_i \Phi_i,$$

where k is the acute dose-conversion factor. The acute dose-conversion factor is used because the SPE event occurs during a short time interval (90 min). Using the problem statement data, the absorbed dose for each proton energy group is

$$D_1 = \left(50 \frac{\text{pGy cm}^2}{\text{proton}}\right)\left(3.2 \times 10^{11} \frac{\text{proton}}{\text{cm}^2}\right)\left(\frac{1 \text{ Gy}}{10^{12} \text{ pGy}}\right) = 16 \text{ Gy},$$

$$D_2 = \left(100 \frac{\text{pGy cm}^2}{\text{proton}}\right)\left(5.0 \times 10^{10} \frac{\text{proton}}{\text{cm}^2}\right)\left(\frac{1 \text{ Gy}}{10^{12} \text{ pGy}}\right) = 5 \text{ Gy},$$

$$D_3 = \left(500 \frac{\text{pGy cm}^2}{\text{proton}}\right)\left(2.2 \times 10^{9} \frac{\text{proton}}{\text{cm}^2}\right)\left(\frac{1 \text{ Gy}}{10^{12} \text{ pGy}}\right) = 1.1 \text{ Gy},$$

$$D = (16 + 5 + 1.1) \text{ Gy} = 22.1 \text{ Gy}.$$

(d) The biological effects of the maximum SPE dose are significant. This absorbed dose exceeds the $LD_{50,30}$ value (3–4 Gy) and is typical of radiation accident absorbed dose values. Even with medical intervention, the prognosis of a 22-Gy absorbed dose is grave. These absorbed dose values are similar to those received by personnel involved in the 1999 Tokai Mura Criticality accident. Even with significant medical intervention, the two highest dose operators died from multiple organ failures. Unless the sphere's occupants have made significant medical advances, they will suffer a significant health detriment.

(e) Given the technological status of the Type II civilization, it is likely that the sphere is designed to accommodate the massive SPE event, and that appropriate defense mechanisms are readily available. If the planet's inhabitants only have 10-min warning, the design features developed to mitigate the SPE would need to be automatically activated. These design features would likely include multiply redundant population center deflection devices to redirect the SPEs charged particles. As the SPE lasts for 18 h, some of the population could be directed to seek refuge in a shielded shelter for a portion of the event.

The administration of radioprotective chemicals is one approach to mitigate the SPE. Additional mitigative measures include nanotechnology and gene

Solutions for Chapter 7 | 447

therapy to minimize the health detriment. These mitigative measures are assumed to be readily available, so that the planet's communication systems facilitate the rapid dissemination of information and that the population is capable of accomplishing or administering the mitigative measures in a timely manner.

07-11

(a) The total fluence (Φ) reaching Cleveland is the sum of the fluence values from each star (A, B, and C) that experiences an SPE:

$$\Phi = \Phi_A + \Phi_B + \Phi_C.$$

The fluence from each star is determined by its output rate ($\dot{O}$), the event duration (T), and distance from Cleveland (r):

$$\Phi_A = \frac{\dot{O}_A T}{4\pi r_A^2} = \frac{(1 \times 10^{30} \text{ particles/s})(2000 \text{ s})}{4\pi (1.5 \text{ AU})^2} = 7.1 \times 10^{31} \frac{\text{particles}}{\text{AU}^2},$$

$$\Phi_B = \frac{\dot{O}_B T}{4\pi r_B^2} = \frac{(3 \times 10^{31} \text{ particles/s})(5000 \text{ s})}{4\pi (10 \text{ AU})^2} = 1.2 \times 10^{32} \frac{\text{particles}}{\text{AU}^2},$$

$$\Phi_C = \frac{\dot{O}_C T}{4\pi r_C^2} = \frac{(5 \times 10^{34} \text{ particles/s})(1000 \text{ s})}{4\pi (400 \text{ AU})^2} = 2.5 \times 10^{31} \frac{\text{particles}}{\text{AU}^2},$$

$$\Phi = (7.1 \times 10^{31} + 1.2 \times 10^{32} + 2.5 \times 10^{31}) \frac{\text{particles}}{\text{AU}^2} = 2.2 \times 10^{32} \frac{\text{particles}}{\text{AU}^2}.$$

(b) Seeding Stars A, B, and C would decrease their fluence by a factor of 5 and decrease their distance to Cleveland by a factor of 1.25. The result of seeding these stars is

$$\Phi_A = \frac{\dot{O}_A T}{4\pi r_A^2} = \frac{(\tfrac{1}{5})(1 \times 10^{30} \text{ particles/s})(2000 \text{ s})}{4\pi (1.5 \text{ AU}/1.25)^2} = 2.2 \times 10^{31} \frac{\text{particles}}{\text{AU}^2},$$

$$\Phi_B = \frac{\dot{O}_B T}{4\pi r_B^2} = \frac{(\tfrac{1}{5})(3 \times 10^{31} \text{ particles/s})(5000 \text{ s})}{4\pi (10 \text{ AU}/1.25)^2} = 3.7 \times 10^{31} \frac{\text{particles}}{\text{AU}^2},$$

$$\Phi_C = \frac{\dot{O}_C T}{4\pi r_C^2} = \frac{(\tfrac{1}{5})(5 \times 10^{34} \text{ particles/s})(1000 \text{ s})}{4\pi (400 \text{ AU}/1.25)^2} = 7.8 \times 10^{30} \frac{\text{particles}}{\text{AU}^2}.$$

The results of the selected seeding are

$$\Phi_{\text{Seed-A}} = (2.2 \times 10^{31} + 1.2 \times 10^{32} + 2.5 \times 10^{31}) \frac{\text{particles}}{\text{AU}^2} = 1.7 \times 10^{32} \frac{\text{particles}}{\text{AU}^2},$$

$$\Phi_{\text{Seed-B}} = (7.1 \times 10^{31} + 3.7 \times 10^{31} + 2.5 \times 10^{31}) \frac{\text{particles}}{\text{AU}^2} = 1.3 \times 10^{32} \frac{\text{particles}}{\text{AU}^2},$$

$$\Phi_{\text{Seed-C}} = (7.1 \times 10^{31} + 1.2 \times 10^{32} + 7.8 \times 10^{30}) \frac{\text{particles}}{\text{AU}^2} = 2.0 \times 10^{32} \frac{\text{particles}}{\text{AU}^2}.$$

The minimum fluence is achieved by seeding Star B. If this approach were credible, the gravitational effects of the seeding evolution should be carefully considered as they could have a dramatic effect on Cleveland's environment.

07-12

It can be demonstrated that the geodesics of flat space–time in a two-dimensional plane are straight lines. To accomplish this, let indexes A and B run over the values 1 and 2 and label the two polar coordinates as $x^1 = r$ and $x^2 = \theta$. The components of the tangent vector ($\vec{u}$) to the path of motion are

$$u^A = \frac{dx^A}{ds}. \tag{07-12.1}$$

The vector components of u are obtained by dividing the line element by ds^2:

$$ds^2 = dr^2 + r^2 d\theta^2. \tag{07-12.2}$$

$$\frac{ds^2}{ds^2} = 1 = \frac{dr^2}{ds^2} + \frac{r^2 d\theta^2}{ds^2} = \left(\frac{dr}{ds}\right)^2 + r^2 \left(\frac{d\theta}{ds}\right)^2. \tag{07-12.3}$$

Define a vector (Killing vector) ξ to have components

$$\xi^r = 0 \quad \text{and} \quad \xi^\theta = 1. \tag{07-12.4}$$

A conserved quantity (L) (constant) is obtained by forming the dot product of ξ and u:

$$L \equiv \vec{\xi} \cdot \vec{u} = g_{AB} \xi^A u^B = r^2 \frac{d\theta}{ds}. \tag{07-12.5}$$

Note that this result is a restatement of the second differential equation provided in the problem statement:

$$\frac{d}{ds}\left(r^2 \frac{d\theta}{ds}\right) = 0. \tag{07-12.6}$$

Inserting the value for $d\theta/ds$ from Equation 07-12.5 into Equation 07-12.3 yields

$$1 = \left(\frac{dr}{ds}\right)^2 + r^2\left(\frac{d\theta}{ds}\right)^2 = \left(\frac{dr}{ds}\right)^2 + \frac{L^2}{r^2}, \qquad (07\text{-}12.7)$$

$$\frac{dr}{ds} = \left(1 - \frac{L^2}{r^2}\right)^{1/2}. \qquad (07\text{-}12.8)$$

If Equation 07-12.5 is divided by Equation 07-12.8, θ as a function of r is obtained:

$$\frac{d\theta}{dr} = \frac{d\theta/ds}{dr/ds} = \frac{L/r^2}{\left(1 - \frac{L^2}{r^2}\right)^{1/2}}. \qquad (07\text{-}12.9)$$

Equation 07-12.9 can be integrated to obtain the geodesic equation:

$$\int d\theta = \int \frac{(L/r^2)dr}{\left(1 - \frac{L^2}{r^2}\right)^{1/2}} \qquad (07\text{-}12.10)$$

Using a change of variables, $x = L/r$ simplifies the integration of Equation 07-12.10 to provide

$$\theta = \theta_o + \cos^{-1}\left(\frac{L}{r}\right), \qquad (07\text{-}12.11)$$

where θ_o is a constant of integration. Equation 07-12.10 is solved for L to provide the shape of the geodesic:

$$L = r\cos(\theta - \theta_o). \qquad (07\text{-}12.12)$$

The cosine is expanded using the identity

$$\cos(\theta - \theta_o) = \cos\theta\cos\theta_o + \sin\theta\sin\theta_o. \qquad (07\text{-}12.13)$$

Using Equations 07-12.12 and 07-12.13 and the polar coordinate relationships,

$$x = r\cos\theta \qquad y = r\sin\theta \qquad (07\text{-}12.14)$$

yields

$$\begin{aligned}L &= r(\cos\theta\cos\theta_o + \sin\theta\sin\theta_o) = (r\cos\theta)\cos\theta_o + (r\sin\theta)\sin\theta_o, \\ &= x\cos\theta_o + y\sin\theta_o.\end{aligned} \qquad (07\text{-}12.15)$$

Equation 07-12.15 is the general equation for a straight line:

$$y = mx + b, \qquad (07\text{-}12.16)$$

where m is the slope and b is the y intercept. As both θ_o and L are constant, Equation 07-12.15 is written in the standard linear form (i.e., $y = mx + b$):

$$(\sin\theta_o)y = -(\cos\theta_o)x + L, \qquad (07\text{-}12.17)$$

$$y = -(\cot\theta_o)x + \frac{L}{\sin\theta_o}, \qquad (07\text{-}12.18)$$

with $m = -\cot\theta_o$ and $b = L/\sin\theta_o$.

This result demonstrates that the geodesic in a plane is a straight line. This example illustrates the complexity of solving geodesic equations and why symbolic algebra codes (e.g., Mathematica® and Maple®) are quite useful when three-, four-, or higher dimensional cases are evaluated in general space–time geometries.

07-13

(a) The unshielded absorbed dose (D_o) is obtained from the relationship

$$D_o = \frac{f\Phi(0)}{r^2}k,$$

where r is the distance from the spacecraft's location to the burst (750 kpc), $\Phi(0)$ is the total γ burst output ($1.9 \times 10^{64}\,\gamma$), the flux-to-dose-conversion factor (k) at 2 MeV is 3.27×10^{-8} Gy/h per γ/cm^2 s, and the fraction of the burst intercepted by the spacecraft (f) is 0.2%. Using these values, the unattenuated absorbed dose is determined:

$$D_o = \frac{(0.002)\left(1.9 \times 10^{64}\,\frac{\gamma}{\text{s}}\right)\left(3.27 \times 10^{-8}\,\frac{\text{Gy cm}^2\,\text{s}}{\text{h}\;\gamma}\right)\left(\frac{1\,\text{h}}{3600\,\text{s}}\right)}{[(750 \times 10^3\,\text{pc})(3.09 \times 10^{18}\,\text{cm/pc})]^2} = 64.3\,\text{Gy}.$$

(b) The shielded absorbed dose (D) is

$$D = D_o B e^{-(\mu/\rho)\rho z},$$

where the shield thickness (z) is 20 cm of water equivalent, the attenuation coefficient (μ/ρ) is $0.05\,\text{cm}^2/\text{g}$, the density of water ($\rho$) is $1.0\,\text{g/cm}^3$, and the buildup factor (B) of 1.8 is applicable to this situation. These values provide the attenuated absorbed dose (D):

$$D = (64.3\,\text{Gy})(1.8)\,e^{-(0.05\,\text{cm}^2/\text{g})(1\text{g/cm}^3)(20\,\text{cm})} = 42.6\,\text{Gy}.$$

(c) Unless significant medical advances have been made at the time of the mission, this absorbed dose is fatal even with medical intervention.

07-14

(a) The luminosity of the gas giant/star (L) is written in terms of its temperature (T) and radius (R) compared to the same quantities as for the Sun:

$$\frac{L}{L_{Sun}} = \left(\frac{R}{R_{Sun}}\right)^2 \left(\frac{T}{T_{Sun}}\right)^4.$$

$$L = L_{Sun} \left(\frac{R}{R_{Sun}}\right)^2 \left(\frac{T}{T_{Sun}}\right)^4.$$

Assuming the gas giant is a sphere, its radius is determined from the mass (M) and volume (V) using the density (ρ) relationship

$$\rho = \frac{M}{V} = \frac{M_{Sun}}{\left(\frac{4\pi R_{Sun}^3}{3}\right)} = \frac{aM_{Sun}}{\left(\frac{4\pi R^3}{3}\right)},$$

where aM_{Sun} is the mass of the gas giant star in terms of solar masses. This equation is solved for the radius of the gas giant star:

$$R = \left(\frac{aM_{Sun}}{\frac{4\pi\rho}{3}}\right)^{1/3} = a^{1/3} R_{Sun},$$

where

$$R_{Sun} = \left(\frac{M_{Sun}}{\frac{4\pi\rho}{3}}\right)^{1/3}.$$

The luminosity of the gas giant is

$$L = L_{Sun} \left(\frac{R}{R_{Sun}}\right)^2 \left(\frac{T}{T_{Sun}}\right)^4,$$

$$L = L_{Sun} \left(\frac{a^{1/3} R_{Sun}}{R_{Sun}}\right)^2 \left(\frac{bT_{Sun}}{T_{Sun}}\right)^4 = a^{2/3} b^4 L_{Sun},$$

where b is a multiplier that expresses the gas giant's temperature in terms of the Sun's temperature. Using the values provided in the problem, the gas giant luminosities are

$$L_{Aries\text{-}6} = L_{Sun} a_{Aries\text{-}6}^{2/3} b_{Aries\text{-}6}^4 = (10)^{2/3} (4)^4 L_{Sun} = 1.19 \times 10^3 L_{Sun},$$

$$L_{Cetus\text{-}4} = L_{Sun} a_{Cetus\text{-}4}^{2/3} b_{Cetus\text{-}4}^{4} = (14)^{2/3}(5)^{4} L_{Sun} = 3.63 \times 10^{3} L_{Sun},$$

$$L_{Bootes\text{-}3} = L_{Sun} a_{Bootes\text{-}3}^{2/3} b_{Bootes\text{-}3}^{4} = (3)^{2/3}(2)^{4} L_{Sun} = 33.3 L_{Sun},$$

(b) As the radiation from the gas giant star decreases as r^{-2}, the luminosity for its companion Earth-like planet is obtained from the point-source relationship. The point-source relationship is applied because the luminosity calculated in (a) is assumed to occur at a distance (r) of 1 AU:

$$L_{Aries\text{-}6}^{Earth\text{-}like\,Planet} = L_{Aries\text{-}6}\left(\frac{1\,AU}{r_{Earth\text{-}like\,Planet}}\right)^{2} = 1.19 \times 10^{3} L_{Sun}\left(\frac{1\,AU}{8\,AU}\right)^{2} = 18.6 L_{Sun},$$

$$L_{Cetus\text{-}4}^{Earth\text{-}like\,Planet} = L_{Cetus\text{-}4}\left(\frac{1\,AU}{r_{Earth\text{-}like\,Planet}}\right)^{2} = 3.63 \times 10^{3} L_{Sun}\left(\frac{1\,AU}{58\,AU}\right)^{2} = 1.08 L_{Sun},$$

$$L_{Bootes\text{-}3}^{Earth\text{-}like\,Planet} = L_{Bootes\text{-}3}\left(\frac{1\,AU}{r_{Earth\text{-}like\,Planet}}\right)^{2} = 33.3 L_{Sun}\left(\frac{1\,AU}{4.6\,AU}\right)^{2} = 1.57 L_{Sun},$$

where L_{Sun} represents the luminosity at the Earth from the output of the Sol.

(c) Colonization of the three Earth-like planets needs to be carefully analyzed. The Aries-6 planet is likely excluded from colonization because it receives a factor 18.6 times the intensity of Sol. Unless its atmosphere is sufficiently thick, it is not a likely planet for colonization by humans.

The Cetus-4 planet has a sun that is 8% more intense than Earth. This increase in intensity requires evaluation. The increase in UV intensity and higher solar loading has an impact on crops and animals transplanted from Earth. The atmospheric thickness and its impact on the amount of radiation reaching the planet's surface also require evaluation.

Bootes-3 planet colonization with its 57% increase in luminosity also merits evaluation. It is a poorer candidate than the Cetus-4 planet.

For these new stars, there is also a concern with possible solar particle events. The magnitude of these events depends on the characteristics of the created star. As the stars were created from gas giants, their characteristics are difficult to predict. This is especially true in their early years post creation, because the new stars require time to reach a state of equilibrium.

The periodicity of the solar cycles of these new stars and the variation in intensity also merit attention. The luminosity values calculated in parts (a) and (b) may have significant variation that would impact their colonization potential.

VI
Appendixes

The 12 appendixes provided in this part are an integral part of *Health Physics in the 21st Century*. These appendixes provide supplementary information and enrich the discussion in Parts Two, Three, and Four. Some of the material may be unfamiliar to health physicists (e.g., the Standard Model of Particle Physics and General Theory of Relativity), but these areas are critical to a complete understanding of emerging health physics issues. Each appendix should be carefully reviewed to gain the maximum benefit from this text.

Appendix A
Significant Events and Important Dates in Physics and Health Physics

This appendix provides a chronological summary of significant events in physics and health physics. The chronology includes historical events and events that will shape the direction of 21st century health physics. The items summarized in Table A.1 are somewhat subjective and reflect the author's opinion of the relevance and importance of events. Omissions are not intended to slight the significance of the omitted work, but reflect limited space to recognize all relevant discoveries.

The events selected for inclusion in Table A.1 also provide background to the historical foundations of the chapters presented in this book. In addition, the events, their timing, and their significance illustrate the rapid development of the science supporting the health physics arena.

There is also a historical trend from the discovery of isolated concepts to the unification of concepts into more integrated and comprehensive theories. These theories facilitate the development of technologies, further experimentation, and improved theories. The coupling of experiment and theory repeats and leads to the rapid advancement of contemporary science compared to slower progress in earlier time periods.

Table A.1 also lists projected events such as the operation of the Generation IV fission reactors and the International Thermonuclear Experimental Reactor. Projected events could be advanced significantly through scientific discovery. If another scientific revolution that parallels the early part of the twentieth century occurs, the transition from a Type 0–I to a Type II civilization could occur with a rapid acceleration of discovery. During this period, events such as travel to other star systems would transition from science fiction to science fact. The beauty of the future is the unrestrained possibilities for growth with associated scientific and human advancement.

Health Physics in the 21st Century. Joseph John Bevelacqua
Copyright © 2008 WILEY-VCH Verlag GmbH & Co. KGaA, Weinheim
ISBN: 978-3-527-40822-1

Table A.1 Significant events and important dates in health physics and physics.

Circa 3000 B.C.	Babylonians measure time
Circa 500 B.C.	Pythagoras develops a prototype periodic table with four elements: earth, air, fire, and water and proves his fundamental theorem of geometry
Circa 350 B.C.	Aristotle writes the first physics textbook
Circa 250 B.C.	Euclid develops plane geometry
Circa 100 B.C.	Ptolemaic theory assumes that the Earth is the fixed center of the universe
Circa 1200	Al-Hazen (Arabia) formulates the basic elements of optics
Circa 1500	Copernicus' *De Revolutionibus Orbis Terranum* presents a new view of the Solar System
Circa 1502	daVinci formulates basic elements of physics, chemistry, astronomy, and geology
1609	Galileo confirms Copernicus' theory
	Kepler formulates laws of planetary motion
1632	Galileo's *Systems of the World* is condemned by the inquisition
1658	Huygens develops wave theory of light
1687	Newton publishes *Principia Mathematica*
1736	Euler formulates analytical mechanics
1738	Bernoulli formulates molecular theory of gases
1750	Franklin draws atmospheric electricity to a conductor
1808	Dalton establishes atomic theory
1811	Avogadro develops kinetic theory of gases
1812	Laplace devises probability theory
1819	Oersted discovers electromagnetism
1826	Ohm's law for electrical conductors is developed
1831	Faraday produces magnetically induced electrical current
1832	Henry discovers electrical self-induction
1838	Bessel measures the distance to a fixed star
1845	Faraday formulates electromagnetic wave theory of light
1847	Thompson (Lord Kelvin) defines absolute temperature
1850	Foucault measures the speed of light in air and in water
1856	Helmholtz writes *Physiological Optics*
1859	Bunsen and Kirchhoff establish the field of spectroscopy
1868	Angstrom maps the solar spectrum
1869	Mendeleev, Meyer, and Newlands find that properties of elements are periodic functions of atomic masses
1873	Maxwell's Theory of Electromagnetic Radiation is presented
1877	Lord Rayleigh publishes *Treatise on Sound*
1879	Edison invents incandescent electric lamp
1887	Michelson–Morley experiment invalidates either theory
	Balmer and Rydberg discover laws of spectral series
1888	Hertz generates and detects electromagnetic waves
1895	Röentgen discovers X-rays
	Lorentz formulates theory of the electron
1896	Becquerel discovers radioactivity
	X-ray images used in court as evidence
1897	Thompson discovers the electron
1898	Curies isolate polonium and radium
1899	Rutherford discovers alpha and beta radiation emitted from uranium

Table A.1 (*Continued*)

1900	Villard discovers gamma rays
	Thompson proposes "plum pudding" atomic model
	Planck suggests that radiation is produced in discrete quantities
	American Röentgen Ray Society (ARRS) is founded
1901	First report of death due to X-ray exposure
	First Nobel Prize in physics awarded to Röentgen
	Marconi generates radio waves that are detected across the Atlantic Ocean
1903	Becquerel and Curies receive the Nobel Prize for their study on radioactivity
	Tsiolkovsky introduces the concept of space travel
1905	Einstein formulates Special Theory of Relativity
	Einstein formulates the explanation of the photoelectric effect
1906	Bergonnie and Tribondeau formulate basics of radiobiology
1907	Boltwood estimates the Earth's age to be 2×10^9 yr, far greater than previous estimates
1908–1913	Hertzsprung and Russel correlate the energy emitted from a star to its temperature
1909	Millikan oil drop experiment yields a precise value of electronic charge
1910	Soddy establishes the existence of isotopes
1911	Rutherford discovers the atomic nucleus
	Wilson develops cloud chamber
1912	von Laue demonstrates interference of X-rays
	Hess discovers cosmic rays
1913	Coolidge applies for X-ray tube patent
	Bohr advances the theory of the hydrogen atom
	Einstein completes the General Theory of Relativity
1914	Franck–Hertz experiment demonstrates discreet atomic energy levels in collisions with electrons
	Goddard initiates experimental rocketry
1915	British Röentgen Society adopts X-ray protection recommendations
1916	Millikan measures Planck's constant
1917	Rutherford produces first artificial nuclear transmutation
	Mount Wilson telescope begins operations
1918	*Noether's Theorem* establishes a relationship between symmetries and conservation laws that was crucial to the later development of quantum gauge field theory and string theory
1919	Aston detects isotopes
	Proton discovered by Rutherford
	Prediction of General Theory of Relativity regarding the gravitational deflection of light is confirmed
1920	ARRS establishes a standing committee for radiation protection
1921	British X-ray and Radium Protection Committee presents its first radiation protection rules
	Kaluza publishes his ideas about unifying gravity with electromagnetism by adding an extra dimension of space
1922	First US radium-related dial painter death
	Compton effect reported
	ARRS adopts British radiation rules
	American Registry of X-ray Technicians founded

(*continued*)

Table A.1 (*Continued*)

	General Theory of Relativity predicts an expanding universe
1923	Szamatolski links dial painter injuries to radium
	Hubble measures the distance to the Andromeda Galaxy
1924	Uhlenbeck and Goudsmit identify electron spin of $\hbar/2$
	de Broglie formulates particle wavelength–momentum relationship
1925	First International Congress of Radiology held, the forerunner to the International Commission on Radiation Units and Measurements (ICRU)
	Pauli exclusion principle formulated
	Heisenberg publishes paper on quantum mechanics
	Mutscheller puts forth tolerance dose for X-rays
	Eddington formulates a relationship between a star's mass and its energy output
	Schröedinger formulates wave mechanics
1926	Dirac develops the basis for quantum electrodynamics
1927	Heisenberg develops the uncertainty principle
	Mueller discovers that ionizing radiation produces genetic mutations
	Davisson and Germer demonstrate that matter has wave properties
	Lemaître formulates the *big bang* theory
1928	International Commission on Radiological Protection (ICRP) formed
	Dirac develops relativistic wave equation for the electron, which established the theoretical basis for antiparticles
1929	Advisory Committee on X-Ray and Radium Protection (ACXRP) is formed in the United States. This was a precursor of the National Council on Radiation Protection and Measurement (NCRP)
	Hubble establishes the expansion of the universe
	Röentgen is adopted as a unit for X-ray radiation by ICRU
1930	Bethe advances quantum-mechanical stopping-power theory
1931	NBS Handbook 15 *X-Ray Protection* published by ACXRP and established the first US guidelines for radiation protection
	Pauli proposes the neutrino to explain the conservation of energy in beta decay
	Lawrence and Livingston construct the first cyclotron
1932	Anderson discovers the positron
	Chadwick reports the discovery of the neutron
	Chandrasekhar calculates stellar collapse to the white dwarf, neutron star, or black hole states
1933	Slizard postulates the nuclear chain reaction
	Zwicky suggests the existence of dark matter
1934	Curie and Joliot produce artificial radioisotopes
	ACXRP recommends daily tolerance dose of 0.1 R
	First whole-body count performed by Evans (MIT)
	Fermi formulates a theory of beta decay
1935	Yukawa predicts the existence of mesons, responsible for the short-range nuclear interaction
1936	Bragg–Gray principle is formulated
1937	First use of a radioisotope (^{32}P) in therapy
	Muons detected in cosmic radiation
1938	Hahn, Meitner, Strassmann, and Fermi study nuclear fission
	Bethe explains stellar energy production in terms of fusion
1939	Meitner and Frisch formulate a fission model
1940	Kerst operates the first betatron

Table A.1 (*Continued*)

1941	Plutonium discovered by Seaborg's research team
	ACXRP recommends first permissible body burden for radium
1942	Manhattan Engineer District created to develop an atomic weapon
	CP-1 uranium/graphite pile achieves first controlled nuclear chain reaction
1942–1945	V-2 Rocket is tested and used in warfare
1943	Oak Ridge's X-10 Clinton Pile achieves criticality
1944	Hanford's B Reactor achieves criticality
1945	First nuclear detonation at Trinity Site
	Nuclear weapons detonated at Hiroshima and Nagasaki, Japan
	First Los Alamos criticality accident
1946	Second Los Alamos criticality accident
	Atomic Energy Act creates the *Atomic Energy Commission*
1947	The Atomic Bomb Casualty Commission (ABCC) is established by the US Academy of Sciences to initiate long-term studies of A-bomb survivors in Hiroshima and Nagasaki
	Pion discovered
	First strange particle (kaon) discovered
1948	Transistor invented by Shockley, Bardeen, and Brittain
	Feynman, Schwinger, and Tomonaga introduce renormalization to eliminate divergence issues in the quantum gauge field theory of electrodynamics
1949	Soviet Union detonates nuclear weapon
	NCRP publishes recommendations and introduces the risk/benefit concept
1951	First cobalt teletherapy treatment
	First reactor to produce electricity by design (EBR-1)
1952	First thermonuclear (fusion) detonation
	Radiation Research Society formed
	Townes formulates the laser concept
1953	President Eisenhower announces Atoms for Peace program
	International Commission on Radiological Units and Measurements (ICRU) introduces concept of absorbed dose
1954	Atomic Energy Act signed
	First power reactor achieves criticality (Obninsk)
	USS Nautilus (first nuclear-powered submarine) launched
	Society of Nuclear Medicine founded
1955	Fermi and Slizard patent CP-1 pile
	Decision to form Health Physics Society (US)
	Arco, ID, becomes the first city to be powered by nuclear power
	First United Nations Conference on peaceful uses of atomic energy
	United Nations Scientific Committee on the Effects of Atomic Radiation (UNSCEAR) established
	Antiproton discovered
1956	First Biological Effects of Atomic Radiation report published
	Lee and Yang discover nonconservation of parity in beta decay
	Neutrino detected
	Health Physics Society founded
1957	UK Windscale accident leads to the release of radioactive material to the environment
	First US commercial power reactor at Shippingport, PA, achieves criticality

(*continued*)

Table A.1 (*Continued*)

	NCRP introduces age proration for occupational doses and recommends nonoccupational exposure limits
	US Congressional Joint Committee on Atomic Energy initiates hearings on radiation hazards with an initial review of weapons test fallout
	First orbiting spacecraft (Sputnik) is launched by the Soviet Union
	International Atomic Energy Agency (IAEA) founded under the United Nations
	Explosion at underground high-level waste reprocessing storage tank at the Mayak Chemical Complex (USSR) released 7.5×10^{10} MBq
1958	Discovery of van Allen radiation belts
	First United Nations Scientific Committee on the Effects of Atomic Radiation Report (UNSCEAR) addresses a study of exposure sources and biological hazards
1959	Nuclear merchant ship, Savannah, launched
	Federal Radiation Council established
	ICRP recommends limitation of genetically significant dose
	Dyson sphere concept introduced
1960	First successful laser
	American Association of Physicists in Medicine formed
	American Board of Health Physics begins certification
	First Biological Effects of Atomic Radiation Report issued by the US National Academy of Sciences
1960–1961	First two reports from the Federal Radiation Council provide radiation protection guides that introduce the concept of biological risks/benefits of radiation dose
1960–present	A diverse group of theories (e.g., string theory, M theory, quantum gravity, D-branes, various gauge theories, superstring theories, supersymmetry, and theory of everything) are proposed to unify the four fundamental interactions. To date, none have been verified
1961	First nuclear-powered aircraft carrier (USS Enterprise) commissioned
	SL-1 reactor of US Army undergoes a prompt criticality accident at the Idaho National Engineering Laboratory that results in the death of three workers
	Federal Radiation Regulations adopted in United States in Title 10, Code of Federal Regulations, Part 20
1962	Differences are noted between the electron neutrino and the muon neutrino
1963	Limited Nuclear Test Ban Treaty signed
1964	International Radiation Protection Association formed
	Quark model introduced by Gell-Mann and Zweig
	US satellite disintegrates over Madagascar and releases 6.3×10^{8} MBq of plutonium into the atmosphere
	NCRP incorporated by Act of Congress
	Kardashev civilization type scale introduced
1965	First nuclear reactor in space
	Temporary Dosimetry System 1965 (T65D) developed for A-bomb survivors
1966	Fermi 1 Atomic Power Plant undergoes a partial fuel melting event
1967	Salam, Weinberg, and Glashow propose theories that unify the weak and electromagnetic interactions
1968	Nuclear Nonproliferation Treaty signed
1969	During the first manned moon landing, Apollo 12 deploys SNAP-27 nuclear generator

Table A.1 (*Continued*)

	Electron–proton scattering at the Stanford Linear Accelerator Center reveals the existence of structures that are interpreted to be the up, down, and strange quarks/partons
	Gofman and Tamplin at the Lawrence Livermore National Laboratory report that no safe threshold exists for radiation dose
1972	Biological Effects of Ionizing Radiation (BEIR I) Report published using a linear model for risk estimates
	First computerized tomography scan performed
1974	The J/Ψ particle is discovered, demonstrating the existence of the charm quark
	A mechanism proposed to explain the energy emission by black holes
1975	Atomic Bomb Casualty Commission replaced by the binational Radiation Effects Research Foundation (RERF) to continue studies of A-bomb survivors
	Tau lepton discovered
	The existence of dark matter is confirmed
1977	ICRP Publication 26, *Recommendations of the International Commission on Radiological Protection*, introduces the stochastic and nonstochastic effects and the dose equivalent concept
	Commercial fuel reprocessing deferred in the United States
	The upsilon particle is discovered, demonstrating the existence of the bottom quark
	Voyager 2 is launched and its electricity is generated from the decay heat of plutonium
1978	Penzias and Wilson awarded Nobel Prize for the discovery of 2.7 K microwave radiation permeating space that is presumed to be a remnant of the *big bang* event that occurred about 13 billion years ago
	ICRP 30, *Limits for Intakes of Radionuclides by Workers*, published
	Standard Model of Particle Physics is accepted as the vehicle for the unification of the strong, weak, and electromagnetic interactions
1979	Three Mile Island Unit-2 accident occurs with minimal iodine release following a small-break loss-of-coolant accident with partial core melt
	Gluons are observed indirectly from three-jet events at the Deutsches Elektronen-Synchrotron (DESY)
1980	Biological Effects of Ionizing Radiation III Report that uses linear-quadratic models for risk estimates published
	Theory of hormesis proposed
1981	First dedicated synchrotron light source becomes operational at Daresbury Laboratory (UK)
1983	Field quanta of the weak interaction are discovered at the European Laboratory for Particle Physics (CERN)
	Nuclear Waste Policy Act (US) establishes a research and development program for the disposal of high-level radioactive waste and spent nuclear fuel
1986	Chernobyl Accident (Ukraine) occurs with a major release of fission products. There are 31 fatalities from the event
	Dosimetry System 1986 (DS86) developed by RERF for A-bomb survivors
1987	Nuclear Waste Policy Amendments Act designates Yucca Mountain, Nevada, as the site for the United States's first geological repository for high-level radioactive waste and spent nuclear fuel
	Neutrinos and γ-rays are detected from Supernova 1987A in the Large Magellanic Cloud

(*continued*)

Table A.1 (Continued)

1988	US National Academy of Sciences publishes the BEIR IV Report, *Health Effects of Radon and Other Internally Deposited Alpha Emitters-BEIR IV*
	Sources, Effects and Risks of Ionizing Radiation published by UNSCEAR
1989	The World Wide Web is launched as a networked information project at CERN
1990	The Hubble Space Telescope becomes operational
	Human Genome Project begins
	US National Academy of Sciences publishes the BEIR V Report, *Health Effects of Exposure to Low Levels of Ionizing Radiation-BEIR V*
1991	IAEA reports on health effects of the 1986 Chernobyl accident
	ICRP Publication 60, *1990 Recommendations of the International Commission on Radiological Protection*, published
1993	The Tokamak reactor at Princeton University generates megawatts of power for 1 s through thermonuclear fusion of hydrogen isotopes
1994	Protocols developed for joint US, Ukraine, Belarus 20-year study of child thyroid disease following the 1986 Chernobyl accident
	ICRP 66, *Human Respiratory Tract Model for Radiological Protection*, published
1995	Researchers use the Tevatron at the Fermi National Accelerator Laboratory to detect the top quark, the sixth and last member of the quark family of fundamental particles proposed by the Standard Model of Particle Physics
	Galileo spacecraft explores Jupiter and its moons at close range
1996	The first Generation III fission reactor goes online (Japan)
1997	The Joint European Torus achieves a world record peak fusion power of 16 MW for less than a second
1999	US National Academy of Sciences BEIR VI Report, *Health Effects of Exposure to Radon-BEIR VI*, published
	Three Japanese workers receive 17, 10, and 3 Gy from a criticality event in a fuel fabrication facility at Tokai Mura (Japan)
2000	Relativistic Heavy Ion Collider at Brookhaven National Laboratory (US) begins operation
2001	NCRP Report No. 136, *Evaluation of the Linear-Nonthreshold Dose-Response Model for Ionizing Radiation*, reviews the linear no-threshold model and recommends its continued use in radiation protection
2002	Dosimetry System 2002 (DS02) developed by RERF for A-bomb survivors
2003	NASA launches two rovers to explore the Martian surface
2004	Westinghouse AP1000 Generation III fission reactor design is certified in the United States by the Nuclear Regulatory Commission
2005	Initial free-electron laser output from DESY
	France selected as the host country for the International Thermonuclear Experimental Reactor (ITER)
	A D-D fusion reaction was achieved through the pyroelectric effect
	D-D fusion achieved through sonoluminescence
2006	BEIR VII, *Health Risks from Exposure to Low levels of Ionizing Radiation, BEIR VII Phase 2*, published
	The 2006 International Astronomical Union assigned Pluto as a minor planet. Ceres, Charon, and Xena were also assigned minor planet status in the 2006 Sol system reclassification
	The largest known supernova (Supernova 2006gy) occurs in the peculiar galaxy NGC 1260. It was about two orders of magnitude larger than any previously known supernova event

Table A.1 (*Continued*)

2007	ICRP Publication 103, *The 2007 Recommendations of the International Commission on Radiological Protection*, published
2008	Large Hadron Collider (CERN) begins operations
2008–2015	International Thermonuclear Experimental Reactor (ITER) construction period
2009	Estimated start date for the Linac Coherent Laser Source (LCLS), an X-ray free-electron laser, at the Stanford Linear Accelerator Center
2010	China projects landing an unmanned vehicle on the surface of the Moon
	Estimated start date for Japan's SPring-8 Compact Self-Amplified Spontaneous Emission X-ray free-electron laser
	Scheduled shutdown of the Fermi National Accelerator Laboratory's Tevatron
	Scheduled shutdown of the Stanford Linear Accelerator Center's B Factory
	Scheduled launch of the Laser Interferometer Space Antenna, an orbital gravitational wave observatory
	NASA's scheduled launch of Juno Mission to orbit Jupiter
2012	Scheduled completion for an X-ray free-electron laser at DESY in Germany
2012–2019	Projected International Linear Collider Construction Period (optimistic projection)
2015	First ITER plasma operations
	New Horizons Spacecraft to approach the vicinity of Pluto and Charon
2015–2020	National Aeronautics and Space Administration's (NASA's) goal of returning humans to the moon
2015–2025	Projected date for operation of first Generation IV fission reactor
2015–2036	ITER operations
2016	Scheduled launch of a nuclear-powered, ion-propelled spacecraft toward the Neptune system by NASA
2024	China projects a manned landing on the Moon
2030	Projected start date for a demonstration fusion power plant
2036–2041	ITER decontamination operations
2050	Projected time frame for an unmanned mission to the Oort Cloud (10 000 AU from Earth) as a logical precursor to subsequent missions to Alpha Centauri and beyond
	Projected start date for a fusion power plant

References

A Century of Physics, The American Physical Society, College Park, MD. http://timeline.aps.org/APS/Timeline/ (accessed on January 15, 2007).

Turner, J.E. (1995) *Atoms, Radiation, and Radiation Protection*, 2nd edn., John Wiley & Sons, Inc., New York.

Weaver, J.E., III, (2003) A Brief Chronology of Radiation and Protection, Michigan Section, American Nuclear Society. http://local.ans.org/mi/Teacher_CD/Historical%20Info/Radiation_history.doc (accessed on January 15, 2007).

Weber, R.L., Manning, K.V. and White, M.W. (1965) *College Physics*, McGraw-Hill, New York.

Appendix B
Production Equations in Health Physics

B.1
Introduction

The assumption that radioactive material enters a system at a constant rate leads to a set of production equations that describe a broad class of phenomena encountered by health physicists. Equations governing activation, buildup of radioactive material on a filter or demineralizer, deposition of material on a surface from a radioactive plume, and release of material into a room are examples of phenomena described consistently by production equations. This appendix describes production equations and their applications in a wide variety of health physics areas.

B.2
Theory

In health physics applications, the rate of change of radioactive material in a system is described by first-order linear differential equations that have exponential solutions. Since exponential forms appear throughout the field, it is not unexpected that phenomena describing the accumulation of radioactive material have a similar mathematical structure. This text refers to these structures as production equations.

To formulate a general form of production relationship, consider the time rate of change of activity $\dot{A}$ associated with the continuous introduction of a radionuclide into a system or structure. For a given radionuclide

$$\dot{A} = Pe^{-Kt}, \tag{B.1}$$

where P is the production term or the rate at which activity is added to the system (e.g., room, accelerator target, or filter paper), K is the total removal rate of the radionuclide from the system, and t is the time from the start of production. To simplify the equation resulting from the integration of Equation B.1, P is assumed to be constant. The production term has units of activity per unit time (Bq/s). Examples of the production term for a variety of physical phenomena are provided in Table B.1.

Health Physics in the 21st Century. Joseph John Bevelacqua
Copyright © 2008 WILEY-VCH Verlag GmbH & Co. KGaA, Weinheim
ISBN: 978-3-527-40822-1

Appendix B

Table B.1 Examples of production terms in health physics applications.

Physical phenomena	P (Bq/s)	Definition of terms (units)
Activation of material in an accelerator	$N\sigma\varphi\lambda$	N = number of target atoms of the nuclide being activated (atoms) σ = activation cross section for the specific activation reaction (b/atom or cm^2/atom) φ = activating flux of a beam of particles (particles/cm^2 s) λ = radioactive disintegration constant (s^{-1})
Activation of material in a reactor	$N\sigma\varphi\lambda$	N = number of target atoms of the nuclide being activated (atoms) σ = activation cross section for the specific activation reaction (b/atom or cm^2/atom) φ = activating flux of neutrons (neutrons/cm^2 s) λ = radioactive disintegration constant (s^{-1})
Deposition of radioactive material in a demineralizer bed	CFe	C = influent activity concentration of an isotope entering the demineralizer (Bq/m^3) F = flow rate of fluid through the demineralizer (m^3/s) e = isotope specific removal efficiency of the demineralizer bed
Deposition of radioactive material in a filter	CFe	C = influent activity concentration of an isotope entering the filter (Bq/m^3) F = flow rate of fluid through the filter (m^3/s) e = isotope specific removal efficiency of the filter
Surface deposition from a radioactive plume	wS	w = ground deposition rate (Bq/m^2 s) S = surface area of the deposition (m^2)
Inhalation of radioactive material	Cr	C = air concentration of radioactive material (Bq/m^3) r = breathing rate (m^3/s)
Surface deposition from a leaking radioactive fluid	CF	C = activity concentration of the isotope in the fluid leaking onto the surface (Bq/m^3) F = leak rate of the fluid onto the surface (m^3/s)
Airborne entry of ^{222}Rn into a home	CF	C = air concentration of ^{222}Rn entering the home (Bq/m^3) F = air infiltration rate entering the home (m^3/s)

Table B.1 (Continued)

Physical phenomena	P (Bq/s)	Definition of terms (units)
Release of radioactive material from a stack	CF	C = air concentration of radioactive material being released (Bq/m³) from a stack F = stack flow rate (m³/s)
Release of radioactive material into a room	Q	Q = release rate of airborne radioactive material into the room (Bq/s)

When using Equation B.1, it is important that the production equation be applied separately for each radionuclide of interest. The quantities P and K depend on the radionuclide half-life as well as on its physical and chemical properties.

The total removal rate has numerous components. The most common components are derived from radioactive decay (λ), biological decay (λ_b), or ventilation (λ_v). Explicit forms for these removal rates are

$$\lambda = \ln(2)/T_{1/2}, \tag{B.2}$$

$$\lambda_b = \ln(2)/T_{1/2}^b, \tag{B.3}$$

$$\lambda_v = F/V, \tag{B.4}$$

where $T_{1/2}$ is the physical half-life, $T_{1/2}^b$ is the biological half-life, F is the ventilation flow rate of the system, and V is the free air volume of the system. The total removal rate

$$K = \lambda + \lambda_b + \lambda_v + \cdots \tag{B.5}$$

is the sum of the individual removal rates as they apply to the problem of interest. Not all terms in Equation B.5 appear in each application. The specific application of removal rates is addressed in subsequent discussion.

Equation B.1 can be integrated with respect to time from $t = 0$ to $t = T$ where the time T is the end of the production interval:

$$\int_0^T \dot{A}\,dt = A(T) = \int_0^T P e^{-Kt}\,dt = P\int_0^T e^{-Kt}\,dt. \tag{B.6}$$

In Equation B.6, we assume that no activity is initially present in the system ($A(0) = 0$). Using this condition leads to the result

$$A(T) = \frac{P}{K}(1 - e^{-KT}). \tag{B.7}$$

Equation B.7 provides a relationship describing the buildup of activity during the time that the production term is active. For $KT \gg 1$, the system activity reaches its maximum value. Accordingly, Equation B.7 is written as

$$A(\infty) = A_{eq} = \frac{P}{K}. \tag{B.8}$$

The saturation or equilibrium activity is the maximum activity that can be achieved in the system.

If T is defined as the time during which the production term is active and t describes the time after the production ceases, Equation B.7 is rewritten to describe the activity variation following the production interval and during the subsequent decay period:

$$A(t) = \frac{P}{K}(1-e^{-KT})e^{-kt} \tag{B.9}$$

where k is the total removal rate postproduction, that is, during the decay time t. As a matter of specificity, $t = 0$ corresponds to the time when production ceases.

B.3
Examples

A number of examples are provided to illustrate the utility of the general production equation. These examples include (1) the activation of a target by an accelerator beam or reactor neutron source, (2) buildup of activity on a filter or demineralizer, (3) buildup of activity in a pond, and (4) release of activity into a room.

B.3.1
Activation

Activation is a process described by the reaction C(c, d)D, during which the radiation of type c strikes a target nucleus C and produces a radioactive nucleus D and radiation of type d. Examples of activation reactions include $^{59}\text{Co}(n, \gamma)^{60}\text{Co}$, $^{16}\text{O}(n, p)^{16}\text{N}$, $^{27}\text{Al}(n, \alpha)^{24}\text{Na}$, and $^{3}\text{H}(p, n)^{3}\text{He}$.

Using the generalized production equation (Equation B.9) and the production term from Table B.1 leads to a relationship that describes the activity in the target as a function of time:

$$A = N\sigma\varphi[1-e^{-\lambda T}]e^{-\lambda t}, \tag{B.10}$$

where N, σ, and φ are defined in Table B.1. For nongaseous products, the removal rates (K and k) are equal to the physical decay constant (λ). T is the irradiation time, that is, the time the target is irradiated by the accelerator's beam or the time the material to be activated is exposed to the reactor's neutron fluence rate (flux). The time after the reactor is shut down or the accelerator beam is terminated is t. The steady-state (saturation) or equilibrium activity is $N\sigma\varphi$.

The application of Equation B.10 is further illustrated by considering the activation of ^{59}Co by thermal neutrons. In this example, N is the number of ^{59}Co atoms in the target, σ is the microscopic cross section for the $^{59}\text{Co}(n_{thermal}, \gamma)^{60}\text{Co}$ reaction, φ is the number of thermal neutrons per cm² s, and λ is the ^{60}Co decay constant. Equation B.10 is applied separately for each activated species.

B.3.2
Demineralizer Activity

Ion exchange is a process used in a variety of nuclear facilities to reduce the radioactive ion content of water by removing radioactive ions and replacing them with nonradioactive ions. The device in which the ion exchange occurs is commonly called a demineralizer.

The activity that accumulates within a demineralizer bed is also obtained from Equation B.9 and Table B.1:

$$A = \frac{CFe}{\lambda}[1-e^{-\lambda T}]e^{-\lambda t}. \tag{B.11}$$

Equation B.11 is also to be applied individually for each isotope trapped in the demineralizer bed. In Equation B.11, C, F, and e are defined in Table B.1, λ is the physical decay constant of the trapped material, T is the time the demineralizer in online (valved in) and removing radioactivity from the influent stream, and t is the time after the demineralizer is no longer in service (valved out). For the demineralizer application, the total removal rate is just the physical decay constant.

Equation B.11 also applies to filters. The saturation activity for both filters and demineralizers is CFe/λ.

B.3.3
Surface Deposition

The deposition of radioactive material onto a surface from an airborne plume is also described by a production equation. Again, using Table B.1 and Equation B.9, the activity deposited onto a surface is

$$A = \frac{wS}{K}[1-e^{-KT}]e^{-kt}, \tag{B.12}$$

and wS is defined in Table B.1. The removal rates k and K are discussed below. Equation B.12 is used to illustrate the versatility of the production equation.

Assuming that there is a continuous release of radioactive material from a plume and that an equilibrium has been reached, an expression for the equilibrium activity that has been removed from the plume and deposited on a surface of area S is written as

$$A_{eq} = \frac{wS}{K}. \tag{B.13}$$

If it is also assumed that the material deposits on the surface of a stationary body of water, such as a pond, then Equation B.13 still applies and

$$K = \lambda + \lambda_b, \tag{B.14}$$

where λ_b is the biological removal rate from the pond.

If the radionuclide deposited onto the surface of the pond is also soluble in the pond water, and instantaneous mixing of the radionuclide within the pond occurs, then the equilibrium concentration C_{eq} of the radionuclide in the pond water is determined from the relation

$$C_{eq} = \frac{A_{eq}}{V}, \tag{B.15}$$

where V is the volume of water in the pond.

Determine the equilibrium concentration of a radionuclide in a pond using Equations B.13 and B.15:

$$C_{eq} = \frac{wS}{KV}. \tag{B.16}$$

The production concept can also be extended to calculate the equilibrium concentration in an organism, such as a fish, living in the pond. The equilibrium activity concentration per unit mass (Bq/kg) in the fish ($C_{eq\text{-fish}}$) is written as

$$C_{eq\text{-fish}} = \frac{IC_{eq}}{K'}, \tag{B.17}$$

where I is the intake of pond water by the fish (m³/kg(fish) s) and K' is the total removal rate of the isotope from the fish

$$K' = \lambda + \lambda'_b, \tag{B.18}$$

where λ'_b is the biological removal rate from the fish. A careful examination of Equation B.18 indicates that the term $C_{eq}I$ is just P per unit mass of the fish. Equation B.18 is another application of the production equation, Equation B.9.

B.3.4
Release of Radioactive Material into a Room

The release of airborne radioactive material into a room is obtained from Equation B.9 and Table B.1:

$$A = \frac{Q}{K}[1 - e^{-KT}]e^{-kt}, \tag{B.19}$$

where the removal of radioactive material includes both physical decay and ventilation terms

$$K = k = \lambda + \frac{F}{V}. \tag{B.20}$$

In Equation B.20, the ventilation rate is assumed to be constant during the production and postproduction periods.

B.4
Conclusions

The use of production equations has been shown to provide a unified explanation for a wide variety of phenomena encountered in health physics. The specific application determines the P, K, and k values, but the form of the equation remains the same. The use of production equations greatly simplifies the understanding of a variety of health physics concepts that appear to involve dissimilar phenomena.

References

Bevelacqua, J.J. (1995) *Contemporary Health Physics: Problems and Solutions*, John Wiley & Sons, Inc., New York.

Bevelacqua, J.J. (1999) *Basic Health Physics: Problems and Solutions*, John Wiley & Sons, Inc., New York.

Bevelacqua, J.J. (2003) Production Equations in Health Physics. *Radiation Protection Management*, **20** (6), 9.

Cember, H. (1996) *Introduction to Health Physics*, 3rd edn, McGraw-Hill, New York.

Turner, J.E. (1995) *Atoms, Radiation, and Radiation Protection*, 2nd edn., John Wiley & Sons, Inc., New York.

Appendix C
Key Health Physics Relationships

This appendix provides a summary of important relationships encountered in a number of areas, including external dosimetry, electromagnetic theory, classical mechanics, quantum mechanics, ionizing radiation, and nonionizing radiation. Internal dosimetry relationships and commentary are provided in Appendix D. The equations represent a set of key health physics relationships that are utilized throughout the book. Applications of these equations are provided in the problems given in Chapters 2–7.

Within this appendix, the following notation is used:

A	Source activity
	Laser beam area
	Hot particle activity
B	Magnetic induction
	Buildup factor
C	Capacitance
C_a	Activity per unit area
C_v	Activity per unit volume
D	Displacement current
	Divergence angle
DF	Duty factor
	Hot particle dose factor
DRCF	Dose rate conversion factor
E	Electric field strength
	Effective dose
	Energy
	Irradiance
$\dot{E}$	Effective dose rate
E_0	Rest mass
F	Force
GB	Gaussian beam
H	Effective dose (used to avoid confusion when the energy appears in an equation defining the effective dose)

	Magnetic field strength
	Radiant exposure
$\dot{H}$	Effective dose rate
I	Current
	Measured radiation quantity (e.g., absorbed dose, flux, exposure, effective dose, and dose)
J	Current density
L	Inductance
M	Magnetization
MPE	Maximum permissible exposure
N	Number of atoms
OD	Optical density
P	Polarization
	Power
	Pressure
PRF	Pulse repetition frequency
PW	Pulse width
Q	Heat
	Release rate
R	Idea gas constant
	Resistance
	Radius of disk source
S	Source strength
	Poynting vector
STP	Standard temperature and pressure
T	Kinetic energy
	Temperature
V	Voltage
	Volume
	Potential energy
W	Work
Y	Yield
Z	Impedance
a	Acceleration
	Aperture radius
	Area
c	Speed of light
	Heat capacity
d	Daughter
	Distance from laser aperture
e	Energy stored in an electric field per unit volume
	Energy (used to avoid confusion when the energy appears in an equation defining a related quantity, such as the radiant exposure)
h	Release height
	Distance from disk source (on-axis)
	Energy stored in a magnetic field per unit volume

i	Summation index	
k	Conversion factor (value depends on the units elected and the particular relationship)	
l	Angular momentum	
m	Mass	
m_0	Rest mass	
n	Number of moles	
p	Momentum	
	Parent	
q	Charge	
r	Radius of circular orbit	
	Radius of laser beam	
	Distance from radiation source	
s	Distance	
t	Shield thickness	
	Time	
	Thickness of disk source	
u	Mean wind speed	
v	Velocity	
x	Shield thickness	
	Vector cross product $[\vec{A} \times \vec{B} = |\vec{A}||\vec{B}|\sin\theta]$	
y	Cross-wind distance	
w	Perpendicular distance from line source	
$\vec{\nabla}$	Gradient operator	
Γ	Dose factor or gamma constant (Sv m^2/MBq h)	
ΔE	Uncertainty in energy or the width of an energy level	
Δt	Uncertainty in time or the lifetime of an energy level	
Δp	Uncertainty in momentum	
Δx	Uncertainty in position	
β	Velocity relative to the speed of light	
γ	Lorentz factor	
ε	Permittivity	
θ	Angle between the two vectors involved in the cross product	
	Included the angle that the point of interest makes with the ends of a line source	
μ	Attenuation coefficient	
	Permeability of a medium	
μ_{en}	Energy absorption coefficient	
ν	Frequency	
λ	Disintegration constant	
	Wavelength	
ρ	Density	
	Charge density	
	Gas density in an ionization chamber	
	Physical density	
σ_y	Horizontal standard deviation	

σ_z Vertical standard deviation
χ Concentration of radioactive material in a plume
χ/Q Dispersion factor (s/m³)
$\chi u/Q$ Dispersion factor (m⁻²)
ψ Wave function
ω Angular frequency
$\vec{a}$ An arrow over a variable indicates it is a vector quantity

Key relationships:

- **Activation**

    See Appendix B.

- **Activity**

$$A = \lambda N,$$

$$A(t) = A(0) e^{-\lambda t},$$

$$A_d(t) = \frac{\lambda_d A_p(0)}{\lambda_d - \lambda_p} \left(e^{-\lambda_p t} - e^{-\lambda_d t} \right).$$

- **Attenuation**

$$I(x) = I(0) B e^{-\mu x}.$$

$$B(\text{Fe}) = 1 + \mu x \text{ for small } \mu x.$$

$$B(\text{Pb}) = 1 + \mu x/3 \text{ for small } \mu x.$$

- **Duty factor**

$$\text{DF} = \frac{I_{\text{average}}}{I_{\text{peak}}} = \frac{P_{\text{average}}}{P_{\text{peak}}} = \text{PW} \times \text{PRF}.$$

- **External dosimetry**

    –Dose – Point source

$$\dot{E} = \frac{A\Gamma}{r^2},$$

$$\dot{H} = \frac{S}{4\pi r^2} \left(\frac{\mu_{\text{en}}}{\rho} \right) \sum_i E_i Y_i.$$

  –Dose – Line source

$$\dot{E} = \frac{A\Gamma\theta}{w}.$$

–Dose – Thin disk source

$$\dot{E} = \pi C_a \Gamma \ln\left(\frac{R^2+h^2}{h^2}\right).$$

–Dose – Thick disk source

$$\dot{E} = \frac{\pi C_v \Gamma (1-\exp(-\mu t))}{\mu} \ln\left(\frac{R^2+h^2}{h^2}\right).$$

–Gamma constant or dose factor

$$\Gamma = k \sum_i E_i Y_i.$$

–Hot particle absorbed dose

$$D = \frac{A(\mathrm{DF})t}{a}.$$

–Ionization chamber dose–current relationship

$$I = \rho V \frac{T_{\mathrm{STP}}}{T} \frac{P}{P_{\mathrm{STP}}} \dot{H}.$$

- **Internal dosimetry**

  See Appendix D.

- **Dispersion relationships**

  –Dispersion theory – Pasquill–Gifford Equation

  $$\chi = \frac{Q}{\pi \sigma_y \sigma_z u} \exp\left[-\frac{1}{2}\left(\frac{y^2}{\sigma_y^2} + \frac{h^2}{\sigma_z^2}\right)\right].$$

  –Dispersion

  $$\dot{E} = Q \left(\frac{\chi u}{Q}\right) \frac{1}{u} (\mathrm{DRCF}),$$

  $$\dot{E} = Q \left(\frac{\chi}{Q}\right) (\mathrm{DRCF}).$$

- **Electromagnetic relationships**

  –Constants

  Permittivity of free space: $\varepsilon_0 = 8.854 \times 10^{-12}$ F/m
  Permeability of free space: $\mu_0 = 4\pi \times 10^{-7}$ N/A^2
  Speed of light: $c = 3.0 \times 10^8$ m/s $= (\varepsilon_0 \mu_0)^{-1/2}$

Impedance of free space: $Z = (\mu_0/\varepsilon_0)^{1/2} = 376.7$ Ohms
Charge: 1 C = 1 A/s
Potential: 1 V = 1 J/C
Magnetic field: 1 T = 1 N/A m = 1.0×10^4 gauss (G)

–Capacitance

$$C = \frac{q}{V}.$$

–Constitutive equations

$$\vec{D} = \varepsilon_0 \vec{E} + \vec{P},$$

$$\vec{H} = \frac{\vec{B}}{\mu_0} - \vec{M}.$$

–Constitutive equations in a linear medium

$$\vec{D} = \varepsilon \vec{E},$$

$$\vec{H} = \frac{\vec{B}}{\mu}.$$

–Current

$$I = \frac{q}{t}.$$

–Electric field strength

$$E = \frac{F}{q}.$$

–Energy

$$E = qV.$$

–Energy stored in an electromagnetic field per unit volume

$$e = \frac{1}{2}\varepsilon_0 E^2,$$

$$h = \frac{1}{2}\mu_0 H^2.$$

–Forces

–Electric force

$$\vec{F} = q\vec{E},$$

$$F = \left(\frac{1}{4\pi\varepsilon_0}\right) \frac{q_1 q_2}{s^2}.$$

–Magnetic force

$$\vec{F} = q\vec{v} \times \vec{B} = q|\vec{v}||\vec{B}|\sin\theta,$$

$$\vec{B} = \mu\vec{H}.$$

–Lorentz force

$$\vec{F} = q(\vec{E} + \vec{v} \times \vec{B}).$$

–Impedance (alternating current)

$$V = ZI,$$

$$V = V_0 \sin\omega t,$$

$$Z = \left[R^2 + \left(\omega L - \frac{1}{\omega C}\right)^2\right]^{1/2}.$$

–Ohm's law (direct current)

$$V = IR.$$

–Power

$$P = IV = I^2 R.$$

–Poynting vector

$$\vec{S} = \vec{E} \times \vec{H} = |\vec{E}||\vec{H}|\sin\theta.$$

–Maxwell equations

$$\vec{\nabla} \cdot \vec{D} = \rho,$$

$$\vec{\nabla} \times \vec{H} - \frac{\partial \vec{D}}{\partial t} = \vec{J},$$

$$\vec{\nabla} \cdot \vec{B} = 0,$$

$$\vec{\nabla} \times \vec{E} + \frac{\partial \vec{B}}{\partial t} = 0.$$

- **General theory of relativity**

  See Appendix L.

- **Mechanics relationships**
  –Angular momentum

  $$l = mvr.$$

–Centrifugal force
$$F = \frac{mv^2}{r}.$$

–Force
$$F = ma.$$

–Heat
$$Q = mc\Delta T.$$

–Ideal gas
$$PV = nRT.$$

–Kinetic energy
$$T = \frac{1}{2}mv^2 = \frac{p^2}{2m}.$$

–Momentum
$$\vec{p} = m\vec{v} = \gamma m_0 \vec{v}.$$

–Total energy
$$E = mc^2 = \gamma m_0 c^2,$$
$$E^2 = p^2 c^2 + m_0^2 c^4 = (m_0 c^2 + T)^2.$$

–Relativistic mass
$$m = m_0 \gamma.$$

–Relativistic notation
$$\beta = \frac{v}{c}.$$
$$\gamma = \frac{1}{\sqrt{1-\beta^2}}.$$

–Rest energy
$$E_0 = m_0 c^2.$$

–Wavelength
$$c = v\lambda.$$

−Work
$$W = Fs.$$

- **Nonionizing radiation relationships**
  −Gaussian beam radius
  $$r_{\text{GB}} = \left(a^2 + d^2 \frac{D^2}{4}\right)^{1/2}.$$

  −Irradiance
  $$E = \frac{P}{A}.$$

  −Nominal ocular hazard distance (NOHD)
  $$\text{NOHD} = \frac{2}{D}\sqrt{\frac{P}{\pi(\text{MPE})} - a^2}.$$

  −Optical density
  $$\text{OD} = \log_{10}\left[\frac{H}{\text{MPE}}\right] = \log_{10}\left[\frac{E}{\text{MPE}}\right].$$

  −Radiant exposure
  $$H = \frac{e}{A}.$$

- **Production equations**

  See Appendix B.

- **Quantum mechanics**
  −Schröedinger equation
  $$\left(-\frac{\hbar^2}{2m}\nabla^2 + V\right)\psi = E\psi.$$

  −Uncertainty relationships
  $$\Delta E \Delta t \geq \hbar,$$
  $$\Delta p \Delta x \geq \hbar.$$

- **Special Theory of Relativity**

  See Appendix F.

References

Bevelacqua, J.J. (1995) *Contemporary Health Physics: Problems and Solutions*, John Wiley & Sons, Inc., New York.

Bevelacqua, J.J. (1999) *Basic Health Physics: Problems and Solutions*, John Wiley & Sons, Inc., New York.

Bevelacqua, J.J. (2003) Production equations in health physics. *Radiation Protection Management*, **20** (6), 9.

Bevelacqua, J.J. (2005) Internal dosimetry primer. *Radiation Protection Management*, **22** (5), 7.

Cember, H. (1996) *Introduction to Health Physics*, 3rd edn, McGraw-Hill, New York.

Turner, J.E. (1995) *Atoms, Radiation, and Radiation Protection*, 2nd edn, John Wiley & Sons, Inc., New York.

Appendix D
Internal Dosimetry

D.1
Introduction

Internal dosimetry can be overwhelming and sometimes confusing because there is a wealth and diversity of models and terminology. International Commission on Radiological Protection (ICRP) and Medical Internal Radiation Dose (MIRD) methodologies are the most commonly used internal dosimetry models. The various ICRP and MIRD models are similar in terms of their assumptions and defining equations. This similarity is obscured by difference in the terminology and notation. These differences contribute to the confusion and can limit a full understanding of these models. Emphasizing the definition of absorbed dose and using this definition to illustrate the terminology and notation of the MIRD and ICRP minimize the confusion.

Contemporary internal dosimetry models began with the single-compartment models of ICRP 2 and 10 (ICRP 2/10). The MIRD methodology and ICRP 26 and 30 (ICRP 26/30) developed the concept of source and target organs. ICRP 60 and supporting publications, including ICRP 66 (ICRP 60/66), continue to refine the internal dosimetry methodology. In this appendix, we use the notation ICRP 2/10, ICRP 26/30, and ICRP 60/66 to refer to the defining internal dosimetry publications and supporting documents. Additional refinement is planned as a part of the 2007 ICRP recommendations and supporting publications.

In this appendix, the essential elements of internal dosimetry are presented. The presentation begins by defining the key elements of the MIRD and ICRP models in terms of the absorbed dose. With the key elements established, the MIRD and ICRP methodologies are presented in additional detail.

D.2
Overview of Internal Dosimetry Models

As an introduction to the MIRD and ICRP internal dosimetry models, the absorbed dose rate following the intake of radioactive material is calculated. If an isolated (single-compartment) organ having a mass m contains an activity $q(t)$ of radioactive material that emits radiation of energy E per disintegration, then the initial absorbed dose rate ($\dot{D}_0$) to this organ is

$$\dot{D}_0 = k\frac{q(0)E}{m}, \tag{D.1}$$

where k is a constant and $q(0)$ is the initial activity in the organ. If $q(0)$ is expressed in μCi, E in MeV/disintegration, and m in grams, then

$$k = 2.13 (\text{rad/h})(\text{g dis/MeV Ci}). \tag{D.2}$$

Equation D.2 is presented in traditional US units because it has historical roots and is still used in literature in the United States.

The dose rate as a function of time t is written in terms of the initial absorbed dose rate:

$$\dot{D}(t) = \dot{D}_0 \exp(-\lambda_{\text{eff}} t), \tag{D.3}$$

where λ_{eff} is the effective removal rate from the organ and

$$\lambda_{\text{eff}} = \lambda_{\text{p}} + \lambda_{\text{b}}. \tag{D.4}$$

In Equation D.4, λ_{p} is the physical removal rate (disintegration constant) and λ_{b} is the biological removal rate. The removal rates are related to their respective half-lives (T) through the relation

$$\lambda = \ln(2)/T. \tag{D.5}$$

The absorbed dose (D) is the integral of the dose rate with respect to time. Using Equations D.1 and D.3 leads to the following expression for the absorbed dose:

$$D = \int_0^T \dot{D}(t) dt = \int_0^T \dot{D}_0 \exp(-\lambda_{\text{eff}} t) dt = \int_0^T k\frac{q(0)E}{m} \exp(-\lambda_{\text{eff}} t) dt. \tag{D.6}$$

Equation D.6 simplifies by recognizing that only the activity in the organ varies with time:

$$D = \frac{kE}{m} \int_0^T q(0) \exp(-\lambda_{\text{eff}} t) dt. \tag{D.7}$$

Equation D.7 can be compared with the basic equations for internal dose within the MIRD (Equation D.8) and ICRP (Equation D.9) methodologies:

$$\bar{D} = \tilde{A}\mathcal{S} \tag{D.8}$$

$$H_{50,T} = 1.6 \times 10^{-10} \frac{\text{Sv g}}{\text{MeV}} U_{\text{S}} \text{SEE}, \tag{D.9}$$

where $\bar{D}$ is the mean absorbed dose, $\tilde{A}$ is the total cumulated activity, $\mathcal{S}$ is the mean dose per unit cumulated activity, $H_{50,T}$ is the 50-year committed dose equivalent, U_{S} is

Table D.1 Comparison of the MIRD and ICRP models.

Equation D.7 term	Corresponding quantities	
	MIRD	ICRP
D	$\bar{D}$	$H_{50,T}$
kE/m	S	$1.6 \times 10^{-10} \dfrac{\text{Sv g}}{\text{MeV}}$ SEE
T	∞	50 yr
$\int_0^T q(0)\exp(-\lambda_{\text{eff}}t)\,\mathrm{d}t$	$\tilde{A}$	U_S

the number of transformations in the source organ over 50 years, and SEE is the specific effective energy. The constant 1.6×10^{-10} in Equation D.9 is the product of conversion factors 1.6×10^{-13} J/MeV and 1000 g/kg, and its units include the definition that a Sv is equivalent to a J/kg.

Comparing Equation D.7 with Equations D.8 and D.9 leads to the explicit identifications summarized in Table D.1. Table D.1 illustrates that the ICRP and MIRD methodologies are essentially equivalent. With the exception of the terminology, the major difference is in the upper limit of integration (i.e., $T = 50$ years for the ICRP and $T = \infty$ for MIRD).

Equations D.7–D.9 and the comparisons of Table D.1 illustrate the inherent consistency of the internal dosimetry models. With this basic consistency established, model-specific aspects are presented. These aspects should be periodically reviewed with regard to Table D.1 to simplify and unify the presented concepts.

D.3
MIRD Methodology

The Committee on Medical Internal Radiation Dose of the Society of Nuclear Medicine developed a methodology to perform radiation absorbed dose calculations. These calculations are performed to assess the risks associated with the administration of radiopharmaceuticals for medical studies, including imaging, therapy, and metabolic applications.

The MIRD technique is a computational methodology that facilitates absorbed dose calculations for specified target organs from radioactive decays that occur in source organs. The source organs contain the radioactive material, and the target is the organ in which the dose is calculated. The target and source organs can be the same tissue. In subsequent discussion, the terms tissue and organ are used interchangeably.

To specify the MIRD methodology, it is necessary to define several terms. The mean energy emitted per transition (Δ), in Gy kg/Bq s, is given as the product of the mean particle energy (E) in MeV or joules and the number of particles emitted per nuclear transformation (n):

$$\Delta = KEn, \tag{D.10}$$

where K is a conversion factor. Within the MIRD methodology, particles are defined to be photons, beta particles, or positrons. These are the radiation types that are used most frequently in nuclear medicine procedures. Recent work is expanding this set to include alpha particles to minimize the dose to healthy tissue.

The cumulated activity or the total number of nuclear transitions occurring within the source organ from time $t=0$ to time T is

$$\tilde{A} = \int_0^T A(t) \, dt. \tag{D.11}$$

The activity as a function of time is

$$A(t) = A(0) \exp(-\lambda_{\text{eff}} t). \tag{D.12}$$

Using Equation D.12, the cumulated activity is simplified if the MIRD upper integration limit of $T = \infty$ is selected. In this case, the total cumulated activity is

$$\tilde{A} = \frac{A(0)}{\lambda_{\text{eff}}} = \frac{A(0) T_{\text{eff}}}{\ln(2)} = 1.44 T_{\text{eff}} A(0), \tag{D.13}$$

where

$$T_{\text{eff}} = \frac{T_p T_b}{T_p + T_b}. \tag{D.14}$$

The initial activity in the organ, $A(0)$, is related to the intake activity $q(0)$:

$$A(0) = f_2 q(0), \tag{D.15}$$

where f_2 is the fraction of the intake reaching the organ of interest.

The total energy emitted by the source organ is the product of Δ and the cumulated activity. However, only a fraction (f) of this energy is deposited in the target organ, which is the location of interest in the dose calculation. With these quantities and knowledge of the mass of the target organ (m), the mean absorbed dose $\bar{D}$ is

$$\bar{D} = \frac{\tilde{A} \Delta f}{m}. \tag{D.16}$$

The MIRD methodology also defines the specific absorbed fraction (F):

$$F = \frac{f}{m}, \tag{D.17}$$

where f is the energy absorbed by the target divided by the energy emitted by the source. The specific absorbed fraction represents the mean target dose per unit energy emitted by the source. Therefore, the mean absorbed dose is written as

$$\bar{D} = \widetilde{A}\Delta F. \tag{D.18}$$

The MIRD methodology defines the mean dose to the target (T) per unit cumulated activity in the source (S) in mGy/MBq s:

$$\mathcal{S}(T \leftarrow S) = \frac{\Delta f}{m} = \Delta F. \tag{D.19}$$

Equations D.18 and D.19 permit the expected MIRD dose relationship to be expressed as

$$\bar{D} = \widetilde{A}\mathcal{S}(T \leftarrow S). \tag{D.20}$$

In Equation D.20, the metabolic factors are contained in the $\widetilde{A}$ term, which depends on the uptake by the source organ and biological elimination of the radiopharmaceutical by the source organ. The $\mathcal{S}$ factor represents the physical decay characteristics of the radionuclide, the range of the emitted radiations, and the size and configuration of the organ. If a standard anatomy is utilized, $\mathcal{S}$ can be calculated and tabulated for a variety of radionuclides and source–target combinations. MIRD Pamphlet 11 provides a tabulation of these $\mathcal{S}$ factors.

D.4
ICRP Methodology

The ICRP internal dosimetry models are based in part on evolving assessments of the biological effects of ionizing radiation. These assessments affect the selection of the organs/tissues of the models and their weighting factors. The biological data and organ models drive the calculated doses that lead to recommendations regarding occupational exposures. Each of these ICRP model aspects is reviewed in the subsequent sections.

The specific ICRP recommendations are incorporated into the national and international regulations. For example, ICRP 26/30 form the basis for the US ionizing radiation regulations (10CFR20 and 10CFR835) and ICRP 60/66 for the current international regulations. The recently approved ICRP 2007 recommendations will soon replace the ICRP 60/66 recommendations.

D.5
Biological Effects

The ICRP models should be viewed in their historical context. The models continue to evolve and incorporate the available data regarding the biological effects of ionizing radiation.

Table D.2 Comparison of the basis for recent ICRP models.

ICRP model	Basis	Dose–response relationship[a]		Risk model
		Solid tumors	Leukemia	
26/30	BEIR III	LQ	LQ	Absolute
60/66	BEIR V	L	LQ	Relative
2007 recommendations	BEIR VII	L	LQ	Various[b]

[a] L = linear; LQ = linear quadratic.
[b] See Table D.4.

A portion of the scientific basis for ICRP 26/30 and ICRP 60/66 are summarized in Table D.2. ICRP 26/30 are based in part on the Biological Effects of Ionizing Radiation (BEIR) III Report. In BEIR III, the dose–response relationships for both solid tumors and leukemia are defined to have a linear-quadratic (LQ) dose–response relationship:

$$f(d) = ad + bd^2, \tag{D.21}$$

where $f(d)$ is the effect of the radiation dose, d is the dose equivalent, and a and b are the risk coefficients. BEIR III based its preferred age-specific cancer model on the absolute (additive) risk model:

$$r(d) = r_0 + f(d)g(\beta), \tag{D.22}$$

where $r(d)$ is the number of cancers of a specific type in the population group, r_0 is the natural incidence of the specific cancer type, and $g(\beta)$ is the excess risk function that contains the time dependence.

BEIR V forms a portion of the basis for ICRP 60/66. In BEIR V, the dose–response model is linear (L) for solid tumors:

$$f(d) = cd, \tag{D.23}$$

and linear-quadratic for leukemia. In Equation D.23, c is a risk coefficient. In contrast to BEIR III, BEIR V uses a relative (multiplicative) risk model:

$$r(d) = r_0[1 + f(d)g(\beta)]. \tag{D.24}$$

Both BEIR III and BEIR V assume that dose–response models have no threshold, that is, any dose no matter how small has an effect (detriment).

There are significant differences between the BEIR III and BEIR V estimates. Table D.3 illustrates the variation in both leukemia and nonleukemia (solid tumor) cancers. The solid tumors include respiratory, digestive, breast, and other cancer types. For leukemia, BEIR V leads to a factor of 4–5 greater risk. A similar increase of about 3–5 occurs for nonleukemia cancers if BEIR III relative risk models are considered.

Considerably larger factors of 11–19 occur for nonleukemia cancers if the BEIR III absolute risk model is compared to BEIR V's relative risk model. BEIR VII supports a combination of absolute and relative risk models and is compared with BEIR III and BEIR V in Table D.4.

Table D.3 Lifetime cancer risk estimates (deaths per 100 000 persons).[a]

Cancer type	Continuous lifetime exposure 1 mGy/yr		Instantaneous exposure 0.1 Gy	
	Male	Female	Male	Female
Leukemia				
BEIR III	15.9	12.1	27.4	18.6
BEIR V	70	60	110	80
BEIR V/BEIR III	4.4	5.0	4.0	4.3
Nonleukemia				
BEIR III (absolute)	24.6	42.4	42.1	66.5
BEIR III (relative)	92.9	118.5	192	213
BEIR V (relative)	450	540	660	730
BEIR V/BEIR III (relative)	4.8	4.6	3.4	3.4
BEIR V/BEIR III (absolute)	18.3	12.7	15.7	11.2

[a] Derived from Bevelacqua (1995).

The BEIR VII Report is consistent with BEIR V. The key elements of BEIR VII and their comparison with BEIR III and BEIR V are summarized in Table D.4.

The BEIR VII risk estimates of total cancer mortality and leukemia from radiation exposure have not changed significantly from BEIR V. The risk estimates of BEIR VII are based on expanded epidemiological data, including cancer mortality data and 15

Table D.4 Comparison of BEIR III, V, and VII.

Parameter/quantity	BEIR III (1980)	BEIR V (1990)	BEIR VII (2006)
Dose–response model – solid tumors	LQ[a]	L[a]	L
Dose–response model – leukemia	LQ	LQ	LQ
Preferred risk model	Absolute	Relative	Various[b,c]
Dosimetry system[d]	T65D	DS86	DS02
DDREF[e] (range)	—	2–10	1.1–2.3
DDREF (adopted)	—	—	1.5 for linear models

[a] L = linear; LQ = linear quadratic.
[b] For solid cancers other than lung, breast, and thyroid, the preferred risk model is a weighted average (on a logarithmic scale) of the relative and absolute risk models with the relative risk given a weight of 0.7 and the absolute risk a weight of 0.3. These weights are reversed for lung cancer. The preferred breast cancer model is based on the absolute risk model. The preferred thyroid cancer model is based on the relative risk model.
[c] For leukemia, the preferred risk model is a weighted average (on a logarithmic scale) of the relative and absolute risk models with the relative risk given a weight of 0.7 and the absolute risk a weight of 0.3.
[d] T65D = Tentative 1965 Dosimetry; DS86 = Dosimetry System 1986; DS02 = Dosimetry System 2002.
[e] Dose and dose rate effectiveness factor.

years of additional mortality follow-up for atomic bomb survivors. Studies involving occupational and environmental exposure were also evaluated.

In formulating its risk models, the BEIR VII Report used the revised Dosimetry System 2002 (DS02) for atomic bomb survivors as the basis for the evaluation of the dependence of risk on dose. The risk models were developed from atomic bomb survivors and persons exposed to radiation for medical reasons.

BEIR VII also reviewed the dose–response model and its functional dependence, the emergence of hormesis as a positive consequence of the radiation dose, and the existence of a threshold for radiation-induced effects. According to BEIR VII, the updated molecular and cellular data from the studies of radiation exposure do not support the postulate that low doses of low-LET radiation are more harmful than predicted by the LNT model. That is, the contention that the dose–response curve exhibits supralinearity is not supported. In addition, the updated molecular and cellular data from the studies of radiation exposure do not support hormesis. BEIR VII reaffirms the LNT hypothesis and concludes that there is cellular-level evidence for the LNT. Thresholds were considered, but not endorsed, as representing the best scientific view of low-dose risk.

BEIR VII also noted that a number of effects, although small, were observed to exist. In particular, BEIR VII concluded that the genetic risks of low-dose, low-LET radiation are very small when compared to the baseline frequencies of genetic disease. In addition, a dose response for noncancer mortality in atomic bomb survivors has been demonstrated. However, the data are not sufficient to determine if this effect exists at low doses and dose rates. BEIR VII does not provide risk estimates for noncancer mortality.

Reports such as BEIR VII are important because they refine the internal dosimetry models and affect the risk estimates. Consequently, conclusions of BEIR VII carry significant weight and are ideally clear, unambiguous, and widely accepted.

The conclusion of the BEIR VII Report on the LNT hypothesis has been challenged by a number of professional organizations including two French academies. The author views the LNT hypothesis as an expedient regulatory model, but the scientific evidence has not yet resolved this issue.

D.6
ICRP 26/30 and ICRP 60/66 Terminology

ICRP 26/30 and ICRP 60/66 utilize different terminologies to describe similar quantities. Table D.5 summarizes the terminology appropriate to each model. The specific terms are defined in the subsequent sections.

D.7
ICRP 26 and ICRP 60 Recommendations

Prior to reviewing specific ICRP internal dose formalism, the ICRP 26 and ICRP 60 recommendations are outlined. This is important because these recommendations

Table D.5 Terminology utilized in recent ICRP models.

ICRP model	Terminology	
	Organ dose	Whole body dose
26/30	Committed dose equivalent ($H_{50,T}$)	Effective dose equivalent[a] (H_E)
60/66	Equivalent dose (H_T)	Effective dose (E)

[a] US Regulations use the term committed effective dose equivalent.

and the internal dose formulation are closely related. The ICRP recommendations are based on the following two general principles:

- Prevent the occurrence of clinically significant radiation-induced deterministic effects.
- Limit the risk of stochastic effects to a reasonable level.

The National Council on Radiation Protection and Measurement (NCRP) also adopts these two general principles. In addition, the NCRP recommends that the risk be limited over a working lifetime to be no greater than the risk of accidental death in a safe industry.

The deterministic effects have a threshold. The term "deterministic effect" was introduced in ICRP 60. Deterministic effects include erythema, cataracts, impairment of fertility, and depletion of blood-forming cells in bone marrow. These effects occur only in irradiated individuals. By keeping the dose below the threshold for the deterministic effect, it is eliminated. The severity of the deterministic effect varies with dose. ICRP 26 refers to deterministic effects as nonstochastic effects.

Stochastic effects include cancer and hereditary effects. These effects occur in the general population as well as in irradiated individuals. With stochastic effects, the probability of the effect increases with increasing dose without threshold.

With ICRP 26, these recommendations are implemented by limiting the effective dose equivalent and committed dose equivalent, and by establishing stochastic and non-stochastic annual limits on intake (ALIs). Considering the purpose of this appendix, the applicable recommendations are summarized in Table D.6. In Table D.6, the annual doses (deep-dose equivalent), eye-dose equivalent, and skin-dose equivalent are evaluated at 1000, 300, and 7 mg/cm^2, respectively.

In ICRP 60, the restrictions on effective dose are sufficient to ensure the avoidance of deterministic effects in all body tissues except the lens of the eye and the skin. The limits for the eye and skin preclude deterministic effects. Therefore, only a stochastic ALI is needed in the ICRP 60/66 internal dosimetry formulation.

D.8
Calculation of Internal Dose Equivalents Using ICRP 26/30

Internal dose equivalents are calculated in a variety of ways. These include the use of the ALI, derived air concentration (DAC), SEE, and U_S values.

Table D.6 Applicable ICRP 26 and ICRP 60 recommendations.

Dose recommendation	Dose (mSv)	
	ICRP 26	ICRP 60
Annual	50[a]	50 maximum[b]
Cumulative	None	100 over 5 yr[b]
		20 per yr average[b]
Eye	150[c]	150[d]
Skin, hands, and feet	500[c]	500[d]

[a]Effective dose equivalent.
[b]Effective dose.
[c]Committed dose equivalent.
[d]Equivalent dose.

Within the ICRP 26/30 methodology, the stochastic and nonstochastic recommendations for internal dose equivalents are developed in terms of the ALI. Following ICRP 26/30, the ALI is defined to be the largest value of intake that satisfies the inequalities of both Equations D.25 and D.26. In Equations D.25 and D.26, ALI_S is the stochastic ALI and ALI_{NS} is the nonstochastic ALI:

$$ALI_S \sum_T w_T H'_{50,T} \leq 0.05 \text{ Sv} \quad \text{for stochastic effects,} \tag{D.25}$$

$$ALI_{NS} H'_{50,T} \leq 0.5 \text{ Sv} \quad \text{for nonstochastic effects,} \tag{D.26}$$

where w_T is the ICRP 26/30 organ/tissue weighting factor and $H'_{50,T}$ is specified as the dose per unit intake (Sv/Bq) that yields the correct units for the ALI. The organ/tissue weighting factors for ICRP 26/30 and 60/66 are summarized in Table D.7.

Table D.7 Weighting factors for recent ICRP models.

Organ or tissue	ICRP 26/30	ICRP 60/66
Gonads	0.25	0.20
Breast	0.15	0.05
Red bone marrow	0.12	0.12
Lung	0.12	0.12
Thyroid	0.03	0.05
Bone surfaces	0.03	0.01
Stomach	—	0.12
Colon	—	0.12
Esophagus	—	0.05
Bladder	—	0.05
Skin	—	0.01
Liver	—	0.05
Remainder	0.30[a]	0.05[b]

[a]Five other highest organs.
[b]Adrenals, brain, small intestine, spleen, kidneys, muscle, pancreas, upper large intestine, thymus, and uterus.

ICRP 26/30 form the basis of the current US regulations embodied in 10CFR20 for US Nuclear Regulatory Commission licensees and 10CFR835 for US Department of Energy licensees. These regulations require the calculation of individual organ doses (i.e., the committed dose equivalent (CDE)) and the committed effective dose equivalents (CEDE). The dose limits are based on the risk of dose to the various organs/tissues included in the ICRP 26/30 model. The CDE and CEDE are calculated in terms of the intake (I) as follows:

$$\text{CDE} = H_{50,T} = \frac{I}{\text{ALI}_{\text{NS}}} 0.5\,\text{Sv} = 1.6 \times 10^{-10}\,\frac{\text{Sv g}}{\text{MeV}} U_S \text{SEE}(T \leftarrow S), \quad (D.27)$$

$$\text{CEDE} = H_E = \sum_T w_T H_{50,T} = \frac{I}{\text{ALI}_S} 0.05\,\text{Sv}. \quad (D.28)$$

Equations D.27 and D.28 can also be rewritten in terms of the DAC

$$\text{DAC} = \text{ALI}/2400\,\text{m}^3. \quad (D.29)$$

D.9
Calculation of Equivalent and Effective Doses Using ICRP 60/66

Within the ICRP 60/66 formalism, a new dose terminology is introduced, including the equivalent and effective doses. The equivalent dose (H_T) is defined as

$$H_T = \sum_R w_R D_{T,R}, \quad (D.30)$$

where w_R is the radiation-weighting factor and $D_{T,R}$ is the average absorbed dose in tissue T because of the radiation of type R. The ICRP 60/66 radiation-weighting factors are provided in Table D.8.

The effective dose (E) is defined as

$$E = \sum_T w_T H_T. \quad (D.31)$$

Using Equation D.30, the effective dose is

$$E = \sum_R w_R \sum_T w_T D_{T,R} = \sum_T w_T \sum_R w_R D_{T,R}. \quad (D.32)$$

Within ICRP 60/66, only one ALI is defined. The committed effective dose $E(50)$ is

$$E(50) = \frac{I}{\text{ALI}} 0.02\,\text{Sv} = \sum_{T=1}^{12} w_T H_T(50) + w_{\text{remainder}} \frac{\sum_{T=13}^{22} m_T H_T(50)}{\sum_{T=13}^{22} m_T}, \quad (D.33)$$

where $H_T(50)$ is the committed equivalent dose, m_T is the mass of the remainder tissue, and $w_{\text{remainder}} = 0.05$. In Equation D.33, the first sum is over the 12 organs/

Table D.8 Radiation-weighting factors.[a]

Type and energy range[b]	Radiation-weighting factor
Photons (all energies)	1
Electrons and muons (all energies)[c]	1
Neutrons	
<10 keV	5
10–100 keV	10
>100 keV–2 MeV	20
>2–20 MeV	10
>20 MeV	5
Protons other than recoil protons (>2 MeV)	5
Alpha particles, fission fragments, and heavy nuclei	20

[a] All values relate to the radiation incident on the body or, for internal sources, emitted from the source.
[b] The choice of values for other radiation types is discussed in Annex A, ICRP 60.
[c] Excluding Auger electrons emitted from nuclei bound to DNA.

tissues with assigned weighting factors (see Table D.7) and the second sum is over the 10 remainder organs/tissues (i.e., adrenals, brain, small intestine, spleen, kidneys, muscles, pancreas, upper large intestine, thymus, and uterus). The right-hand side of Equation D.33 is applicable whenever one of the 12 organs with assigned weighting factors has the largest committed equivalent dose. In the exceptional case, in which one of the remainder organs receives a committed equivalent dose in excess of the highest committed equivalent dose in any of the 12 organs for which a weighting factor is assigned, a weighting factor of 0.025 is applied to the remainder organ or tissue. A weighting factor of 0.025 is also assigned to the average dose in the rest of the remainder, and in the exceptional case the $E(50)$ equation has the following form:

$$E(50) = \sum_{T=1}^{12} w_T H_T(50) + 0.025 H_{T'}(50) + 0.025 \frac{\sum_{T=13}^{22} m_T H_T(50) - m_{T'} H_{T'}(50)}{\sum_{T=13}^{22} m_T - m_{T'}},$$

(D.34)

where $m_{T'}$ is the mass of the remainder tissue or organ in which the committed equivalent dose is calculated to be higher than in any of the 12 specified tissues/organs with assigned weighting factors, and $H_{T'}(50)$ is the committed equivalent dose in that remainder tissue/organ.

A careful reader will note that the first term in Equation D.33 contains no ALI subscript because the ICRP 60/66 formulation only utilizes a stochastic ALI. The 0.02 Sv (20 mSv) multiplier is a direct consequence of the cumulative effective dose recommendation given in Table D.6.

D.10
Model Dependence

Equations D.25–D.34 and Tables D.3, D.4, and D.7 illustrate the model dependence of ICRP 26/30 and 60/66. The selection of the tissues, governed by a host of inherent model assumptions and historical data, dictates the dose result. An examination of Table D.7 and the differences in the number of listed tissues, their associated weighting factors, and the treatment of the remainder illustrate the evolving nature of the ICRP internal dosimetry models.

D.11
Conclusions

Using the concept of absorbed dose, the MIRD and ICRP internal dosimetry models are found to be quite similar. The ICRP general principles and the supporting biological effects of ionizing radiation publications affect the specific model formulations. With the exception of terminology, the ICRP methodology remains consistent with the definition of absorbed dose with model refinements being governed by evolving assessments of the biological effects of ionizing radiation.

References

10CFR20 (2007) Standards for Protection Against Radiation, National Archives and Records Administration, U.S. Government Printing Office, Washington, DC.

10CFR835 (2007) Occupational Radiation Protection, National Archives and Records Administration, U.S. Government Printing Office, Washington, DC.

Bevelacqua, J.J. (1995) *Contemporary Health Physics: Problems and Solutions*, John Wiley & Sons, Inc., New York.

Bevelacqua, J.J. (1999) *Basic Health Physics: Problems and Solutions*, John Wiley & Sons, Inc., New York.

Bevelacqua, J.J. (2005) Internal dosimetry primer. *Radiation Protection Management*, **22** (5), 7.

ICRP Publication 2 (1959) *Permissible Dose for Internal Radiation*, Pergamon Press, Oxford, England.

ICRP Publication 10 (1968) *Evaluation of Radiation Doses to Body Tissues from Internal Contamination due to Occupational Exposure*, Pergamon Press, Oxford, England.

ICRP Publication 10A (1971) *The Assessment of Internal Contamination Resulting from Recurrent or Prolonged Uptakes*, Pergamon Press, Oxford, England.

ICRP Publication 26 (1977) *Recommendations of the International Commission on Radiological Protection*, Pergamon Press, Oxford, England.

ICRP Publication 30 (1979) *Limits for Intakes of Radionuclides by Workers*, Pergamon Press, England.

ICRP Publication 60 (1991) *Recommendations of the ICRP*, Pergamon Press, Oxford, England.

ICRP Publication 66 (1994) *Human Respiratory Tract Model for Radiological Protection*, Pergamon Press, Oxford, England.

ICRP Publication 103 (2007) *The 2007 Recommendations of the International*

Commission on Radiological Protection, Elsevier, Amsterdam.

Loevinger, R., Budinger, T.F. and Watson, E.E. (1988) *MIRD Primer for Absorbed Dose Calculations*, The Society of Nuclear Medicine, New York.

National Research Council (1980) *Committee on the Biological Effects of Ionizing Radiation, The Effects on Populations of Exposures to Low Levels of Ionizing Radiation (BEIR III)*, National Academy Press, Washington, DC.

National Research Council (1990) *The Health Effects of Exposures to Low Levels of Ionizing Radiation, BEIR V*, National Academy Press, Washington, DC.

National Research Council (2006) *Health Risks from Exposure to Low Levels of Ionizing Radiation, BEIR VII Phase 2*, National Academy Press, Washington, DC.

NCRP Report No. 116 (1993) *Limitation of Exposure to Ionizing Radiation*, National Council on Radiation Protection and Measurement, Bethesda, MD.

Snyder, W.S., Ford, M.R., Warner, G.G. and Watson, S.B. (1975) *"S", Absorbed Dose per Unit Cumulated Activity for Selected Radionuclides and Organs*, MIRD Pamphlet No. 11, The Society of Nuclear Medicine, New York.

Tubiana, M. and Aurengo, A. (2005) Dose–effect relationship and estimation of the carcinogenic effects of low doses of ionizing radiation: the Joint Report of the Académie des Sciences (Paris) and the Académie Nationale de Médecine. *International Journal of Low Radiation*, **2** (3/4), 1.

Appendix E
The Standard Model of Particle Physics

E.1
Overview

The theoretical formulation of high-energy phenomena is embodied in the Standard Model of Particle Physics. This model also provides a basis for the unification or consistent description of the strong, electromagnetic, and weak interactions. However, it does not include gravity. The Standard Model forms the foundation for subsequent theoretical development of the accelerator and space sections of this text.

E.2
Particle Properties and Supporting Terminology

In this appendix, the basic particle properties are reviewed. Terminology that supports their characterization and facilitates the discussion is also presented. The properties of particles (e.g., muons and kaons) that are less familiar to some applied health physicists are compared with particles of more familiarity (e.g., neutrons, protons, and electrons).

As background material for the effective dose calculations, an overview of the four fundamental interactions (strong, electromagnetic, weak, and gravitational) and the basic conservation laws, governing particle interactions, is presented in the context of the Standard Model. This overview provides a foundation for a discussion of the characteristics of particle decays and particle interactions, and the resultant radiation types.

E.2.1
Terminology

Specific terminology is introduced to facilitate presentation of the basic physics associated with the Standard Model. These terms include the following:

- *baryon* – A heavy particle normally composed of three quarks. Protons and neutrons are baryons. Baryons can be electrically charged or uncharged.

- *boson* – A particle having integer spin. The mediators or carriers of each of the four fundamental interactions are bosons. Bosons include both mediators and other integer spin particles. The photon and pions are bosons. Bosons can be electrically charged or uncharged.

- *charge* – A general term used to assign a particular property to a particle or field quanta. Health physicists are most familiar with the electric charge that influences processes, such as ionization, and governs the electromagnetic force. There are other types of charges, including color charge that governs the strong interaction and weak charge that manifests itself either as a charged or neutral weak current. These currents govern the weak interaction.

- *fermion* – A particle having half-integer spin. Neutrons, protons, and electrons are examples of fermions. Fermions can be electrically charged or uncharged.

- *flavor* – A designation for the type of quark. Within the Standard Model, the flavors are d, u, s, c, b, and t. These designations are further defined in subsequent discussion.

- *hadron* – A particle that interacts primarily through the strong interaction. Mesons and baryons are hadrons.

- *lepton* – A fundamental particle that interacts primarily through the weak interaction. The electron and the electron neutrino are examples of leptons. Leptons can be electrically charged or uncharged.

- *meson* – A middle-weight particle normally composed of a quark and an antiquark. The charged and neutral pions are examples of mesons.

- *quark* – A particle having a fractional charge that interacts through the strong, electromagnetic, and weak interactions. Quarks were initially inferred from the high-energy electron–proton (e–p) scattering. The e–p scattering cross section indicates the presence of point-like structures inside the proton that have been interpreted as quarks.

E.3
Basic Physics

E.3.1
Basic Particle Properties

Table E.1 provides a summary of the properties of selected low-energy particles that will likely be of concern to health physicists in the twenty-first century. These properties include the particle mass, mean lifetime, and dominant decay mode, and are provided for the neutrinos (electron, muon, and tau), electron (e^-) and its antiparticle (e^+), muon (μ^-) and its antiparticle (μ^+), three pions (π^+, π°, and π^-), three kaons (K^+, K°, and K^-), proton (p) and its antiparticle ($\bar{p}$), and the neutron (n) and its antiparticle ($\bar{n}$).

Table E.1 Properties of selected low-energy particles.

Particle	Mass (MeV)	Mean lifetime	Dominant decay mode
ν_e	<0.000002	>300 s/eV[a]	[b]
$\bar{\nu}_e$	<0.000002	>300 s/eV[a]	[b]
ν_μ	<0.19	>15.4 s/eV[a]	[b]
$\bar{\nu}_\mu$	<0.19	>15.4 s/eV[a]	[b]
ν_τ	<18.2	Not yet determined[a]	[b,c]
$\bar{\nu}_\tau$	<18.2	Not yet determined[a]	[b,c]
e^-	0.511	$>4.6\times 10^{26}$ yr	Stable
e^+	0.511	$>4.6\times 10^{26}$ yr	Stable
μ^-	105.7	2.2×10^{-6} s	$\mu^- \to e^- + \nu_\mu + \bar{\nu}_e$
μ^+	105.7	2.2×10^{-6} s	$\mu^+ \to e^+ + \bar{\nu}_\mu + \nu_e$
τ^-	1777	2.9×10^{-13} s	Multiple decay modes
τ^+	1777	2.9×10^{-13} s	Multiple decay modes
π^-	139.6	2.6×10^{-8} s	$\pi^- \to \mu^- + \bar{\nu}_\mu$
π^0	135.0	8.4×10^{-17} s	$\pi^0 \to \gamma + \gamma$
π^+	139.6	2.6×10^{-8} s	$\pi^+ \to \mu^+ + \nu_\mu$
K^-	493.7	1.24×10^{-8} s	$K^- \to \mu^- + \bar{\nu}_\mu$
K^0	497.6	[d]	$K^0 \to \pi^+ + \pi^-$
K^+	493.7	1.24×10^{-8} s	$K^+ \to \mu^+ + \nu_\mu$
p	938.3	$>2.1\times 10^{29}$ yr	Stable
$\bar{p}$	938.3	$>2.1\times 10^{29}$ yr	Stable
n	939.6	885.7 s	$n \to p + e^- + \bar{\nu}_e$
$\bar{n}$	939.6	885.7 s	$\bar{n} \to \bar{p} + e^+ + \nu_e$

[a]The Particle Data Group (2004) quoted specific lifetime values, or noted that the lifetime was not yet determined. In 2006, the Particle Data Group did not quote specific lifetime values, but did note that the measured quantities depend upon the mixing parameters of the Standard Model and to some extent on the experimental conditions (e.g., energy resolution).
[b]Dependent on the degree of neutrino mixing.
[c]Decay mode not yet determined.
[d]The K^0 particle is a superposition of two states K_S^0 and K_L^0; $K^0 = \frac{1}{\sqrt{2}}(K_S^0 + K_L^0)$ with lifetimes of $K_S^0 = 8.95\times 10^{-11}$ s and $K_L^0 = 5.11\times 10^{-8}$ s.

Neutrinos are neutral leptons, once believed to be massless, but now evidence suggests that they have a nonzero mass. There are three known varieties (also known as flavors or generations) of neutrinos (ν) and their corresponding antiparticles (antineutrinos, $\bar{\nu}$), specifically including the electron neutrino (ν_e) and its antiparticle ($\bar{\nu}_e$), the muon neutrino (ν_μ) and its antiparticle ($\bar{\nu}_\mu$), and the tau neutrino (ν_τ) and its antiparticle ($\bar{\nu}_\tau$). The electron and muon neutrinos are well studied, but much less is known about tau neutrinos.

The leptons (electrons, muons, and neutrinos) appearing in Table E.1 are fundamental and have no discernable substructure. This is not true of the mesons and baryons that have an underlying quark structure. The properties of these quarks and the composition of selected baryons and mesons are summarized in Tables E.2–E.4, respectively.

Table E.2 summarizes the properties of quarks within the Standard Model. The quark masses are listed for the theoretical value of an isolated (bare) quark flavor, and

Table E.2 Properties of quarks within the standard model.[a]

Generation	Flavor	Charge[b] (e)	Bare	Effective In baryons	Effective In mesons
				Mass (MeV)	
First	d	$-\frac{1}{3}$	7.5	363	310
First	u	$+\frac{2}{3}$	4.2	363	310
Second	s	$-\frac{1}{3}$	150	538	483
Second	c	$+\frac{2}{3}$	1 100	1 500	1 500
Third	b	$-\frac{1}{3}$	4 200	4 700	4 700
Third	t	$+\frac{2}{3}$	>23 000	>23 000	>23 000

[a] Derived from Griffiths (1987).
[b] Charge (Q) has units of electric charge. For example, $Q = -\frac{1}{3}$ means one-third unit of negative charge or $Q = -\frac{1}{3}e$.

for the effective mass of the quark flavor when it appears as a part of the baryon or meson structure. These values are model dependent and may not be experimentally measurable. These results are based on the currently accepted quark interaction spatial dependence having the properties of quantum chromodynamic confinement and asymptotic freedom.

Tables E.3 and E.4 provide the properties of selected baryons and mesons, respectively. The quark structure, electric charge, mass, and lifetime are provided.

Table E.3 Properties of selected baryons within the standard model.[a]

Baryon	Quark structure	Charge[b] (e)	Mass (MeV)	Lifetime
p	uud	+1	938.3	2.1×10^{29} yr
n	udd	0	939.6	885.7 s
Λ	uds	0	1115.6	2.63×10^{-10} s
Σ^+	uus	+1	1189.4	0.80×10^{-10} s
Σ^0	uds	0	1192.5	6×10^{-20} s
Σ^-	dds	−1	1197.3	1.48×10^{-10} s
Ξ^0	uss	0	1314.9	2.90×10^{-10} s
Ξ^-	dss	−1	1321.3	1.64×10^{-10} s
Λ_c^+	udc	+1	2281	2×10^{-13} s

[a] Derived from Griffiths (1987).
[b] Charge (Q) has units of electric charge (e.g., $Q = -1$ means one unit of negative charge or $Q = -e$.).

Table E.4 Properties of selected mesons within the standard model.[a]

Meson	Quark structure	Charge[b] (e)	Mass (MeV)	Lifetime (s)
π^+	$u\bar{d}$	+1	139.6	2.6×10^{-8}
π^-	$d\bar{u}$	−1	139.6	2.6×10^{-8}
π^0	$(u\bar{u}-d\bar{d})/\sqrt{2}$	0	135.0	8.4×10^{-17}
K^+	$u\bar{s}$	+1	493.7	1.24×10^{-8}
K^0	$d\bar{s}$	0	497.6	[c]
K^-	$s\bar{u}$	−1	493.7	1.24×10^{-8}

[a] Derived from Griffiths (1987).
[b] Charge (Q) has units of electric charge (e.g., $Q=-1$ means one unit of negative charge or $Q=-e$).
[c] See Table E.1.

E.3.2
Fundamental Interactions

Four fundamental interactions or forces describe the phenomena observed in the universe. These are the strong, electromagnetic, weak, and gravitational interactions, and their properties are summarized in Table E.5. The unique aspects of each interaction govern particle decays and interactions, which influence the health physics consequences of the resulting radiation types.

In Table E.5, the field boson is the mediator or the carrier of the force. For example, the electromagnetic interaction is mediated by photons. It is the photon that is exchanged between the two particles involved in an electromagnetic interaction. The field mediators have been directly observed or inferred from observed phenomena. All mediators are based on significant experimental evidence, with the exception of the graviton that is inferred from gravitational field theory.

Table E.5 Fundamental interactions and their properties.

Property	Fundamental interaction			
	Gravitational	Electromagnetic	Weak	Strong
Field bosons	Graviton	Photon	W^+, W^-, and Z^0	Eight gluons
Mass of field boson (GeV/c^2)	0	0	$M_W = 80.4$ $M_Z = 91.2$	0
Range of the interaction (m)	∞	∞	10^{-18}	$\leq 10^{-15}$
Source of the interaction	Mass	Electric charge	Weak charge	Color charge
Strength (relative to the strong interaction)	10^{-39}	10^{-2}	10^{-5}	1
Typical cross section (m^2)	[a]	10^{-33}	10^{-39}	10^{-30}
Typical lifetime (s)	[a]	10^{-20}	10^{-10}	10^{-23}

[a] In view of the range and source of the gravitational interaction, the cross section and lifetime are not well-defined quantities.

These field bosons give unique properties to the various fundamental interactions. Although the photon is a well-known type of radiation, it is has a much deeper physical significance, because its exchange defines the electromagnetic interaction. In a similar fashion, the exchange of gluons (of which there are 8) defines the strong interaction. The weak interaction is also complex because there are three particles (i.e., W^+, W^-, and Z^0) that are exchanged. Properties of the field bosons give each fundamental interaction a distinctive character. For example, the distinctive character of the weak interaction manifests itself in the magnitude and dose profile of the neutrino effective dose. The field boson mass also exhibits a distinctive nature.

The photon, gluons, and graviton are all massless. In contrast, the weak interaction field bosons have masses in the 80–90 GeV/c^2 range. The field boson mass does not uniquely determine the nature of the fundamental interaction. It is the collective nature of the mass of field boson, charge, number of allowed states, lifetime, and coupling constant that determines its unique characteristics.

In Table E.5, the source of the interaction refers to the basic physical quantity that gives rise to the force. The four fundamental interactions arise from very different physical constructs. For example, the gravitational and electromagnetic interactions are derived from mass and electric charge, respectively. The concepts of mass and electric charge are well known to health physicists. However, weak charge and color charge are not.

It is well known from classical physics that a moving charge produces a current. Therefore, weak charges in motion generate a weak current. Weak currents produce weak forces that govern lepton interactions. Leptons have no color charge, and consequently do not participate in the strong interaction. Neutrinos have no electric charge, so they experience no electromagnetic force, but they do participate in the weak interaction.

The color charge produces the strong interaction. However, the color charge is considerably more complex than the electric charge. The color charge is a property assigned to a quark or gluon, and it has three colors (states) (i.e., red, white, and blue). There are eight gluons governing the strong interaction instead of one photon for the electromagnetic interaction. Because the gluons themselves carry a color charge, they can directly interact with the other gluons. This possibility is not available with the electromagnetic force because photons do not have electric charge. Therefore, it is not surprising that the strong and electromagnetic forces have different properties.

In Table E.5, the interaction strength is the magnitude of the force as measured over its effective range. The term interaction strength is intrinsically ambiguous because it depends on the measurement distance from the source. Accordingly, the strength values listed in Table E.5 may be quoted with different values by other authors. Table E.5 provides the strength relative to the strong interaction. In terms of decreasing strength, nature orders these interactions as follows: strong, electromagnetic, weak, and gravitational.

The cross section describes the probability of a typical interaction that is solely governed by one of the fundamental interactions. The lifetime represents the time over which an interaction occurs, assuming that the interaction is governed solely by that fundamental force. For example, strong interactions typically create particles with cross sections in the mb range that have lifetimes on the order of 10^{-23} s. The cross section and lifetime of a created particle are often clear indications of the type of force that governs it.

Neutrino interaction cross sections, governed by the weak interaction, are orders of magnitude smaller than the typical strong or electromagnetic interaction cross sections. The magnitude of the weak interaction cross section makes neutrino detection difficult.

The gravitational interaction is an interaction affecting massive objects, such as planets, Solar Systems, and galaxies. The terms cross section and lifetime are not clearly defined within the context of the gravitational interaction.

The gravitational force was not a health physics concern in the twentieth century, and may not be a health physics issue in the twenty-first century. It is too ineffectual to play a major role in the contemporary health physics phenomena, but plays a key role in cosmology and astrophysics. However, the gravitational interaction could affect the health physics aspects of nuclear decays through the emergence of extra dimensions. The emergence of these dimensions is speculative and may arise from spatial anomalies. Although spatial exploration has been limited, an anomaly has already been detected (e.g., the Pioneer Anomaly, addressed in Chapter 6).

E.4
Fundamental Interactions and Their Health Physics Impacts

The strong interaction binds quarks into baryons, and is also responsible for binding nucleons in the nucleus. It arises from the exchange of gluons between quarks, and governs a number of commonly observed processes, including fission, fusion, and activation. The radiation hazards from these processes are well known to health physicists.

The electromagnetic force results from the exchange of photons. It governs much of the physics encountered in our daily lives; for example, atomic physics and molecular chemistry are governed by the electromagnetic interaction. This interaction also influences nuclear reactions and competes with the strong force in nuclear processes. The electromagnetic interaction depends on the electric charge of the interacting particles. As a practical example, ions can be accelerated because they have an electric charge and the electromagnetic force governs their final energy.

The weak force arises from the exchange of particles known as intermediate vector bosons. These include the W^+, W^-, and Z^0 that mediate weak processes, such as beta decay and positron decay. The weak interaction also governs the behavior of leptons that includes muons and neutrinos.

Although the fundamental interactions are distinct phenomena, they often appear collectively in Nature. As an example, consider the beta decay of ^{60}Co:

$$^{60}\text{Co} \rightarrow {}^{60}\text{Ni} + e^- + \bar{\nu}_e. \tag{E.1}$$

The nuclear energy levels in the ^{60}Co and ^{60}Ni nuclei are determined by the strong and electromagnetic interactions. The relative position of the energy levels in the ^{60}Co and ^{60}Ni nuclei, their specific properties (spin and parity), and conservation laws determine if the transition between a specific set of energy levels produces a beta particle.

During the beta decay, a neutron single particle level in ^{60}Co(^{59}Co+n) transitions to a proton single particle level in ^{60}Ni(^{59}Co+p) with the emission of an electron (beta particle) and antielectron neutrino.

From a nuclear transformation perspective, beta decay is described by

$$n \rightarrow p + e^- + \bar{\nu}_e. \tag{E.2}$$

Beta decay is described within the Standard Model as a sequential process through the W^- field boson as follows

$$n \rightarrow p + W^-, \tag{E.3}$$

$$W^- \rightarrow e^- + \bar{\nu}_e. \tag{E.4}$$

Although Equations E.1 through E.4 represent the same physical process, they differ in terms of the type of model utilized in the description of the neutron decay process.

Equations E.1 through E.4 may be accepted on face value, but the reader should question why these are the physical beta decay modes. Are the other modes equally likely, and why do the listed modes preferentially occur? The following section illustrates conservation laws and how these laws govern the decay characteristics of a particle.

E.4.1
Conservation Laws

Fundamental physics is governed by basic symmetries that are expressed in terms of a set of conservation laws that permit certain reactions and forbid others. In this section, we examine the specific conservation laws that facilitate an understanding of the processes that lead to radiation types of concern in health physics. Four conservation laws are useful in understanding the underlying physics. These specific conservation laws include the following:

(1) *Conservation of electric charge* – All three of the fundamental interactions governing health physics applications (strong, electromagnetic, and weak) conserve electric charge. Many particles participating in the various fundamental interactions contain electric charge (e.g., protons, pions, muons, and electrons).

(2) *Conservation of color charge* – The electromagnetic and weak interactions do not affect color charge. Color charge is conserved in strong interactions. Within the Standard Model, physical particles (e.g., baryons and mesons) are normally considered to be colorless. This means mesons contain a quark of one color (red, white, or blue) and an antiquark of the same anticolor (antired, antiwhite, or antiblue). Baryons consist of three quarks, each of a different color.

(3) *Conservation of baryon number* – The total number of quarks is a constant. Because the baryons are composed of three quarks, the baryon number is just the quark

number divided by 3. There is no corresponding conservation of meson number because the mesons, composed of quark–antiquark pairs, carry zero baryon number.

(4) *Conservation of electron number, muon number, and tau number* – A strong interaction does not affect leptons. In an electromagnetic interaction, the same particle comes out (accompanied by a photon) that went in. The weak interaction only mixes together the leptons from the same generation. Therefore, the lepton number, muon number, and tau number are all conserved.

These conservation laws provide a key input to understanding the decay schemes summarized in Table E.1. An examination of the health physics consequences of particle decays and their associated radiation is possible when these conservation laws are combined with an understanding of the Standard Model of particle physics.

The leptons, for example, interact primarily through the weak interaction, and electrically charged leptons also experience the effects of the electromagnetic force. They are not affected by the strong interaction. There are six leptons, classified according to their electric charge (Q), electron number (L_e), muon number (L_μ), and tau number (L_τ). The leptons are naturally grouped into three families or generations, as summarized in Table E.6.

There are also six antileptons, with all the signs in Table E.6 reversed (i.e., $+$ to $-$ and $-$ to $+$). The positron, for example, has an electric charge of $+1$ and an electron number of -1. Considering both particles and antiparticles, there are a total of 12 leptons.

In a similar manner, there are six types (flavors) of quarks (u, d, s, c, t, and b) that are classified according to their electric charge, upness (U), downness (D), strangeness (S), charm (C), bottomness (B), and topness (T). These labels are historical and have no underlying physical meaning. The quarks also fall into three generations, as summarized in Table E.7. Again, all signs are reversed on a table of antiquarks. Since each quark and antiquark comes in three colors, there are 36 distinct quarks.

Table E.5 and the subsequent discussion noted eight mediators for the strong interaction (gluons), the photon for the electromagnetic interaction, and three mediators for the weak interaction (W^+, W^-, and Z^0). This yields a total of 12 mediators for the Standard Model.

Table E.6 Lepton classification.

Generation	Lepton	Q^a	L_e	L_μ	L_τ
First	e^-	-1	1	0	0
	ν_e	0	1	0	0
Second	μ^-	-1	0	1	0
	ν_μ	0	0	1	0
Third	τ^-	-1	0	0	1
	ν_τ	0	0	0	1

[a] Q has units of electric charge. For example, $Q = -1$ means one unit of negative charge or $Q = -e$.

Table E.7 Quark classification.

Generation	Quark	Q^a	D	U	S	C	B	T
First	d	$-\frac{1}{3}$	−1	0	0	0	0	0
	u	$\frac{2}{3}$	0	1	0	0	0	0
Second	s	$-\frac{1}{3}$	0	0	−1	0	0	0
	c	$\frac{2}{3}$	0	0	0	1	0	0
Third	b	$-\frac{1}{3}$	0	0	0	0	−1	0
	t	$\frac{2}{3}$	0	0	0	0	0	1

[a]Q has units of electric charge. For example, $Q = -\frac{1}{3}$ means one-third unit of negative charge or $Q = -\frac{1}{3}$ e.

A careful reader notes that one of the shortcomings of the Standard Model is the number of free parameters or elementary particles that it requires: 12 leptons, 36 quarks, and 12 mediators. There is also at least one other particle (the Higgs particle) required to complete the theory. Therefore, there is a minimum of 61 parameters to be addressed. The Standard Model has been remarkably successful, but mounting evidence (e.g., indication that neutrinos have mass and recent publications, regarding evidence for four quark mesons and five quark baryons) suggests that physics beyond the Standard Model is required to explain the recent experimental results. However, the Standard Model is sufficient for the purposes of this text.

E.4.2
Consequences of the Conservation Laws and the Standard Model

With knowledge of conservation laws and the Standard Model, we will illustrate how these laws are satisfied for simple beta decay (Equation E.2) and muon decay (Table E.1). Tables E.8 and E.9 summarize beta decay and muon decay,

Table E.8 Beta decay ($n \rightarrow p + e^- + \bar{\nu}_e$).

	Initial state	Final state		
Conservation law	n	p	e^-	$\bar{\nu}_e$
Baryon number	1	1	0	0
Lepton number $(L_e)^a$	0	0	1	−1
Lepton number $(L_\mu)^a$	0	0	0	0
Electric charge	0	e	−e	0
Color charge	0	0	0	0

[a]See Table E.6.

Table E.9 Muon decay ($\mu^- \to e^- + \nu_\mu + \bar{\nu}_e$).

	Initial state	Final state		
Conservation law	μ^-	e^-	ν_μ	$\bar{\nu}_e$
Baryon number	0	0	0	0
Lepton number (L_e)[a]	0	1	0	−1
Lepton number (L_μ)[a]	1	0	1	0
Electric charge	−e	−e	0	0
Color charge	0	0	0	0

[a] See Table E.6.

respectively. These tables illustrate that the decays summarized in Table E.1 are not arbitrary, and are governed by the conservation laws that follow from the symmetries underlying the Standard Model.

Tables E.8 and E.9 illustrate the application of conservation laws to predict a particle's decay and its associated radiation types. These laws and the Standard Model are sufficient to predict the radiation types that occur in particle decay and interaction processes of interest in health physics applications.

Conservation laws are also implied by the fundamental interactions and their underlying symmetry properties. Noether's Theorem provides a mathematical proof of the relationship between a symmetry and its conservation law. For the purpose of this appendix, we state without proof that the symmetries are expressed in terms of group properties. As an example, the electromagnetic, weak, and strong interaction field quanta are represented by the generators of the unitary group of dimension 1 [U(1)], the special unitary group of dimension 2 [SU(2)], and the special unitary group of dimension 3 [SU(3)], respectively.

Within the Standard Model, the number of generators (N) of a group of dimension n is given by

$$N = n^2 - 1 \quad \text{for} \quad n > 1 \tag{E.5}$$

These generators are equivalent to the number of field bosons summarized in Table E.5. Therefore, it is expected from Equation E.5 that the electromagnetic ($n = 1$), weak ($n = 2$), and strong ($n = 3$) interactions have 1, 3, and 8 generators (field quanta), respectively. This prediction is observed experimentally with one field boson (photon) for the electromagnetic interaction, three field bosons (W^+, W^-, and Z^0) for the weak interaction, and eight field bosons (eight gluons) for the strong interaction. The prediction of the number and characteristics of the field bosons for the electromagnetic, weak, and strong interactions is an impressive success of the Standard Model of particle physics, and provides additional confidence in its ability to predict the radiation types and their intensity resulting from the decay and interaction of fundamental particles.

E.5
Cross-Section Relationships for Specific Processes

Cross-section relationships are most readily obtained from evaluating the appropriate Feynman diagram. A Feynman diagram is a method of depicting interaction processes in a manner that reveals the underlying physics and simplifies the computation of the interaction probability in quantitative mathematical terms. With Feynman diagrams, interaction processes are defined by linking expressions corresponding to each component of the diagram. Unfortunately, Feynman diagrams are beyond the scope of this appendix. The interested reader is referred to the appendix references for a more detailed description of these diagrams and their application.

The complexity of Standard Model cross-section formulas is illustrated for electron–positron scattering, leading to the production of point-like, spin-1/2 fermions through a virtual photon. Additional cross-section relationships are found in the references to this appendix.

For point-like, spin-1/2 fermions ($\bar{f}$), the differential cross section in the center-of-mass (CM) for $e^+ + e^- \to \gamma \to f + \bar{f}$ is

$$\frac{d\sigma}{d\Omega} = N_c \frac{\alpha^2}{4s} \beta [1 + \cos^2\theta + (1 - \beta^2)\sin^2\theta] Q_f^2, \tag{E.6}$$

where $N_c = 1$ if f is a charged lepton and $N_c = 3$ if f is a quark, α is the fine structure constant, s is the square of the CM energy, θ is the CM scattering angle, β is the v/c value for the final state fermion in the CM, and Q_f is the charge of the fermion. Equation E.6 can be integrated to provide the total cross section. In the limit of $\beta \to 1$,

$$\sigma = N_c \frac{4\pi\alpha^2}{3s} Q_f^2 = N_c \frac{(86.8 \text{ GeV}^2 \text{ nb}) Q_f^2}{s[\text{GeV}^2]}. \tag{E.7}$$

In Equation E.7, the cross section is expressed in nb when s has units of GeV^2.

References

Bevelacqua, J.J. (2004) Muon colliders and neutrino dose equivalents: ALARA challenges for the 21st century. *Radiation Protection Management*, **21** (4), 8.

Griffiths, D. (1987) *Introduction to Elementary Particles*, John Wiley & Sons, Inc., New York, NY.

Halzen, F. and Martin, A.D. (1984) *Quarks and Leptons: An Introductory Course in Modern Particle Physics*, John Wiley & Sons, Inc., New York, NY.

Particle Data Group (2004) Review of particle physics. *Physics Letters*, **B592**, 1.

Particle Data Group (2006) Review of particle physics. *Journal of Physics G: Nuclear and Particle Physics*, **33**, 1.

Appendix F
Special Theory of Relativity

The Special Theory of Relativity describes the relationship of physical quantities (e.g., energy and mass) in inertial reference frames, which are mathematical constructs that describe uniform motion between two coordinate systems. In inertial reference frames, the coordinate systems move apart at a constant velocity and their relative acceleration is zero.

The basis of the special relativity is contained in two postulates:

(1) The laws of physical phenomena are the same in all inertial reference frames. Only the relative motion of inertial frames can be measured.
(2) The velocity of light (in free space) is a constant, independent of the motion of the source.

Using these postulates, Einstein constructed the Special Theory of Relativity.

In this appendix, special relativity is addressed from a health physics perspective. In addition to the traditional treatment of mass, length, and time, applications of relevance to accelerator and space health physics are provided.

F.1
Length, Mass, and Time

The transformation relationships for length, mass, and time are obtained by considering two inertial reference frames, K_o and K. The frame K is fixed in the Earth and frame K_o moves at a constant velocity relative to frame K. For specificity, all motion is in the vertical or z-direction. Observers in the K_o and K frames are assumed to be at rest.

If an observer in frame K measures a vertical distance D between two points along the z-axis, an observer in the K_o frame measures a different distance D_o between the same two points. The relationship between the distances D and D_o is

$$D = \gamma D_o, \tag{F.1}$$

where

$$\gamma = (1-\beta^2)^{-1/2} \tag{F.2}$$

and

$$\beta = v/c. \tag{F.3}$$

In Equation F.3, v is the relative velocity of the two frames and c is the speed of light in a vacuum. Equation F.1 is often referred to as "length contraction" because an observer in the K_o frame perceives a given distance or length to be shorter by a factor of $1/\gamma$ than the length observed by an observer in the K frame.

In a similar manner, intervals of time (e.g., particle lifetimes) do not have the same value when measured by observers in the K and K_o frames. If τ and τ_o are the lifetimes of a particle as measured by an observer in the K and K_o frames, respectively, then the lifetimes are related by the relationship

$$\tau = \gamma \tau_o. \tag{F.4}$$

Equation F.4 is often referred to as "time dilation" because the lifetime measured by the observer in the K frame is longer than that measured by the observer in K_o.

A striking example of time dilation is afforded by observations made on cosmic ray muons. Muons are produced in large numbers by cosmic-ray particles interacting with nuclei in the upper atmosphere. Although muons are unstable particles that eventually decay, the fact that they are observed at the surface of the Earth implies that their lifetime in the K frame is well in excess of their physical half-life. This apparent increase in lifetime is in accordance with the Special Theory of Relativity (Equation F.4).

In a similar fashion, the particle mass (m_o) increases as noted in Equation F.5:

$$m = \gamma m_o, \tag{F.5}$$

where m_o is the mass of the body (rest mass) when it is at rest with respect to the observer. Equation F.5 notes that the particles mass increases without bound as the particle's velocity approaches c. The fact that a particle's mass increases markedly with relativistic velocity affects the cyclotron design. In particular, the magnetic field strength of cyclotron must be modified as the particle's mass increases. Additional impacts of relativistic mass are noted below.

F.1.1
Cosmic Ray Muons and Pions

As a specific application of Special Relativity, consider the dynamics of muons propagation to the Earth from their birth in the upper atmosphere. Muons are part of the natural background radiation environment of the Earth. Two of the more commonly created particles arising from cosmic-ray interactions in the atmosphere are the muon and pion whose properties are summarized in Appendix E. On the basis of their lifetimes, neither the pion nor the muon should reach the surface of the Earth.

Consider the creation of pions and muons in the atmosphere and their subsequent journey to the Earth. This process is effectively described by considering the two inertial reference frames, K_o and K, defined previously.

If a particle travels a distance D relative to the Earth, it appears to travel a distance D_o when observed from the K_o frame. Following Equations F.1–F.3, the distance traveled as perceived by an observer in the K_o frame would be shorter than that perceived by the observer in the K frame. In addition, the lifetime (τ), as noted by an observer in the K frame, is longer than τ_o when following Equation F.4.

To provide an explanation for muons reaching the Earth's surface, we use parameters appropriate to particles born in the atmosphere from the interaction of cosmic rays with nuclei. Typical values are used in the discussion. However, the results hold for any physical altitude and particle velocity.

Consider a muon born at an altitude of 8000 m above the Earth that has a velocity (v) of 0.998 c appropriate for high-energy cosmic ray collisions. Given the muon lifetime τ_o, it should only travel a distance D_o in the K_o frame before it decays:

$$D_o = v\tau_o = D_o = (0.998)(3.0 \times 10^8 \text{ m/s})(2.2 \times 10^{-6} \text{ s}) = 659 \text{ m}. \quad (F.6)$$

Recall from Equation F.1 that the distance D_o is contracted relative to D, the distance that the muon is observed to travel in the frame K:

$$D = \frac{659 \text{ m}}{\sqrt{1-\left(\frac{0.998c}{c}\right)^2}} = 1.04 \times 10^4 \text{ m}. \quad (F.7)$$

As the muon was born at 8000 m in frame K and would travel a distance 1.04×10^4 m before decaying, it reaches the Earth.

The extension of the muon lifetime, as observed on the Earth, is specified by Equation F.4:

$$\tau = \frac{2.2 \times 10^{-6} \text{ s}}{\sqrt{1-\left(\frac{0.998c}{c}\right)^2}} = 3.48 \times 10^{-5} \text{ s}. \quad (F.8)$$

Therefore, the muon lifetime appears to be increased by over an order of magnitude, and thus it can reach the Earth.

Special Relativity also explains why pions born in the atmosphere do not reach the Earth. From Appendix E, a charged pion has a lifetime of 2.6×10^{-8} s. Equations F.6 and F.7 lead to the following values for the charged pion born at 8000 m with a velocity of 0.998 c:

$$D_0 = (0.998)(3.0 \times 10^8 \text{ m/s})(2.6 \times 10^{-8} \text{ s}) = 7.78 \text{ m}, \quad (F.9)$$

$$D = \frac{7.78 \text{ m}}{\sqrt{1-\left(\frac{0.998c}{c}\right)^2}} = 123 \text{ m}. \quad (F.10)$$

As $D <$ 8000 m, pions do not reach the Earth. This simple example provides a physical explanation for the fact that cosmic-ray-induced muons reach the Earth's surface, but pions do not.

F.2
Energy and Momentum

The total energy (W) of a relativistic particle is the sum of the rest energy (E_o) and kinetic energy (E):

$$W = E + E_o, \tag{F.11}$$

$$E_o = m_o c^2. \tag{F.12}$$

The total energy can also be expressed in terms of the relativistic mass (m) and momentum (p):

$$W = \gamma m_o c^2 = mc^2 = \sqrt{m_o c^2 + p^2 c^2} = \frac{m_o c^2}{\sqrt{1-\beta^2}}. \tag{F.13}$$

Equations F.11–F.13 lead to an expression for the particle's kinetic energy

$$E = W - E_o = m_o c^2 \left(\frac{1}{\sqrt{1-\beta^2}} - 1 \right), \tag{F.14}$$

which can be solved for β:

$$\beta = \left[1 - \left(\frac{m_o c^2}{E + m_o c^2} \right)^2 \right]^{1/2} = \left[1 - \left(\frac{E_o}{W} \right)^2 \right]^{1/2}. \tag{F.15}$$

This relationship is useful because β is defined in terms of the particle's rest mass and total energy.

Table F.1 Values of β for selected particles expected to be accelerated in twenty-first century accelerators.

Kinetic energy (MeV)	β for indicated particles			
	e^- and e^+	μ^- and μ^+	p and $\bar{p}$	^{208}Pb
0.1	0.5482	0.0435	0.0146	0.0010
1	0.9411	0.1366	0.0461	0.0032
10	0.9988	0.4067	0.1448	0.0101
10^2	1.0000	0.8579	0.4282	0.0320
10^3	1.0000	0.9954	0.8750	0.1008
10^4	1.0000	0.9999	0.9963	0.3083
10^5	1.0000	1.0000	1.0000	0.7500
10^6	1.0000	1.0000	1.0000	0.9866
10^7	1.0000	1.0000	1.0000	0.9998
$\geq 10^8$	1.0000	1.0000	1.0000	1.0000

A particle initially at rest, having a charge q, and accelerated through a potential difference V has a kinetic energy of

$$E = qV. \tag{F.16}$$

Equations F.14 and F.16 lead to a relationship for β in terms of the terminal potential of an accelerator:

$$\beta = \left[1 - \left(\frac{qV}{m_o c^2} + 1\right)^{-2}\right]^{1/2}. \tag{F.17}$$

Table F.1 illustrates the values of β for particles that will be accelerated in twenty-first century accelerators. In particular, Table F.1 includes electrons, positrons, protons, antiprotons, charged muons, and heavy ions (e.g., ^{208}Pb) with values derived from Equation F.17.

References

Bevelacqua, J.J. (1999) *Basic Health Physics: Problems and Solutions*, John Wiley & Sons, Inc., New York.

Goldstein, H., Poole, C.P. and Safko, J.L. (2002) *Classical Mechanics*, 3rd edn, Prentice-Hall, Upper Saddle River, NJ.

Griffiths, D. (1987) *Introduction to Elementary Particles*, John Wiley & Sons, Inc., New York, NY.

Halzen, F. and Martin, A.D. (1984) *Quarks and Leptons: An Introductory Course in Modern Particle Physics*, John Wiley & Sons, Inc., New York, NY.

Jackson, J.D. (1999) *Classical Electrodynamics*, 3rd edn, John Wiley & Sons, Inc., New York, NY.

Particle Data Group (2006) Review of particle physics. *Journal of Physics G: Nuclear and Particle Physics*, **33**, 1.

Appendix G
Muon Characteristics

G.1
Overview

Muons are leptons that have a rest mass of 105.7 MeV and a mean lifetime of 2.2×10^{-6} s. They have a relatively large rest mass relative to that of the electron (0.511 MeV), and to first order these particles do not participate in the strong interaction. Muons can penetrate long distances into matter and are less susceptible to radiative effects compared to electrons. Over a broad energy range, the dominant energy loss mechanism is that of ionization. This makes the shielding of muons and the knowledge of their range considerably important at high-energy accelerators and in other high-energy applications. Muon radiation becomes more important as the accelerator energy increases.

G.2
Stopping Power and Range

The mean stopping power for high-energy muons in a material is described by the relationship

$$-\frac{dE}{dx} = (a(E) + b(E)E), \tag{G.1}$$

where E is the total energy, $a(E)$ is the electronic stopping power, and $b(E)$ is due to radiative processes including bremsstrahlung, pair production, and photoelectric interactions. The quantities $a(E)$ and $b(E)$ are slowly varying functions of E at the high energies where radiative contributions are important. The term $b(E)E$ is less than 1% of $a(E)$ for $E \leq 100$ GeV for most materials. For example, $a(E)$ is ≈ 0.002 GeV cm^2/g in iron, and $b(E)$ is the radiative coefficient in GeV that has values of about 1, 3, 5.5, 7.5, 8, and 8.4×10^{-6} cm^2/g for 1, 10, 100, 1000, 10 000, and 100 000 GeV muon energy, respectively.

Table G.1 Muon stopping power and range in water.[a]

Muon energy (GeV)	$\frac{1}{\rho}\left|\frac{dE}{dx}\right|$ (MeV cm²/g)	CSDA range (g/cm²)
1	2.109	4.706×10^2
10	2.507	4.260×10^3
100	3.020	3.629×10^4
1000	6.014	2.426×10^5
10^4	36.462	7.787×10^5
10^5	353.358	1.428×10^6

[a]Derived from Groom, Mokhov and Striganov (2001).

Table G.2 Muon stopping power and range in polyethylene.[a]

Muon energy (GeV)	$\frac{1}{\rho}\left|\frac{dE}{dx}\right|$ (MeV cm²/g)	CSDA range (g/cm²)
1	2.191	4.512×10^2
10	2.584	4.119×10^3
100	3.065	3.544×10^4
1000	5.584	2.487×10^5
10^4	30.902	8.607×10^5
10^5	295.995	1.634×10^6

[a]Derived from Groom, Mokhov and Striganov (2001).

Table G.3 Muon stopping power and range in air.[a]

Muon energy (GeV)	$\frac{1}{\rho}\left|\frac{dE}{dx}\right|$ (MeV cm²/g)	CSDA range (g/cm²)
1	2.021	5.077×10^2
10	2.642	4.204×10^3
100	3.240	3.409×10^4
1000	6.196	2.307×10^5
10^4	36.207	7.634×10^5
10^5	348.500	1.421×10^6

[a]Derived from Groom, Mokhov and Striganov (2001).

The muon range can be calculated using the continuous-slowing-down approximation (CSDA):

$$R(E) = \int_{E_o}^{E} \left(-\frac{dE'}{dx}\right)^{-1} dE' = \int_{E_o}^{E} (a(E') + B(E')E')^{-1} dE', \tag{G.2}$$

where E_o is sufficiently small, so that the range is insensitive to its exact value. At high energies, where a and b are essentially constant, the range is

$$R(E) \approx \frac{1}{b} \ln\left(1 + \frac{E}{E_{\mu c}}\right), \tag{G.3}$$

Table G.4 Muon stopping power and range in concrete.[a]

Muon energy (GeV)	$-\frac{1}{\rho}\left|\frac{dE}{dx}\right|$ (MeV cm²/g)	CSDA range (g/cm²)
1	1.834	5.460×10^2
10	2.216	4.855×10^3
100	2.775	4.038×10^4
1000	6.645	2.433×10^5
10^4	46.625	6.848×10^5
10^5	459.676	1.186×10^6

[a]Derived from Groom, Mokhov and Striganov (2001).

Table G.5 Muon stopping power and range in standard rock.[a]

Muon energy (GeV)	$-\frac{1}{\rho}\left|\frac{dE}{dx}\right|$ (MeV cm²/g)	CSDA range (g/cm²)
1	1.808	5.534×10^2
10	2.188	4.920×10^3
100	2.747	4.084×10^4
1000	6.615	2.453×10^5
10^4	46.586	6.877×10^5
10^5	459.512	1.189×10^6

[a]Derived from Groom, Mokhov and Striganov (2001).

where

$$E_{\mu c} = a(E_{\mu c})/b(E_{\mu c}) \tag{G.4}$$

is the muon critical energy, which is the energy at which electronic and radiative losses are equal. The critical energies for muons incident on water, polyethylene, air, concrete, and standard rock are 1.03, 1.28, 1.11, 0.70, and 0.693 TeV, respectively. At high energies (>100 GeV), the distribution of the ranges of individual muons about the mean range becomes severe.

The muon mass stopping power and CSDA range are provided for a variety of materials in Tables G.1–G.5. Tables G.1–G.5 summarize these quantities for muons incident on water, polyethylene, air, concrete, and standard rock.

Tables G.6 and G.7 provide fractional energy loss and comparisons of muon ranges in soil at high energies for different physical mechanisms. The dominant energy loss mechanisms are ionization, bremsstrahlung, pair production, and deep inelastic scattering. As the energy increases, the contribution of ionization decreases. Of the three remaining processes, bremsstrahlung dominates up to about 1000 GeV and then pair production dominates. The emergence of these processes occurs because additional energy facilitates accessing these reaction channels.

Table G.7 illustrates the importance of including all physical processes in model calculations. This becomes more important as the muon energy increases. In addition, straggling is important as shielding calculations based upon using the mean range values can lead to significant underestimates of the fluence of muons that penetrate the shield.

Table G.6 Fractional energy loss of muons in soil with a density of 2.0 g/cm^3 [a].

	Fractions of the total energy loss due to dominant energy loss mechanisms			
Energy (GeV)	Ionization	Bremsstrahlung	Pair production	Deep inelastic nuclear scattering
10	0.972	0.037	8.8×10^{-4}	9.7×10^{-4}
100	0.888	0.086	0.020	0.0093
1000	0.580	0.193	0.168	0.055
10 000	0.167	0.335	0.388	0.110

[a] Derived from Fermilab Report TM-1834 (2004), Van Ginneken et al. (1987), and Schopper (1990).

Table G.7 Comparison of muon ranges (meters) in heavy soil with a density of 2.24 g/cm^3 [a].

			Energy mean ranges from dE/dx in heavy soil (m)		
Energy (GeV)	Mean range (m)	Standard deviation (m)	All processes	Coulomb losses only	Coulomb and pair production losses
10	22.8	1.6	21.4	21.5	21.5
30	63.0	5.6	60.3	61.1	60.8
100	188	23	183	193	188
300	481	78	474	558	574
1000	1140	250	1140	1790	1390
3000	1970	550	2060	5170	2930
10 000	3080	890	3240	16 700	5340
20 000	3730	1070	[b]	[b]	[b]

[a] Derived from Fermilab Report TM-1834 (2004), Van Ginneken et al. (1987), and Schopper (1990).
[b] Not provided.

References

Cossairt, J.D. (2004) *Radiation Physics for Personnel and Environmental Protection*, Fermilab Report TM-1834, Revision 7, Fermi National Accelerator Laboratory, Batavia, IL.

Groom, D.E., Mokhov, N.V. and Striganov, S.I. (2001) Muon stopping power and range. *Atomic Data and Nuclear Data Tables*, **78**, 183.

Particle Data Group (2006) Review of particle physics. *Journal of Physics G: Nuclear and Particle Physics*, **33**, 1.

Schopper, H. (ed.) (1990) A. Fassò, K. Goebel, M. Höfert, J. Ranft, and G. Stevenson, Landolt–Börnstein numerical data and functional relationships in science and technology new series; Group I: nuclear and particle physics volume II: shielding against high energy radiation (O. Madelung, Editor-in-Chief, Springer-Verlag, Berlin, Heidelberg).

Van Ginneken, A., Yurista, P. and Yamaguchi, C. (1987) *Shielding Calculations for Multi-TeV Hadron Colliders*, Fermilab Report FN-447, Fermi National Accelerator Laboratory, Batavia, IL.

Appendix H
Luminosity

H.1
Overview

It is desirable to maximize the number of interacting particles in the design of accelerators. Accelerator design attempts to maximize the luminosity with minimum emittance and amplitude functions. This appendix defines these quantities, specifies their interrelationships, and relates them to accelerator design.

H.2
Accelerator Physics

Luminosity is a parameter that relates the event rate R in a collider and the interaction cross section (σ_{int}) through the relationship

$$R = L\sigma_{\text{int}}. \tag{H.1}$$

Luminosity is often expressed in units of $1/\text{cm}^2\,\text{s}$, and tends to be a large number.

For the case of colliding beams, the luminosity is written in terms of the number of particles in each of the beams. The collection of grouped particles in a beam is called a bunch. If two bunches containing n_1 and n_2 particles collide with frequency f, then the luminosity is approximately given by

$$L = \frac{fn_1 n_2}{4\pi\sigma_x\sigma_y}, \tag{H.2}$$

where σ_x and σ_y characterize the Gaussian transverse beam profiles in the horizontal and vertical directions. Even if a Gaussian shape only approximates the initial particle distribution at the source, the normal form is a good particle distribution approximation in the high-energy limit. The Gaussian shape at high energies is a consequence of the central limit theorem and the diminished importance of space charge effects.

If the bunches are assumed to be contained in a cylindrical shape of cross-sectional area A and length l, a simplified expression for the luminosity is achieved. To obtain

Health Physics in the 21st Century. Joseph John Bevelacqua
Copyright © 2008 WILEY-VCH Verlag GmbH & Co. KGaA, Weinheim
ISBN: 978-3-527-40822-1

this simplification, consider the probability (P) of a single collision of the two bunches containing n_1 and n_2 particles:

$$P = n_1 \sigma \left(\frac{n_2}{Al}\right) l = \sigma \frac{n_1 n_2}{A}. \tag{H.3}$$

The quantity $n_2/(Al)$ is the density of particles in bunch 2. The rate at which collisions occur (R) is

$$R = Pf = L\sigma. \tag{H.4}$$

Combining Equations H.3 and H.4 provides a simplified relationship for the luminosity:

$$L = \frac{fn_1 n_2}{A}, \tag{H.5}$$

which is critically dependent on the beam size.

The beam size is expressed in terms the transverse emittance (ε) and the amplitude function (β). Transverse emittance is related to beam quality and reflects bunch preparation. Beam optics, particularly the magnet configuration, determines the amplitude function. The transverse emittance is written in terms of the amplitude function and the Gaussian transverse beam profile:

$$\varepsilon_x = \frac{\pi \sigma_x^2}{\beta}, \tag{H.6}$$

$$\varepsilon_y = \frac{\pi \sigma_y^2}{\beta}. \tag{H.7}$$

From a design perspective, it is desirable to focus the beam and achieve the minimum physical size at the interaction point. This is achieved by making the amplitude function at the interaction point (β^*) as small as possible. Equations H.6 and H.7 are used to recast Equation H.2 in terms of emittance and amplitude functions:

$$L = \frac{fn_1 n_2}{4\sqrt{\varepsilon_x \beta_x^* \varepsilon_y \beta_y^*}}. \tag{H.8}$$

Equation H.8 suggests that achieving high luminosity requires high population bunches of low emittance to collide at high frequency at locations where the beam optics provides minimum values of the amplitude functions.

In storage rings, the luminosity is expected to degrade over time as a result of defocusing and radiative processes. The luminosity degrades primarily owing to particles leaving the established energy region as a result of a variety of effects (e.g., radiation losses including bremsstrahlung, bunch charge density effects, and beam instability). In general, stored particles are intentionally removed from the ring when the luminosity drops to the point where a refill with new bunches improves the integrated luminosity.

References

Chao, A.W. and Tigner, M. (eds)(2002) *Handbook of Accelerator Physics and Engineering*, World Science Publishing, Co., Singapore.

Edwards, D.A. and Syphers, M.J. (2004) *An Introduction to the Physics of High Energy Accelerators*, Wiley-VCH Verlag GmbH & Co. KGaA, Weinheim, Germany.

Particle Data Group (2006) Review of Particle Physics. *Journal of Physics G: Nuclear and Particle Physics*, **33**, 1.

Appendix I
Dose Factors for Typical Radiation Types

I.1
Overview

Dose factors are a convenient method to calculate the effective dose and compare the relative detriment of various radiation types. This appendix provides a summary of dose factors for typical radiation types over a range of energies.

I.2
Dose Factors

The effective dose (H) is written in terms of a dose factor (k) through a simple relationship:

$$H = k\Phi, \tag{I.1}$$

where Φ is the particle fluence. The primary particle fluence depends on the accelerator characteristics. Secondary particle fluence varies with the reactions under investigation and their associated cross sections.

High-energy accelerators and space applications involve a variety of radiation types with a wide range of energies. Given this complex situation, it is important to conceptually define the effective dose for an ensemble of particles and energies. For a radiation field containing a mixture of n different components (e.g., different particle types), the effective dose is defined as

$$H = \sum_{i=1}^{n} \int_{E_{\min}}^{E_{\max}} dE\, k_i(E) \Phi_i(E), \tag{I.2}$$

where $\Phi_i(E)$ is the fluence of particles of type i with energy E, and $k_i(E)$ is the effective dose per unit fluence at energy E. A similar relationship can be written for the dose equivalent. The dose equivalent per unit fluence for protons, neutrons, pions, muons, electrons, and photons is summarized in Table I.1.

Health Physics in the 21st Century. Joseph John Bevelacqua
Copyright © 2008 WILEY-VCH Verlag GmbH & Co. KGaA, Weinheim
ISBN: 978-3-527-40822-1

Table I.1 Dose equivalent per unit fluence for various radiation types as a function of energy[a,b].

Energy (GeV)	Dose equivalent per fluence (μSv cm^2/particle)						
	p	n	π^+	π^-	$\mu^\pm$	e	γ
10^{-10}	c	1×10^{-5}	c	c	c	c	c
10^{-9}	c	1×10^{-5}	c	c	c	c	c
10^{-8}	c	9×10^{-6}	c	c	c	c	c
10^{-7}	c	7×10^{-6}	c	c	c	c	c
10^{-6}	c	6×10^{-6}	c	c	c	c	c
10^{-5}	c	9×10^{-6}	c	c	c	c	3×10^{-8}
10^{-4}	c	1×10^{-4}	c	c	c	2×10^{-3}	3×10^{-7}
0.001	7×10^{-3}	4×10^{-4}	c	c	c	6×10^{-4}	4×10^{-6}
0.01	7×10^{-3}	4×10^{-4}	4×10^{-3}	7×10^{-2}	3×10^{-3}	4×10^{-4}	3×10^{-5}
0.1	8×10^{-3}	5×10^{-4}	3×10^{-3}	3×10^{-3}	3×10^{-3}	4×10^{-4}	2×10^{-4}
1	2×10^{-3}	1×10^{-3}	2×10^{-3}	2×10^{-3}	4×10^{-4}	7×10^{-4}	4×10^{-4}
10	4×10^{-3}	3×10^{-3}	3×10^{-3}	3×10^{-3}	4×10^{-4}	1×10^{-3}	6×10^{-4}
100	1×10^{-2}	6×10^{-3}	6×10^{-3}	6×10^{-3}	5×10^{-4}	c	c
1000	c	c	1×10^{-2}	1×10^{-2}	7×10^{-4}	c	c

[a]Derived from Fermilab Report TM-1834 (2004) and Schopper (1990).
[b]The values for muons are valid for both μ^+ and μ^-.
[c]Value was not provided.

I.3
Dose Terminology

The use of dose equivalent is prevalent in the United States because its regulations follow the ICRP 26/30 methodology. Much of the world derives its radiation protection standards from the recommendations of ICRP 60 that uses effective dose. Table I.1 is derived from a US source, and it expresses dose in terms of the dose equivalent.

References

Cossairt, J.D. (2004) Radiation Physics for Personnel and Environmental Protection, Fermilab Report TM-1834, Revision 7, Fermi National Accelerator Laboratory Batavia, IL.

ICRP Publication 26 (1977) *Recommendations of the International Commission on Radiological Protection*, Pergamon Press, Oxford, England.

ICRP Publication 30 (1979) *Limits for Intakes of Radionuclides by Workers*, Pergamon Press, Oxford, England.

ICRP Publication 60 (1991) *1990 Recommendations of the ICRP*, Pergamon Press, Oxford, England.

Schopper H. (ed.) (1990) A. Fassò, K. Goebel, M. Höfert, J. Ranft, and G. Stevenson, Landolt-Börnstein numerical data and functional relationships in science and technology new series; Group I: nuclear and particle physics volume II: shielding against high energy radiation (O. Madelung, Editor-in-Chief, Springer-Verlag, Berlin, Heidelberg).

Appendix J
Health Physics Related Computer Codes

J.1
Code Overview

This appendix summarizes a selected listing of computer codes and supporting data used in health physics applications. The listing represents only a sample of the broad scope of available codes. Additional codes are summarized in National Council on Radiation Protection and Measurements Report No. 144.

The codes summarized in this appendix are utilized in fission power reactor, fusion reactor research, accelerator, photon light source, and space applications. This appendix contains a brief summary of the code, its applications, and a web address that provides additional information.

J.1.1
EGS Code System (http://www.irs.inms.nrc.ca/EGS4/get_egs4.html)

EGS (Electron Gamma Shower) is a Monte Carlo code that simulates the transport of electrons and photons in arbitrary geometries. It was originally developed at the Stanford Linear Accelerator Center (SLAC) for high-energy physics applications and has been extended with the help of the National Research Council of Canada and the High Energy Research Organization in Japan (KEK) to apply to lower energy applications. The applicable energy range of EGS4 is approximately 1 keV–0.1 TeV. EGS4 has been extensively benchmarked for medical physics applications.

J.1.2
ENDF (http://www.nndc.bnl.gov/exfor3/endf00.htm)

The Evaluated Nuclear Data File (ENDF) library includes a nuclear reaction database containing evaluated (recommended) cross sections, spectra, angular distributions, fission product yields, photo-atomic data, and thermal scattering data. The emphasis of the data set is on neutron-induced reactions. The data were analyzed to produce recommended libraries for one of the national (United States, European, Japanese,

Russian, and Chinese) nuclear data projects. All data are stored in the internationally adopted format (ENDF-6).

J.1.3
FLUKA (http://www.fluka.org/)

FLUKA is a fully integrated Monte Carlo simulation package. It has applications in high-energy physics; engineering; shielding, detector, and telescope design; cosmic ray studies; dosimetry; medical physics; and radiobiology.

J.1.4
JENDL (http://wwwndc.tokai-sc.jaea.go.jp/jendl/jendl.html)

The Japanese Evaluated Nuclear Data Library (JENDL) provides a standard library for fast breeder reactors, thermal reactors, fusion reactors, shielding calculations, and other applications. The latest version is JENDL-3.3 (2002), and it contains neutron-induced reaction data for 337 nuclides, in the neutron energy range from 10^{-5} eV to 20 MeV.

J.1.5
MARS (http://www-ap.fnal.gov/MARS/)

MARS is a Monte Carlo code for the simulation of three-dimensional hadronic and electromagnetic cascades. Its applications include muon, heavy-ion and low-energy neutron transport in accelerators; detector development and evaluation; spacecraft shielding design; and a variety of shielding applications. MARS is applied to energies spanning the eV to 100 TeV range.

J.1.6
MCNP (http://mcnp-green.lanl.gov/index.html)

MCNP is a general-purpose Monte Carlo N-Particle code used for neutron, photon, electron, or coupled neutron/photon/electron transport. Applications include radiation protection and dosimetry, radiation shielding, radiography, medical physics, nuclear criticality safety, detector design and analysis, well logging, accelerator target design, fission and fusion reactor design, and decontamination and decommissioning.

J.1.7
MCNPX (http://mcnpx.lanl.gov/)

MCNPX (Monte Carlo N-Particle code extended) is a three-dimensional, time-dependent, and general-purpose Monte Carlo radiation transport code for modeling radiation interactions in a wide variety of situations. It extends the capabilities of MCNP4C3 to many particle types and over a wide energy range. MCNPX is applicable to a diverse set of applications including Earth orbit and planetary space radiation evaluations, oil exploration, nuclear medicine, nuclear safeguards, accelerator applications, and nuclear criticality safety.

J.1.8
MicroShield® (http://www.radiationsoftware.com/mshield.html)

MicroShield® is a photon/gamma-ray shielding and dose assessment program. It has applications in the health physics, waste management, radiological-related design, and radiological engineering applications. MicroShield® has a relatively simple input format.

J.1.9
MicroSkyshine® (http://www.radiationsoftware.com/mskyshine.html)

MicroSkyshine® calculates the photon dose from sky scattered gamma radiation, and its method of solution is based on the use of "beam functions" for a point source as developed for the US Nuclear Regulatory Commission. MicroSkyshine® has been used to evaluate conformance with US Regulations (e.g., 10CFR50, Appendix I ALARA requirements, and 40CFR190 fuel cycle exposure criteria). Typical applications include scattering in boiling water reactor turbine buildings, radioactive waste storage facilities, and waste disposal sites.

J.1.10
SCALE 5 (http://www-rsicc.ornl.gov/codes/ccc/ccc7/ccc-725.html)

The Standardized Computer Analyses for Licensing Evaluation (SCALE) system was developed for the US Nuclear Regulatory Commission to provide a method of analysis for the evaluation of nuclear fuel facilities and package designs. The system has the capability to perform criticality safety, shielding, radiation source term, spent fuel depletion/decay, and heat transfer analyses.

The criticality safety analysis sequences (CSAS) control module calculates the neutron multiplication factor for one-dimensional (1D) (XSDRNPM S) and multidimensional (KENO V.a) system models. It also has the capability to perform criticality searches (optimum, minimum, or specified values of k_{eff}) on geometry dimensions or nuclide concentrations in KENO V.a.

The SAS2H module uses ORIGEN S to perform a one-dimensional fuel depletion analysis. This module can be used to characterize spent fuel and generate source terms.

Four shielding analysis sequence (SAS) codes are provided. General one-dimensional shielding problems can be analyzed using XSDRNPM S. Shielding analysis using the MORSE SGC Monte Carlo code is also available. The SAS4 module can be used to perform a Monte Carlo shielding analysis for cask-type geometry. The QADS module analyzes three-dimensional gamma-ray shielding problems via the point kernel code, QAD CGGP.

The thermal analysis module HTAS1 performs a two-dimensional thermal analysis for a specific class of spent fuel casks during normal, fire, and postfire conditions.

J.1.11
SKYSHINE-KSU (http://www-rsicc.ornl.gov/codes/ccc/ccc6/ccc-646.html)

SKYSHINE-KSU was developed at Kansas State University to form a comprehensive system for calculating gamma-ray scattering from the sky (skyshine). It includes the SKYNEUT 1.1, SKYDOSE 2.2, and MCSKY 2.3 codes plus the DLC-0188/ZZ-SKYDATA library.

SKYNEUT evaluates neutron and neutron-induced secondary gamma-ray skyshine doses from an isotropic, point, neutron source collimated by three simple geometries. These geometries are an open silo; a vertical, perfectly absorbing wall; and a rectangular building. The source may emit monoenergetic neutrons or neutrons with a spectrum of energies.

SKYDOSE evaluates the gamma-ray skyshine dose from an isotropic, monoenergetic, point gamma–photon source collimated by three simple geometries. These are a source in a silo, a source behind an infinitely long, vertical, perfectly absorbing wall, and a source in a rectangular building. In all the three geometries, an optional overhead slab shield may be specified.

MCSKY evaluates the gamma-ray skyshine dose from an isotropic, monoenergetic, point, gamma source collimated into either a vertical cone or a vertically oriented structure with an N-sided polygon cross section. An overhead laminate shield composed of two different materials is assumed.

J.1.12
SPAR (http://www-rsicc.ornl.gov/codes/ccc/ccc2/ccc-228.html)

SPAR (Stopping Powers and Ranges) is a legacy code that still provides useful information. It computes the stopping powers and ranges for muons, pions, protons, and heavy ions in any nongaseous medium for energies up to several hundred GeV.

J.2
Code Utilization

Computer code users need to exercise caution in using any numerical algorithm. Users must clearly understand the limitations and capabilities of a code to address the problem of interest. This caution is more encompassing than the old adage "GARBAGE IN – GARBAGE OUT." It involves the interpretation of results and understanding the inherent limits and assumptions of the code package.

As an example, I cite a series of shielding design calculations that the author recently performed. The problem involved the scattering of photons from a source to an elevated penetration. Before evaluating MCNP, I performed a scaling (hand) calculation based on data that I obtained in similar circumstances at other facilities. The next step was to perform a hand calculation using Rockwell's methodology. Next, a deterministic shielding computer code was executed that provided a refined calculation. Finally, MCNP was evaluated.

At each step, differences from the previous step were evaluated and assessed for credibility. If MCNP had been run without the other steps and without any internal benchmarking, how would a user know if the results were credible? Errors could include input/geometry errors, misinterpreting MCNP caution or error flags, or applying MCNP to a problem that was outside its zone of applicability. In the case cited, all codes and hand calculations provided a consistent solution that suggested a reasonable degree of confidence in the MCNP results. The message to any code user is to be cautious and to perform internal benchmarking to improve confidence in the code's final results.

References

NCRP Report No. 144 (2003) *Radiation Protection for Particle Accelerator Facilities*, National Council on Radiation Protection and Measurements, Bethesda, MD.

Rockwell, T., III (1956) *Reactor Shielding Design Manual*, McGraw-Hill Book Company, Inc., New York.

Appendix K
Systematics of Heavy Ion Interactions with Matter

K.1
Introduction

Heavy ion interactions with matter are important health physics considerations. These interactions are encountered in accelerator and therapy applications, during deep space missions, and during planetary missions within our Solar System. Heavy ion interactions also affect fusion energy facilities that operate under sustained plasma conditions.

The interaction of heavy ions with matter is described by well-known relationships. These relationships provide the stopping power, range, and dosimetric information of ions over a wide range of heavy-ion energies. This appendix provides an analytical framework for these calculations. Appendix J summarized the computer codes that provided numerical algorithms for heavy ion calculations.

K.2
Overview of External Radiation Sources

The interaction of heavy ions with matter is an important health physics consideration, but it is particularly important at the high energies encountered in the space and accelerator applications. Both of these areas include a range of energies and a diversity of radiation types. Accelerator health physics issues are primarily associated with the shielding of the generated radiation, and the shielding is designed to ensure that the radiation levels meet applicable standards and requirements. External radiation in space is more diverse and offers a more complex challenge.

External radiation encountered in space applications arises from trapped radiation, galactic cosmic rays, and Solar particle events that involve a variety of radiation types, including photons, electrons, protons, and heavy ions. For photon radiation, scattering and attenuation reduce the photon fluence as it penetrates the spacecraft shielding.

With electrons, the density builds to an equilibrium value inside the shield such that the electron fluence rises to a maximum and then decreases with increasing

Health Physics in the 21st Century. Joseph John Bevelacqua
Copyright © 2008 WILEY-VCH Verlag GmbH & Co. KGaA, Weinheim
ISBN: 978-3-527-40822-1

depth into the shield. Electron backscatter increases the surface fluence and is considered in the shielding analysis. The depth of the maximum fluence increases with the increase in electron energy. With electrons, the primary particles slow down in the shield and produce high ionizations per unit length as they reach their maximum range. For depths beyond the maximum range, the electron fluence decreases very rapidly to a value of only a few percent of the maximum value. Similar comments apply to electrons that penetrate the shield and reach tissue.

The stopping power for high-energy electrons is about 2 MeV/cm in tissue and about twice this value in bone. For the electron energies below 1 MeV, the maximum effective dose occurs near the skin surface. As the electron energy increases from 4 to 20 MeV, the shape of the effective dose curve shifts from a surface peak to a broader plateau extending into the tissue. Beyond 20 MeV, the plateau expands and the additional tissue is at risk.

Protons have a range that varies with energy. Proton beams produce a relatively low constant fluence that terminates in a narrow Bragg peak at the end of the range of the particle. The methods for determining the proton stopping power and range are provided below. As a matter of reference, Table K.1 summarizes the range of electrons and protons in water as a function of energy.

K.3
Physical Basis for Heavy Ion Interactions with Matter

Using relativistic quantum mechanics, Bethe derived the following equation for the stopping power ($-dE/dx$) in a uniform medium for a heavy charged particle or heavy ion:

$$-\frac{dE}{dx} = \frac{4\pi k^2 z^2 e^4 n}{mc^2 \beta^2} \left[\ln \frac{2mc^2 \beta^2}{I(1-\beta^2)} - \beta^2 \right], \tag{K.1}$$

where k is an electric constant (8.99×10^9 N m^2/C^2), z is the atomic number of the heavy ion, e is the magnitude of the electric charge, n is the number of electrons per unit volume in the medium interacting with the heavy ion, m is the electron rest mass, c is the velocity of light in a vacuum, β is the velocity of the ion relative to the

Table K.1 Range of protons and electrons in water.

Kinetic energy (MeV)	Range (g/cm²)	
	Protons	Electrons
0.01	0.00003	0.0002
0.1	0.0001	0.0140
1	0.002	0.430
10	0.118	4.88
100	7.57	32.5
1000	321	101

Derived from Turner (1995).

speed of light (v/c), v is the velocity of the heavy ion, and I is the mean excitation energy of the medium interacting with the heavy ion.

Using relativistic mechanics, β is determined from the total energy (W) and rest energy (E_o):

$$W = E + E_o, \tag{K.2}$$

$$E_o = m_o c^2, \tag{K.3}$$

$$W = \frac{m_o c^2}{\sqrt{1-\beta^2}}, \tag{K.4}$$

where E is the kinetic energy and m_o is the rest mass of ion. Equations K.2–K.4 lead to the following expression for the ion's kinetic energy:

$$E = W - E_o = m_o c^2 \left(\frac{1}{\sqrt{1-\beta^2}} - 1 \right). \tag{K.5}$$

Equation K.5 is solved for β:

$$\beta = \left[1 - \left(\frac{m_o c^2}{E + m_o c^2} \right)^2 \right]^{1/2} = \left[1 - \left(\frac{E_o}{W} \right)^2 \right]^{1/2}. \tag{K.6}$$

The mean excitation energy I can be represented by the following empirical formulas for an element with atomic number Z:

$$I \cong 19.0 \, \text{eV}, \quad Z = 1, \tag{K.7}$$

$$I \cong (11.2 + 11.72 Z) \, \text{eV}, \quad 2 \leq Z \leq 13, \tag{K.8}$$

$$I \cong (52.8 + 8.71 Z) \, \text{eV}, \quad Z > 13. \tag{K.9}$$

Once the stopping power is known, it is possible to calculate the range of ions. The range of a charged particle is the distance it travels before coming to rest. The reciprocal of the stopping power is the distance traveled per unit energy loss. Therefore, the range $R(E)$ of a charged particle having kinetic energy E is the integral of the reciprocal of the negative stopping power from the initial kinetic energy E_i to the final kinetic energy of a stopped particle ($E = 0$):

$$R(E) = \int_{E_i}^{0} (dE/dx)^{-1} dE. \tag{K.10}$$

Equation K.10 is often written in terms of the stopping power as given below:

$$R(E) = \int_{0}^{E_i} (-dE/dx)^{-1} dE. \tag{K.11}$$

As the heavy ion beam loses energy, it broadens in a variety of ways, including its energy, width, and angular dispersion. For example, the Bragg peak spreads in energy and has a distinctive width. Each of these spreading mechanisms affects the energy delivered to the medium. Accordingly, energy straggling, range straggling, and angle straggling are briefly addressed.

For a beam of heavy ions, the width of the Bragg peak is caused by the summation of multiple scattering events that yield a Gaussian energy loss distribution, often referred to as energy straggling:

$$\frac{N(E)dE}{N} = \frac{1}{\alpha \pi^{1/2}} \exp\left[-\frac{(E-\bar{E})^2}{\alpha^2}\right]. \tag{K.12}$$

Energy straggling represents the specific number $N(E)$ of particles having energies in the range E to $E + dE$ divided by the number of particles N, with mean energy $\bar{E}$ after traversing a thickness x_o of absorber. The distribution parameter or straggling parameter (α) expresses the half-width at the $(1/e)$th height and is given by the expression:

$$\alpha^2 = 4\pi z^2 e^4 n Z x_o \left[1 + \frac{KI}{mv^2} \ln\left(\frac{2mv^2}{I}\right)\right], \tag{K.13}$$

where K is a constant that depends on the electron shell structure of the absorber and has a value between 2/3 and 4/3, and Z is the atomic number of the absorber. It is also possible to recast Equation K.13 to represent the full width at half-maximum (FWHM) height.

In an analogous manner, the range straggling, expressed as the number of particles $N(R)$ with ranges R to $R + dR$ divided by the total number of particles of the same initial energy, is given by the equation

$$\frac{N(R)dR}{N} = \frac{1}{\alpha \pi^{1/2}} \exp\left[-\frac{(R-\bar{R})^2}{\alpha^2}\right], \tag{K.14}$$

where $\bar{R}$ is the mean range.

Upon entering a medium of thickness x_o, a collimated beam experiences multiple collisions that broaden the beam and cause it to diverge. This phenomena is called angle straggling, and the mean divergence angle $(\bar{\theta})$ is given by the following relationship:

$$\bar{\theta}^2 = \frac{2\pi z^2 e^4}{\bar{E}^2} n Z^2 x_o \ln\left(\frac{\bar{E} a_o}{z Z^{4/3} e^2}\right), \tag{K.15}$$

where a_o is the Bohr radius

$$a_o = \frac{\hbar^2}{kme^2}, \tag{K.16}$$

and k is a unit specific constant.

K.4
Range Calculations

Heavy ion range calculations are provided in this section. Because the range is being calculated, the relationship between the range and the peak of the dose equivalent distribution needs to be established. To accomplish this, the effects of straggling are briefly considered.

The position of the Bragg peak and straggling full width at half-maximum are summarized in Table K.2. For consistency with the literature, ion energies are expressed in terms of MeV per nucleon (MeV/n).

Ion range and straggling widths are provided for ^{12}C ions with energies between 90 and 330 MeV/n. Table K.2 indicates that the heavy ion range is reasonably approximated by the location of the Bragg peak.

The values in parenthesis in Table K.2 are the results for Stopping Powers and Ranges (SPAR) Code calculations to verify the model used in this appendix. SPAR includes a somewhat different formulation of the stopping power than utilized in this appendix, but it is sufficient to verify the validity of our calculations. The differences in the calculated ranges arise from SPAR's parameterization of the mean ionization and the inclusion of shell-effect and density-effect corrections. In addition, SPAR utilizes approximations that become less valid as the ion's atomic number increases beyond 50.

Because one of the purposes of this appendix is the calculation of the range of heavy ions, Table K.3 provides the results of calculations of the range in water for a number of heavy ions, including ^{4}He, ^{12}C, ^{16}O, ^{20}Ne, ^{40}Ca, ^{63}Cu, ^{92}Mo, ^{107}Ag, ^{142}Nd, ^{172}Hf, ^{184}Os, ^{197}Au, ^{209}Bi, ^{238}U, and ^{236}Np. The ions ranges are evaluated for energies between 90 and 330 MeV/n.

The results of Table K.3 illustrate that the locations of peak irradiation vary considerably with the specific ion and energy combinations. This characteristic complicates the shielding design, particularly for deep space applications. In addition, the inability to target a specific location because of the wide variability in the ion and its energy makes heavy ion dosimetry and the calculation of tissue dose a challenge. A method for the determination of the absorbed dose in tissue is outlined in the next section.

Table K.2 ^{12}C ion range and straggling widths in water.

Energy (MeV/n)	Range @ peak position (cm)[a]	Straggling FWHM (cm)[a]	Range (cm) this work[b]
90	2.13	0.07	2.14 (2.12)
198	8.28	0.23	8.54 (8.45)
270	14.43	0.5	14.5 (14.3)
330	20.05	0.7	20.2 (19.9)

[a] Weber (1996).
[b] Values in parenthesis are based on the SPAR Code (1985).

Table K.3 Heavy ion ranges in water (cm) for selected energies.

Ion	Ion energy (MeV/n)			
	90	198	270	330
^{4}He	6.42	25.6	43.4	60.5
^{12}C	2.14	8.54	14.5	20.2
^{16}O	1.60	6.40	10.8	15.1
^{20}Ne	1.28	5.12	8.67	12.1
^{40}Ca	0.64	2.56	4.34	6.05
^{63}Cu	0.48	1.92	3.25	4.53
^{92}Mo	0.34	1.34	2.26	3.15
^{107}Ag	0.31	1.24	2.10	2.93
^{142}Nd	0.25	1.01	1.71	2.38
^{172}Hf	0.21	0.85	1.44	2.01
^{184}Os	0.20	0.82	1.38	1.93
^{197}Au	0.20	0.81	1.37	1.91
^{209}Bi	0.19	0.78	1.32	1.83
^{238}U	0.18	0.72	1.22	1.70
^{236}Np	0.18	0.70	1.18	1.65

K.5
Tissue Absorbed Dose from a Heavy Ion Beam

For a tissue volume irradiated by a parallel beam of particles, the absorbed dose (D) as a function of penetration distance x is given by

$$D(x) = \frac{1}{\rho}\left(-\frac{dE}{dx}\right)\Phi(x), \tag{K.17}$$

where ρ is the density of the material (tissue) attenuating the heavy ion, $-dE/dx$ is the stopping power, and Φ is the heavy ion fluence. The particle fluence varies with the penetration distance according to the following relationship:

$$\Phi(x) = \Phi(0)\exp(-\mu x), \tag{K.18}$$

where $\Phi(0)$ is the entrance fluence and μ is the macroscopic reaction cross section (linear attenuation coefficient). The linear attenuation coefficient is defined as

$$\mu = N\sigma, \tag{K.19}$$

where N is the number of atoms of absorbing material per unit volume and σ is the total microscopic reaction cross section for the heavy-ion–tissue interaction.

In principle, the dose distribution from each heavy ion in the beam is summed to obtain the total dose distribution. However, in performing this sum, the absorbed dose must be modified by an energy dependent radiation weighting factor.

When calculating the effective dose to a complex medium, such as tissue, the methodology must be modified. In particular, modifications to the linear attenuation coefficient and stopping power are required.

For a medium, such as tissue composed of hydrogen (5.98×10^{22} atoms/cm^3), oxygen (2.45×10^{22} atoms/cm^3), carbon (9.03×10^{21} atoms/cm^3), and nitrogen (1.29×10^{21} atoms/cm^3), the linear attenuation coefficient is the sum of the product of the attenuation coefficient and number density for each element:

$$\mu = \sum_i \mu_i N_i. \tag{K.20}$$

In a similar fashion, the stopping power for a medium, composed of a number of elements i having charge Z_i, number density N_i, and mean excitation I_i, is obtained through a modification of Equation K.1. In particular, for a complex medium, the following substitution is made in Equation K.1:

$$n/\ln I \rightarrow \sum_i N_i Z_i / \ln I_i. \tag{K.21}$$

The actual dosimetry situation involved in spacecraft situations is more complex than assumed in Equations K.17–K.21. In particular, the heavy ion beam is shielded by spacecraft structures prior to impinging on tissue. For that case, the primary fluence is modified to account for the attenuation of the heavy ion beam. In addition, secondary particle fluence is generated from interactions of the primary particles and shielding materials.

K.6
Determination of Total Reaction Cross Section

Equation K.19 uses the total microscopic reaction cross section to obtain the total macroscopic reaction cross section. The total microscopic reaction cross section is obtained from parameterizations or the use of nuclear optical model codes, such as DWUCK or MERCURY.

The parametric models fit available cross-section data, using the established relationships, including trends in nuclear radii, reaction kinematics, and energy dependence. The optical model codes require parameterization of the entrance and exit channels, nuclear structure information for the transferred particles, spectroscopic information, and specification of kinematic information related to the reaction under investigation. Each of these approaches has its inherent shortcomings, and these must be clearly understood. The best practice is to use measured data. However, the use of models is often required because a complete set of cross sections is often not available.

References

Bethe, H. (1930) Zur Theorie des Durchgangs schneller Korpuskularstrahlung durch Materie. *Annalen Der Physik (Leipzig)*, 5, 325.

Bevelacqua, J.J. (1995) *Contemporary Health Physics: Problems and Solutions*, John Wiley & Sons, Inc., New York.

Bevelacqua, J.J. (1999) *Basic Health Physics: Problems and Solutions*, John Wiley & Sons, Inc., New York.

Bevelacqua, J.J. (2005) Systematics of Heavy Ion Radiotherapy. *Radiation Protection Management*, **22** (6), 4.

Bevelacqua, J.J., Stanley, D. and Robson, D. (1977) Sensitivity of the Reaction $^{40}\text{Ca}(^{13}\text{C},^{14}\text{N})^{39}\text{K}$ to the $^{39}\text{K}+\text{p}$ Form Factor. *Physical Review*, **C15**, 447.

Charlton, L.A. (1973) Specific New Approach to Finite-Range Distorted-Wave Born Approximation. *Physical Review*, **C8**, 146.

Charlton, L.A. and Robson, D. (1973) MERCURY, Florida State University Technical Report No. 5, Tallahassee, FL.

Kraft, G. (2000) Tumor Therapy with Heavy Charged Particles. *Progress in Particle and Nuclear Physics*, **45**, S473.

Kunz, P.D. (1986) *Computer Code DWUCK*, University of Colorado, Bolder, Co; available as a part of Radiation Safety Information Computational Center Computer Code Collection: Code Package PSR-235 DWUCK, Nuclear Model Computer Codes for Distorted Wave Born Approximation and Coupled Channel Calculations, Oak Ridge National Laboratory, Oak Ridge, TN. http://rsicc.ornl.gov (accessed on January 16, 2007).

Marmier, P. and Sheldon, E. (1969) *Physics of Nuclei and Particles*, Vol. I, Academic Press, New York.

Radiation Safety Information Computational Center Computer Code Collection: Code Package CCC-228 SPAR. (1985) *Calculation of Stopping Powers and Ranges for Muons, Charged Pions, Protons, and Heavy Ions*, Oak Ridge National Laboratory, Oak Ridge, TN. http://rsicc. ornl.gov (accessed on January 16, 2007).

Shen, W.Q., Wang, B., Feng, J., Zhan, W.L., Zhu, Y.T. and Feng, E.P. (1989) Total reaction cross section for heavy-ion collisions and its relation to the neutron excess degree of freedom. *Nuclear Physics A*, **491**, 130.

Turner, J.E. (1995) *Atoms, Radiation, and Radiation Protection*, 2nd edn, John Wiley & Sons, Inc., New York.

Weber, U. (1996) Volumenkonforme Bestrahlung mit Kohlenstoffionen, PhD thesis, Universität Gh Kassel, Germany.

Appendix L
Curvature Systematics in General Relativity

L.1
Introduction

This appendix provides an overview of elements of the general theory of relativity that are relevant to planetary and deep space health physics applications. In particular, it compiles basic connection coefficients and tensors for a number of representative spacetime geometries. This compilation and associated discussion is intended to facilitate an understanding of a portion of the physics encountered in general relativity, and provides additional insight into various spacetime geometries and their associated physical content. These connection coefficients also facilitate the calculation of geodesics that govern the trajectory of spacecraft in deep space after they leave our Solar System. Connection coefficients also determine the characteristics of wormholes that permit options for traveling more efficiently than the geodesic pathway.

L.2
Basic Curvature Quantities

There are a number of quantities that can be used to describe spacetime geometries. These include the metric tensor, inverse metric tensor, affine connection coefficients or Christoffel symbols, the Riemann curvature tensor, the Ricci tensor, scalar curvature, and the Einstein tensor. Each of these is well defined once the spacetime geometry is specified. For each geometry, a specific coordinate system is provided. The various tensors and connection coefficients are defined in terms of these coordinates.

The metric tensor $g_{\mu\nu}$ is defined in terms of the specified coordinates. From a given metric $g_{\mu\nu}$, we compute the components of the following: the inverse metric, the Christoffel symbols or affine connection coefficients, the Riemann curvature tensor, the Ricci tensor, the scalar curvature, and the Einstein tensor.

The Christoffel symbols are defined in terms of the inverse metric tensor and partial derivatives of the metric tensor:

$$\Gamma^\lambda_{\mu\nu} = \frac{1}{2} g^{\lambda\sigma}(\partial_\mu g_{\sigma\nu} + \partial_\nu g_{\sigma\mu} - \partial_\sigma g_{\mu\nu}), \tag{L.1}$$

where ∂_α stands for the partial derivative $\partial/\partial x^\alpha$, and repeated indexes are summed. An examination of Equation L.1 reveals that the Christoffel symbols are symmetric in the lower two indexes:

$$\Gamma^\lambda_{\mu\nu} = \Gamma^\lambda_{\nu\mu}. \tag{L.2}$$

The Christoffel symbols are uniquely related to the equation for time-like geodesics:

$$\frac{d^2 x^\lambda}{d\tau^2} + \Gamma^\lambda_{\mu\nu} \frac{dx^\mu}{d\tau} \frac{dx^\nu}{d\tau} = 0, \tag{L.3}$$

where x^λ are the coordinates in the four-dimensional basis and τ is the proper time. As noted by Misner et al. (1973) in describing geodesic motion on the Earth: "...the connection coefficients serve as 'turning coefficients' to tell how fast to "turn" the components of a vector in order to keep that vector constant (against the turning influence of the base vectors)."

The Christoffel symbols are also an important ingredient of the equation of geodesic deviation:

$$R^\lambda_{\mu\nu\sigma} = \partial_\nu \Gamma^\lambda_{\mu\sigma} - \partial_\sigma \Gamma^\lambda_{\mu\nu} + \Gamma^\eta_{\mu\sigma} \Gamma^\lambda_{\eta\nu} - \Gamma^\eta_{\mu\nu} \Gamma^\lambda_{\eta\sigma}. \tag{L.4}$$

The quantity $R^\lambda_{\mu\nu\sigma}$ is a rank-four tensor called the Riemann curvature tensor or Riemann curvature. It represents a measure of spacetime curvature.

An examination of Equation L.4 reveals the antisymmetry of the Riemann curvature tensor under the exchange of the first two indexes and the last two indexes:

$$R^\lambda_{\mu\nu\sigma} = -R_\mu{}^\lambda{}_{\nu\sigma}, \tag{L.5}$$

$$R^\lambda_{\mu\nu\sigma} = -R^\lambda_{\mu\sigma\nu}. \tag{L.6}$$

By summing the first and third indexes of the Riemann curvature tensor, the rank-two Ricci tensor $R_{\mu\nu}$ is obtained:

$$R_{\mu\nu} = R^\lambda_{\mu\lambda\nu}. \tag{L.7}$$

The Ricci tensor can be expressed in terms of the Christoffel symbols

$$R_{\mu\nu} = \frac{\partial \Gamma^\gamma_{\mu\nu}}{\partial x^\gamma} - \frac{\partial \Gamma^\gamma_{\mu\gamma}}{\partial x^\nu} + \Gamma^\gamma_{\mu\nu} \Gamma^\delta_{\gamma\delta} - \Gamma^\gamma_{\mu\delta} \Gamma^\delta_{\nu\gamma}. \tag{L.8}$$

An inspection of Equation L.8 reveals that the Ricci tensor is symmetric in μ and ν.

The scalar curvature (R) is defined in terms of the inverse metric and the Ricci tensor:

$$R = g^{\mu\nu} R_{\mu\nu}. \tag{L.9}$$

Finally, the Einstein curvature tensor ($G_{\mu\nu}$) is defined in terms of the Ricci tensor, metric tensor, and the scalar curvature:

$$G_{\mu\nu} = R_{\mu\nu} - \frac{1}{2}g_{\mu\nu}R. \tag{L.10}$$

For completeness, we note that the Einstein curvature tensor, describing the spacetime geometry, is related to the stress–energy tensor $T_{\mu\nu}$:

$$G_{\mu\nu} = 8\pi T_{\mu\nu}, \tag{L.11}$$

where $T_{\mu\nu}$ is the measure of the matter energy density.

In the following section, a summary of connection coefficients and tensors for common spacetime geometries is provided. Only nonzero components are presented, and symmetry properties are utilized to minimize the number of listed components.

L.3
Tensors and Connection Coefficients

A number of commonly encountered spacetime geometries are investigated to illustrate their impact on their associated derived quantities. For each, we provide all nonzero Christoffel symbols, the scalar curvature, and nonzero elements of the Riemann curvature tensor, the Ricci tensor, and the Einstein tensor. These quantities are provided for flat spacetime, the Schwarzchild geometry, the Morris–Thorne (MT) wormhole geometry, the Friedman–Robertson–Walker (FRW) geometry, and a generalized Schwarzchild geometry. In the subsequent discussion, spherical coordinates $\{r, \theta, \varphi, t\}$ are utilized in the description of all spacetime geometries.

The use of spherical coordinates provides internal consistency between the various metrics utilized in this appendix. It is worth noting that a specific orthonormal basis could provide a simpler expression for a highly symmetric metric or one that is not singular in spherical coordinates. However, these bases would depend on the specific spacetime geometry. An orthonormal basis would also simplify the solution the Einstein equation, but solutions of this equation are beyond the scope of this appendix.

Geometrized units are used in the subsequent discussion. These units are convenient for general relativity, and utilize a system in which mass, length, and time all have units of length. In these units, the speed of light and the gravitational constant have unit value.

L.3.1
Flat Spacetime Geometry

The coordinates used to define the flat spacetime geometry are $\{r, \theta, \varphi, t\}$. The metric tensor $(g_{\mu\nu})$ and inverse metric tensor $(g^{\mu\nu})$ are:

metric tensor:

$$g_{\mu\nu} = \begin{bmatrix} 1 & 0 & 0 & 0 \\ 0 & r^2 & 0 & 0 \\ 0 & 0 & r^2\sin^2\theta & 0 \\ 0 & 0 & 0 & -1 \end{bmatrix}, \tag{L.12}$$

inverse metric tensor:

$$g^{\mu\nu} = \begin{bmatrix} 1 & 0 & 0 & 0 \\ 0 & \dfrac{1}{r^2} & 0 & 0 \\ 0 & 0 & \dfrac{\csc^2\theta}{r^2} & 0 \\ 0 & 0 & 0 & -1 \end{bmatrix}. \tag{L.13}$$

For flat spacetime, it is expected that the scalar curvature will be zero. Zero curvature also suggests the tensors associated with its definition (e.g., the Ricci tensor and the Riemann curvature tensor) have few, if any, nonzero elements. In a similar fashion, the Einstein curvature in flat spacetime is expected to have few, if any, nonzero elements. This qualitative argument is supported by calculation of the elements of these tensors.

All the affine coefficients are not expected to be zero. An inspection of the flat spacetime metric suggests that the Christoffel symbols involving t as an index are zero because no metric coefficients are time dependent. Because there is an interrelationship between r, θ, and φ, it is expected that some of the Christoffel symbols having these elements are nonzero. This is in fact the case. A listing of these flat spacetime connection coefficients and tensors is as follows:

Christoffel symbols:

$$\begin{aligned} &\Gamma^r{}_{\theta\theta} = -r \quad \Gamma^r{}_{\varphi\varphi} = -r\sin^2\theta \\ &\Gamma^\theta{}_{\theta r} = \frac{1}{r} \quad \Gamma^\theta{}_{\varphi\varphi} = -\cos\theta\sin\theta \\ &\Gamma^\varphi{}_{\varphi r} = \frac{1}{r} \quad \Gamma^\varphi{}_{\varphi\theta} = \cot\theta. \end{aligned} \tag{L.14}$$

Riemann curvature tensor:

All elements of the Riemann curvature tensor are zero within the flat spacetime geometry:

$$R^\lambda{}_{\mu\nu\sigma} = 0. \tag{L.15}$$

Ricci tensor:

All elements of the Ricci tensor are zero within the flat spacetime geometry:

$$R_{\mu\nu} = 0. \tag{L.16}$$

Scalar curvature:

The scalar curvature is zero within the flat spacetime geometry:

$$R = 0. \tag{L.17}$$

Einstein tensor:

All elements of the Einstein tensor are zero within the flat spacetime geometry:

$$G_{\mu\nu} = 0. \tag{L.18}$$

L.3.2
Schwarzschild Geometry

The simplest curved spacetimes of general relativity are those that are the most symmetric. One of the most useful spacetime geometries is the Schwarzschild geometry that describes empty space outside a spherically symmetric source of curvature (e.g., a spherical star). In addition, the Schwarzschild geometry is a solution of the vacuum Einstein equation or the equation describing spacetime devoid of matter.

The coordinates used to define the Schwarzschild metric are $\{r, \theta, \varphi, t\}$, and the metric tensor and its inverse are as follows:

metric tensor:

$$g_{\mu\nu} = \begin{bmatrix} \dfrac{1}{1-\dfrac{2m}{r}} & 0 & 0 & 0 \\ 0 & r^2 & 0 & 0 \\ 0 & 0 & r^2\sin^2\theta & 0 \\ 0 & 0 & 0 & -1+\dfrac{2m}{r} \end{bmatrix}, \quad (L.19)$$

inverse metric tensor:

$$g^{\mu\nu} = \begin{bmatrix} 1-\dfrac{2m}{r} & 0 & 0 & 0 \\ 0 & \dfrac{1}{r^2} & 0 & 0 \\ 0 & 0 & \dfrac{\csc^2\theta}{r^2} & 0 \\ 0 & 0 & 0 & \dfrac{r}{2m-r} \end{bmatrix}. \quad (L.20)$$

The Schwarzschild metric has the following properties:

- The metric is independent of time.
- The metric is spherically symmetric. The geometry of a surface of constant t and constant r has the symmetries of a sphere of radius r with respect to changes in the angles θ and φ.
- The coordinate r is not the distance from any center. It is related to the area (A) of a two-dimensional sphere of fixed r and t, $r = (A/4\pi)^{1/2}$.
- The constant m can be identified as the total mass of the source of curvature.
- The geometry becomes interesting at $r = 0$ and $r = 2\,\text{m}$. The $r = 2\,\text{m}$ value is called the Schwarzschild radius and is the characteristic length scale for curvature in the Schwarzschild geometry. The surface of a static star (i.e., a star not undergoing gravitational collapse) lies well outside $r = 0$ and $r = 2\,\text{m}$.

- At large r ($r \gg 2\,m$), the Schwarzschild spacetime approaches flat spacetime.
- For small m ($m \to 0$), the Schwarzschild spacetime approaches flat spacetime.

An examination of Equations L.19 and L.20 indicates that the Schwarzschild geometry is identical to flat spacetime in the limit that $m \to 0$ or $r \gg 2\,m$ (Equations L.12 and L.13). This limit serves as a natural check on the affine connection coefficients and curvature tensors presented below.

Christoffel symbols:

$$\begin{aligned}
&\Gamma^{r}{}_{rr} = \frac{m}{2mr - r^2} & &\Gamma^{\theta}{}_{\theta r} = \frac{1}{r} \\
&\Gamma^{r}{}_{\theta\theta} = 2m - r & &\Gamma^{\theta}{}_{\varphi\varphi} = -\cos\theta \sin\theta \\
&\Gamma^{r}{}_{\varphi\varphi} = (2m - r)\sin^2\theta & &\Gamma^{\varphi}{}_{\varphi r} = \frac{1}{r} \\
&\Gamma^{r}{}_{tt} = \frac{m(-2m + r)}{r^3} & &\Gamma^{\varphi}{}_{\varphi\theta} = \cot\theta. \\
&\Gamma^{t}{}_{tr} = \frac{m}{-2mr + r^2} & &
\end{aligned} \qquad (\text{L.}21)$$

Examination of Equation L.21 supports the requirement that the Christoffel symbols derived from the Schwarzschild geometry reduce to those derived from flat spacetime in the $m \to 0$ limit or the $r \gg 2\,m$ limit.

Riemann curvature tensor:

The Riemann curvature tensor has a number of nonzero elements within the Schwarzschild geometry. In the $m \to 0$ limit or the $r \gg 2\,m$ limit, the flat spacetime results (Equation L.15) are obtained. The nonzero Schwarzschild Riemann curvature tensor elements are

$$\begin{aligned}
&R^{r}{}_{\theta\theta r} = \frac{m}{r} & &R^{\varphi}{}_{r\varphi r} = \frac{m}{(2m - r)r^2} \\
&R^{r}{}_{\varphi\varphi r} = \frac{m \sin^2\theta}{r} & &R^{\varphi}{}_{\theta\varphi\theta} = \frac{2m}{r} \\
&R^{r}{}_{ttr} = \frac{2m(-2m + r)}{r^4} & &R^{\varphi}{}_{tt\varphi} = \frac{m(2m - r)}{r^4} \\
&R^{\theta}{}_{r\theta r} = \frac{m}{(2m - r)r^2} & &R^{t}{}_{rtr} = \frac{2m}{r^2(-2m + r)} \\
&R^{\theta}{}_{\varphi\varphi\theta} = -\frac{2m \sin^2\theta}{r} & &R^{t}{}_{\theta t\theta} = -\frac{m}{r} \\
&R^{\theta}{}_{tt\theta} = \frac{m(2m - r)}{r^4} & &R^{t}{}_{\varphi t\varphi} = -\frac{m \sin^2\theta}{r}
\end{aligned} \qquad (\text{L.}22)$$

Ricci tensor:

All elements of the Ricci tensor are zero within the Schwarzschild geometry:

$$R_{\mu\nu} = 0. \qquad (\text{L.}23)$$

Scalar curvature:

The scalar curvature is zero within the Schwarzschild geometry:

$$R = 0. \tag{L.24}$$

Einstein tensor:
All elements of the Einstein tensor are zero within the Schwarzschild geometry:

$$G_{\mu\nu} = 0. \tag{L.25}$$

The Schwarzschild geometry exhibits a discontinuity as $r \to 2\,m$. This condition may be viewed as a Schwarzschild wormhole or conduit that connects two distinct regions of a single asymptotically flat universe.

Other spacetime geometries also exhibit wormhole characteristics. Accordingly, we will further pursue the wormhole concept in the following section.

L.3.3
MT Wormhole Geometry

To further illustrate the wormhole concept, the Morris–Thorne wormhole geometry is reviewed. The coordinates used to define the MT wormhole geometry are $\{r, \theta, \varphi, t\}$, and the metric tensor and its inverse are:

metric tensor:

$$g_{\mu\nu} = \begin{bmatrix} 1 & 0 & 0 & 0 \\ 0 & b^2+r^2 & 0 & 0 \\ 0 & 0 & (b^2+r^2)\sin^2\theta & 0 \\ 0 & 0 & 0 & -1 \end{bmatrix}, \tag{L.26}$$

inverse metric tensor:

$$g^{\mu\nu} = \begin{bmatrix} 1 & 0 & 0 & 0 \\ 0 & \dfrac{1}{b^2+r^2} & 0 & 0 \\ 0 & 0 & \dfrac{\csc^2\theta}{b^2+r^2} & 0 \\ 0 & 0 & 0 & -1 \end{bmatrix}, \tag{L.27}$$

where b is a constant having the dimensions of length. An examination of the wormhole geometry indicates that it reduces to flat spacetime (Equations L.12 and L.13) in the limit $b \to 0$.

At the present time, the MT wormhole geometry does not represent a physically realistic spacetime. Except for the $b = 0$ metric, the geometry is not flat but is curved. For $b \neq 0$, an embedding of the (r, φ) slice of the wormhole geometry produces a surface with two asymptotically flat regions connected by a region of minimum radius b. This region resembles a tunnel or wormhole connecting the two asymptotically flat regions.

Appendix L

An insight into the geometry of the MT wormhole can be gained if the spherical symmetry and static nature of the metric are considered. For simplicity, the discussion is limited to the equatorial plane ($\theta = \pi/2$) at a fixed instant of time. Using the coordinate transformation:

$$R^2 = b^2 + r^2, \tag{L.28}$$

the metric in the plane $\theta = \pi/2$, $t = $ constant is

$$d\sigma^2_{2-\text{surface}} = \frac{1}{1-\left(\frac{b}{R}\right)^2} dR^2 + R^2 d\varphi^2. \tag{L.29}$$

This 2-surface can be imbedded in a three-dimensional Euclidean space which is represented by the cylindrical coordinates (R, φ, z) by identifying this surface with the surface $z = z(R)$. The metric of the surface in Euclidean space is written as

$$d\sigma^2_{\text{Euclidean}} = \left[1 + \left(\frac{dz}{dR}\right)^2\right] dR^2 + R^2 d\varphi^2. \tag{L.30}$$

The comparison of Equations L.29 and L.30 and integration with respect to R leads to the shape of the embedding diagram

$$z(R) = \pm b \ln\left[\frac{R}{b} + \left(\left(\frac{R}{b}\right)^2 - 1\right)^{1/2}\right]. \tag{L.31}$$

The embedding space has no physical meaning. The structure of Equation L.31 is an upper universe connected by a throat of radius b to a lower universe. The impression of a tube (throat) suggested by Equation L.31 is misleading. There is no tube in spacetime, because the regions with radial coordinate $R < b$ are not part of the spacetime. The throat has a spherical topology and becomes important only for geodesics that spiral in the direction of decreasing R (like water flowing down a drain).

The wormhole geometry cannot be produced from smooth distortions of flat spacetime. The creation of a wormhole geometry not only has a different geometry from flat spacetime but also a different topology.

In addition to the previous discussion, the MT wormhole metric has the following properties:

- The metric is independent of time.
- The metric is spherically symmetric because a surface of constant r and t has the geometry of a sphere.
- At very large r ($r \gg b$), the MT spacetime approaches flat spacetime.

The MT wormhole geometry also reduces to flat spacetime in the $b \to 0$ limit. This consistency check is indeed observed for the MT wormhole connection coefficients and curvature tensors presented below.

Christoffel symbols:

$$\Gamma^r{}_{\theta\theta} = -r \qquad \Gamma^r{}_{\varphi\varphi} = -r\sin^2\theta$$

$$\Gamma^\theta{}_{\theta r} = \frac{r}{b^2+r^2} \qquad \Gamma^\theta{}_{\varphi\varphi} = -\cos\theta\sin\theta \qquad \text{(L.32)}$$

$$\Gamma^\varphi{}_{\varphi r} = \frac{r}{b^2+r^2} \qquad \Gamma^\varphi{}_{\varphi\theta} = \cot\theta.$$

Riemann curvature tensor:

$$R^r{}_{\theta\theta r} = \frac{b^2}{b^2+r^2} \qquad R^r{}_{\varphi\varphi r} = \frac{b^2\sin^2\theta}{b^2+r^2}$$

$$R^\theta{}_{r\theta r} = -\frac{b^2}{(b^2+r^2)^2} \qquad R^\theta{}_{\varphi\varphi\theta} = -\frac{b^2\sin^2\theta}{b^2+r^2} \qquad \text{(L.33)}$$

$$R^\varphi{}_{r\varphi r} = -\frac{b^2}{(b^2+r^2)^2} \qquad R^\varphi{}_{\theta\varphi\theta} = \frac{b^2}{b^2+r^2}.$$

Ricci tensor:
Only the R_{rr} element is nonzero within the MT wormhole geometry:

$$R_{rr} = -\frac{2b^2}{(b^2+r^2)^2}. \qquad \text{(L.34)}$$

Scalar curvature:
The scalar curvature is nonzero within the MT wormhole geometry:

$$R = -\frac{2b^2}{(b^2+r^2)^2}. \qquad \text{(L.35)}$$

Einstein tensor:
The diagonal elements of the Einstein tensor are nonzero within the MT wormhole geometry:

$$G_{rr} = -\frac{b^2}{(b^2+r^2)^2}$$

$$G_{\theta\theta} = \frac{b^2}{b^2+r^2}$$

$$G_{\varphi\varphi} = \frac{b^2\sin^2\theta}{b^2+r^2} \qquad \text{(L.36)}$$

$$G_{tt} = -\frac{b^2}{(b^2+r^2)^2}.$$

L.3.4
Generalized Schwarzchild Geometry

The coordinates used to define the generalized Schwarzchild geometry are $\{r, \theta, \varphi, t\}$. The rr and tt metric tensor elements of the generalized Schwarzchild geometry are functions of r, namely exponential functions of $\lambda(r)$ and $\varphi(r)$. The metric tensor and its inverse are the following:

metric tensor:

$$g_{\mu\nu} = \begin{bmatrix} e^{2\lambda(r)} & 0 & 0 & 0 \\ 0 & r^2 & 0 & 0 \\ 0 & 0 & r^2\sin^2\theta & 0 \\ 0 & 0 & 0 & -e^{2\varphi(r)} \end{bmatrix}, \qquad (L.37)$$

inverse metric tensor:

$$g^{\mu\nu} = \begin{bmatrix} e^{-2\lambda(r)} & 0 & 0 & 0 \\ 0 & \dfrac{1}{r^2} & 0 & 0 \\ 0 & 0 & \dfrac{\csc^2\theta}{r^2} & 0 \\ 0 & 0 & 0 & -e^{-2\varphi(r)} \end{bmatrix}. \qquad (L.38)$$

In the subsequent discussion, the derivative with respect to r is indicated by a prime. That is $\lambda' = d\lambda/dr$; similarly, $\varphi' = d\varphi/dr$.

The generalized Schwarzchild geometry reduces to the flat spacetime geometry in the limit $\lambda(r) \to 0$ and $\varphi(r) \to 0$. This consistency check is verified by examining the tensors and connection coefficients noted below.

Christoffel symbols:

$$\begin{aligned}
&\Gamma^r{}_{rr} = \lambda'(r) &&\Gamma^r{}_{\theta\theta} = -re^{-2\lambda(r)} \\
&\Gamma^r{}_{\varphi\varphi} = -e^{-2\lambda(r)} r\sin^2\theta &&\Gamma^r{}_{tt} = e^{-2\lambda(r)+2\varphi(r)}\varphi'(r) \\
&\Gamma^\theta{}_{\theta r} = \frac{1}{r} &&\Gamma^\theta{}_{\varphi\varphi} = -\cos\theta\sin\theta \\
&\Gamma^\varphi{}_{\varphi r} = \frac{1}{r} &&\Gamma^\varphi{}_{\varphi\theta} = \cot\theta. \\
&\Gamma^t{}_{tr} = \varphi'(r)
\end{aligned} \qquad (L.39)$$

Riemann curvature tensor:

$$\begin{aligned}
&R^r{}_{\theta\theta r} = -e^{-2\lambda(r)}r\lambda'(r) &&R^r{}_{\varphi\varphi r} = -e^{-2\lambda(r)}r\sin^2\theta\,\lambda'(r) \\
&R^r{}_{ttr} = e^{-2\lambda(r)+2\varphi(r)}(\lambda'(r)\varphi'(r)-\varphi'^2(r)-\varphi''(r)) \\
&R^\theta{}_{r\theta r} = \frac{\lambda'(r)}{r} &&R^\theta{}_{\varphi\varphi\theta} = (-1+e^{-2\lambda(r)})\sin^2\theta \\
&R^\theta{}_{tt\theta} = -\frac{e^{-2\lambda(r)+2\varphi(r)}\varphi'(r)}{r} &&R^\varphi{}_{r\varphi r} = \frac{\lambda'(r)}{r} \\
&R^\varphi{}_{\theta\varphi\theta} = 1-e^{-2\lambda(r)} &&R^\varphi{}_{tt\varphi} = -\frac{e^{-2\lambda(r)+2\varphi(r)}\varphi'(r)}{r} \\
&R^t{}_{rtr} = \lambda'(r)\varphi'(r)-\varphi'^2(r)-\varphi''(r) &&R^t{}_{\theta t\theta} = -e^{-2\lambda(r)}r\varphi'(r). \\
&R^t{}_{\varphi t\varphi} = -e^{-2\lambda(r)}r\sin^2\theta\,\varphi'(r)
\end{aligned}$$

$$(L.40)$$

Ricci tensor:
Only the diagonal Ricci tensor elements are nonzero within the generalized Schwarzchild geometry:

$$
\begin{aligned}
R_{rr} &= \frac{\lambda'(r)(2+r\varphi'(r))-r(\varphi'^2(r)+\varphi''(r))}{r} \\
R_{\theta\theta} &= e^{-2\lambda(r)}(-1+e^{2\lambda(r)}+r\lambda'(r)-r\varphi'(r)) \\
R_{\varphi\varphi} &= e^{-2\lambda(r)}\sin^2\theta(-1+e^{2\lambda(r)}+r\lambda'(r)-r\varphi'(r)) \\
R_{tt} &= \frac{e^{-2\lambda(r)+2\varphi(r)}((2-r\lambda'(r))\varphi'(r)+r\varphi'^2(r)+r\varphi''(r))}{r}.
\end{aligned}
\tag{L.41}
$$

Scalar curvature:
The scalar curvature is nonzero within the generalized Schwarzchild geometry:

$$
R = \frac{1}{r^2}(2e^{-2\lambda(r)}(-1+e^{2\lambda(r)}-2r\varphi'(r)-r^2\varphi'^2(r)+r\lambda'(r)(2+r\varphi'(r))-r^2\varphi''(r)).
\tag{L.42}
$$

Einstein tensor:
The diagonal elements of the Einstein tensor are nonzero within the generalized Schwarzchild geometry:

$$
\begin{aligned}
G_{rr} &= \frac{1-e^{2\lambda(r)}+2r\varphi'(r)}{r^2} \\
G_{\theta\theta} &= e^{-2\lambda(r)}r(\varphi'(r)+r\varphi'^2(r)-\lambda'(r)(1+r\varphi'(r))+r\varphi''(r)) \\
G_{\varphi\varphi} &= e^{-2\lambda(r)}r\sin^2\theta(\varphi'(r)+r\varphi'^2(r)-\lambda'(r)(1+r\varphi'(r))+r\varphi''(r)) \\
G_{tt} &= \frac{e^{-2\lambda(r)+2\varphi(r)}(-1+e^{2\lambda(r)}+2r\lambda'(r))}{r^2}.
\end{aligned}
\tag{L.43}
$$

L.3.5
Friedman–Robertson–Walker (FRW) Geometry

The FRW geometry describes the time evolution of a homogeneous, isotropic space that expands in time as $a(t)$ increases and contracts as $a(t)$ decreases. The function $a(t)$ contains all information about the temporal evolution of the universe.

In addition to the scaling factor $a(t)$, a constant k is included in the FRW metric. The constant k determines the classification of the universe (i.e., $k=+1$ indicates a closed universe, $k=0$ indicates a flat universe, and $k=-1$ indicates an open universe). Although the conventional terminology flat, closed, and open are used to distinguish the three possible homogeneous and isotropic geometries of space, it is more physical to distinguish these features in terms of their spatial curvature.

Homogeneity requires that the spatial curvature be the same at each point in these geometries. The flat case has zero spatial curvature everywhere. The closed and open cases have constant positive and constant negative curvature, respectively.

Following the previous discussion of the MT wormhole geometry, embedding can be constructed for the possible homogeneous and isotropic geometries for the FRW metric. If the $t=$ constant, $\theta=\pi/2$ 2-surface is considered, the flat and closed embedding diagrams correspond to a plane and a sphere, respectively. These are constant zero-curvature and positive-curvature surfaces, respectively. A $t=$ constant, $\theta=\pi/2$ slice of the open FRW geometry cannot be embedded as an axisymmetric surface in the flat three-dimensional space. That surface has a constant negative curvature.

The coordinates used to define the FRW geometry are $\{r, \theta, \varphi, t\}$, and the metric tensor and its inverse are the following:

metric tensor:

$$g_{\mu\nu} = \begin{bmatrix} \frac{a^2(t)}{1-kr^2} & 0 & 0 & 0 \\ 0 & r^2 a^2(t) & 0 & 0 \\ 0 & 0 & r^2 a^2(t)\sin^2\theta & 0 \\ 0 & 0 & 0 & -1 \end{bmatrix}, \quad (L.44)$$

inverse metric tensor:

$$g^{\mu\nu} = \begin{bmatrix} \frac{1-kr^2}{a^2(t)} & 0 & 0 & 0 \\ 0 & \frac{1}{r^2 a^2(t)} & 0 & 0 \\ 0 & 0 & \frac{\csc^2\theta}{r^2 a^2(t)} & 0 \\ 0 & 0 & 0 & -1 \end{bmatrix}. \quad (L.45)$$

The FRW geometry reduces to the flat spacetime geometry in the limit $a \to 1$ and $k \to 0$. This consistency check is verified by examining the tensors and connection coefficients noted below.

Christoffel symbols:

$$\Gamma^r_{rr} = \frac{kr}{1-kr^2}$$

$$\Gamma^r_{\theta\theta} = (-1+kr^2)r \quad \Gamma^r_{\varphi\varphi} = (-1+kr^2)r\sin^2\theta$$

$$\Gamma^r_{tr} = \frac{\dot{a}(t)}{a(t)} \quad \Gamma^\theta_{\theta r} = \frac{1}{r}$$

$$\Gamma^\theta_{\varphi\varphi} = -\cos\theta\sin\theta \quad \Gamma^\theta_{t\theta} = \frac{\dot{a}(t)}{a(t)} \quad (L.46)$$

$$\Gamma^\varphi_{\varphi r} = \frac{1}{r} \quad \Gamma^\varphi_{\varphi\theta} = \cot\theta$$

$$\Gamma^\varphi_{t\varphi} = \frac{\dot{a}(t)}{a(t)} \quad \Gamma^t_{rr} = \frac{a(t)\dot{a}(t)}{1-kr^2}$$

$$\Gamma^t_{\theta\theta} = r^2 a(t)\dot{a}(t) \quad \Gamma^t_{\varphi\varphi} = r^2 a(t)\dot{a}(t)\sin^2\theta.$$

Riemann curvature tensor:

$$R^r{}_{\theta\theta r} = -r^2(\dot{a}^2(t)+k) \qquad R^r{}_{\varphi\varphi r} = -r^2(\dot{a}^2(t)+k)\sin^2\theta$$

$$R^r{}_{ttr} = \frac{\ddot{a}(t)}{a(t)} \qquad R^\theta{}_{r\theta r} = \frac{\dot{a}^2(t)+k}{1-kr^2}$$

$$R^\theta{}_{\varphi\varphi\theta} = -r^2(k+\dot{a}^2(t))\sin^2\theta$$

$$R^\theta{}_{tt\theta} = \frac{\ddot{a}(t)}{a(t)} \qquad R^\varphi{}_{r\varphi r} = \frac{\dot{a}^2(t)+k}{1-kr^2} \qquad \text{(L.47)}$$

$$R^\varphi{}_{\theta\varphi\theta} = r^2(k+\dot{a}^2(t)) \qquad R^\varphi{}_{tt\varphi} = \frac{\ddot{a}(t)}{a(t)}$$

$$R^t{}_{rtr} = \frac{a(t)\ddot{a}(t)}{1-kr^2} \qquad R^t{}_{\theta t\theta} = r^2 a(t)\ddot{a}(t).$$

$$R^t{}_{\varphi t\varphi} = r^2 a(t)\ddot{a}(t)\sin^2\theta$$

Ricci tensor:
Only the diagonal Ricci tensor elements are nonzero within the FRW geometry:

$$R_{rr} = \frac{2k+2\dot{a}^2(t)+a(t)\ddot{a}(t)}{1-kr^2}$$
$$R_{\theta\theta} = r^2(2k+2\dot{a}^2(t)+a(t)\ddot{a}(t))$$
$$R_{\varphi\varphi} = r^2(2k+2\dot{a}^2(t)+a(t)\ddot{a}(t))\sin^2\theta \qquad \text{(L.48)}$$
$$R_{tt} = -\frac{3\ddot{a}(t)}{a(t)}.$$

Scalar curvature:
The scalar curvature is nonzero within the FRW geometry:

$$R = \frac{6(k+\dot{a}^2(t)+a(t)\ddot{a}(t))}{a^2(t)}. \qquad \text{(L.49)}$$

Einstein tensor:
The diagonal elements of the Einstein tensor are nonzero within the FRW geometry:

$$G_{rr} = \frac{k+\dot{a}^2(t)+2a(t)\ddot{a}(t)}{(-1+kr^2)}$$
$$G_{\theta\theta} = -r^2(k+\dot{a}^2(t)+2a(t)\ddot{a}(t))$$
$$G_{\varphi\varphi} = -r^2\sin^2\theta(k+\dot{a}^2(t)+2a(t)\ddot{a}(t)) \qquad \text{(L.50)}$$
$$G_{tt} = \frac{3(k+\dot{a}^2(t))}{a^2(t)}.$$

L.4 Conclusions

Using spherical coordinates, the affine connection coefficients, the Riemann curvature tensor, the Ricci tensor, scalar curvature, and the Einstein tensor are determined for flat spacetime, the Schwarzchild geometry, the Morris–Thorne wormhole geometry, the Friedman–Robertson–Walker geometry, and a generalized Schwarzchild geometry. This approach provides a logical and consistent treatment of the basic quantities and spacetime geometries associated with general relativity, gravitation, and differential geometry. In addition, the approach provides a physical description of these quantities and their interrelationships that will be useful in obtaining geodesics and spacetime properties relevant to twenty-first century health physics applications.

References

Abdulezer, L. (2006) The Einstein Field Equations and Their Applications to Stellar Models (using Mathematica), http://www.evolvingtech.com/etc/math/GR1ma.pdf (accessed on November 22, 2006).

Bevelacqua, J.J. (2006) Curvature systematics in general relativity. *FIZIKA*, **A15**, 133.

Carroll, S.M. (2004) *An Introduction to General Relativity: Spacetime and Geometry*, Addison-Wesley, New York.

Friedmann, A. (1922) Über die Krümmung des Raumes. *Zeitschrift Fur Physik*, **10**, 377.

Fuller, R.W. and Wheeler, J.A. (1962) Causality and multiply connected spacetime. *Physical Reviews*, **128**, 919.

Hartle, J.B. (2003) *Gravity: An Introduction to Einstein's General Relativity*, Addison-Wesley, New York.

Misner, C.W., Thorne, K.S. and Wheeler, J.A. (1973) *Gravitation*, Freeman, San Francisco.

Morris, M.S. and Thorne, K.S. (1988) Wormholes and supersymmetry. *American Journal of Physics*, **56**, 395.

Müller, T. (2004) Visual appearance of a Morris–Thorne wormhole. *American Journal of Physics*, **72**, 1045.

Schwarzschild, K. (1916) On the gravitational field of a mass point according to Einstein's theory, Sitzungsberichte der Königlich Preussischen Akademie der Wissenschaften 1, 189.

Schwarzschild, K. (1916) On the gravitational field of a sphere of incompressible fluid according to Einstein's theory. *Sitzber Deut Akad Wiss Berlin, Kl Math -Phys Tech.*, 424.

Visser, M. (1995) *Lorentzian Wormholes: From Einstein to Hawking*, American Institute of Physics, Woodbury, NY.

Index

a

abnormal operations 61
absorbed dose 57–58, 101, 234, 264–269, 271–273, 280, 397–400, 427–431, 486–487
accelerator 1, 3, 4, 94, 131, 133–143, 145, 147, 149, 154–155, 157–158, 160–172, 175–178, 180–192, 194–195, 199, 202–204, 215–216, 218, 221, 225–227, 229–232, 234–238, 241, 243, 386, 388–390, 395, 416, 461–463, 468, 497, 509, 512–513, 515, 519, 523, 525–527, 531
– beam loss events 139, 152, 157, 159, 163
accumulator ring 138, 139, 202
activation
– air 142, 163, 179–180, 184, 227
– soil 142, 163, 179–180, 184, 227
– water 142, 163, 179–180, 184, 227
activation product 10, 17, 27–28, 42–43, 47–49, 85, 103–105, 238
active safety system 22, 28–29
advanced boiling water reactor (ABWR) 22, 27, 64
advanced Canadian deuterium reactor (ACR) 22, 24
advanced passive pressurized water reactor (APPWR) 22
advanced pressurized water reactor (APWR) 22–23, 62–63
affine connection coefficients 539, 544, 552
air ejector 16, 17, 368
ALARA 21, 49, 59, 61–62, 71, 101–103, 117–120, 167, 178
ALICE 151–152
Alpha Centauri 303, 319, 321–322, 327, 333–334
alpha particles 94, 100–101, 110, 325, 370
aluminum 184–185, 188, 286–288, 298–299, 331–332, 345
^{241}Am 64, 358
amplification factor 232–233
Andromeda Galaxy 311, 314–315

annihilation 152–153
ANSI-Z136.1 235, 239–240, 411–413
anthracene 110
antimatter annihilation propulsion 323
antineutrons 163, 439
antiprotons 150, 152–153, 323, 513
aperture 165, 239, 242–243
Apollo lunar mission 263, 266–267
^{41}Ar 48, 99
argon 103, 378
asymptotic freedom 500
ATLAS 151
attenuation length 167, 172–173
^{197}Au 187–188, 535–536
^{198}Au 187
auxiliary building 36, 51, 60
auxiliary system accident 90
aversion response 240

b

^{11}B 72, 325
Barnard's Star 307, 321
barred spiral galaxy 316
baryon 497–500, 504–507
^{7}Be 72, 96, 98, 184–185, 188, 235, 414
^{8}Be 254, 304
beam containment system 190
beam emittance 203–204, 222, 420
beam line 157, 158, 227, 238
beam loss event 139, 157, 159–160, 163, 236–237
beam stop 182, 184, 238
bending magnet 202–206, 209, 234
beta 324, 364, 377–379, 400–401, 503–504
beyond design basis event 49, 53–55, 90
BF_3 detector 111, 189
big bang 257, 304, 317
binary star 305
bioassay 65–66, 109, 125, 377
biological half-life 467
biomedical intervention 334–335

blackbody radiation 317
black dwarf 305
black hole 306, 311–312, 321
blanket assembly 98, 103
blood forming organs 252, 270, 392
blowdown 16
– radiation monitor 16
blue shift 336–337
boiling water reactor 9, 17, 95, 126
boson 165, 181, 498, 501–502
Bragg peak 532, 534–535
BRAHMS 158
bremsstrahlung 101, 137, 141–143, 145, 199–201, 203, 515, 517
brightness 199, 203–208, 315–316
buildup factor 58, 64, 348, 405, 450
buildup time 221, 419–420
bunching 216–218, 223

c

^{11}C 98, 184–185, 235
^{12}C 254, 292, 304, 310, 535
^{47}Ca 34, 127
calorimetry 113
Canadian deuterium (CANDU) reactor 13, 22, 94, 126
carbon 259, 304–305, 537
carcinogenic inhibitors 334
centrifugal force 480
CERN 134, 142, 149, 160–161, 174, 181, 207
charge
– color 498, 501–502, 504, 506, 507
– electric 74, 75, 500–507
– weak 498, 501–502, 504
charged current 165, 167
Christoffel symbol 321, 539–542, 544, 547–548, 550
circular collider 133, 141, 168, 175
classical mechanics 4, 473
CMS 151
^{56}Co 185, 414
^{57}Co 13, 47, 185, 414
^{58}Co 10, 13, 21, 47–49, 97, 99, 103, 127
^{59}Co 47, 64, 97, 356–357, 468–469
^{60}Co 10, 11, 21, 47–49, 64, 66, 97, 103, 105, 126, 357, 381
cobalt 48–49, 64, 305–306
Code of Federal Regulations 66, 117, 365
– 10CFR20 65, 118, 493
– 10CFR835 493
cold fusion 121, 122
collider 133–135, 138, 140–144, 149–151, 158–183, 192–194
colliding beam accelerator 191

committed dose equivalent 65, 484–485, 491–493
composition measurement 112, 113
Compton backscatter 199, 227–228
condenser 16, 17, 29, 67, 368
– air ejector 46, 354, 368
consequence level 236–237
containment building 17, 26, 45, 50–51, 83
continuous-slowing-down-approximation 516
control rod 16–18, 51
core 10, 15–19, 25–29, 31–37, 45, 47, 51, 55–57, 157, 271, 304–307, 310, 312, 316, 329
coronal mass ejection 257
Coulomb barrier 136, 190, 384
Coulomb force 77
Coulomb interaction 254
Coulomb losses 518
^{50}Cr 105
^{51}Cr 104–106, 185
critical energy 143–144, 147–149, 408
criticality 60, 84, 527
critical wavelength 207, 241, 406–408
cross section 44, 73, 146, 169–170, 187–188, 371, 503, 508, 537
– macroscopic 331, 536–537
– microscopic 331, 468–469, 536–537
^{137}Cs 12, 47, 48
CsI(Tl) detector 110, 111
^{60}Cu 185
^{61}Cu 185
^{62}Cu 185
^{63}Cu 105, 535–536
^{64}Cu 105, 106, 185, 235
cumulated activity 484, 486
curvature 320, 539, 543, 544, 546

d

D_2 113, 118
damping ring 161–163
dark energy 160, 163, 181, 311–312, 441
dark matter 160, 163, 181, 311–312, 441
D-D fusion 73, 84, 119, 122
Debye length 77, 78
decontamination factor 62–63, 66, 354–355, 366
deep inelastic nuclear scattering 518
deep space 255, 259, 317–318, 339
– mission 249, 251, 255, 260, 267–268, 270, 307, 318–319, 321, 334
defense-in-depth system 140
deflector efficiency 294, 427
demineralizer 15–16, 43, 44, 67, 361
demonstration fusion reactor 84, 102
design basis event 17, 49, 52–54, 57, 87, 92

design feature 118, 120, 446
differential geometry 320, 552
diffused junction detector 110, 111
dipole 205, 207, 210, 407
dipole magnet 147, 157, 162, 205, 241
dispersion factor 92, 375, 476
distributed source 80, 81, 125, 375
divergence angle 165–166, 222, 412, 420
Doppler effect 314
dose equivalent 65–66, 95, 236–237, 252–253, 261–263, 268–274, 278–288, 294, 296, 363, 491–493, 523–524, 535
dose factor 46, 65, 66, 127, 193, 386, 523
dosimeter 184, 237, 265–268
DT 324, 330–332, 345, 370, 439, 445
D-T fusion 73, 79–81, 94–97, 100, 119
DTO 85
DWUCK code 537
Dyson sphere 326–327

e

eccentricity 281
economic and simplified boiling water reactor (ESBWR) 22
effective dose 10, 59, 65, 66, 80, 97
effective removal rate 484
Einstein tensor 539, 541, 545, 547, 551
electric field 73, 74, 161–162, 217–219, 229–230, 478
electric force 291, 478
electromagnetic cascade 142–147, 155, 190, 526
electromagnetic deflection 269, 288–289, 440–441
– engineering requirements 295
– field radial requirements 293
– field requirements 292
electromagnetic field 121, 215, 258, 282, 323, 397
electromagnetic force 89, 498, 502, 503
electromagnetic interaction 501–503, 505, 507
electron 75, 78, 83, 89, 103, 131, 133, 139–155, 160–164, 183, 187, 199–231, 234, 239–245, 257–259, 271, 278, 305–306, 323, 370, 388–389, 397, 403–405, 414, 421, 424–425, 434, 497–499, 508, 513, 515, 517, 525–526, 531–532, 534
electron accelerator 155, 183, 216, 231, 234
Electron Gamma Shower (EGS) code 405, 525
electron–positron collider 133, 140–144, 149, 163–164, 300
elliptical galaxy 315–316
emergency core cooling system 17, 19
emergency operations 56, 61

emittance 222, 519–520
energy 142, 143, 299, 304–305, 307, 309–311, 397, 434, 487, 536
energy acceptance 223–224, 421
energy level 136, 233, 339–340
energy loss 143, 144, 515, 518
energy recovery linac 204
engineered safety system 17
Epsilon Eridani 321
European pressurized water reactor (EPWR) 22
Evaluated Nuclear Data File (ENDF) library 525
evaporation 17, 152, 154–155
event horizon 306, 311
exposure time 178, 239
external bremsstrahlung 143, 145
external dose 40, 114, 360
external hazard 38, 234, 440, 441
extranuclear cascade 146, 156
extra spatial dimensions 253, 339
extravehicular activity 261, 295, 334
extremely unlikely events 92, 93, 319
eye dose 253, 273, 274, 491
– limit 274

f

^{18}F 185
faster than light travel 336
^{54}Fe 47, 104, 105, 356
^{55}Fe 48, 99, 103, 414
^{56}Fe 124, 306, 310, 371, 414
^{58}Fe 13, 47, 243
^{59}Fe 13, 106, 185, 243, 356, 414
Fermi National Accelerator Laboratory 135, 462, 463
– CDF detector 397
fermion 181, 498, 508
filter 41
fission 7, 9–22, 24, 30, 32, 36, 38, 41, 43, 47, 51, 232, 323, 378, 379, 401
fission chamber 100, 140, 184
fission gas 11, 12, 43, 51
fission neutron 10, 42, 97, 102, 107, 195, 378
fission power reactor 9, 101, 107, 123, 384, 525
fission product 50–52, 55, 224, 361, 525
fission product barrier 29, 30, 36, 38, 45–48, 50–52
fission propulsion 324, 325, 330
fixed target accelerator 133
flat space–time 314, 320, 321, 338, 339, 347, 448, 541, 542, 544, 546, 548, 550, 552
flavor 500
fluence 187, 259, 272, 298, 299, 345, 347, 524

FLUKA code package 526
folding space–time 336, 338
free electron laser 4, 131, 134, 199, 202, 204, 205, 211, 215, 218, 224–226, 228, 231, 239, 241, 421
Friedman–Robertson–Walker geometry 541, 549, 552
fuel handling accident 49, 51, 53
full width at half-maximum 204, 534, 535
fusion 7, 71–76, 79–81, 83–85, 87, 88, 94, 95, 101–103, 106, 107, 109, 118, 121–123, 125, 304, 310, 322–324, 327, 345, 375, 378, 379
fusion accident scenario 117,
fusion gamma 115, 118
fusion neutron 97, 103, 104, 115, 118–119, 379
fusion power reactor 75, 101, 102, 109, 110, 114, 117, 120
fusion process 71, 72, 79–85, 97, 101, 102, 121, 123, 304, 324, 375, 377
fusion product 71, 84, 87, 97, 109, 115–117
fusion propulsion 324, 325, 326

g

gain 24, 116, 217, 220, 221, 223–225, 231, 240, 245, 418–421, 453
galactic cosmic radiation 252, 259, 296, 300, 318, 321, 329, 440
– doses 269
galaxy 282, 299, 303, 307, 312, 314, 316–318, 321, 326, 462
Galileo 339, 456, 462
gamma constant 62, 124, 126, 344, 345, 367
gamma-ray 20, 42, 193, 194, 195, 202, 224, 226, 228, 232, 312, 313, 314, 342, 347, 385, 443, 528
gamma-ray burst 309, 312–314, 318, 327, 441
gamma-ray free electron laser 202, 226
gamma-ray laser 199, 227, 232, 233
gamma spectrometer 187
gas bremsstrahlung 234
gas-cooled fast reactors (GFR) 30–32
gas turbine–modular helium reactor (GT-MHR) 22
Gaussian plume model 127, 383, 384
Geiger-Mueller detector 392
Gemini mission 263–265, 267
generalized Schwarzschild geometry 541, 547, 548, 549, 552
general relativity 3, 320, 336, 539, 541, 543, 552
generation I reactor 13
generation II reactor 13, 17, 20, 22, 37, 57
generation III reactor 17, 20, 22, 24, 26, 52, 56
generation IV reactor 30, 38, 40, 50, 368
genetic enhancement 334, 335

genetic screening 334, 335
geodesic 319, 320, 336, 345, 448, 450, 539, 540, 546
gluon 151, 152, 461, 501–503, 505, 507
GRASER 199, 232–234
gravitational anomaly 318
gravitational collapse 310
gravitational force 264, 305, 314, 315, 503
gravitational interaction 253, 275, 304, 305, 311, 317, 339, 501, 503
gray equivalent 252, 253, 269
ground deposition 92, 466

h

^{3}H 10, 13, 47, 72, 95–97, 102, 119, 189, 190, 385, 414, 468
hadron 131, 133, 140–142, 150, 158, 163, 165–167, 526
Hadron collider 131, 133, 140–141, 150, 151, 152, 161, 163–164, 190, 397, 463
hadronic cascade 152, 154–156, 170
half divergence angle 173, 175, 176, 222, 420
half-life 11, 12, 20, 51, 58, 64, 66, 97, 124, 126, 193, 324, 357, 366, 416, 467, 510
hands-on maintenance 116, 117
HD 109
HDO 109, 113
^{3}He 72, 73, 96, 97, 101, 104, 105, 124, 183, 189
^{4}He 33, 47, 72, 73, 82, 85, 88, 95, 96, 101, 102, 124
^{5}He 72, 95, 96, 375
heavy ion 5, 100, 106, 114, 131, 136, 152, 153, 156–158, 181, 183, 186, 193, 229–231
heavy ion accelerator 157, 182, 230, 389
heavy ion interaction 114, 531–533
helium 10, 14, 19, 22, 38, 48, 87, 89, 257, 304, 305, 317, 375
helium-cooled fast reactor 38
^{178}Hf 231, 232
^{178}Hf^{m2} 231
hibernation 334, 335
high dose rate components 102
high mass star 305, 306
high temperature gas cooled reactor 19
Higgs boson 151, 160, 163, 181
Hohmann orbit 252, 276–279, 281, 283, 319
hot particle 42, 43, 46, 60, 114, 116, 127, 283
HT 91, 363, 493
HTO 38, 65, 66, 79, 85, 91, 94, 108, 113, 363, 384
Hubble's law 315, 316, 458, 462
hypernova 312, 318

hypothetical events 92, 93
HZE particles 257, 259, 273, 279, 289, 295, 296, 334, 335

i

^{131}I 12, 38, 47, 63, 65, 354, 355, 358–361
^{115}In 187
^{116}In 187
inclination 252, 265, 266
inertial confinement 78–80, 119, 125,
inertial reference frame 509, 511
infrared radiation 239
ingestion 46, 93, 108, 109, 363
inhalation 46, 108, 109, 363, 466
inorganic scintillation detector 110, 111
insertion device 201, 203, 205, 208–213, 218, 234, 240
inspection and maintenance doses 59, 60
interaction
– electromagnetic 503
– gravitational 501–503
– strong 139, 147, 152, 155
– weak 164–165, 177, 425, 461, 497, 498, 502–505, 507
interlock 140, 154, 157, 160, 191, 214, 227, 238, 391
internal dose 45, 237, 360, 484, 490, 491
internal dosimetry 44, 473, 483, 485, 490, 495
internal hazard 11, 12, 71, 79, 101, 108, 123, 440
internal intake 79, 97, 108, 109, 113, 118, 119, 121
International Atomic Energy Agency (IAEA) 22, 460
International Commission on Radiological Protection (ICRP) 25, 458, 483
– ICRP 2 483, 495
– ICRP 10 483
– ICRP 26 65, 66, 192, 360, 483, 487, 488, 490–493, 524
– ICRP 30 461, 483, 492
– ICRP 60 194, 331, 382, 400, 483, 487, 488, 490–494, 524
– ICRP 66 462
International Experimental Thermonuclear Reactor (ITER) 81–87, 90–94, 102, 104–106, 117, 119, 387, 463
– effective dose values 10, 114, 164, 178
International Linear Collider 131, 134, 135, 138, 140, 150, 160, 161, 163, 193, 463
International Space Station 251, 263, 266
interstellar ramjet 323, 324
intranuclear cascade 146, 152, 155, 156
ion chamber tritium-in-air monitor 112, 364

ionization 65, 75, 110, 112, 137, 143, 144, 147, 154, 163, 184, 186, 187, 191, 239, 387–389, 477, 517, 532, 535
ionization chamber 65, 110, 111, 163, 184, 187, 191, 237, 387, 388, 389, 475, 477
iron 10, 103, 195, 243, 255, 259, 305, 306, 309, 310, 324, 397, 401, 414, 515
irregular galaxy 316
isomer 231
isomeric transition 199, 227, 231–232

j

Japanese Evaluated Nuclear Data Library (JENDL) 526
j-value 221, 418

k

kaon 154–156, 163, 183, 234, 389, 397, 459, 497, 498
Kardashev civilization type scale 460
Keppler's third law 275, 276
klystron 149, 220, 223, 236, 245, 420
^{85}Kr 11, 47, 51, 52, 67, 368
^{88}Kr 11, 47
krypton 10, 45, 103, 112, 378

l

large electron positron collider 140, 142, 149
Large Hadron Collider 134, 135, 138, 150, 151, 161, 163, 181, 190, 397, 463
laser
– continuous wave 239, 411
– infrared 239
– pulsed 240
– ultraviolet 215
– visible 240, 412
laser accelerator 199, 227, 229, 230
laser ion acceleration 199, 227, 230
laser Raman spectrometer 113
laser safety calculation 239–240
Lawson criterion 79, 119
lead-bismuth-cooled fast reactor 30–32
length contraction 510
LEO missions 260, 266, 269
lepton 131, 133, 141, 146, 147, 154, 160, 162, 163, 176, 181, 183, 184, 187, 389, 414, 461, 498
lepton collider 141, 160, 163
letdown 15, 45, 51, 52, 361
LHCb 151
^{6}Li 13, 33, 47, 48, 72, 95, 96, 98, 102, 189, 378
^{7}Li 72, 96
lifetime 22, 32, 37, 55, 92, 102, 106, 152, 169, 191, 231–233, 238, 244, 427, 510, 511

light source 199–204, 206–209, 234, 236, 238, 242, 244, 461, 525
LiI(Eu) detector 111
likely events 92, 318–319
limiting aperture 239, 412, 413
linear accelerator (LINAC) 138
linear collider 131, 134, 135, 138, 140, 141, 150, 160, 161, 163, 168, 171, 193
linear energy transfer 165, 252, 259, 261
line element 448
line source approximation 44, 362, 395
liquid metal fast breeder reactor 9, 20, 64
liquid scintillation counting 112, 364
lithium drifted germanium detector 111
lithium drifted silicon detector 110
Lorentz factor 169, 201, 212, 218, 222, 240, 289, 403, 408, 411, 418, 475
loss of coolant accident 18, 20, 50, 51, 54, 86, 87, 461
loss of cryogen 87, 89
loss of flow accident 87–88
loss of power event 53
loss of vacuum accident 87, 88
low-Earth orbit 1, 251, 263, 268, 296
low mass star 304, 305
luminosity 134, 135, 137, 141, 152, 157, 163, 191, 304, 307, 309, 348, 451, 519, 520
lunar missions 251, 260, 266, 268

m

magnetic confinement 78–81, 84, 121
magnetic fault transient 89
magnetic field 74, 79, 81, 85, 89, 138, 147, 211, 217–219, 221, 259, 263, 287, 290, 293–295, 512
magnetic force 79, 205, 209, 370, 422
magnetic induction 73, 209, 293, 296
magnetic rigidity 271, 287
magnetosphere 255, 257, 269, 279, 296
main sequence star 307, 318
main steamline break event 53
maintenance operations 114, 116
mapping spacetime 338
Mars 251, 277–283, 285–288, 298, 526
MARS code 528
Mars mission dose 278, 280, 296
Martian atmosphere 285
mass extinction 311, 313
mass spectrometer 113
maximum permissible exposure 239, 413
Maxwell equations 3, 74, 75, 220
mean dose per unit cumulated activity 484
mean lifetime 152, 416, 423, 499
mechanical handling equipment 114

medical internal radiation dose (MIRD) 483
MERCURY code 537
mercury mission 267, 281
meson 152, 182–184
meteorology 92, 354, 375
metric tensor 320, 338, 541–545, 547, 548-550
MicroShield® code 527
MicroSkyshine® code 527
microwave radiation 205
Milky Way Galaxy 303, 307, 313, 321
mission risk 296, 298
^{52}Mn 185
^{52m}Mn 185
^{54}Mn 47, 48, 64, 105, 414
^{55}Mn 97, 124, 370–372
^{56}Mn 64, 97, 103, 124, 370–372
^{92}Mo 536
^{98}Mo 105
^{99}Mo 105
^{100}Mo 105
molten salt epithermal reactor (MSR) 31, 36
Monte Carlo method 167
Monte Carlo N-particle (MCNP) code 526
Monte Carlo N-particle extended (MCNPX) code 526–527
Morris–Thorne wormhole geometry 338, 541, 545, 552
MORSE code 145, 527
muon 121, 136, 146–147, 190, 227, 334, 391, 395, 399, 414, 425, 510–511
– range 154, 516–518
– stopping power 515, 517
muon catalyzed fusion 121
muon collider 161, 165–168, 170, 172, 174, 175–180

n

^{13}N 413–416
^{16}N 10, 16, 17, 28, 38, 48, 59, 101, 106–107, 119, 375
^{22}Na 414
^{23}Na 20, 392, 414, 415
^{24}Na 20, 38, 101, 116, 244, 392, 415, 468
NaI(Tl) detector 111
nanotechnology 335, 446
NASA-Mir space station 251
^{92m}Nb 35, 104–106
NCRP 157, 182
NCRP 132 Leo dose recommendations 263
– blood forming organ dose limit 253–255, 298, 392
– ten year career dose limit 256
– whole-body exposure limit 255–256, 427
NCRP 142 262

NCRP   144 155, 187, 189, 252, 255
^{20}Ne   310, 536
neon   254, 305
net cavity gain   221
neutral current   165, 167
neutrino   165–169, 170–180, 310, 502–506
neutrino attenuation length   167, 173
neutrino effective dose   164, 168–180, 395
neutrino fluence-to-dose conversion factor   172, 176
neutrino interaction cross section   173, 503
neutron   81, 85, 102, 122, 123, 136–138, 142, 146, 154, 187, 188, 190, 192, 246, 299, 334, 348, 381, 389, 393, 401, 403, 425, 426, 440, 441, 447, 460, 468, 506, 528
neutron capture   21, 48, 116–117, 235, 308, 529, 530
neutron poison   15, 18
neutron star   306, 312
Newton's laws   3, 209, 291
^{57}Ni   105–106, 185, 235
^{58}Ni   13, 47, 99, 105
^{59}Ni   48, 99, 103
^{60}Ni   99, 105, 503–504
^{61}Ni   105
^{62}Ni   105
^{63}Ni   48, 99, 103
^{65}Ni   185
nickel   10, 21, 48, 305
nitrogen   48, 89, 163, 189, 297, 425
nitrous oxides   239
noble gas   46–47, 62, 368–379
Noerther's theorem   507
normal operations   61
nuclear cascade   155, 389
nuclear matter   100, 152
nuclear propulsion   319, 330
nucleosynthesis   253, 255

o
^{15}O   393–394, 415–416
^{16}O   48, 106–107, 192, 253, 310, 414, 468, 535
Oort Cloud   322
operational events   92
operational restrictions   335
optical cavity   218, 228, 232–234
optical gain   220
optical klystron   223, 420
optimized spatial bubble   337
orbital dynamics   275
orbital period   264, 275, 281
organic scintillation counter   110
outage   15, 27, 41, 60, 61, 115, 362
outage operations   60

outer planet mission   286–288
oxides of nitrogen   163, 189, 244
oxygen   36, 39, 254, 259, 297, 305, 313, 425, 537
ozone   239, 244, 313–314, 417

p
^{32}P   188, 457
^{234m}Pa   101
pair production   143, 146, 425, 515, 517
particulates   85, 109, 114
passive safety system   17, 23, 28
pebble bed modular reactor (PBMR)   22
peculiar galaxy   310, 316
phase bunching   216–218
phase slip   222, 418
PHOBOS   159
photochemical hazard   240, 411
photochemical reaction   243, 414
photoelectric effect/interaction   232, 515
photokaon   146
photon   81, 100, 101, 112, 131, 141–144, 146
photoneutron   146
photon generating device   161, 199
photopion   146
pion   136, 153–158, 227, 323, 389, 425, 439, 511, 523
pioneer   322, 339
planetary atmospheric attenuation   285
planetary mission   251, 252, 287, 296, 441, 531
planetary nebula   305
plasma   71–83, 87–90, 97, 102, 121, 229–230, 324, 375
plasma facing components   85, 87, 88
plasma transient   88
plume   58, 164, 167, 177, 180, 354, 384, 465, 469
^{147}Pm   12
point source   80–81, 195, 341–344, 375–378, 401, 527
point source approximation   44, 283, 331, 357, 367, 441, 443–444
Poisson's equation   76
positron   140–141, 143–145, 147–149, 154, 161–164, 205, 211, 403, 414, 486
power   9, 24, 30, 37, 39, 41–42, 44, 46, 49, 55, 56, 57, 62, 64, 66, 71, 75, 80, 81, 89, 106, 107, 109, 117, 120, 127, 138, 145, 147, 148, 180, 187, 190, 202, 207, 213, 220, 221, 224, 230, 244, 312, 313, 326, 388, 435, 518, 533
power output   19, 84, 181, 213–221, 326
pressurized water reactor   9, 14, 22, 26, 63
primary beam   163, 227, 234
primary coolant   15–17, 19, 21, 27, 30, 36, 38, 45–48, 50–51, 63, 65, 66, 354, 361

probabilistic risk assessment 56
probability level 236
production equations 44, 465, 471
proportional counter 184, 266
proton 106, 111, 131, 134, 136, 139, 141, 151-156, 158, 160, 164, 181, 182, 230, 258, 259, 261, 262, 271, 272, 274, 287, 292, 299, 300, 303, 306, 389, 391, 394, 401, 422, 425, 446, 457, 459, 461, 532
proton accelerator 141, 154–155, 182, 191
proton–antiproton collider 192
Proxima Centauri 321, 344, 345
^{239}Pu 10, 20, 43, 64, 377
^{241}Pu 20, 43, 64
pulsar 306, 311
pyroelectric process 122

q

quadrupole magnet 157, 162
quark 141, 152, 163, 181, 245, 499, 500–508
quark-gluon phase 152
quark-gluon plasma 158–159
Q-value 73, 83, 94, 310

r

radiance 207
radiant flux 413
radiation radiation countermeasures 334
radiation damage 82, 85, 88, 102, 106, 335
radiation length 143–145
radiation safety system 190
radiation types
 – alpha 9, 100, 165
 – beta 9, 101, 324, 486
 – gamma 9, 97, 314, 327, 401
 – heavy ion 106, 183, 186, 399, 533
 – kaon 154, 163, 184, 234
 – muon 154, 165, 183, 190, 191, 234, 414
 – neutrino 441
 – neutron 9, 97, 102, 155, 164, 186, 187, 188, 190, 324, 386, 394, 402, 415
 – proton 131, 155, 164, 397, 400, 416, 532
 – X-ray 139, 236, 311
radioactive gas 58, 112, 240–241, 246, 366, 418
radiofrequency cavity 138, 206
radiofrequency sources 235
radioiodine 10, 17, 67, 117
radioprotective chemical 269, 334–335, 446
radio wave 306, 312
range 535
^{88}Rb 11
reactor 9, 10, 13–15, 18–23, 26, 30–32, 36–38, 40–41, 45–53, 61–63, 71, 100, 108, 126, 332
reactor building 17

reactor building spray 17
reactor vessel 10, 15, 17–20, 29, 47, 62, 107
recession velocity 314–315
red giant 304–305, 318
red shift 314
red supergiant 305–306, 309
reference frame 288, 509, 511
relative biological effectiveness 252–253
relativistic heavy ion collider 135, 181, 462
 – radiation levels 180
remote handling equipment 114
resonant energy 216–218
Ricci tensor 539–542, 544, 547, 549, 551
Riemann curvature tensor 539–542, 544, 547–549, 551–552
risk 56, 92, 236–237, 255, 270, 274, 295–296, 318, 335, 427, 432, 485, 488–491, 493, 532
 – analysis 236–237
 – level 236–237

s

safety analysis report 117, 365, 391
safety envelope 159, 160
safety system 189–190, 238
saturation activity 97, 107, 415, 469
^{44}Sc 185
^{44m}Sc 185
^{46}Sc 30, 34, 48, 104, 185
^{47}Sc 34, 104, 185
^{48}Sc 34, 104–106, 185
scalar curvature 539–540, 542, 545, 547, 551
Schröedinger's equation 339–340
Schwarzchild geometry 541, 543–545, 547–549
secondary beam 238
secondary coolant 16, 46, 51, 55
self amplified spontaneous emission 216, 224, 226
semiconductor detector 110, 111, 186
semi-infinite cloud model 57–58
shadow shielding 273, 334, 335
shielding 331–332
shielding materials 288, 387, 391, 537
simplified boiling water reactor (SBWR) 22
skin absorption 108, 363–364
skin dose 46, 271, 274, 382
skin dose limit 274
Skylab 251, 263, 265, 267
skyshine 189, 528
SKYSHINE-KSU code system 528
slippage 217
social trauma 334
sodium-cooled fast reactors (SFR) 30, 31, 37
Sol 252, 263, 276, 281, 309, 321, 328

solar cycle   252, 261, 271, 452
solar flare   257, 260, 271, 287, 330
solar maximum   252, 269, 278, 286, 329
solar minimum   252, 279, 286, 329
solar particle event   259–260, 297, 306, 327, 330, 440
– August   1972 271, 286
– Carrington Flare   1859 271
– September   29, 1989 271, 280
solar system   78, 251, 257–259, 268, 278, 282–284, 303, 314, 319–322, 325, 327–328, 339, 344–345, 438, 440, 442, 503, 531, 539
solar wind   252, 263, 282, 283
sonoluminescence   121–122
South Atlantic Anomaly   258, 329
space radiation biology   295
spacetime
– geometry   319–320, 336, 343, 539, 541, 548, 550
Space Transport Shuttle   251, 263, 265
spallation   138, 155, 183–184, 229
spallation neutron source   138
special theory of relativity   134, 333, 457, 481, 509–510
spectral flux density   207, 241, 406–407
spectral type   307–308, 330
– A   307–308
– B   307
– F   307
– G   307–308
– K   307–308
– M   307
– O   307–308
spiral galaxy   309, 316
square well potential   340
^{90}Sr   11–12, 47
stability class   63, 354
stainless steel   13, 15, 18, 45, 47–48, 64, 66, 103-105, 184–185, 356–357
Standardized Computer Analyses for Licensing Evaluation (SCALE) code system   527
standard model of particle Physics   163, 453, 461–462, 497, 505, 507
star   158, 201, 249, 251, 303-308, 317, 315, 318–319, 321, 327–330, 333, 345–347, 382, 418, 451–452, 457–458, 543
STAR detector   158
steam generator   10, 14–17, 19–20, 26–27, 36, 42, 45–46, 52, 54, 57, 60–61, 87, 354, 362, 367
– tube rupture   51, 53, 57, 67, 354
steam generator tubes   16, 45, 49, 51, 61
steam line monitor   16
Stefan–Boltzmann law   317
stellar radiation   303

stopping power   515–517, 531–533, 536
Stopping Powers and Ranges (SPAR) code   528, 535
storage ring   158, 167, 174, 202–203, 216, 218, 520
straggling   517, 534–535
– angle   534
– energy   534
– range   534
sulfur oxides   239
supercritical water-cooled reactors (SWCR)   30, 31, 37
supernova
– type I   305, 309–310
– type II   306, 309–310
supersymmetry   150–151, 160, 181
supersymmetry particle   151, 160, 181, 193, 397
surface barrier detector   110, 111
synchrotron   159, 199, 208, 216, 407–408
synchrotron radiation   141–143, 147–149, 156, 181, 201–206

t

target area   182, 184, 190, 192, 227, 238, 286, 393
Tau Ceti   321, 345, 442
^{149}Tb   188
^{98}Tc   105
^{99m}Tc   105, 106
^{232}Th   10, 20, 43, 194
thallium   306
thermal injury   240
thermal method   112, 113
thermoluminescent dosimeter   184, 237, 265, 266
thick disk source   477
^{48}Ti   34, 105
time dilation   333, 334, 336, 510
T_2O 85
Tokai Mura criticality   446
tokamak   79, 81, 84, 85, 90, 91, 125, 370, 462
total removal rate   238, 358, 359, 393, 467–470
TOTEM   151, 152
toxic gas   238, 239, 416
trajectory   139, 142, 151, 154, 162, 200, 203–205, 215, 218, 252, 275, 281, 337
transmission factor   356
trapped radiation   257, 258, 260, 266, 269, 275, 278, 279, 281, 282, 285, 296, 297, 328, 329, 441
tritium bioassay   65, 125, 377
tritium bubbler   112, 364
tritium plant event   87, 89
true count rate   186

u

^{233}U   10, 22, 36, 43
^{235}U   10, 43, 377
^{238}U   10, 12, 20, 31, 43, 100, 101, 190, 358, 377, 384, 535, 536
ultraviolet radiation   311, 314
Ulysses   339, 348
undulator   161, 202–210, 214–217, 223, 224
undulator harmonic   409
undulator wavelength   409–411
unique nuclear reaction propulsion   323, 325
unlikely events   92, 319
unrestricted linear energy transfer   252, 261
uranium   10, 15, 19, 20, 32, 36–38
urinalysis   109, 386
US Department of Energy   159, 236, 493
US Nuclear Regulatory Commission   52, 493, 527
US Occupational Safety and Health Administration   121

v

^{50}V   104, 105
^{51}V   103, 104, 105
vacuum vessel   81–83, 85, 87–90, 95, 100–104, 106–109, 114, 117, 124, 126–127, 234, 379
vanadium   102–103, 121
van Allen belts
– inner belt   258
– outer belt   258
very high temperature helium-cooled thermal reactors (VHTR)   38
Very Large Hadron Collider   138, 161, 181
visible radiation   319

w

wake-field acceleration   229
warp bubble   336, 337
warp drive metric   336
warping spacetime   336
waste gas decay tank   41, 43
waste gas decay tank rupture   49, 51
waveguide   149, 235
wavelength   165, 199, 207–208, 211–212, 215–220, 224–226, 232, 239–242, 245, 314, 405–411, 420
white dwarf   305, 309, 311, 318
whole body dose   58, 253
width   139, 146, 204, 213, 222, 224, 226, 230, 232–233, 240, 411, 534–535
wiggler
– helical   245, 418
– parameter   214, 219–220, 242, 245, 411, 421
– period   219–220, 222, 411
– planar   220
wormhole, 321, 336, 338, 339, 539, 541, 545, 546

x

^{133}Xe   12, 47
^{135}Xe   12, 47
xenon   10, 46, 103, 112, 354
X-ray   1, 109, 193, 202, 204, 206, 224–226, 234, 236, 239, 241, 243, 313, 405
X-ray free electron laser   202, 224–226, 231
X-ray induced isomeric transition   199, 227, 232
X-ray laser   224
X-ray tube   203–206, 208, 406

y

^{90}Y   12, 34, 47
yield   63, 75, 76, 156, 174, 254, 291, 344, 396

z

^{63}Zn   185
^{64}Zn   105
^{65}Zn   185, 235
^{95}Zr   13, 35, 47